The Facts On File

DICTIONARY

of

BIOLOGY

D0209819

The Facts On File
DICTIONARY
of
BIOLOGY

Fourth Edition

Edited by
Robert Hine

Facts On File, Inc.

The Facts On File Dictionary of Biology
Fourth Edition

Copyright © 2005, 1999 by Market House Books Ltd

Facts On File, Inc.
132 West 31st Street
New York NY 10001

Library of Congress Cataloging-in-Publication Data
The Facts On File dictionary of biology / edited by Robert Hine.— 4th ed.
    p. cm.
  Includes bibliographical references.
  ISBN 0-8160-5647-1 (hc)
  1. Biology—Dictionaries. I. Title: Dictionary of biology. II. Hine, Robert. III. Facts On File, Inc.

  QH302.5.F38 2005
  570'.3—dc22

                                                                    2005040698

Compiled and typeset by Market House Books Ltd, Aylesbury, UK

Printed in the United States of America

MP PKG 10 9 8 7 6 5 4 3 2 1

This book is printed on acid-free paper.

PREFACE

This dictionary is one of a series designed for use in schools. It is intended for students of biology, but we hope that it will also be helpful to other science students and to anyone interested in science. Facts On File also publishes dictionaries in a variety of other disciplines, including chemistry, physics, mathematics, forensic science, weather and climate, marine science, and space and astronomy.

The Facts On File Dictionary of Biology was first published in 1980 and the third edition was published in 1999. This latest edition of the dictionary has been extensively revised and extended. The dictionary now contains over 3,700 headwords covering the terminology of modern biology. A totally new feature of this edition is the inclusion of over 2,500 pronunciations for terms that are not in everyday use. An Appendix has been included at the end of the book containing useful charts together with a list of Web sites and a bibliography. A guide to using the dictionary has also been added.

We would like to thank all the people who have cooperated in producing this book. A list of contributors is given on the acknowledgments page. We are also grateful to the many people who have given additional help and advice.

ACKNOWLEDGMENTS

Editor *(First and Second Editions)*

Elizabeth Tootill B.Sc.

Contributors

E. K. Daintith B.Sc.
Dennis J. Taylor B.Sc., Ph.D.
Sue O'Neill B.Sc.
Roderick Fischer B.Sc.
E. A. Martin M.A.
M. M. Richards B.Sc.
Owen Bishop B.Sc.
R. A. Prince M.A.
M. R. Ingle B.Sc., M.I.Biol., Ph.D., F.L.S.
Anne Moorhead B.Sc.

Additional contributions by

S. Pain B.Sc.
Lynne Mayers, B.Sc. D.T.A.
Mary Myles B.Sc.
Ruth D. Newell M.Sc., D.I.C.
J. Cohen B.Sc., Ph.D.
B. King B.Sc., Ph.D.
Lesley A. Bradnam B.Sc., M.Sc.
Malcolm P. Hart B.Sc.
Kathryn J. Green B.Sc.
Derek Cooper B.Sc., Ph.D., F.R.I.C.
P. R. Mercer B.A., T.C.
K. R. Dixon B.Ed.

Pronunciations

William Gould B.A.

CONTENTS

Preface   v

Acknowledgments   vi

Guide to Using the Dictionary   viii

Pronunciation Key   x

Entries A to Z   1

Appendixes

    I.   The Animal Kingdom   400

    II. The Plant Kingdom   401

    III. Amino Acids   402

    IV. Web Sites   404

Bibliography   406

GUIDE TO USING THE DICTIONARY

The main features of dictionary entries are as follows.

Headwords
The term being defined is in bold type:

> **absorption** The uptake of liquid by cells. Most digested food is absorbed in the small intestine, ...

Plurals
Irregular plurals are given in brackets after the headword.

> **stimulus** (*pl.* **stimuli**) A change in the external or internal environment of an organism that elicits a response in the organism.

Variants
Sometimes a word has a synonym or alternative spelling. This is placed in brackets after the headword, and is also in bold type:

> **vitamin C** (**ascorbic acid**) A water-soluble vitamin, which is widely required in metabolism. ...

Here, 'ascorbic acid' is another name for 'vitamin C'. Generally, the entry for the synonym consists of a simple cross-reference:

> **ascorbic acid** *See* vitamin C.

Abbreviations
Abbreviations for terms are treated in the same way as variants:

> **Biochemical Oxygen Demand** (BOD) The standard measurement for determining the level of organic pollution ...

The entry for the abbreviation consists of a simple cross-reference:

> **BOD** *See* Biochemical Oxygen Demand.

If the abbreviation is more common than the full form (e.g. DNA) then the main entry is at the headword for the abbreviation.

Multiple definitions
Some terms have two or more distinct senses. These are numbered in bold type

> **capsule** /**kap**-sewl/ **1.** A dry dehiscent fruit that is formed from several fused carpels. **2.** The mucilaginous covering often found around the cell membrane in bacteria. **3.** The structure within which spores of the

sporophyte generation of mosses and liverworts are formed....

4. (*Zoology*) A protective or supportive
sheath or envelope that surrounds an
organ or part of the body....

Cross-references

These are references within an entry to other entries that may give additional useful information. Cross-references are indicated in two ways. When the word appears in the definition, it is printed in small capitals:

> **annular thickening** Rings of thickening
> laid down on the inner wall of protoxylem
> vessels and tracheids. Such thickening al-
> lows extension of the XYLEM between the
> rings so that ...

In this case the cross-reference is to the entry for 'xylem'.

Alternatively, a cross-reference may be indicated by 'See', 'See also', or 'Compare', usually at the end of an entry:

> **aerobe** /**air**-ohb/ An organism that can
> live and grow only in the presence of free
> oxygen, i.e. it respires aerobically (*see* aer-
> obic respiration). *Compare* anaerobe.

Sometimes it is convenient to define one term within the entry for another term:

> **sere** A plant succession in which each
> community itself effects changes in the
> habitat that determine the nature of the fol-
> lowing stage. The successive stages are
> known as *seral stages*. ...

Here, 'seral stage' is a hidden entry under 'sere', and is indicated by italic type. The entry
for 'seral stage' consists of a simple cross-reference:

> **seral stage** *See* sere.

Pronunciations

Where appropriate pronunciations are indicated immediately after the headword, enclosed in forward slashes:

> **serology** /si-**rol**-ŏ-jee/ The *in vitro* study
> of reactions between antigens and antibod-
> ies in the blood serum....

Note that simple words in everyday language are not given pronunciations. Also headwords that are two-word phrases do not have pronunciations if the component words are
pronounced elsewhere in the dictionary.

Pronunciation Key

Bold type indicates a stressed syllable. In pronunciations, a consonant is sometimes doubled to prevent accidental mispronunciation of a syllable resembling a familiar word; for example, /**ass**-id/ (acid), rather than /**as**-id/ and /ul-tră- **sonn**-iks/ (ultrasonics), rather than /ul-tră-**son**-iks/. An apostrophe is used: (a) between two consonants forming a syllable, as in /**den**-t'l/ (dental), and (b) between two letters when the syllable might otherwise be mispronounced through resembling a familiar word, as in /**th'e**-ră-pee/ (therapy) and /tal'k/ (talc). The symbols used are:

/a/ *as in* back /bak/, active /**ak**-tiv/

/ă/ *as in* abduct /ăb-**dukt**/, gamma /**gam**-ă/

/ah/ *as in* palm /pahm/, father /**fah**-*th*er/,

/air/ *as in* care /kair/, aerospace /**air**-ŏ-spays/

/ar/ *as in* tar /tar/, starfish /**star**-fish/, heart /hart/

/aw/ *as in* jaw /jaw/, gall /gawl/, taut /tawt/

/ay/ *as in* mania /**may**-niă/ ,grey /gray/

/b/ *as in* bed /bed/

/ch/ *as in* chin /chin/

/d/ *as in* day /day/

/e/ *as in* red /red/

/ĕ/ *as in* bowel /**bow**-ĕl/

/ee/ *as in* see /see/, haem /heem/, caffeine /**kaf**-een/, baby /**bay**-bee/

/eer/ *as in* fear /feer/, serum /**seer**-ŭm/

/er/ *as in* dermal /**der**-măl/, labour /**lay**-ber/

/ew/ *as in* dew /dew/, nucleus /**new**-klee-ŭs/

/ewr/ *as in* epidural /ep-i-**dewr**-ăl/

/f/ *as in* fat /fat/, phobia /**foh**-biă/, rough /ruf/

/g/ *as in* gag /gag/

/h/ *as in* hip /hip/

/i/ *as in* fit /fit/, reduction /ri-**duk**-shăn/

/j/ *as in* jaw /jaw/, gene /jeen/, ridge /rij/

/k/ *as in* kidney /**kid**-nee/, chlorine /**klor**-een/, crisis /**krÿ**-sis/

/ks/ *as in* toxic /**toks**-ik/

/kw/ *as in* quadrate /**kwod**-rayt/

/l/ *as in* liver /**liv**-er/, seal /seel/

/m/ *as in* milk /milk/

/n/ *as in* nit /nit/

/ng/ *as in* sing /sing/

/nk/ *as in* rank /rank/, bronchus /**bronk**-ŭs/

/o/ *as in* pot /pot/

/ô/ *as in* dog /dôg/

/ŏ/ *as in* buttock /**but**-ŏk/

/oh/ *as in* home /hohm/, post /pohst/

/oi/ *as in* boil /boil/

/oo/ *as in* food /food/, croup /kroop/, fluke /flook/

/oor/ *as in* pruritus /proor-**ÿ**-tis/

/or/ *as in* organ /**or**-găn/, wart /wort/

/ow/ *as in* powder /**pow**-der/, pouch /powch/

/p/ *as in* pill /pil/

/r/ *as in* rib /rib/

/s/ *as in* skin /skin/, cell /sel/

/sh/ *as in* shock /shok/, action /**ak**-shŏn/

/t/ *as in* tone /tohn/

/th/ *as in* thin /thin/, stealth /stelth/

/*th*/ as in then /*th*en/, bathe /bay*th*/

/u/ *as in* pulp /pulp/, blood /blud/

/ŭ/ *as in* typhus /**tÿ**-fŭs/

/û/ *as in* pull /pûl/, hook /hûk/

/v/ *as in* vein /vayn/

/w/ *as in* wind /wind/

/y/ *as in* yeast /yeest/

/ÿ/ *as in* bite /bÿt/, high /hÿ/, hyperfine /**hÿ**-per-fÿn/

/yoo/ *as in* unit /**yoo**-nit/, formula /**form**-yoo-lă/

/yoor/ *as in* pure /pyoor/, ureter /yoor-**ee**-ter/

/ÿr/ *as in* fire /fÿr/

A band *See* sarcomere.

abaxial /ab-**aks**-ee-ăl/ In lateral organs, such as a leaf, the lower surface, i.e. the side facing away from the main axis. Abaxial is synonymous with dorsal when the latter term is being applied to lateral organs. *Compare* adaxial.

abdomen 1. The section of the body cavity of vertebrates that contains the stomach, intestines, and other viscera.
2. The posterior section of the body in arthropods. Primitively it consists of a series of similar segments.

abducens nerve /ab-**dew**-sĕnz/ (**cranial nerve VI**) One of the pair of nerves that arises from the anterior end of the medulla oblongata in the vertebrate brain to supply the posterior rectus muscle of each eyeball. It contains chiefly motor nerve fibers. *See* cranial nerves.

abiogenesis /ay-bÿ-oh-**jen**-ĕ-sis/ The development of living from nonliving matter, as in the origin of life.

abiotic environment /ay-bÿ-**ot**-ik/ The nonliving factors of the environment that influence ecological systems. Abiotic factors include climate, chemical pollution, geographical features, etc.

abomasum /ab-ŏ-**may**-sŭm/ The fourth and last region of the specialized stomach of ruminants (e.g. the cow). It is lined with normal mucosa containing gastric pits, producing hydrochloric acid, pepsin, and rennin, and is the true digestive stomach.

ABO system /ay-bee-**oh**/ *See* blood groups.

abscisic acid /ab-**siss**-ik/ (**ABA**) A plant hormone that is involved in seed development and in regulating closure of leaf pores in response to water deficit. ABA is thought to induce the formation of storage proteins in developing seeds, and might also act to prevent precocious germination of seeds. In a quite different role, ABA is produced in wilting leaves and diffuses to GUARD CELLS in the epidermis to induce closure of the associated leaf pores (stomata), thereby reducing further water loss. ABA was first isolated from sycamore, in which it is associated with the onset of bud dormancy (hence its original name 'dormin'), but there is now doubt over the general applicability of this effect. Also, it is no longer thought to play a significant role in the shedding (abscission) of flowers and fruits, as originally proposed.

abscission /ab-**sizh**-ŏn, -**sish**-/ The organized loss of part of a plant, usually a leaf, fruit, or unfertilized flower. An *abscission zone* occurs at the base of the organ. Here a separation layer (*abscission layer*) is formed by breakdown or separation of cells and final severance occurs when the vascular bundles are broken mechanically, e.g. by wind. Abscission is normally suppressed by the continual flow of the plant hormone AUXIN through the leaf stalk or other connecting structure. But a decline in the flow of auxin, e.g. due to aging or injury, triggers abscission and consequent separation of the part.

absorption The uptake of liquid by cells. Most digested food is absorbed in the small intestine, the inner surface of which is lined with finger-like projections (villi): the liquid products of digestion are absorbed through the villi into the blood and lymphatic systems. In plants, water and mineral salts are absorbed mainly by the root hairs, just behind the root tips.

absorption spectrum A plot of the absorbance by a substance of radiation at different wavelengths, usually of ultraviolet, visible, or infrared radiation. It can give information about the identity or quantity of a substance. Chlorophylls, for example, have absorption peaks in the red and blue (and therefore reflect green light).

abyssal /ă-**bis**-ăl/ Inhabiting the portion of the ocean deeper than 2000 m and shallower than 6000 m. The abyssal realm is the largest environment on earth. Abyssal fauna tend to be black or gray, unstreamlined, and delicately structured. Many can produce light (*see* bioluminescence).

Acarina /ak-ă-**rÿ**-nă/ A subclass or order of arthropods belonging to the class ARACHNIDA that contains the mites and ticks. Roughly 50 000 species of mites have been described, and about 850 tick species. Both mites and ticks have rounded bodies lacking a waist, and have mouthparts variously modified for crushing, cutting, or piercing. Mites are usually tiny, less than 1.5 mm long, and occur in most habitats worldwide. The majority are free living, feeding on dead plant and animal matter in the soil or in water, but some are predators on small arthropods, and others are parasitic on animals or plants. For example, itch mites feed on the blood of humans and pets, causing skin irritation. Ticks are generally larger than mites, with the largest reaching nearly 30 mm when fully expanded. They are blood-sucking parasites of other animals, including reptiles, birds, and mammals, and can transmit certain diseases, including Lyme disease to humans and tick fever to cattle and sheep.

accessory cell (**subsidiary cell**) One of a number of specialized epidermal cells of a plant that are found adjacent to the guard cells, and may help in opening and closing the stomata.

accessory nerve (**cranial nerve XI**) One of the pair of nerves that arises from the posterior region of the medulla oblongata in the brain of higher vertebrates. It carries motor nerve fibers and merges with the adjacent vagus nerve close to its root. In fishes and amphibians it is considered an integral part of the vagus. *See* cranial nerves.

accessory pigment *See* photosynthetic pigments.

acclimatization /ă-klÿ-mă-ti-**zay**-shŏn/ Physiological, biochemical, or anatomical changes in an organism that occur in response to changes in the natural environment during the lifetime of the organism, and better equip it for living in the altered or new environment. For example, humans who move from a low altitude to live in a mountainous region undergo a gradual increase in the concentration of their red blood cells, enabling them to breathe normally in the rarefied atmosphere. When changes are the result of imposed laboratory conditions, the process is called *acclimation*. Whether the result of natural or laboratory conditions, such changes are reversible and not heritable, and are distinct from the process of ADAPTATION.

accommodation 1. The reflex process in the eye by which an image is focused on the retina. In humans, the eye at rest is focused on infinity, with the lens and cornea flattened. To focus on a near object the ciliary muscles contract, and the lens, being elastic, becomes more convex and thus of shorter focal length; the cornea also becomes more convex and assists in the focusing.

Although the structure of the eye is similar in all vertebrates, the method of accommodation varies from group to group. Fishes and amphibians have muscles that move the lens (of fixed focal length) backwards for distant vision and forwards for near vision.

Reptiles and birds accommodate by action of the ciliary muscles, which on contraction actually push the lens into a convex shape.

2. *See* adaptation.

acellular /ay-**sell**-yŭ-ler/ Denoting relatively large tissues or organisms that are not composed of discrete cells and are, in effect, unicellular. Examples are aseptate fungal hyphae and muscle fibers. The term is used in preference to unicellular to distinguish such structures (which are often multinucleate) from conventional cells and show their equivalence to multicellular structures. *See also* syncytium.

acentric /ay-**sen**-trik/ Denoting a chromosome or fragment of a chromosome that lacks a centromere.

acetabulum /ass-ĕ-**tab**-yŭ-lŭm/ A cup-shaped socket on each side of the pelvic girdle in tetrapods that holds the rounded head of the thigh bone (femur) to form the hip joint. This 'ball-and-socket' arrangement allows for great stability and a wide range of movement.

acetic acid /ă-**see**-tik, ă-**set**-ik/ (**ethanoic acid**) A carboxylic acid, CH_3COOH, obtained by the oxidation of ethyl alcohol. Acetic acid is a component of vinegar (which is obtained by bacterial oxidation of wine waste).

acetylcholine /ass-ĕ-t'l-**koh**-leen, ă-see-t'l-, -lin/ (**ACh**) A NEUROTRANSMITTER found in both vertebrates and invertebrates in various parts of the nervous system and at neuromuscular junctions. It can have either excitatory or inhibitory effects, depending on the location. Nerves that produce acetylcholine are called *cholinergic nerves*.

acetyl CoA /**ass**-ĕ-t'l koh-**ay**, ă-/ (**acetyl coenzyme A**) An important intermediate compound in the metabolism of all organisms, especially in the oxidation of sugars, fatty acids, and amino acids, and in certain biosynthetic pathways. *See* Krebs cycle.

ACh *See* acetylcholine.

achene /ă-**keen**/ A dry indehiscent fruit formed from a monocarpellary ovary containing a single seed. Different types of ach-

enes include the caryopsis, cypsela, nut, and samara. *See also* etaerio.

acid A substance that gives rise to hydrogen ions (or H_3O^+) when dissolved in water. An acid in solution will have a pH below 7. This definition does not take into account the competitive behavior of acids in solvents and it refers only to aqueous systems. The *Lowry-Brønsted theory* defines an acid as a substance that exhibits a tendency to release a proton, and a base as a substance that tends to accept a proton. Thus, when an acid releases a proton, the ion formed is the *conjugate base* of the acid. Strong acids (e.g. HNO_3) react completely with water to give H_3O^+, i.e. HNO_3 is stronger than H_3O^+ and the conjugate base NO_3^- is weak. Weak acids (e.g. CH_3COOH and C_6H_5COOH) are only partly dissociated because H_3O^+ is a stronger acid than the free acids and the ions CH_3COO^- and $C_6H_5COO^-$ are moderately strong bases.

acid–base balance Maintenance of the optimum pH of body fluids by regulating the acid:base ratio. This usually involves a buffer system (*see* buffer). For example mammalian blood must be maintained at a pH of 7.4, which requires a ratio of carbonic acid to bicarbonate of 1:20; any serious deviation would result in the conditions of acidosis or alkalosis. The optimum pH for higher plants is around 6.7, and they tend to be somewhat more tolerant to pH changes.

acidic stain *See* staining.

acid rain The deposition of acids by natural precipitation (e.g. rain, snow, fog), leading to acid pollution. The very dilute acids are formed by reaction of gaseous waste products (e.g. sulfur dioxide, nitrogen oxide, or carbon monoxide), with moisture present in the air. This has led to environmental damage in many areas. Unpolluted rain is slightly acidic, having a pH of 5.0–5.6; damage will occur to sensitive ecosystems when the pH falls below 4.6.

acid value A measure of the free acid present in fats, oils, resins, plasticizers, and solvents, defined as the number of milligrams of potassium hydroxide required to neutralize the free acids in one gram of the substance.

acoelomate /ay-**see**-lŏ-mayt, -mit/ Describing an animal that lacks a body cavity, apart from that of the gut and other organs. Acoelomate animals include the flatworms and nemertines (ribbon worms).

acoustic nerve *See* auditory nerve.

acoustico-lateralis system /ă-koos-ti-koh-lat-er-**al**-is/ A system found in fish and certain amphibians consisting of the neuromast cells in various tracts on or near the surface of the body (i.e. the *lateral-line system*) and the inner ear (*membranous labyrinth*). Neuromasts and receptor cells of the inner ear are similar, consisting of groups of cells with sensory processes innervated by nerves from the medulla oblongata of the brain. They are responsible for the detection of sounds in water.

acquired characteristics Changes in the structure or function of an organ or system during the life of an organism, brought about by the use or disuse of that organ or system, or by environmental influences. For example, sportsmen may develop strong muscles, and plants growing near coasts show adaptations to the drying effects of sea air. Acquired characteristics are not inherited. *See also* Lamarckism.

acquired immune deficiency syndrome *See* AIDS.

acraniate /ay-**kray**-nee-ayt/ Any chordate animal with a notochord and lacking a brain and skull. Acraniates include the urochordates and cephalochordates. *See* Cephalochordata; Urochordata.

acropetal /ă-**krop**-ĕ-t'l/ Developing or moving from the base upwards. An acropetal sequence of development is seen in flower formation in which the calyx forms first and the carpels form last. *Compare* basipetal. *See also* centrifugal; centripetal.

acrosome /**ak**-rŏ-sohm/ A membrane-bound structure in the anterior head region of a spermatozoon, usually forming a cap over the nucleus. It contains enzymes that are released on contact with the egg at fertilization and break down the egg coats, enabling the spermatozoon nucleus to enter the egg.

ACTH *See* corticotropin.

actin /**ak**-tin/ A protein that, with other proteins such as MYOSIN, is responsible for many types of cell movement and contraction, including muscle contraction. It also forms the MICROFILAMENTS of the internal scaffolding (CYTOSKELETON) of cells. Changes in cell shape involve extension or shortening of these microfilaments, achieved by the assembly or disassembly of their actin subunits. Actin polymers also form the thin filaments in muscle myofibrils (*see* sarcomere). The alternating and overlapping thick filaments (myosin) and thin filaments interact during muscle contraction to form actomyosin complexes. Actin and myosin also interact in nonmuscle cells, notably in accomplishing cell cleavage during cytokinesis, and producing cytoplasmic streaming in certain algal and plant cells.

actinomorphy /ak-tă-nŏ-**morf**-ee/ *See* radial symmetry.

actinomycetes /ak-tă-nŏ-**mỹ**-seets/ (**filamentous bacteria**) A group of bacteria characterized by a mycelial funguslike growth habit and belonging to the phylum Actinobacteria. They are numerous in the topsoil and are important in soil fertility. Some may cause infections in animals and humans. Most of the antibiotics (e.g. streptomycin, actinomycin, and tetracycline) are obtained from actinomycetes.

actinostele /ak-**tin**-ŏ-steel, -stee-lee/ A type of protostele in which the xylem is star shaped and the phloem lies between the points of the star. *See* stele.

action potential The transitory change in electrical potential that occurs across the membrane of a nerve or muscle fiber during the passage of a nervous impulse. The degree of change is independent of the strength of the impulse; it either occurs or does not (*see* all-or-none). In the absence of an impulse a *resting potential* of about −70 mV exists across the membrane (inside negative) as a result of the unequal distribution of ions between the intracellular and extracellular media (*see* sodium pump). However, during the passage of an impulse the resistance of the membrane falls and allows an inward current of sodium ions, which makes the inside less negative and eventually alters the membrane potential to about +30 mV (inside positive). An outward flow of potassium ions follows shortly afterward and restores the resting potential. Local currents flow ahead of the action potential that in turn give rise to further action potentials. In this way a unidirectional wave of activity is propagated along the membrane to produce the nerve impulse.

action spectrum A graph showing the effect of different wavelengths of radiation, usually light, on a given process. It is often similar to the absorption spectrum of the substance that absorbs the radiation and can therefore be helpful in identifying that substance. For example, the action spectrum of photosynthesis is similar to the absorption spectrum of chlorophyll. *See* absorption spectrum.

activation energy Symbol: E_a The minimum energy a particle, molecule, etc., must acquire before it can react; i.e. the energy required to *initiate* a reaction regardless of whether the reaction is exothermic or endothermic. Activation energy is often represented as an energy barrier that must be overcome if a reaction is to take place. *See* Arrhenius equation.

active site The region of an enzyme molecule that combines with and acts on the substrate. It consists of amino acids arranged in a configuration specific to a particular substrate or type of substrate. Binding of an inhibiting compound elsewhere on the enzyme molecule may change this configuration and hence the efficiency of the enzyme activity.

active transport The transport of molecules or ions across a cell membrane against a concentration gradient, with the expenditure of energy. Anything that interferes with the provision of energy will interfere with active transport. The mechanism typically involves a carrier protein that spans the cell membrane and transfers substances in or out of the cell, or cell organelle, by changing shape. *See also* AT-Pase.

actomyosin /ac-tŏ-**mÿ**-ŏ-sin/ A protein complex found in muscle, formed between molecules of actin and myosin present in adjacent thick and thin muscle filaments. These actomyosin complexes are involved in the process of muscle contraction; their formation is important in the concerted mechanism that pulls the thin filaments past the thick ones.

adaptation 1. The extent to which an organism, or a physiological or structural characteristic of an organism, is suited to a particular environment. Organisms that have become highly adapted to one environment are then often not so adaptable as less specialized organisms and are at a disadvantage in a changing environment (adaptation versus adaptability).
2. (**accommodation**) The decrease with time of the frequency of response of sensory receptors subjected to a constant intensity of stimulus. A more intense stimulus is then required to elicit a response.

Receptors vary greatly in the rate and degree of adaptation they show depending on their function. Many monitor relatively static stimuli, such as internal temperature, and exhibit no adaptation, while others monitoring rapidly changing stimuli, such as touch sensation in the skin, rapidly adapt to a constant intensity stimulus.

adaptive enzyme (**inducible enzyme**) An enzyme that is produced by a cell only in

5

the presence of its substrate. *Compare* constitutive enzyme.

adaptive radiation The gradual formation through evolution of a number of different varieties or species from a common ancestor, each adapted to a different ecological niche. A classic example of adaptive radiation is illustrated by DARWIN'S FINCHES.

adaxial /ad-**aks**-ee-ăl/ In lateral organs, such as a leaf, the upper surface, i.e. the side facing towards the main axis. Adaxial is synonymous with ventral when the latter term is being applied to lateral organs. *Compare* abaxial.

adenine /**ad**-ĕ-neen, -in, -ÿn/ A nitrogenous base found in DNA and RNA. It is also a constituent of certain coenzymes, e.g. NAD and FAD, and when combined with the sugar ribose it forms the nucleoside adenosine found in AMP, ADP, and ATP. Adenine has a purine ring structure. *See illustration at* DNA.

adenohypophysis /ad-ĕn-oh-hÿ-**poff**-ă-sis/ *See* pituitary gland.

adenosine /ă-**den**-ŏ-seen, -sin, **ad**-'n-ŏ-seen/ (**adenine nucleoside**) A nucleoside formed from adenine linked to D-ribose with a β-glycosidic bond. Adenosine triphosphate (ATP) is a nucleotide derived from adenosine.

adenosine diphosphate /dÿ-**fos**-fayt/ *See* ADP.

adenosine monophosphate /**mon**-oh-fos-fayt/ *See* AMP.

adenosine triphosphate /trÿ-**fos**-fayt/ *See* ATP.

adenovirus /ad-ĕn-oh-**vÿ**-rŭs/ One of a group of DNA-containing viruses, about 80 nm in diameter, causing respiratory diseases and tumors in humans and animals. *See* oncogenic.

adenylate cyclase /ă-**den**-ă-layt sÿ-klayss/ An enzyme that catalyses the conversion of ATP to CYCLIC AMP (cAMP) within living cells. cAMP is an important regulator of numerous cellular processes, and its formation is a key event in signaling pathways inside cells. Adenylate cyclase is closely associated with the plasma membrane, and its activity is regulated by G PROTEINS in response to the binding of hormones and other signal molecules to receptors on the outer surface of the cell.

ADH (**antidiuretic hormone**) *See* vasopressin.

adipose tissue /**ad**-ă-pohs/ (**fatty tissue**) A type of CONNECTIVE TISSUE consisting of closely packed cells (*adipocytes*) containing fat. Adipose tissue is found in varying amounts in the dermis of the skin and around the kidneys, heart, and blood vessels. It provides an energy store, heat insulation, and mechanical protection. The fat is deposited as tiny droplets scattered in the cytoplasm, but as these enlarge they fuse to form one large drop with a thin layer of cytoplasm and a nucleus squashed to one side. The distended cells are pressed against each other, so that they appear angular.

Most adipose tissue is in the form of *white fat*, but hibernating and newborn animals have deposits of darker colored *brown fat*, rich in unsaturated fatty acids, which is well supplied with blood vessels and acts as a readily mobilizable source of heat energy (*see* thermogenesis). Some theories on the cause of obesity in humans postulate a lack of brown fat in those affected.

See illustration at skin.

ADP (**adenosine diphosphate**) A nucleotide consisting of adenine and ribose with two phosphate groups attached. *See* ATP.

adrenal glands /ă-**dree**-năl/ A pair of glands that are situated above each kidney and produce various hormones, notably epinephrine. Secretion is controlled by the nervous system. The adrenal glands have a central medulla and an outer cortex, each behaving independently. The former produces norepinephrine and epinephrine. The adrenal cortex is rich in vitamin C and cholesterol, and produces three types of hormones: aldosterone, cortisol, and sex hormones.

adrenaline /ă-**dren**-ă-lin/ *See* epinephrine.

adrenergic /ad-rĕ-**ner**-jik/ Designating the type of nerve fiber that releases epinephrine or norepinephrine from its ending when stimulated by a nerve impulse. Vertebrate sympathetic motor nerve fibers are adrenergic. *Compare* cholinergic.

adrenocorticotropic hormone /ă-dree-noh-kor-ti-koh-**trop**-ik/ *See* corticotropin.

adventitious /ad-ven-**tish**-ŭs/ Describing plant organs that arise in unexpected places, for example the development of adventitious roots from stems, and adventitious buds from leaves.

aerenchyma /air-**enk**-ă-mă/ Modified parenchyma with large air spaces to facilitate diffusion of gases and, in aquatic plants, to provide buoyancy.

aerobe /**air**-ohb/ An organism that can live and grow only in the presence of free oxygen, i.e. it respires aerobically (*see* aerobic respiration). *Compare* anaerobe.

aerobic respiration Repiration in which free oxygen is used to oxidize organic substrates to carbon dioxide and water, with a high yield of energy. The reaction overall is:

$$C_6H_{12}O_6 + 6O_2 = 6CO_2 + 6H_2O + \text{energy}$$

It occurs in a number of stages, the first of which (glycolysis) also occurs in anaerobic respiration in the cell cytoplasm. With glucose as the substrate, a sequence of reactions results in the formation of pyruvate. The remaining stages, which do not occur in anaerobic respiration, take place in the mitochondria. Pyruvate is converted to acetyl coenzyme A, which enters a cyclic series of reactions, the Krebs cycle, with the production of carbon dioxide and hydrogen atoms. These and other hydrogen atoms produced at earlier stages are now passed to the ELECTRON-TRANSPORT CHAIN (involving cytochromes and flavoproteins). Here they combine with atoms of free oxygen to form water. Energy released at each stage of the chain is used to form ATP during a coupling process (*see* oxidative phosphorylation). Current opinion is that aerobic respiration gives a net yield of 31 or 29.5 ATP per glucose molecule oxidized, depending on the shuttle mechanism by which NADH in the cytoplasm is transported into the mitochondrion. This compares with a maximum net yield of 38 ATP per glucose molecule quoted in some texts. The lower totals are based on revised calculations of the number of protons pumped across the inner mitochondrial membrane by the electron-transport chain, and the number required by ATP SYNTHETASE to form each molecule of ATP.

aerotaxis /air-ŏ-**taks**-is/ (aerotactic movement) A taxis in response to an oxygen concentration gradient. For instance, motile aerobic bacteria are positively aerotactic, whereas motile obligate anaerobic bacteria are negatively aerotactic. *See* taxis.

afferent /**af**-ĕr-ĕnt/ Conveying (impulses, blood, etc.) from the outer regions of an organ or body towards the center. For example, afferent nerves convey nerve impulses from sense organs to the central nervous system. *Compare* efferent.

aflatoxin /af-lă-**toks**-in/ *See* Aspergillus.

afterbirth The placenta, umbilical cord, and membranes discharged from the body shortly after the birth of the young.

after-ripening The collective name for processes that are necessary before germination can take place in certain seeds, even though external conditions may be suitable. A period of dormancy is thus imposed on the seed, preventing premature germination before an unfavorable season such as winter. After-ripening is common in the family Rosaceae.

aftershaft *See* contour feathers.

agamospermy /ă-**gam**-ŏ-sper-mee/ *See* apomixis.

agar /**ay**-gar, **ah**-gar/ A gelling agent prepared from seaweed, used to set liquid nutrients. Agar gels are extensively used for growing microorganisms.

agglutination /ă-gloo-tă-**nay**-shŏn/ The clumping together of red blood cells or bacteria as a result of the action of antibodies. Agglutination may occur in transfusion if blood of the wrong group is given. The surfaces of the donor's red blood cells contain antigen molecules that are attacked by antibody molecules in the serum of the recipient, which causes the red cells to clump together. These clumps may block capillaries, causing fatal damage to the heart or brain. Agglutination of bacteria by antibodies causes them to disintegrate. *See also* antibody; blood groups.

aggression The type of animal behavior involving threats and actual attacks on other animals. Aggression is normally a response to opposition (e.g. defense of territory) and results in the displacement of the opponent, which is usually of the same species. It operates in the interests of survival of the individual or species.

Agnatha /**ag**-nă-thă/ The superclass containing the earliest and most primitive vertebrates, characterized by the absence of jaws. The only living class is the Cyclostomata, which includes lampreys (e.g. *Petromyzon*) and hagfish (e.g. *Myxine*) – aquatic fishlike animals lacking the paired fins typical of true fishes. There are also several extinct Paleozoic groups, whose members had a heavy armor of bony plates and scales. *Compare* Gnathostomata. *See* Cyclostomata.

agonism /**ag**-ŏ-niz-ăm/ *See* agonistic behavior.

agonist A substance, such as a drug or hormone, that binds to a cell's receptors and elicits a response.

agonistic behavior (**agonism**) A type of animal behavior exhibiting features of both aggression and avoidance. For example, the establishment of a territory involves both attack and escape behavior; if two neighbors meet at the boundary of their territories, conflicting tendencies to attack and flee are aroused, resulting in conflicting responses. Some forms of agonistic behavior have become modified into threat displays, which tend to intimidate rivals and reduce actual fighting.

agranulocyte /ay-**gran**-yŭ-lŏ-sÿt/ A white blood cell (leukocyte) that does not contain granules in its cytoplasm. There are two types: LYMPHOCYTES and MONOCYTES (comprising 25% and 4%, respectively, of all leukocytes). Both have large nuclei and a small amount of clear cytoplasm.

Agrobacterium /ag-roh-bac-**teer**-ee-ŭm/ A genus of soil bacteria, the species *A. tumefaciens* being the causative agent of crown gall, a type of tumor in plants. A segment of DNA (transferred DNA, T-DNA) from a plasmid in the bacterium is transferred into the host DNA and induces tumor formation. Since the plasmid is capable of independent replication in host cells of many dicotyledonous plants, it has been used as a cloning vector in genetic engineering. Once the desired segment of DNA, for example a gene, has been spliced into the T-DNA, the plasmid can be introduced into certain plant cultures and entire plants with the desired characteristic can be produced. Unfortunately, the bacterium does not infect monocotyledonous plants, which include important cereal crops. *See also* gene cloning; tissue culture.

AIDS (**acquired immunodeficiency syndrome**) A viral disease of humans characterized by destruction of the lymphocytes responsible for cell-mediated immunity (*see* T-cell), the patient consequently succumbing to opportunistic fatal infections, cancers, etc. The causative agent – HIV (human immunodeficiency virus) – is transmitted in blood, semen, or vaginal fluid; it is a retrovirus and can remain inactive in its host for several years without causing serious symptoms. Such carriers, described as *HIV-positive*, can nevertheless transmit the virus.

air bladder *See* swim bladder.

air sac 1. Any of several interconnected tissue sacs in birds that assist in ventilating the lungs by acting like bellows. Some of the air sacs penetrate the hollow internal cavities of the bones.
2. An expanded chamber that forms part of the network of breathing tubes (*see* trachea) in certain insects and increases the area available for gaseous exchange.
3. An inflatable pouch used for display in certain animals, such as that hanging from the throat of an orang-utan.

alanine /al-ă-neen, -nÿn/ *See* amino acids.

albinism /al-bă-niz-ăm/ The absence of pigmentation in the eyes, skin, feathers, hair, etc., of some animals. It is a hereditary condition in vertebrates, resulting from abnormalities in production or function of the pigment cells. Albinism in humans is due to a recessive gene and results in the absence of a dark brown pigment (melanin). The complete albino has milk-white skin and hair and the irises of the eyes appear pink.

albumen /al-byoo-měn/ *See* albumin.

albumin /al-byoo-min/ One of a group of simple proteins, which are usually deficient in glycine. They are water-soluble and when heated coagulate. Albumins are the products of plants and animals; e.g. legumelin in peas and the serum albumin of blood. The protein in egg white (*albumen*) is a mixture of albumins. These also contain carbohydrate groups and consequently are classified as conjugated proteins.

albuminous cell /al-byoo-mă-nŭs/ A vertically elongated parenchyma cell, found in

groups in the rays of the secondary phloem in gymnosperms.

alcohol A type of organic compound of the general formula ROH, where R is a hydrocarbon group. Examples of simple alcohols are methanol (CH_3OH) and ethanol (C_2H_5OH).

Alcohols have the –OH group attached to a carbon atom that is not part of an aromatic ring: C_6H_5OH, in which the –OH group is attached to the ring, is thus a phenol. Phenylmethanol ($C_6H_5CH_2OH$) does have the characteristic properties of alcohols.

Alcohols can have more than one –OH group; those containing two, three or more such groups are described as *dihydric, trihydric*, and *polyhydric* respectively (as opposed to those containing one –OH group, which are *monohydric*). Examples are ethane-1,2-diol (ethylene glycol; ($HOCH_2CH_2OH$) and propane-1,2,3-triol (glycerol; $HOCH_2CH(OH)CH_2OH$).

Alcohols are further classified according to the environment of the –C–OH grouping. If the carbon atom is attached to two hydrogen atoms, the compound is a *primary alcohol*. If the carbon atom is attached to one hydrogen atom and two other groups, it is a *secondary alcohol*. If the carbon atom is attached to three other groups, it is a *tertiary alcohol*.

aldehyde /al-dě-hÿd/ A type of organic compound with the general formula RCHO, where the –CHO group (the aldehyde group) consists of a carbonyl group attached to a hydrogen atom. Simple examples of aldehydes are methanal (formaldehyde, HCHO) and ethanal (acetaldehyde, CH_3CHO).

Aldehydes are formed by oxidizing a primary alcohol. In the laboratory potassium dichromate(VI) is used in sulfuric acid. They can be further oxidized to carboxylic acids. Reduction (using a catalyst or nascent hydrogen from sodium amalgam in water) produces the parent alcohol.

Aldehydes undergo a number of reactions:
1. They act as reducing agents, being oxidized to carboxylic acids in the process.

These reactions are used as tests for aldehydes using such reagents as Fehling's solution and Tollen's reagent (silver-mirror test).

2. They form addition compounds with hydrogen cyanide to give 'cyanohydrins'. For example, propanal gives 2-hydroxybutanonitrile:

$$C_2H_5CHO + HCN \rightarrow$$
$$C_2H_5CH(OH)CN$$

3. They form addition compounds with the hydrogensulfate(IV) ion (hydrogensulfite; HSO_3^-):

$$RCHO + HSO_3^- \rightarrow RCH(OH)(HSO_3)$$

4. They undergo condensation reactions with such compounds as hydrazine, hydroxylamine, and their derivatives.
5. With alcohols they form hemiacetals and acetals.
6. They polymerize readily. Polymethanal or methanal trimer can be formed from methanal depending on the conditions. Ethanal gives ethanal trimer or ethanal tetramer.

See also acetal; Cannizzaro reaction; condensation reaction; ketone.

aldohexose /al-doh-**heks**-ohs/ An aldose sugar with six carbon atoms. *See* sugar.

aldopentose /al-doh-**pen**-tohs/ An aldose sugar with five carbon atoms. *See* sugar.

aldose /**al**-dohs/ A sugar containing an aldehyde (CHO) or potential aldehyde group. *See* sugar.

aldosterone /al-**dos**-tĕ-rohn/ A steroid hormone secreted by the cortex of the adrenal glands. It stimulates the reabsorption of sodium ions from urine by the kidney tubules, and hence causes the reabsorption of water, thereby promoting water retention. This raises blood pressure. It also promotes the excretion of potassium and hydrogen ions by the kidneys.

aleurone grain /**al**-yŭ-rohn / (**aleurone body**) A modified vacuole found in the embryo and endosperm of seeds and containing mostly reserve proteins, but also phytic acid and various enzymes associated with mobilization (digestion) of these reserves.

aleurone layer The outermost protein-rich layer of the endosperm of grass fruits (e.g. cereal grains). At germination, the embryo produces gibberellin, which stimulates the aleurone layer to synthesize enzymes, especially α-amylase. The latter causes hydrolysis of the starch in the endosperm. The enzymes are synthesized from the amino acids supplied by breakdown of aleurone grains.

aleuroplast /**al**-yŭ-roh-plast/ A type of proteoplast in which protein is stored in the form of aleurone grains. They are common in seeds; for example, in the endosperm of castor oil.

algae /**al**-jee/ A large mixed group of photosynthesizing organisms, now usually placed in the kingdom Protoctista. They often resemble plants and are found mainly in marine or fresh-water habitats, although some algae are terrestrial. Algae differ from plants in lacking any real differentiation of leaves, stems, and roots, and in not having an embryo stage in their life cycle. Algae can be unicellular (e.g. *Chlamydomonas*), colonial (e.g. *Volvox*), filamentous (e.g. *Spirogyra*), or thalloid (e.g. *Fucus*). All algae contain chlorophyll but this may be masked by various accessory pigments, these being one of the major characteristics used to divide the algae into their various phyla. Other characters used to classify the algae are the nature of storage products, the type of cell wall, the form and number of undulipodia (flagella), ultrastructural cell details, and reproductive processes. *See also* Chlorophyta; Phaeophyta; Rhodophyta; Xanthophyta.

alimentary canal /al-ă-**men**-tă-ree/ A tube (8–9 m long in humans) through which food is passed for digestion and absorption into the bloodstream of an animal. In most animals, it leads from the mouth to the anus, and different parts are modified for digestion and absorption of soluble food. Numerous glands pass secretions, containing enzymes, into the alimentary canal and these digest the food as it is moved along by peristalsis.

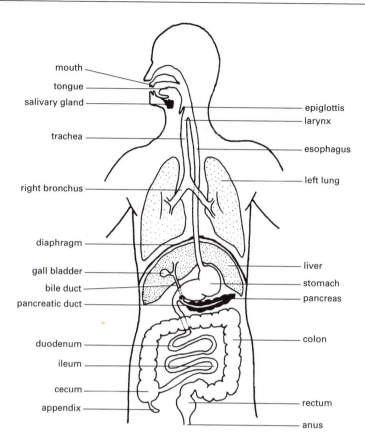

mouth
tongue
salivary gland
trachea
right bronchus
diaphragm
gall bladder
bile duct
pancreatic duct
duodenum
ileum
cecum
appendix

epiglottis
larynx
esophagus
left lung
liver
stomach
pancreas
colon
rectum
anus

The alimentary canal

The wall of the alimentary canal of vertebrates consists of a number of coats. Starting from the inside they are:

1. The mucous membrane – epithelium and underlying connective tissue.
2. Muscularis mucosa – a thin band of muscle.
3. Submucous coat of areolar connective tissue with blood vessels and lymphatics.
4. The muscle coat – circular and longitudinal, and in the stomach, oblique.
5. The peritoneum – a thin layer of connective tissue with an outer squamous epithelium.

aliphatic compound /al-ă-**fat**-ik/ An organic compound with properties similar to those of the alkanes, alkenes, and alkynes and their derivatives. Most aliphatic compounds have an open chain structure but some, such as cyclohexane and sucrose, have rings. The term is used in distinction to AROMATIC COMPOUNDS, which are similar to benzene.

alkaloid /**al**-kă-loid/ One of a group of organic compounds found in plants, that are poisonous insoluble crystalline compounds. They contain nitrogen and usually occur as salts of acids such as citric, malic, and succinic acids. Their function in plants remains obscure, but it is suggested that they may be nitrogenous end-products of metabolism, or they may have a protective function against herbivores. Important examples are quinine, nicotine, atropine, opium, morphine, codeine, and strychnine. They occur mainly in the poppy family, the buttercup family, and the nightshade family of plants.

allantois /ă-**lan**-toh-is/ One of the three embryonic membranes of reptiles, birds, and mammals. It first appears as a ventral outgrowth of the embryo's gut or a 'bubble' ventrally (in primates). It nearly always expands into the coelom under the hindgut, then across the extra-embryonic coelom to fuse with the CHORION, forming the chorio-allantoic membrane (birds) or PLACENTA (mammals). In reptiles and birds, excreted uric acid is stored in the allantoic cavity. *See* cleidoic egg. *See also* amnion.

allele /ă-**leel**/ (allelomorph) One of the possible forms of a given gene. The alleles of a particular gene occupy the same positions (*loci*) on homologous chromosomes. A gene is said to be *homozygous* if the two loci have identical alleles and *heterozygous* when the alleles are different. When two different alleles are present, one (the *dominant* allele) usually masks the effect of the other (the *recessive* allele). The allele determining the normal form of the gene is usually dominant while mutant alleles are usually recessive. Thus most mutations only show in the phenotype when they are homozygous. In some cases one allele is not completely dominant or recessive to another allele. Thus an intermediate phenotype will be produced in the heterozygote. *See also* co-dominance; multiple allelism.

allele frequency (gene frequency) The number of copies of an allele as a proportion of all the alleles of a particular gene in a given population of organisms. It is expressed as a decimal fraction. Hence, if a gene has just two equally common alleles, each has an allele frequency of 0.5.

allelochemical /ă-lee-lŏ-**kem**-ă-kăl, / (allelochemic) A substance produced by organisms of one species that affects the behavior, physiology, or growth of members of another species. There are various types, involved in a wide range of interactions. Some plants release chemicals from their roots into the soil to inhibit the growth of other plant species in the surrounding area. This chemical growth inhibition is called *allelopathy*. Animals often exploit chemicals released by other species to locate suitable prey or hosts, and some release pungent chemicals to warn potential predators that they are toxic or taste unpleasant. Sometimes, both producer and recipient benefit. For example, damage by beetles may cause a pine tree to emit terpenes, which attract insects that parasitize the beetles, thereby countering the beetle infestation.

allelomorph /ă-**lee**-lŏ-morf, -**lel**-ŏ-/ *See* allele.

allelopathy /ă-lee-**lop**-ă-th'ee, al-ĕ-/ Inhibition of the germination, growth, or reproduction of an organism effected by a chemical substance released from another organism. This is a common anti-competition mechanism in plants; for example, barley secretes an alkaloid substance from its roots to inhibit competing weeds. *See also* phytoalexin.

allergy A type of abnormal immune response in which the body produces antibodies against substances, such as dust or pollen, that are usually not harmful and are normally removed or destroyed in other individuals. The allergy-triggering substances (*allergens*) stimulate certain lymphocytes (B-cells) to secrete antibodies, termed *reagins*, which bind to mast cells. When an allergen encounters and binds to the mast cell the latter degranulates, discharging histamine and other substances. Histamine dilates blood capillaries in the region and increases their permeability, resulting in inflammation and increased mucus secretion. It also stimulates contraction of smooth muscle, leading to bronchial constriction. Hence, drugs that block histamine (*antihistamines*) are used to relieve the symptoms of hay fever, asthma, and similar allergies.

allogamy /ă-**log**-ă-mee/ Cross-fertilization in plants. *Compare* autogamy.

allograft /al-ŏ-graft, -grahft/ *See* graft.

allometric growth /al-ŏ-**met**-rik/ The growth of different parts of the body of an organism at different rates or at different times. In humans, for example, brain growth stops at about the age of five years, while other parts of the body continue to grow.

allopatric species /al-ŏ-**pat**-rik/ *See* species.

allopolyploidy /al-ŏ-**pol**-i-ploi-dee/ A type of polyploidy involving the combination of chromosomes from two or more different species. *Allopolyploids* usually arise from the doubling of chromosomes of a hybrid between species, the doubling often making the hybrid fertile. The properties of the hybrid, e.g. greater vigor and adaptability, are retained in the allopolyploid in subsequent generations and such organisms are often highly successful. *Compare* autopolyploidy.

all-or-none Designating the type of response shown by certain irritable tissues that occurs either with full strength or not at all. A stimulus will produce no response until it reaches a certain threshold level, when it produces a fixed maximum response, independent of stimulus intensity. No intermediate response is ever elicited. The firing of a nerve impulse is a typical example.

allosteric site /al-ŏ-**ste**-rik/ A part of an enzyme to which a specific effector or modulator can be attached. This attachment is reversible. Allosteric enzymes possess an allosteric site in addition to their active site. This site is as specific in its relationship to modulators as active sites are to substrates.

allotetraploid /al-ŏ-**tet**-ră-ploid/ (**amphidiploid**) An allopolyploid whose chromosomes are derived from two different species and which therefore has four times the haploid number of chromosomes. *See* allopolyploidy.

alpha helix A highly stable structure in which peptide chains are coiled to form a spiral. Each turn of the spiral contains approximately 3.6 amino-acid residues. The R group of these amino-acids extends outward from the helix. Hydrogen bonding between successive coils holds the helix together. If the alpha helix is stretched the hydrogen bonds are broken but reform on relaxation. The alpha helix is found in muscle protein and keratin.

alternation of generations The occurrence of two, or occasionally more, generations during the life cycle of an organism. It is found in some animals, e.g. certain parasites and hydrozoan coelenterates, and in plants, being particularly clear in some ferns where the generations are independent. Most commonly there is an alternation between sexual and asexual generations, which are usually very different from each other morphologically. In nearly all plants there is also an alternation between haploid and diploid stages. Generally the haploid plant produces gametes mitotically and is thus termed the *gametophyte* while the diploid plant produces spores meiotically and is called the *sporophyte*, though many algae do not follow this rule. The gametes fuse to form a zygote, which develops into the sporophyte, and the spores germinate and produce the gametophyte, so forming a cycle. In bryophytes the haploid gametophyte is the dominant phase of the life cycle and the sporophyte is represented only by the capsule, seta, and foot. In vascular plants the diploid sporophyte is the dominant phase and in the ferns, for example, the gametophyte is a small prothallus. The concept of an alternation of generations can be extended to the flowering plants, in which the embryo sac and pollen represent the much reduced female and male gametophyte generations respectively.

altricial /al-**trish**-ăl/ *See* nidicolous.

altruism Behavior by an animal that favors the survival of other animals of the same species at its own expense. The most common example is that of parents putting themselves at risk, and sometimes losing their lives, to protect and save their off-spring. This has been shown to be genetically favorable to the altruist, increasing the chance that its genes will be passed on, particularly if the parent animal has exhausted, or nearly exhausted, its reproductive capacity. Similarly group altruism, in which genetically more distant group members are protected, will favor gene survival in the long term. *See also* kin selection.

alveolus /al-**vee**-ŏ-lŭs/ (*pl.* **alveoli**) **1.** A minute air sac in the lungs of mammals. Alveoli occur in clusters at the ends of each bronchiole. They have thin moist walls and each is surrounded by a network of capillaries, enabling free exchange of gases between the blood in the capillaries and the inspired air in the alveolus. In birds their function is performed by air capillaries leading from the *parabronchi* (branches of the bronchus). *See* lung.
2. The socket in the jaw bone of a mammal that encloses the root of a tooth. The tooth is fixed into the socket by the vascular *alveolar (periodontal) membrane*, which is derived from the periosteal membranes of the jaw bone and the cement covering the root of the tooth. The interlacing fibers of the alveolar membrane pass into both the jaw bone and the cement, allowing a very slight movement between tooth and jaw during chewing.

amebocyte /ă-**mee**-bŏ-sÿt/ A cell that can wander freely in animal tissues. Amebocytes are found, for example, in the walls of sponges and in the blood and body fluids of mammals. They have the general appearance of *Amoeba*, particularly in exhibiting ameboid movement. *See* macrophage.

ameboid movement A type of locomotion shown by amebas and certain other motile cells. It involves the protrusion of the plasma membrane in the forward direction to form finger-like or balloon-like extensions of the cell called pseudopodia, or 'false feet'. Formation of pseudopodia is linked to reversible changes in the cytoplasm, which cycles between a fluid sol state in the cell interior (the endoplasm) and a gel state just beneath the plasma membrane (the ectoplasm). Endoplasm flows to fill a newly forming pseudopodium, and at the leading edge is converted to ectoplasm. At the cell's trailing edge, the reverse occurs, with ectoplasm changing to endoplasm and flowing forward, causing the 'tail' of the cell to retreat. These sol–gel transformations are themselves caused by the assembly and disassembly of actin microfilament networks, which ultimately determine the viscosity of the cytosol.

amensalism /ă=**men**-sˇ-liz-ăm/ An interaction between populations of two species in which one is adversely affected whereas the other is unaffected. For example, a plant may release toxins into the soil that are detrimental to certain soil insects, thereby inhibiting the latter in the plant's vicinity.

amine /ă-**meen**, **am**-in/ A compound containing a nitrogen atom bound to hydrogen atoms or hydrocarbon groups. They have the general formula R_3N, where R can be hydrogen or an alkyl or aryl group. Amines can be prepared by reduction of amides or nitro compounds.

An amine is classified according to the number of organic groups bonded to the nitrogen atom: one, primary; two, secondary; three, tertiary. Since amines are basic in nature they can form the quaternium ion, R_3NH^+. All three classes, plus a quaternium salt, can be produced by the Hofmann reaction (which occurs in a sealed vessel at 100°C):

$$RX + NH_3 \rightarrow RNH_3^+ X^-$$
$$RNH_3^+ X^- + NH_3 \rightleftharpoons RNH_2 + NH_4X$$
$$RNH_2 + RX \rightarrow R_2NH_2^+ X^-$$
$$R_2NH_2^+ X^- + NH_3 \rightleftharpoons R_2NH + NH_4X$$
$$R_2NH + RX \rightarrow R_3NH^+ X^-$$
$$R_3NH^+ X^- + NH_3 \rightleftharpoons R_3N + NH_4X$$
$$R_3N + RX \rightarrow R_4N^+X^-$$

Reactions of amines include:
1. Reaction with acids to form salts:

$R_3N + HX \rightarrow R_3NH^+X^-$

2. Reaction with acid chlorides to give N-substituted acid amides (primary and secondary amines only):

$RNH_2 + R'COCl \rightarrow R'CONHR + HX$

amino acids /ă-**mee**-noh, am-ă-/ Derivatives of carboxylic acids in which a hydrogen atom in an aliphatic acid has been replaced by an amino group. Thus, from acetic acid, the amino acid, glycine, is formed. All are white, crystalline, soluble in water, and with the sole exception of the simplest member, glycine, all are optically active. In the body the various proteins are assembled from the necessary amino acids and it is important therefore that all the amino acids should be present in sufficient quantities. In humans ten of the twenty amino acids can be synthesized by the body itself. Since these are not required in the diet they are known as *non-essential amino acids*. The remaining ten cannot be synthesized by the body and have to be supplied in the diet. They are known as *essential amino acids. See Appendix for structures.*

Various other amino acids fulfill important roles in metabolic processes other than as constituents of proteins. For example, ornithine $(H_2N(CH_2)_3CH(NH_2)COOH)$ and citrulline $(H_2N.CO.NH.(CH_2)_3 CH(NH_2)COOH)$ are intermediates in the production of urea. Some amino acids function as NEUROTRANSMITTERS; notably gamma-aminobutyric acid (GABA), glutamic acid (glutamate), and glycine.

amino sugar A sugar in which a hydroxyl group (OH) has been replaced by an amino group (NH_2). Glucosamine (from glucose) occurs in many polysaccharides of vertebrates and is a major component of chitin. Galactosamine or chondrosamine (from galactose) is a major component of cartilage and glycolipids. Amino sugars are important components of bacterial cell walls.

amitosis /am-ă-**toh**-sis/ Nuclear division characterized by the absence of a nuclear spindle and leading to the production of daughter nuclei with unequal sets of chromosomes. The ordered process of division, duplication of chromosomes, dissolution of nuclear membrane, and production of a spindle as in mitosis is apparently absent. Cells produced amitotically inherit vari-

---

THE AMINO ACIDS MOST COMMONLY FOUND IN PROTEINS

alanine	glycine	proline**
arginine	histidine*	serine
asparagine	isoleucine*	threonine*
aspartic acid	leucine*	tryptophan*
cysteine	lysine*	tyrosine*
glutamic acid	methionine*	valine*
glutamine	phenylalanine*	

* essential amino acids in animal diets
** an imino acid derived from pyrollidine

```
        H
        |
R ----- C ----- COOH
        |
        NH₂
```

Amino acid: the amino acids in proteins are alpha amino acids. The –COOH group and –NH_2 group are on the same carbon atom able numbers of chromosomes. The chances of a daughter cell lacking essential genes are less than may be expected since many cells that characteristically divide amitotically are polyploid, e.g. the endosperm nucleus in angiosperms and the macronucleus of ciliates. *Compare* endomitosis; mitosis.

ammonia /ă-**moh**-nee-ă/ A colorless pungent gas, NH_3, that is soluble in water and toxic to living organisms at high concentrations. It is the form in which nitrogenous waste is excreted in AMMONOTELIC animals, chiefly teleost fishes and aquatic invertebrates. Ammonia is also important in the NITROGEN CYCLE, being produced from atmospheric nitrogen by nitrogen-fixing bacteria in the soil, and by the Haber process, which provides raw material for the manufacture of nitric acid, ammonium nitrate and other industrial chemicals. *See* nitrogen fixation.

ammonites /am-ŏ-nÿts/ An extinct group of cephalopod mollusks (*see* Cephalopoda) that had multi-chambered shells and flourished in the Mesozoic era, dying out at the end of the Cretaceous period some 65 million years ago. The shells were typically spirally coiled, and the animal secreted successive chambers as it grew, living in the chamber nearest the opening. Ammonites underwent distinct changes in form over their long history, for example in the pattern of external shell sutures, and this has enabled a detailed phylogenetic tree to be constructed. Hence, ammonites are often used as index fossils to date rock strata.

ammonotelic /am-ŏ-noh-**tel**-ik/ Describing animals that excrete their nitrogenous waste mainly in the form of ammonia. Because of its toxicity, the ammonia must be kept diluted and its excretion requires large volumes of water. Hence ammonotelic animals are mostly teleost fishes and aquatic invertebrates, in which the ammonia is excreted across the gills. Aquatic birds, such as ducks, can also excrete significant amounts of ammonia, but only small amounts are present in the urine of mammals. *Compare* ureotelic; uricotelic.

amniocentesis /am-nee-oh-sen-**tee**-sis/ A process whereby a sample of amniotic fluid is obtained from a pregnant woman in order to identify an abnormal fetus. It is usually performed between the 16th and 20th weeks of pregnancy, and involves inserting a needle through the abdominal wall and uterus and withdrawing a sample of amniotic fluid. Stray fetal cells in the fluid may be examined for chromosome abnormalities (*see* Down's syndrome), while abnormal concentrations of various substances may indicate other problems, e.g. a significantly raised concentration of alpha-fetoprotein may be due to spina bifida or anencephaly.

amnion /am-nee-ŏn/ One of the three embryonic membranes of reptiles, birds, and mammals. It is the inner layer of membrane enclosing the embryo, first as a hood and later as a complete bubble. The amniotic cavity contains *amniotic fluid*, which may be sampled in pregnant women to diagnose some kinds of genetic abnormality in the fetus (*see* amniocentesis). *See also* chorion.

amniotes /**am**-nee-ohts/ Reptiles, birds, and mammals, i.e. those vertebrates whose embryos always possess an amnion.

Amoeba /ă-**mee**-bă/ The best-known protozoan protoctist. A microscopic organism found universally in fresh water, it has a continually changing shape due to the formation of pseudopodia for locomotion and food capture. Food and water vacuoles carry out digestion and osmoregulation. Reproduction is by binary fission and spore formation takes place in adverse conditions. *See also* protozoa; Rhizopoda.

AMP (adenosine monophosphate) A nucleotide consisting of adenine, ribose, and phosphate. *See* ATP; cyclic AMP.

Amphibia /am-**fib**-ee-ă/ The class of vertebrates that contains the most primitive terrestrial tetrapods – the frogs, toads, newts, and salamanders. Amphibians have four pentadactyl limbs, a moist skin without scales, a pelvic girdle articulating with the sacrum, and a middle-ear apparatus for detecting airborne sounds, but no external ear. They are poikilothermic and the adults have lungs and live on land but their skin, also used in respiration, is thin and moist and body fluids are easily lost, therefore they are confined to damp places. In reproduction, fertilization is external and so they must return to the water to breed. The eggs are covered with jelly and the aquatic larvae have gills for respiration and undergo metamorphosis to the adult. Partial or complete neoteny occurs in some amphibians; for example *Ambystoma* (Mexican axolotl) is permanently aquatic, with larval gills retained in the adult and atrophied lungs. *See also* Anura.

amphicribal /am-fă-**krÿ**-băl/ *See* amphiphloic.

amphidiploid /am-fă-**dÿ**-ploid/ *See* allotetraploid.

amphimixis /am-fă-**miks**-is/ True sexual reproduction by fusion of gametes. *Compare* apomixis.

Amphioxus /am-fee-**oks**-ŭs/ *See* Branchiostoma.

amphiphloic /am-fă-**floh**-ik/ (**amphicribal; periphloic**) Describing a centric vascular bundle in which the tissues are arranged concentrically and an outer ring of phloem completely surrounds a central core of xylem, as seen in the stele of *Selaginella*. *Compare* amphixylic.

amphivasal /am-fă-**vay**-săl/ *See* amphixylic.

amphixylic /am-fiks-**il**-ik/ (**amphivasal; perixylic**) Describing a centric vascular bundle in which the tissues are arranged concentrically and an outer ring of xylem completely surrounds a central core of phloem cells. This arrangement can be seen, for instance, in lily of the valley. *Compare* amphiphloic.

ampulla /am-**pul**-ă -**pûl**-/ (*pl.* **ampullae**) 1. A small enlargement or swelling of a tube or canal.

2. A swelling in each of the three SEMICIRCULAR CANALS of the inner ear that contains sensory hair cells responsible for detecting acceleration of the head.
3. Any of the small sacs in the water vascular system of starfish and other members of the ECHINODERMATA that act as reservoirs for hydraulic expansion of the tube feet.

amylase /am-ă-layss/ A member of a group of closely related enzymes, found widely in plants, animals, and microorganisms, that hydrolyzes starch or glycogen to the sugars maltose, glucose, or dextrin. Both α- and β-amylases occur in plants, the latter particularly in malt (being used in the brewing industry), but only α-amylase is found in animals, in the pancreatic juices and in saliva (*see* ptyalin), having an important role in digestion.

amyloid /am-ă-loid/ A waxy translucent material consisting of protein fibrils that is deposited in tissues of humans and other animals in certain disorders (collectively called amyloidoses). The accumulation of amyloid interferes with tissue function, causing progressive worsening of symptoms.

amylopectin /am-ă-lo-**pek**-tin/ The water-insoluble fraction of starch. *See* starch.

amyloplast /am-ă-lŏ-plast/ A plastid storing starch grains. They are common in storage organs, e.g. the potato tuber. They have a physiological role in the root cap and elsewhere where the starch grains act as statoliths. *See also* geotropism.

amylose /am-ă-lohs/ The water-soluble fraction of starch. *See* starch.

anabolic steroid Any steroid hormone or synthetic steroid that promotes growth and formation of new tissue. They are used in the treatment of wasting diseases, and are sometimes used in agriculture to boost livestock production, and, controversially, in sport to build up athletes' muscles, although this is now generally outlawed. Most androgens have anabolic activity, but when used by women they lead to masculinization.

anabolism /ă-**nab**-ŏ-liz-ăm/ Metabolic reactions in which molecules are linked together to form more complex compounds. Thus, anabolic reactions are concerned with building up structures, storage compounds, and complex metabolites in the cell. Starch, glycogen, fats, and proteins are all products of anabolic pathways. Anabolic reactions generally require energy provided by ATP produced by catabolism. *See also* metabolism.

anaerobe /an-**air**-ohb/ An organism that can live and grow in the absence of free oxygen, i.e. it respires anaerobically (*see* anaerobic respiration). Anaerobes can be facultative, in that they usually respire aerobically but can switch to anaerobic respiration when free oxygen is in short supply, or obligate, in that they never respire aerobically and may even be poisoned by free oxygen. *Compare* aerobe.

anaerobic respiration   Respiration in which oxygen is not involved, found in yeasts, bacteria, and occasionally in muscle tissue. The organic substrate is not completely oxidized and the energy yield is low. In the absence of oxygen in animal muscle tissue, glucose is degraded to pyruvate by glycolysis, with the production of a small amount of energy and also lactic acid, which may be oxidized later when oxygen becomes available (*see* oxygen debt). Fermentation is an example of anaerobic respiration, in which certain yeasts produce ethanol and carbon dioxide as end products. Only two molecules of ATP are produced by this process. *See* anaerobe. *Compare* aerobic respiration.

analogous   Describing structures that are apparently similar (structurally or functionally) but have a different evolutionary origin, and thus a different embryological origin and structure. The wings of birds and insects have a similar function, but are analogous not HOMOLOGOUS.

anamniotes /an-**am**-nee-ohts/ Fishes and amphibians, i.e. those vertebrates whose embryos rarely possess an amnion.

anaphase /an-ă-fayz/ The stage in mitosis or meiosis when chromatids are pulled towards opposite poles of the nuclear spindle. In mitosis the chromatids moving towards the poles represent a single complete chromosome. During anaphase I of meiosis a pair of chromatids still connected at their centromere move to the spindle poles. During anaphase II the centromeres divide and single chromatids are drawn towards the poles.

anaphylaxis /an-ă-fă-**laks**-is/ The severe reaction of an animal to an antigen to which it has previously been exposed. In humans it occurs rarely after injection of antiserum or antibiotics, after bee or wasp stings, etc. It results from the release of histamine and other substances from mast cells, following antigen–antibody interaction.

anatomy 1. The study of the structure of the body of an organism as learned by dissecting it.
2. The organization of the parts of the body and the structural relationships between them.

anatropous /ă-**nat**-rŏ-pŭs/ Describing the position of the ovule in the ovary when the developing ovule has turned through 180°, so that the micropylar end has folded over and lies close to the base of the funicle (stalk). This is the most frequent position of the ovule in flowering plants. *Compare* campylotropous; orthotropous. *See illustration at* ovule.

ancestral trait   *See* plesiomorphy.

androdioecious /an-drŏ-dÿ-**ee**-shŭs/ Describing species in which male and hermaphrodite flowers are borne on separate plants. *Compare* gynodioecious.

androecium /an-**dree**-shee-ŭm/ The collective name in higher plants for the male parts of a plant, i.e. the stamens. It is denoted in the floral formula by a letter *A*. *See also* stamen.

androgen /**an**-drŏ-jĕn/ A male sex hormone (a steroid) that controls the development, function, and maintenance of secondary male characteristics (e.g. facial hair and deepening of the voice), male accessory sex organs, and spermatogenesis. Androgens are produced chiefly by the testis (smaller amounts are produced by the ovary, adrenal cortex, and placenta); the most important is testosterone. Castration (removal of the testes) leads to atrophy of accessory sexual organs; this effect can be prevented by androgen replacement therapy. Androgens are also used in the treatment of diseases in which androgen secretion is reduced or absent (e.g. hypogonadism, hypopituitarism) and in the treatment of certain breast cancers. They also have anabolic activity, promoting growth and formation of new tissue. *See also* anabolic steroid.

andromonoecious /an-drŏ-mŏ-**nee**-shŭs/ Describing species in which both male and hermaphrodite flowers are borne on the same plant. *Compare* gynomonoecious.

androsterone /an-**dros**-tĕ-rohn/ A steroid hormone formed in the liver from the metabolism of testosterone. It has only weak androgenic activity. *See* androgen.

anemophily /an-ĕ-**moff**-ă-lee/ Pollination by wind. Plants pollinated in this manner (e.g. grasses) have insignificant unscented flowers with large, often feathery stigmas.

aneuploidy /an-yŭ-ploi-dee/ The condition, resulting from nondisjunction of homologous chromosomes at meiosis, in which one or more chromosomes are missing from or added to the normal somatic chromosome number. If both of a pair of homologous chromosomes are missing, *nullisomy* results. *Monosomy* and *trisomy* are the conditions in which one or three homologs occur respectively, instead of the normal two. *Polysomy*, which includes trisomy, is the condition in which one or more chromosomes are represented more than twice in the cell. *See* nondisjunction.

angiotensin /an-jee-oh-**ten**-sin/ A polypeptide, existing in two forms in the blood, that raises blood pressure. Angiotensin I is an inactive decapeptide produced by the action of renin, a kidney enzyme released when blood pressure is low, on a globulin in the blood. The active form, angiotensin II, is formed by the enzyme-catalyzed removal of two terminal amino acids from angiotensin I. Angiotensin II raises blood pressure by stimulating the constriction of arterioles and the secretion of the hormone aldosterone. High blood pressure may be treated by inhibiting the enzyme responsible for converting inactive angiotensin I to active angiotensin II.

angstrom /**ang**-strŏm/ Symbol: Å A unit of length equal to 10^{-10} meter. It is still used occasionally for measurements of wavelength or interatomic distance. The unit is named for the Swedish physicist and astronomer Anders Jonas Ångstrom (1814–74).

aniline stains *See* staining.

animal An organism that feeds on other organisms or on organic matter, is often motile, and reacts to stimuli quickly. Animal cells are surrounded by cell membranes. There is no chlorophyll and growth is usually limited. *Compare* plant.

animal pole That part of the surface of animal eggs to which the nucleus (germinal vesicle) is closest. It is usually opposite the vegetal, or yolky, pole.

animal starch *See* glycogen.

anion /**an**-ÿ-ŏn, -on/ A negatively charged ion, formed by addition of electrons to atoms or molecules. In electrolysis anions are attracted to the positive electrode (the anode). *Compare* cation.

anisogamy /an-ÿ-**sog**-ă-mee/ The sexual fusion of non-identical gametes. Anisogamy grades from situations in which the

---

DIFFERENCES BETWEEN PLANTS AND ANIMALS

	Plants	*Animals*
Nutrition	by synthesis. Take in simple substances (carbon dioxide, water, minerals) and, using light energy, convert these into all the compounds needed for growth; chlorophyll normally present.	by breakdown. Take in complex substances (plant or animal tissues) and, by digestion, reduce these to simpler compounds that are absorbed into the body providing energy, or building blocks for growth; chlorophyll never present.
Support	by pressure exerted on rigid cellulose cell walls by water-filled vacuole; additional strengthening tissues, e.g. lignin, are formed in older plants.	by various mechanisms, e.g. internal or external skeletons; cell walls not rigid, cellulose never present.
Sensitivity	response to stimuli slow, and generally only occuring if stimulus maintained over long period.	response to stimuli rapid and generally occurring immediately after application of the stimulus.
Mobility	land plants of necessity immobile as need to withdraw water and nutrients from the soil by roots; some aquatic microscopic plants possess flagella and are mobile.	organism able to move whole body from place to place

---

gametes differ only in size to the extreme of oogamy, in which one gamete is a large immotile ovum and the other a small motile sperm. *Compare* isogamy.

Annelida /ă-**nel**-ă-dă/ A phylum of triploblastic bilaterally symmetrical metamerically segmented invertebrates, the segmented worms, including *Nereis, Lumbricus*, and the leeches. Annelids have a long soft cylindrical body covered by a thin cuticle and most have segmentally arranged chitinous bristles (chaetae), which assist in locomotion. The body wall contains layers of circular and longitudinal muscle and the body cavity is a coelom isolating the gut from the body wall. These features, together with the metamerism, provide an efficient means of locomotion. Many are hermaphrodite. The gut runs from the mouth at the front to the posterior anus. There are well-developed blood and nervous systems and nephridia for excretion. The phylum is divided into three classes (*see* Polychaeta; Oligochaeta; Hirudinea). *See also* archiannelid.

annual A plant that completes its life cycle within a year. Examples are the common field poppy and the sunflower. *Compare* biennial; ephemeral; perennial.

annual ring (growth ring) The annual increase in girth of the stems or roots of woody plants, as a result of cambial activity. The annual rings of plants growing in temperate climates can be seen in cross-section as two consecutive rings of light and dark colored xylem tissue. These are formed from a zone containing larger vessel elements produced by the cambium in the spring (lighter layer), followed by a zone containing smaller vessel elements (darker layer) produced during the late summer. This process is repeated annually so that the number of light or dark rings indicate the age of that part of the plant. *See illustration at* wood.

annular thickening Rings of thickening laid down on the inner wall of protoxylem vessels and tracheids. Such thickening allows extension of the XYLEM between the rings so that it is not ruptured as the surrounding tissues grow.

annulus /an-yŭ-lŭs/ (*pl.* **annuli**) **1.** The ring of tissue surrounding the stalk of the mature fruiting body (toadstool) of the basidiomycete fungi. The annulus is all that remains of the veil, which joined the rim of the pileus to the stalk in the immature toadstool. It is sometimes termed the *velum*.
2. A special arc or ring of cells in the sporangia of ferns that constitutes the mechanism for spore dispersal. The cells of the annulus are thickened except on the outer wall, so that they contract on drying out. This causes the stomium to rupture exposing the spores within the capsule. The capsule wall gradually bends back as the water in the annulus cells continues to evaporate until a point when the remaining water in the cells suddenly turns to vapor. This results in the wall springing back to its original position, the sudden movement dispersing any remaining spores.
3. A ring of large cells separating the epidermis from the operculum in certain bryophytes (e.g. *Funaria*).
4. (*Zoology*) A ring-shaped structure, such as any of the external segments of an annelid worm.

Anoplura /an-ŏ-**ploor**-ă/ The order of insects that contains the sucking lice, e.g. *Pediculus humanus* (the head and body louse). Lice, ectoparasites of mammals, are found universally in unhygienic conditions and are carriers of typhus and other diseases. They lack wings and eyes and have a flattened transparent body with piercing and sucking mouthparts for feeding on the host's blood and prehensile clawed legs for attachment to the host. The eggs (nits) are attached to the host's hair and develop into blood-sucking nymphs.

anoxic /an-**oks**-ik/ Lacking or not needing oxygen.

ANS *See* autonomic nervous system.

antagonism **1.** The interaction of two substances, e.g. drugs or hormones, which have opposing effects in a given system, such that one partially or completely inhibits the effect of the other.
2. The interaction of two types of organisms, such that the growth of one is partially or completely inhibited by the other.
3. The opposing action of two muscles, such that the contraction of one is accompanied by relaxation of the other.

antagonist **1.** (*Anatomy*) A muscle whose contractions oppose the action of another muscle (the agonist). Muscles are commonly arranged in antagonistic combinations to flex and extend joints. For example, in the human lower leg, the tibialis anterior muscle antagonizes the gastrocnemius and soleus muscles; the former flexes (closes) the ankle joint, while the latter two muscles extend (open) the ankle.
2. (*Biochemistry*) An agent, such as a drug or other substance, that opposes the effects of another agent, the agonist (e.g. a neurotransmitter or hormone). Antagonists often work by binding to the cell receptors normally occupied by the agonist, thereby reducing or abolishing cell activation by the agonist.

antenna (*pl.* **antennae**) One of a pair of threadlike appendages on the heads of many arthropods. They generally have a sensory function (touch and smell) but some crustaceans use them for swimming and attachment. *See also* antennule.

antennule /an-**ten**-yool/ One of a pair of small threadlike sensory appendages that occur anterior to the antennae on the heads of Crustacea.

anterior /an-**teer**-ee-er/ **1.** Designating the front end of an animal. In bilaterally symmetrical animals this is the end that leads during locomotion. However, in bipedal animals, such as humans, the anterior side corresponds to the ventral side of other animals.
2. Designating the part of a flower or axillary bud facing away from the inflorescence axis or stem, respectively.
Compare posterior.

anther /an-th'er/ The part of the stamen that produces the pollen. The anther is usually joined to the tip of the filament (stalk) and is made up of two lobes. Each lobe contains two pollen sacs that produce very large quantities of small pollen grains. The pollen is released when the lobes split open longitudinally.

The anther is made up of an outer epidermis, a middle fibrous layer, and an inner nutritive layer, the *tapetum*. The haploid pollen cells develop in the tapetal zone from spore mother cells.
See illustration at flower.

anther culture (**pollen culture**) The generation of haploid plants from immature pollen grains or intact excised anthers. The resultant plants are generally smaller than their diploid counterparts. When single pollen grains are cultured, some anther tissue may have to be present as nurse tissue. Treatment with colchicine can double the ploidy level of a pollen grain and the resulting diploid cells can then be cultured. Rice and tobacco plants have been cultured in this way.

antheridium /an-th'ĕ-**rid**-ee-ŭm/ (*pl.* **antheridia**) The male sex organ of the algae, fungi, bryophytes, and pteridophytes. It may be made up of one cell, or one or many layers of cells. It produces gametes that are usually motile. *Compare* archegonium.

antherozoid /an-th'er-ŏ-**zoh**-id, **an**-th'er-ŏ-zoid/ (**spermatozoid**) The male gamete of algae, fungi, bryophytes, pteridophytes and some of the gymnosperms. It is motile and is produced in an antheridium, except in certain gymnosperms (e.g. *Cycas*, *Ginkgo*) in which antherozoids develop from the generative cells of the pollen tube.

Anthoceratophyta /an-th'oh-se-ră-**toff**-ă-tă/ (**hornworts**) A phylum of simple nonvascular plants known as hornworts, or horned liverworts, and comprising about 400 species. Superficially similar to certain liverworts, hornworts have a thin, flattened leaflike thallus (the gametophyte), from which arise slender elongate sporophytes (the 'horns'). They grow in temperate and tropical regions on moist soil, for example in road cuts and on river banks. In some species the thallus contains both male and female sex organs, while in others the sexes are separate. The sperms are motile, and swim through moisture films to fertilize the egg. The sporophytes are green and able to photosynthesize. When mature they split longitudinally from base to tip to release spores, which germinate to form new gametophytes. The cells of hornworts contain a single large chloroplast, which, uniquely in the plant kingdom, contains a PYRENOID for starch formation, a feature more typical of algae.

anthocyanin /an-th'ŏ-**sȳ**-ă-nin/ One of a group of water-soluble pigments found dissolved in higher plant cell vacuoles. Anthocyanins are red, purple, and blue and are widely distributed, particularly in flowers and fruits where they are important in attracting insects, birds, etc. They also occur in buds and sometimes contribute to the autumn colors of leaves. They are natural pH indicators, often changing from red to blue as pH increases, i.e. acidity decreases. Color may also be modified by traces of iron and other metal salts and organic substances, for example cyanin is red in roses but blue in the cornflower. *See* flavonoid.

Anthophyta /an-**thof**-ă-tă/ (**angiosperms**; **flowering plants**) An extremely important phylum of vascular seed plants characterized by their flowers, which contain the male and female reproductive structures. They differ from conifers and other gymnosperms by having the ovule enclosed within an ovary, which after fertilization develops into a fruit. The female gameto-phyte is represented by the embryo sac, archegonia being absent. Angiosperms are divided into two major groups depending on the number of cotyledons, giving the monocotyledons and dicotyledons. *See* Liliopsida; Magnoliopsida.

Anthozoa /an-th'ŏ-**zoh**-ă/ A class of cnidarians, the sea anemones and corals, in which the polyp is the only form and the medusa is absent. The solitary sea anemone has numerous feathery tentacles. Corals are colonial, with the polyp contained in a gelatinous matrix (the soft corals), a horny skeleton (the horny corals), or a skeleton of calcium carbonate (the stony or true corals). Accumulations of these corals in warm shallow seas form coral reefs.

antibiotic One of a group of organic compounds, varying in structure, that are produced by microorganisms and can kill or inhibit the activities of other microorganisms. One of the best known examples is penicillin, which was discovered by Sir Alexander Fleming. Another example is streptomycin. Antibiotics are widely used in medicine to combat bacterial infections.

antibody A protein molecule formed within the body of an animal in order to neutralize the effect of a foreign invading protein (called an *antigen*). Antibodies are produced by lymphocytes (B-cells) in response to the presence of antigens. Each antibody has a molecule structure that exactly fits the structure of one particular antigen molecule, like a lock and key (i.e. they are specific). Antibody molecules attach themselves to invading antigen molecules (on or from bacteria, transfused red blood cells, or grafted tissue of another animal) and so render them inactive. Some cause agglutination (clumping) of invading cells, so that they disintegrate; others, called *opsonins*, make bacteria more easily engulfed by phagocytic leukocytes. Antibodies are important in defense against infectious diseases and in developing immunities. *See also* agglutination; immunity; lymphocyte.

anticlinal /an-tee-**klÿ**-năl/ Describing a line of cell division at right angles to the surface of the organ. *Compare* periclinal.

anticoagulant /an-tee-koh-**ag**-yŭ-lănt/ A chemical that can prevent blood clotting. An example is heparin, which occurs in the saliva of leeches, mosquitoes, and other blood-sucking animals. Traces are found normally in blood and may help to prevent clotting in the blood. Heparin acts by inhibiting the formation of thrombin. Another anticoagulant is acid sodium citrate, which is added to blood taken for transfusion and prevents clotting by removing the calcium ions (which are necessary for clotting). *See also* blood clotting.

anticoding strand /an-tee-**koh**-ding/ *See* template strand.

anticodon /an-tee-**koh**-don/ A nucleotide triplet on transfer RNA that is complementary to and bonds with the corresponding codon of messenger RNA in the ribosomes. *See* transfer RNA.

antidiuretic hormone /an-tee-dÿ-yŭ-**ret**-ik/ (**ADH**) *See* vasopressin.

antigen /**an**-ti-jĕn/ A substance that induces the production of antibodies that are able to bind to the antigen.

antigenic variation /an-ti-**jen**-ik/ The capacity of some pathogenic organisms, such as viruses, to alter their surface molecules and thereby avoid detection by the immune system of their host. For example, influenza viruses continually change the nature of their coat proteins, and hence antibodies formed after a previous infection no longer recognize and bind to the variant viruses, and are unable to prevent a new infection. The variation is caused by recombination within the genes encoding the coat proteins, in some cases involving genes from different strains, and it presents an immense problem when attempting to develop effective vaccines.

antioxidant /an-tee-**oks**-ă-dănt/ A substance that slows or inhibits oxidation reactions, especially in biological materials or within cells, thereby reducing spoilage or preventing damage. Natural antioxidants include vitamin E and β-carotene. These work by reacting with peroxides or oxygen free radicals, effectively mopping them up. Antioxidants are added to a range of foods, e.g. margarines, and industrial materials, such as plastics.

antipodal cells /an-**tip**-ŏ-dăl/ The three haploid cells found in the embryo sac of seed-bearing plants that migrate to the chalazal end of the sac opposite the micropyle. These nuclei arise as a result of the three meiotic divisions that produce the egg cell, synergid cells, and polar nuclei, but they do not themselves take part in the fertilization process. They are eventually absorbed by the developing embryo, and their function is uncertain.

antisense DNA /**an**-tee-sens/ The DNA strand that is transcribed. In transcription the DNA double helix unwinds and only one of the strands acts as the template for messenger RNA (mRNA) synthesis. The base sequence of the resultant mRNA carries the genetic code, and is the 'sense' molecule. It is complementary to the corresponding segment of transcribed DNA, which is therefore the 'antisense' molecule. Short segments of single-stranded antisense nucleic acid (DNA or RNA) can be synthesized in order to bind to specific regions of messenger RNA and prevent translation (*see* RNA interference). They can also bind to some parts of the DNA double helix, forming a triple helix, and so prevent gene transcription. Antisense nucleic acids can thus suppress the activity of certain genes, and have potential as therapeutic agents for certain genetic disorders and cancers, and as antiviral agents.

antisense RNA A short single-stranded RNA molecule whose base sequence is complementary to that of a messenger RNA (mRNA – the 'sense' molecule) and is able to undergo base pairing with it. Binding of the antisense RNA blocks translation of the mRNA and hence gene expression, and may trigger cleavage of the double-stranded RNA by cellular ribonuclease enzymes. Hence antisense RNA can be used as a means of targeting and blocking the expression of particular genes, for example in genetic engineering or gene therapy. However, double-stranded RNA is now known to be much more effective in silencing the expression of specific genes due to the phenomenon of RNA INTERFERENCE.

antiserum /an-tee-**seer**-ŭm/ A serum containing antibodies against a particular antigen, obtained from a human or animal which has been exposed to the antigen, either naturally (through disease) or through immunization. Antisera are injected to give passive immunity against specific diseases and also used in the laboratory to identify unknown pathogens.

ants *See* Hymenoptera.

anucleate /ay-**new**-klee-ayt -it/ Lacking a nucleus; for example, erythrocytes (red blood cells) are anucleate.

Anura /ăn-**yoor**-ă/ The order of amphibians that contains the frogs and toads. The adults are highly specialized for jumping, having a short backbone, no tail, very long powerful hind limbs, and a strengthened pectoral girdle to absorb the shock of landing. The hind feet are webbed for swimming. Most of their oxygen is absorbed through the skin, supplementing the limited supply drawn into the lungs by the pumping action of the floor of the mouth. Oxygenated and deoxygenated blood are not completely separated in the heart. The eggs (spawn), covered with jelly, are laid in water and hatch into aquatic larvae (tadpoles), which undergo a rapid and extensive metamorphosis in which the tail is absorbed and the gill slits are replaced by lungs. Most frogs (e.g. *Rana*) live in damp places or are aquatic; some are arboreal. Toads (e.g. *Bufo*), which have a dry warty skin, are better adapted to drier habitats.

anus The posterior opening to the alimentary canal, which occurs in nearly all animals. Through it, feces and sometimes semisolid wastes are expelled from the body, often under muscular control. It sometimes opens into a cloaca. Some aquatic animals also use the anus during respiration. *See also* sphincter. *See illustration at* alimentary canal.

aorta In mammals, the large artery arising from the left ventricle of the heart that carries oxygenated blood, via various branches, to all parts of the body. It is divided into an ascending portion, an arch, and a descending portion. The aorta forms the left branch of the systemic arch. *See also* dorsal aorta; ventral aorta. *See illustration at* heart.

aortic arches Six pairs of blood vessels present in all vertebrate embryos, which link the ventral aorta leaving the heart with the dorsal aorta. Arches one and two soon disappear and in adult fish arches three to six lead to the gills. Adult tetrapods lose arch five, arch three becomes the carotid arch supplying the head, arch four becomes the systemic arch supplying the body, and arch six becomes the pulmonary arch supplying the lungs.

aphids /**ay**-fidz/ *See* Hemiptera.

apical dominance /**ap**-ă-kăl, **ay**-pă-/ The phenomenon in which the presence of a growing apical bud on a plant inhibits the growth of lateral buds. It is controlled by the interactions of plant hormones, particularly auxin (produced by the shoot tip).

apical meristem The actively dividing cells constituting the growing point at the tip of the root or stem in vascular plants. New cells are cut off on the lower side to form new stem tissue at the stem apex, and on both sides in the root apex to form root tissue and a protective root cap. The apical meristems in the lower plants consist of one cell only, as in the ferns, but become more complex and consist of groups of cells in the higher plants. *See* histogen theory; tunica–corpus theory.

Apicomplexa /ap-ă-kom-**pleks**-ă/ A phy-

lum of spore-forming protoctists that are parasites of animals. It includes the malaria parasite, *Plasmodium*, and several important pathogens of domestic livestock, such as *Toxoplasma* (causing toxoplasmosis) and *Eimeria* (responsible for coccidiosis). Apicomplexans are named for a collection of fibrils, vacuoles, and organelles (the 'apical complex') visible at one end of the cell. They reproduce sexually and have complex life cycles, often involving several hosts, and can proliferate rapidly by a series of cell divisions (schizogony). Members of the Apicomplexa were traditionally classified as sporozoans.

aplanospore /ă-**plan**-ŏ-spor/ A nonmotile spore, characteristic of the pin molds and green algae. It is an asexual spore formed in a sporangium and is usually thick walled.

apocarpy /**ap**-ŏ-kar-pee/ An ovary made up of unfused carpels, as in the buttercup. *Compare* syncarpy.

apoenzyme /ap-oh-**en**-zȳm/ An enzyme whose cofactor has been removed (e.g. via dialysis) rendering it catalytically inactive. It is the protein part of a conjugate enzyme. When combined with its prosthetic group (coenzyme) it forms a complete enzyme (holoenzyme).

apogamy /ă-**pog**-ă-mee/ In pteridophytes, the development of the sporophyte directly from a cell of the gametophyte, so fusion of gametes is bypassed. It frequently occurs in gametophytes that have been produced aposporously and are thus diploid. The term also describes the development of an unfertilized female gamete into the sporophyte, a phenomenon called PARTHENOGENESIS. *See* apospory; apomixis.

apomixis /ap-ŏ-**miks**-eez/ (**agamospermy**) A modified form of reproduction by plants in which seeds are formed without fusion of gametes. It is comparable to the conditions of APOGAMY and APOSPORY, which are seen in many pteridophytes. Apomixis includes the process whereby a diploid cell of the nucellus develops into an embryo giving a diploid seed with a genetic constitution identical to the parent. Another form of apomixis in which seeds develop from unfertilized gametes can also be termed PARTHENOGENESIS. Seeds produced in this way may be either haploid or diploid depending on whether or not the megaspore mother cell undergoes meiosis. Often, in the process termed *pseudogamy*, entry of the male gamete is required to stimulate the development of the female gamete, even though nuclear fusion does not occur. Such cases of apomixis are difficult to distinguish from true sexual reproduction. *Compare* amphimixis.

apomorphy /**ap**-ŏ-mor-fee/ (**derived trait**) A character, or trait, that is not present in an ancestor but is present in one or more of its descendants. Apomorphies are useful in constructing phylogenetic trees. *See* cladistics; synapomorphy. *Compare* plesiomorphy.

apoplast /**ap**-ŏ-plast/ The system of cell walls extending through a plant body and along which water containing mineral salts, etc. can move passively. It is an important pathway for movement of these substances outside the xylem, for example across the root cortex. *Compare* symplast.

apoptosis (**programmed cell death**) The orderly destruction and disposal of cells within multicellular organisms. Apoptosis is an essential means of sculpting tissues and remodeling organ systems during development, and of eliminating damaged or superfluous cells during later life. The events of apoptosis follow a characteristic pattern. The cell starts to shrink, the organelles break up, and in the nucleus the chromosomes become condensed and fragmented. Then the nucleus and cytoplasm break up, forming fragments known as *apoptotic bodies*, which are taken up and digested by motile phagocytic cells. Apoptosis involves the activation of enzymes called *caspases*. These cleave certain target proteins and thereby activate other 'executioner' enzymes that chop up the cell's DNA, its cytoskeleton, etc. Apoptosis is controlled by various intracellular and ex-

tracellular signals. For example, most cells require continual extracellular survival signals to stay alive. If these signals cease, the cell activates a 'suicide' programme of apoptosis. Apoptosis can also be triggered by specific extracellular signals. This occurs, for instance, in the destruction of nonfunctional lymphocytes, which is prompted by 'murder' signals from other lymphocytes.

aposematic coloration /ap-ŏ-sĕ-**mat**-ik/ *See* warning coloration.

apospory /**ap**-ŏ-spor-ee, -spoh-ree, ă-**pos**-pŏ-ree/ The development of the gametophyte directly from the cell of a sporophyte, thus bypassing meiosis and spore production. Gametophytes produced in this manner are thus diploid instead of haploid. If such gametophytes produce fertile gametes the resulting sporophyte is then tetraploid, and large polyploid series may subsequently be developed. Apospory is found in some bryophytes and pteridophytes. *See also* apogamy; apomixis.

apothecium /ap-ŏ-**th'ee**-shee-ŭm, -see-ŭm/ (*pl.* **apothecia**) *See* ascocarp.

appeasement A type of behavior performed by an animal to prevent attack by another of its species. For example, a female entering a male's territory during courtship may be threatened as he defends his territory. The female performs appeasement gestures to inhibit attack; they may be actions that arouse nonaggressive tendencies in the male, for example, food-begging may arouse a parental response; or the female may adopt postures totally different from those characteristic of a rival male, thus indicating her nonthreatening intentions. In dominance hierarchies, subordinate members of the group will use appeasement behavior to avoid attack by the dominant animals.

appendicular skeleton /ap-en-**dik**-yŭ-ler/ The parts of the skeleton of a vertebrate animal (*see* endoskeleton) that are attached to the vertebral column (i.e. the axial skeleton), comprising the pectoral and pelvic girdles and the limbs (arms, legs, wings, fins).

appendix (vermiform appendix) A blind tube or diverticulum protruding from the cecum. In humans and other primates it is small, whereas in herbivorous mammals, such as the rabbit, it is larger and contains microorganisms concerned with cellulose digestion. The human appendix contains numerous leukocytes within its walls, and it may play a role in immune defenses. However, its removal entails no ill effects. *See illustration at* alimentary canal.

appetitive behavior /ap-ĕ-tÿ-tiv/ The first phase of a series of actions performed by an animal after initial motivation towards a specific goal. For example, the appetitive behavior of a hungry animal is that employed in the search for food.

apposition /ap-ŏ-**zish**-ŭn/ The deposition of successive layers of cellulose on the inner wall of a plant cell, resulting in an increase in thickness of the wall. *Compare* intussusception.

aqueous humor The fluid that fills the space in front of the lens in a vertebrate eye. It is secreted by the ciliary glands and supplies the cornea and lens with nutrients. The pressure of the fluid is maintained at a constant level (10–20 mmHg) to keep the eyeball rigid. A rise in the pressure leads to the condition known as *glaucoma*. *See illustration at* eye.

arachidonic acid /a-ră-kă-**don**-ik/ A polyunsaturated fatty acid, $CH_3(CH_2)_3$-$(CH_2CH:CH)_4(CH_2)_3COOH$, that is an essential dietary component for some mammals, although in humans it can be manufactured in the body from linoleic acid. It is the precursor of various biologically important molecules, notably the PROSTAGLANDINS, and is cleaved from diacylglycerols in the cell's plasma membrane in response to binding of hormones or other extracellular signal molecules. *See* essential fatty acid.

Arachnida /ă-**rak**-nă-dă/ The class of the Arthropoda that contains the mostly terrestrial and carnivorous scorpions and spiders, which typically have spinnerets on the abdomen for web spinning, and the ticks and mites. The body is divided into two parts, the anterior cephalothorax (*prosoma*), and the posterior abdomen (*opisthosoma*), and there are four pairs of walking legs. There are no antennae and the eyes are simple.

The cephalothorax bears prehensile chelicerae and leglike, usually sensory, pedipalps, as well as the legs. Respiration is carried out by lung books and/or tracheae and excretion is by coxal glands and Malpighian tubules. *See* Acarina.

arachnoid membrane /ă-**rak**-noid/ The middle one of the three membranes (meninges) that surround and protect the brain and spinal cord in vertebrates. *See* meninges.

arbovirus One of a group of RNA-containing viruses that cause such serious diseases as yellow fever and encephalitis. The viruses are transmitted from animals to man by insects (arthropods), hence the name (*ar*thropod-*bo*rne viruses).

archaebacteria /ar-kĕ-bak-**teer**-ee-ă/ (**archaea**) A group of bacteria containing many that live in harsh environments, such as hot springs, salt flats, or sea vents, thought to resemble the early earth environment. However, they are distinguished from the vast majority of bacteria (the eubacteria) principally on the basis of biochemical differences. For example, they differ in the nature of their lipid constituents, and lack a peptidoglycan layer in the cell wall. They include methanogic (methane-producing), thermophilic (heat-loving), and halophilic (salt-loving) species. However, they are not restricted to such extreme lifestyles, and are widespread in more congenial settings. The archaebacteria are now considered by many authorities to be so distant from the eubacteria in evolutionary terms that they constitute a separate DOMAIN of prokaryotic organisms.

Archean /ar-**kee**-ăn/ (**Archeozoic**) *See* Precambrian.

archegonium /ar-kĕ-**goh**-nee-ŭm/ (*pl.* **archegonia**) The female sex organ of the bryophytes, pteridophytes, and most gymnosperms. It is a multicellular flask-shaped structure made up of a narrow neck and a swollen base (venter) that contains the female gamete. *Compare* antheridium.

archenteron /ar-**ken**-tĕ-ron/ The earliest gut cavity of most animal embryos. It is produced by an infolding of part of the outer surface of a blastula to form an internal cavity that is in continuity with the outside via the blastopore.

archesporium /ar-kĕ-**spor**-ee-ŭm, -**spoh**-ree-ŭm/ The single cell or group of plant cells in the sporophyte from which spores may eventually develop in a sporangium.

archiannelid /ar-kee-**an**-ĕ-lid/ Any of various small marine annelid worms most of which are scavengers with a protrusible tongue for conveying food to the mouth. They are regarded as remnants of ancestral annelids, and some authorities place them in a separate class, Archiannelida.

Arctogea /ark-tŏ-**jee**-ă/ A major region comprising the four northern zoogeographical regions of the earth: the Palaearctic, Nearctic, Oriental, and Ethiopian regions. The Palaearctic and Nearctic regions are sometimes termed the *Holarctic region* because their fauna is very similar.

arginine /**ar**-jă-nÿn/ *See* amino acids.

aril /**a**-ril/ A brightly colored fleshy outgrowth from the funicle at the base of the ovule which may partly or completely cover the seed. The mace-yielding outgrowth around the fruit of the nutmeg is an example. *See also* caruncle.

aromatic compound An organic compound containing benzene rings in its structure. Aromatic compounds, such as benzene, have a planar ring of atoms linked by alternate single and double bonds. The characteristic of aromatic compounds is that their chemical properties are not those expected for an unsaturated compound; they tend to undergo nucleophilic substitution of hydrogen (or other groups) on the ring, and addition reactions only occur under special circumstances. *Compare* aliphatic compound.

arousal A general level of alertness in an animal, resulting from activity of a particular part of the brain. *See* reticular activating system.

Arrhenius equation /ah-**ren**-ee-ŭs/ An equation relating the rate constant of a chemical reaction and the temperature at which the reaction is taking place:
$$k = A\exp(-E_a/RT)$$
where A is a constant, k the rate constant, T the thermodynamic temperature in kelvins, R the gas constant, and E_a the activation energy of the reaction.

Reactions proceed at different rates at different temperatures, i.e. the magnitude of the rate constant is temperature dependent. The Arrhenius equation is often written in a logarithmic form, i.e.
$$\log_e k = \log_e A - E/2.3RT$$
This equation enables the activation energy for a reaction to be determined.

arteriole /ar-**teer**-ee-ohl/ One of a number of small blood vessels leading from an artery. The arterioles divide further into capillaries.

arteriovenous anastomosis /ar-teer-ee-oh-**vee**-nŭs ă-nas-tŏ-**moh**-sis/ A small muscular blood vessel that carries blood directly from the arterioles to the venules and bypasses the capillary network. By dilating or contracting it can regulate the amount of blood flowing through a particular capillary network at any given time. It is stimulated by sympathetic nerves.

artery A large thick-walled blood vessel that carries blood from the heart to the limbs and organs. All arteries except the pulmonary artery carry oxygenated blood. The artery wall consists of an inner layer of endothelium (*tunica intima*); a thick middle layer (*tunica media*), composed of smooth muscle and elastic tissue; and a thin outer layer (*tunica externa*) of collagen fibers. It is thus well suited to withstand the pressures resulting from the pumping of the heart.

Arthropoda /ar-**throp**-ŏ-dă/ The largest phylum in the animal kingdom and the only invertebrate phylum with aquatic, terrestrial, and aerial members. Arthropods are bilaterally symmetrical segmented animals with a characteristic tough chitinous protective exoskeleton flexible only at the joints; growth is by ecdysis. Each segment typically bears a pair of jointed appendages, which are modified for different functions. The phylum includes the crustaceans, insects, centipedes, millipedes, and spiders. Some authorities divide the arthropods into three separate phyla, the CHELICERATA, MANDIBULATA, and CRUSTACEA.

In the Arthropoda the coelom is reduced and the body cavity is a hemocoel. There is a ventral nerve cord with a pair of cerebral ganglia and paired segmental ganglia. *See also* Arachnida; Chilopoda; Diplopoda; Insecta; Onychophora; Trilobita.

articular /ar-**tik**-yŭ-ler/ A small bone of the lower jaw in bony fish (Osteichthyes), amphibians, and reptiles that forms a hinge joint with the quadrate bone of the upper jaw. The articular is derived from the ossification of Meckel's cartilage.

articulation The surface where two skeletal elements meet, forming a movable joint.

artificial chromosome A VECTOR that resembles a natural chromosome and is introduced to a host cell in order to clone relatively large pieces of foreign DNA, or to alter the cell's genetic make-up. *Bacterial artificial chromosomes* (BACs) are based on a naturally occurring bacterial plasmid, engineered to incorporate foreign DNA up to about 300 kilobase (kb) in length. They also contain bacterial genes needed for replication of the plasmid by the host bacterial cell, and an antibiotic-resistance gene to enable selection of host cells that have successfully taken up the BAC. *Yeast artificial chromosomes* (YACs) are linear DNA molecules capable of accommodating DNA fragments of up to 1000 kb. They have certain basic features of eukaryotic chromosomes, including telomeres (the DNA sequences that form the ends of a chromosome) and a centromere, via which the YAC attaches to the mitotic spindle of its yeast host so that it segregates normally during cell division. YACs are used especially for creating gene libraries of organisms with large genomes, such as mammals. Various other types of eukaryotic artificial chromosomes are being developed, including human artificial chromosomes. These offer the potential for introducing functioning genes to modify genetic expression in a way that will be reliably replicated and transmitted during cell division, for example as a technique for gene therapy.

artificial insemination The artificial introduction of semen into the reproductive tract of a female animal. It is used extensively in breeding animals (e.g. sheep, cattle). Semen collected from a male animal with desirable hereditary characters can be frozen and transported long distances to fertilize numerous females. The method is also used for human females who wish to conceive where the partner is unable to copulate successfully or is sterile.

artificial parthenogenesis *See* parthenogenesis.

Artiodactyla /ar-tee-oh-**dak**-tă-lă/ The order of mammals that contains the even-toed ungulates, in which the weight of the body is supported on the third and fourth digits only. These large herbivorous mammals include sheep, goats, deer, domestic cattle, antelopes, pigs, camels, and giraffes. The cud-chewing cloven-hoofed camels and *ruminants* have three or four chambers in the stomach, food being regurgitated from the first and chewed while the animal is resting before being swallowed again for complete digestion. *Compare* Perissodactyla.

ascocarp /ass-kŏ-karp/ The fruiting body of ascomycete fungi in which the asci are borne. It is formed from sterile hyphae surrounding the asci. If the ascocarp is closed it is termed a *cleistothecium* but if there is a pore through which ascospores may be discharged it is termed a *perithecium* or *pyrenocarp*. An *apothecium* is a cup- or disk-shaped ascocarp.

ascogonium /ass-kŏ-**goh**-nee-ŭm/ The female gametangium of certain ascomycete fungi (e.g. *Erysiphe*, *Eurotium*). The hyphal branches bearing ascogonia become entwined with those bearing antheridia. The contents of the antheridia pass into the ascogonia and the nuclei become paired. Pairs of nuclei later fuse and meiosis takes place to give ascospores.

ascomycete /ass-kŏ-**mӯ**-seet/ Any member of a large phylum of fungi (Ascomycota) characterized by their distinctive reproductive structure, the ascus, in which spores are formed, usually eight in number. Ascomycetes are subdivided according to how the asci are borne, to give the hemiascomycetes (asci naked, i.e. borne directly on the mycelium) and the euascomycetes (asci borne in an ascocarp).

ascorbic acid /ă-**skor**-bik/ *See* vitamin C.

ascus /ass-kŭs/ (*pl.* asci) The spore-producing cell of the ascomycete fungi. It is a saclike structure that is formed either singly or in large numbers in ascocarps. After meiosis the ascus contains four or eight haploid ascospores that are liberated through a pore at the end of the sac.

asexual reproduction /ay-**sek**-shoo-ăl/ The formation of new individuals from a single parent without the production of gametes or special reproductive structures. It occurs in many plants, usually by vegetative propagation or spore formation; in unicellular organisms usually by fission; and in multicellular invertebrates by fission, budding, fragmentation, etc.

asparagine /ă-**spa**-ră-jeen, -jin/ *See* amino acids.

aspartic acid /ă-**spar**-tik/ *See* amino acids.

Aspergillus /as-per-**jil**-ŭs/ A genus of ascomycete fungi including mostly saprophytes and opportunist pathogens. Inhalation of the spores of *A. fumigatus* causes the lung disease *aspergillosis*, while other *Aspergillus* species (including *A. flavus*) produce the poisonous metabolite *aflatoxin*, which may lead to liver diseases when consumed in *Aspergillus*-contaminated nuts and cereals.

aspirin (acetylsalicylic acid) A white crystalline powder, widely used for its analgesic, anti-inflammatory, and antipyretic (fever-relieving) properties, and to prevent thrombosis. It acts by inhibiting the synthesis of prostaglandin, which is released from damaged tissue during inflammation.

assimilation The process of incorporation of simple molecules of food that has been digested and absorbed into living cells of an animal and conversion into the complex molecules making up the organism.

association A climax plant community named according to the dominant type of species. Examples are heath associations and coniferous forest associations. *See also* consociation.

assortative mating /ă-**sor**-tă-tiv/ Reproduction of animals in which the males and females appear not to pair at random but tend to select partners of a similar phenotype.

Astacus /ass-tă-kŭs/ A genus of crawfish. Species are widely distributed in fresh water and are omnivorous. The carapace protecting the cephalothorax is prolonged in front to form a pointed *rostrum*, on each side of which is a stalked mobile compound eye. The thorax bears greatly enlarged pincers (chelae), used in feeding and defense, and four pairs of walking legs. Crawfish can swim rapidly backwards to escape danger by means of abdominal flexions and the use of the tail fan as a paddle. Respiration is by filamentous gills on the base of the legs and sides of the thorax. The eggs are carried over winter on the abdominal appendages of the female until hatching in the spring. *See also* Decapoda.

aster A cluster of radiating microtubules around the centrosome at each end of the spindle at nuclear division, so-called from its starlike appearance in the light microscope. It is generally more apparent in animal cells, which have centrioles, than in plant cells, which lack centrioles. Asters are believed to help locate the spindle in relation to the cell boundaries. They might also participate in cleavage of the cytoplasm following nuclear division by sending signals to the cell's equator.

Asteroidea /ass-tĕ-**roi**-dee-ă/ The class of the Echinodermata that contains the starfish (e.g. *Asterias*), which are often found just below the low-tide mark. A starfish typically has five arms radiating from a central disk, which contains the main body organs and has the mouth on its ventral surface. Chalky plates in the skin form a skeletal test. Suckered tube feet on the underside of the arms are used for locomotion and holding prey.

astrosclereid /ass-trŏ-**skleer**-ee-id/ An irregularly branched sclereid found in the leaves of certain dicotyledons.

atactostele /ă-**tak**-tŏ-steel, -stee-lee/ Describing the distribution of vascular tissue in those angiosperms in which the vascular bundles are scattered in an apparently random fashion in the ground tissue. Each vascular bundle is surrounded by a pericycle and endodermis. This arrangement is typical of the stem structure of monocotyledons but also occurs in some dicotyledons in which a complete network of interconnected bundles arises as the stem thickens with age.

atherosclerosis /ath-ĕ-roh-sklĕ-**roh**-sis/ The accumulation of fatty material, cholesterol and other substances on the interior walls of the arteries. This build-up, known as plaque, reduces the elasticity of the vessel wall ('hardening of the arteries'), and impedes blood flow through the vessel.

However, it can also trigger the formation of blood clots, which can detach and cause blockage in arteries of the heart, brain, or lungs, leading to heart attack or stroke. Similarly, plaque material can break off and travel through the blood system to obstruct a vessel elsewhere. Atherosclerosis is a progressive disease beginning in childhood, but symptoms, including angina and leg pain, do not begin usually until later life. Risk factors include high levels of blood cholesterol (especially low-density lipoproteins, or 'bad' cholesterol), diabetes, high blood pressure, smoking, obesity, and lack of exercise.

atlas A ringlike bone in tetrapods that forms the first neck (cervical) vertebra of the vertebral column, on which the skull rests. In reptiles, birds, and mammals, it articulates with the skull to allow nodding of the head and articulates with the axis to allow rotatory movement.

ATP (**adenosine triphosphate**) The universal energy carrier of living cells. Energy from respiration or, in photosynthesis, from sunlight is used to make ATP from ADP. It is then reconverted to ADP in various parts of the cell, the energy released being used to drive cell reactions. ATP supplies the energy for the synthesis of cell components (e.g. proteins and nucleic acids); it powers the active transport of substances; it drives cell division and growth; and it fuels muscle contraction and other forms of cell movement.

ATP is a nucleotide consisting of adenine and ribose with three phosphate groups attached. Hydrolysis of the terminal phosphate bond releases energy (30.6 kJ mol^{-1}) and is coupled to an energy-requiring process. Further hydrolysis of ADP to AMP sometimes occurs, releasing more energy. The pool of ATP is small, but the faster it is used, the faster it is replenished. *See also* electron-transport chain; respiration.

ATPase Any enzyme that catalyzes the hydrolysis of ATP to yield ADP and inorganic phosphate (Pi), with the release of energy. ATPases are integral components of numerous cellular processes, including active transport and cell motility.

atrioventricular node /ay-tree-oh-ven-**trik**-yŭ-ler/ (**AV node**) A bundle of specialized muscle cells in the wall of the right atrium of the mammalian heart, near the right ventricle. It receives electrical impulses from the cardiac PACEMAKER (the sinoatrial node), and transmits these to the ventricles via conducting fibers of the BUNDLE OF HIS and the Purkinje fibers.

atrium /ay-tree-ŭm/ (*pl.* **atria**) **1.** (**auricle**) A cavity or chamber in the body, especially in the vertebrate heart. In mammals, there are two atria forming the upper chambers of the heart: the left atrium receives oxygenated blood from the lungs; the right atrium receives deoxygenated blood from the body. Other tetrapods also have two atria, while fish have one, forming the second chamber of the heart. This atrium pumps deoxygenated blood from the sinus venosus into the ventricle. *See illustration at* heart.
2. A cavity surrounding the gills of primitive chordates; for example, in *Branchiostoma* the atrium connects with the exterior via a small pore (an *atriopore*).
3. Any cavity opening to the exterior; for example, the genital cavity of *Helix*.

atrophy /at-rŏ-fee/ The shrinking in size of a tissue or an organ.

atropous /at-rŏ-pŭs/ *See* orthotropous.

attenuation /ă-ten-yoo-**ay**-shŏn/ **1.** The loss of virulence of a pathogenic microorganism after several generations of culture *in vitro*. Attenuated microorganisms are commonly used in vaccines.
2. (*Genetics*) A 'back-up' mechanism that helps to regulate gene expression in bacterial OPERONS. It involves an *attenuator*, a site lying downstream of the promoter and operator sites, but before the start of the structural genes. The attenuator is transcribed into messenger RNA, but is able to interact with a substance in the environment (e.g. the product of enzymes encoded by the operon) to curtail translation of the mRNA when the operon's enzymes are not needed.

auditory capsule /**aw**-dă-tor-ee, -toh-ree/ (**otic capsule**) The cartilaginous or bony part of the vertebrate skull that encloses the inner ear. It consists of a single periotic bone in adult mammals. *See also* chondrocranium; neurocranium.

auditory nerve (**vestibulocochlear nerve; acoustic nerve; cranial nerve VIII**) The nerve that supplies the inner ear. It arises from the dorsal region of the medulla oblongata in the vertebrate brain and has two main branches – the *vestibular nerve*, which serves the anterior region, and the *cochlear nerve*, which supplies the posterior part (including the cochlea). *See* cranial nerves.

auricle /**ôr**-ă-kăl/ *See* atrium.

Australasia One of the six main zoogeographical areas, composed of Australia and the islands of its continental shelf, Tasmania, New Guinea, and New Zealand. Marsupial (pouched) and monotreme (egg-laying) mammals are particularly characteristic, but many other unique vertebrates and invertebrates are also found. *See also* Wallace's line.

Australopithecus /ô-stră-loh-**pith**-ĕ-kŭs/ (**southern ape**) An extinct genus of early hominids whose fossils show features intermediate between those of apes and humans. It is thought that its members, the australopithecines, arose some 4 million years ago and from about 2.5 million years ago coexisted with early forms of *Homo* before becoming extinct some 2 million years ago. They were vegetarians and could walk upright.

autecology /aw-tĕ-**kol**-ŏ-jee/ The study of the interactions of an individual organism or a single species with the living and nonliving components of its environment. *Compare* synecology.

autoclave /**aw**-tŏ-klayv/ An apparatus, similar in principle to a pressure cooker, in which materials are placed to be sterilized by steam under pressure. High temperatures, well above the boiling point of water, can be achieved in autoclaves.

autoecious /aw-**tee**-shŭs/ Denoting rust fungi that require only one host species to complete the various stages of their life cycle. An example is *Puccinium antirrhini* found on antirrhinum. *Compare* heteroecious.

autogamy /aw-**tog**-ă-mee/ **1.** (*Zoology*) Reproduction in which the nucleus of an individual cell divides into two and forms two gametes, which reunite to form a zygote. It occurs in some protozoans, e.g. *Paramecium*.
2. (*Botany*) Self-fertilization in plants. *Compare* allogamy.

autogenic movements /aw-tŏ-**jen**-ik/ *See* autonomic movements.

autograft /**aw**-toh-graft, -grahft/ A type of graft involving transplantation of tissue or an organ from one part of an individual to another part of the same individual.

autoimmunity /aw-toh-i-**myoo**-nă-tee/ A diseased state in which antibodies are formed and react against a normal component of the animal's own tissues. Autoimmunity is a contributory factor of a number of diseases (*autoimmune diseases*), such as rheumatoid arthritis and some forms of gout.

autolysis /aw-**tol**-ă-sis/ The self-destruction of cells by digestive enzyme activity. It is the final stage of cell senescence resulting in complete digestion of all cell components. *See* apoptosis; lysosome.

autonomic movements /aw-tŏ-**nom**-ik/ (autogenic movements; spontaneous movements) Movements of plants in response to internal rather than external stimuli. Examples are cytoplasmic streaming, chromosome movement during nuclear division, and growth itself. *Compare* paratonic movements.

autonomic nervous system The division of the vertebrate nervous system that supplies motor nerves to the smooth muscles of the gut and internal organs and to heart muscle. It comprises the SYMPATHETIC NERVOUS SYSTEM, which (when stimulated) increases heart rate, breathing rate, and blood pressure and slows down digestive processes, and the PARASYMPATHETIC NERVOUS SYSTEM, which slows heart rate and promotes digestion. Each organ is innervated by both systems and their relative rates of stimulation determine the net effect on the organ concerned. Many functions of the autonomic nervous system, such as the control of heart rate and blood pressure, are regulated by centers in the medulla oblongata of the brain.

autophagy /aw-**tof**-ă-jee/ The process whereby redundant, faulty, or aging cell organelles are destroyed. The organelle or cell portion is surrounded by a membrane derived either from the endoplasmic reticulum or a vacuole. Lysosomes then fuse with the compartment thus formed, releasing their digestive enzymes and destroying its contents. It is part of the normal turnover of cell constituents, but accelerates during senescence and may be part of a developmental process, such as clearing of cell contents during sieve tube and tracheid formation.

autopolyploidy /aw-toh-**pol**-i-ploi-dee/ A type of polyploidy involving the multiplication of chromosome sets from only one species. *Autopolyploids* may arise from the fusion of diploid gametes that have resulted from the nondisjunction of chromosomes at meiosis. Alternatively, like allopolyploids, they may arise by the nondisjunction of chromatids during the mitotic division of a zygote. *Compare* allopolyploidy.

autoradiography /aw-toh-ray-dee-**og**-ră-fee/ A technique whereby a thin slice of tissue containing a radioactive isotope is placed in contact with a photographic plate or film. The image obtained on development shows the distribution of the isotope in the tissue.

autosomes /**aw**-tŏ-sohmz/ Paired somatic chromosomes that play no part in sex determination. *Compare* sex chromosomes.

autotetraploid /aw-toh-**tet**-ră-ploid/ An autopolyploid that has four times the haploid number of chromosomes. *See* autopolyploidy.

autotomy Self-amputation of a trapped or damaged part of the body (e.g. tail or limb) of certain animals, especially lizards, arthropods, and worms. The lost part is usually regenerated. A special muscular mechanism exists, such that contraction along one of the various lines of weakness results in separation at that point.

autotrophism /aw-tŏ-**trof**-iz-ăm/ A type of nutrition in which the principal source of carbon is inorganic (carbon dioxide or carbonate). Organic materials are synthesized from inorganic starting materials. The process may occur by use of light energy (*photoautotrophism*) or chemical energy (*chemoautotrophism*). Autotrophic organisms (autotrophs) are important ecologically as primary producers, their activities ultimately supplying the carbon requirements of all heterotrophic organisms. *Compare* heterotrophism. *See* chemotrophism; phototrophism.

auxanometer /awks-ă-**nom**-ĕ-ter/ An instrument designed to measure the increase in length of a plant part. A growing plant is attached by thread to the end of a lever that magnifies any growth movement. The opposite end of the lever is used to record a trace on a slowly rotating drum.

auxin /**awks**-in/ Any of a group of plant hormones, the most common naturally occurring one being indole acetic acid, IAA. Auxins are made continually in growing shoot and root tips. They are actively moved away from the tip, having various effects along their route before being inactivated. They regulate the rate of extension of cells in the growing region behind the shoot tip and are involved in phototropic and geotropic curvature responses of shoot tips (but probably not root tips) moving laterally away from light and towards gravity.

Auxins stimulate cell enlargement, probably by stimulating excretion of protons leading to acid-induced wall loosening and thus wall extension. They help maintain apical dominance by inhibiting lateral bud development. Root initiation may be stimulated by auxin from the shoot, and auxins have been shown to move towards the root tips. Pollen tube growth is stimulated by auxin and its production by developing seeds stimulates fruit set and pericarp growth in fleshy fruits. It interacts synergistically with gibberellins and cytokinins in stimulating cell division and differentiation in the cambium. A high auxin:cytokinin ratio stimulates root growth but inhibits regeneration of buds in tobacco pith callus. Auxins suppress the onset of shedding (abscission) of leaves, fruits, etc.

Synthetic auxins, cheaper and more stable than IAA, are employed in agriculture, horticulture, and research. These include indoles and naphthyls: e.g. NAA (naphthalene acetic acid) used mainly as a rooting and fruit setting hormone; phenoxyacetic acids, e.g. 2,4-D (2,4-dichlorophenoxyacetic acid) used as weed-killers and modifiers of fruit development; and more toxic and persistent benzoic auxins, e.g. 2,4,5-trichlorobenzoic acid, also formerly used as herbicides but now widely restricted.

average *See* mean.

Aves /**ay**-veez/ The class of vertebrates that contains the birds, most of whose characteristics are adaptations for flight. The forelimbs are modified as wings with three digits only, the third being greatly elongated. Birds have light strong hollow bones and a rigid skeleton strengthened by bone fusion. The sternum usually has a large keel for attachment of the powerful pectoral muscles, which depress the wings. Birds are homoiothermic. The short deep body is covered with feathers, which provide insulation as well as a large surface area for flight. The jaws form a horny beak and teeth are absent. They lay yolky eggs with a calcareous shell and typically have a well-developed social life, including territorial and courtship displays, nesting, parental care, and song. Many undertake long migrations. There are 28 orders, including the passerines (perching birds), which alone account for about 60% of all birds. The large flightless birds (ratites) tend to be swift runners.

Birds evolved from reptiles in the Jurassic period and retain many reptilian characteristics, such as scaly legs and feet. However, in contrast to reptiles, oxygenated and deoxygenated blood are completely separated in the four-chambered heart. *See also* passerines; ratites.

axenic culture /ay-**zen**-ik, -**zee**-nik/ (**pure culture**) A culture containing only one species of microorganism.

axial skeleton The longitudinal part of the vertebrate skeleton including the skull, vertebral column, and rib cage. *Compare* appendicular skeleton.

axil /**aks**-ăl/ The angle between the upper side of a leaf or branch and the shoot bearing it. Leaf axils are the site of lateral (axillary) buds.

axis A bone in tetrapods that forms the second and strongest neck (cervical) vertebra of the vertebral column. In reptiles, birds, and mammals, it bears a process (the *odontoid peg*), which projects upward into the 'ring' of the atlas and acts as a pivot for rotation of the head.

axon /**aks**-on/ (**nerve fiber**) The part of a NEURON that conveys impulses from the cell body towards a synapse. It is an extension of the cell body and consists of an axis cylinder (*axoplasm*), surrounded in most vertebrates by a fatty (myelin) sheath, outside which is a thin membrane (*neurilemma*).

axoneme /**aks**-ŏ-neem/ The central '9+2' core of microtubules found in cilia and flagella (undulipodia) of eukaryotes, consisting of nine pairs of outer microtubules surrounding two single central microtubules. *See* undulipodium.

Azotobacter /ă-zoh-tŏ-**bak**-ter/ A genus of free-living aerobic nitrogen-fixing bacteria found in limestone soils and water. The cells are plump rods or cocci, surrounded by slime. A multilayered wall may be synthesized around the cell to produce a microcyst resistant to desiccation.

B

Bacillariophyta /bas-i-lair-ee-**off**-ă-tă/ *See* diatoms.

bacillus /bă-**sil**-ŭs/ Any rod-shaped bacterium. Bacilli may occur singly (e.g. *Pseudomonas*), in pairs, or in chains (e.g. *Lactobacillus*). Some are motile.

backbone *See* vertebral column.

back cross The crossing of a hybrid back to the original parent generation. If a homozygous dominant AA is crossed with a homozygous recessive aa, the F_2 generation obtained by selfing the F_1 would be 25% AA, 50% Aa, and 25% aa. To distinguish between the phenotypically identical AA and Aa, these can be back-crossed to the homozygous recessive parent aa (a *test cross*). The offspring from AA × aa will all be identical (Aa) whereas in the cross Aa × aa, 50% will have the dominant phenotype (Aa) and 50% will show the recessive character (aa).

bacteria A large and diverse group of organisms, which, in terms of numbers and variety of habitats, includes the most successful life forms. In nature, bacteria are important in the nitrogen and carbon cycles, and some are useful to man in various industrial processes, especially in the food industry, and in techniques of genetic engineering (*see also* biotechnology). However, there are also many harmful parasitic bacteria that cause diseases such as botulism and tetanus.

Bacterial cells are simpler than those of animals and plants. They lack well-defined membrane-bound nuclei, and do not contain complex organelles such as chloroplasts and mitochondria. They may divide every 20 minutes and can thus reproduce very rapidly. They also form resistant spores. Bacteria include all prokaryotic organisms, and constitute the kingdom Monera (or Bacteria). They are divided into two evolutionary distant groups: the ARCHAEBACTERIA, which often occur in extreme conditions, such as hot springs; and the EUBACTERIA, which include the vast majority of bacteria. *See* actinomycetes; Cyanobacteria; myxobacteria; rickettsiae; spirochetes; sulfur bacteria. *See also* prokaryote.

bactericidal /bak-teer-ă-**sÿ**-d'l/ Used to describe a compound that has a lethal effect on bacteria. A bactericidal compound may act by interfering with a vital biochemical pathway, or by destroying the molecular structure of the cell.

bacteriochlorophyll /bak-teer-ee-oh-**klor**-ŏ-fil, -**kloh**-rŏ-/ Any of several types of chlorophyll found in photosynthetic bacteria, such as the purple bacteria. There are seven forms, designated bacteriochlorophylls *a–g*. All are structurally similar to chlorophyll *a* of plants. The bacteriochlorophylls absorb light at longer wavelengths than chlorophyll *a* enabling far-red and infrared light to be used in photosynthesis. *See* photosynthetic pigments.

bacteriophage /bak-**teer**-ee-ŏ-fayj/ *See* phage.

bacteriostatic /bak-teer-ee-ŏ-**stat**-ik/ Used to describe a compound that prevents the reproduction of bacteria, but does not kill them.

Balbiani ring /bal-bee-**an**-ee, -**ah**-nee/ *See* puff.

baleen /bă-**leen**/ (**whalebone**) Parallel

horny plates that grow down from the sides of the upper jaw in baleen whales. 150–400 plates may occur on each side, each roughly triangular in shape and comprised of hollow fibers, which on the inner edge are separated out into a fringe. They are used to strain their food, krill, from the sea water.

barb 1. A hooked hair or bristle.
2. *See* feathers.

barbule *See* feathers.

bark The outermost tissue of the stem and roots in woody plants. If the same cork cambium functions from year to year, then the bark is smooth and consists only of tissue external to the phloem, mainly cork. If new cork cambia arise every few years then the bark includes dead phloem and cortex as well as cork, and is termed *rhytidome*. The texture and patterning of the bark is often characteristic of the species. *See illustration at* wood.

baroreceptor /ba-rŏ-ri-**sep**-ter/ (**barocep-tor**) A sensory receptor in the walls of the blood vascular system that responds to changes in blood pressure. Its function is to maintain a steady blood pressure. Any increase in pressure distends the wall and stimulates the receptor to fire nervous impulses that result in reflex slowing of the heart and vasodilation. In humans, the major baroreceptors occur in the systemic arch, the carotid sinus, and the left atrium.

basal body /**bayss**-ăl/ 1. (**kinetosome**) A barrel-shaped body found at the base of all eukaryote undulipodia (cilia and flagella) and identical in structure to the centriole. It is essential for formation and anchorage of undulipodia. *See also* centriole; undulipodium.
2. An assembly of proteins found at the base of a bacterial flagellum, embedded in the plasma membrane and cell wall. It consists of disk-shaped structural proteins, which anchor the shaft of the flagellum; motor proteins, which rotate the base of the shaft; and switch proteins, which can reverse the direction of rotation.

basal ganglia *See* cerebral nuclei.

basal lamina (**basement membrane**) A thin sheet of protein fibers, typically 60–100 nm thick, that underlies epithelial and endothelial cells, and also surrounds muscle cells, fat cells, and Schwann cells. It is a component of the EXTRACELLULAR MATRIX, and is secreted by the tissue cells themselves. The main constituents are COLLAGEN fibrils, which form a framework, and laminin, a protein that binds the lamina to adjacent cells. The lamina helps to maintain the structural integrity of the associated tissue cells, and regulates the passage of materials to and from the cells into neighboring blood vessels. The kidney glomerulus has a thick basal lamina formed by fusion of endothelial and epithelial laminae. This acts as the filter through which water and nonprotein components of blood are forced under pressure into the capsule to form urine.

basal metabolic rate (**BMR**) The minimum rate of energy expenditure by an animal necessary to maintain the vital processes, e.g. circulation, respiration, etc. It is expressed as the output of heat in joules (or kilojoules) per square meter of body surface area per hour and is measured either directly from heat production or indirectly from oxygen consumption. The thyroid hormones are the prime regulators of the BMR.

base 1. (*Chemistry*) A compound that reacts with an acid to produce water plus a salt.
2. (*Biochemistry*) A nitrogenous molecule, either a pyrimidine or a purine, that combines with a pentose sugar and phosphoric acid to form a nucleotide, the fundamental unit of nucleic acids. The most abundant bases are cytosine, thymine, and uracil (pyrimidines) and adenine and guanine (purines). *See* purine; pyrimidine.

basement membrane *See* basal lamina.

base pairing The linking together of the two helical strands of DNA by bonds between complementary bases, adenine pairing with thymine and guanine pairing with cytosine. The specific nature of base pairing enables accurate replication of the chromosomes and thus maintains the constant composition of the genetic material. In pairing between DNA and RNA the uracil of RNA pairs with adenine.

base ratio The ratio of adenine (A) plus thymine (T) to guanine (G) plus cytosine (C). In DNA the amount of A is equal to the amount of T, and the amount of G equals the amount of C, but the amount of A + T does not equal the amount of C + G. The A + T : G + C ratio is constant within a species but varies between species.

basic stain *See* staining.

basidiocarp /bă-**sid**-ee-ŏ-karp/ The fruiting body of the basidiomycete fungi, excepting the rusts and smuts. The basidiocarp is the above-ground portion of the fungus that is commonly referred to as the toadstool. It may be of a fleshy, corky, or spongy nature. The basidiocarp contains the basidial hyphae that, after meiosis, give rise to the basidiospores.

basidiomycete /bă-sid-ee-ŏ-**mÿ**-seet/ Any member of a phylum of fungi (Basidiomycota) in which the sexual spores, basidiospores, are borne externally on the end of specialized cells termed basidia. The basidia are formed on a fertile compact layer called the hymenium and each basidium usually contains four basidiospores. There are two groups, the homobasidiomycetes, which contains the edible mushrooms, and the heterobasidiomycetes, which includes the rusts and smuts.

basidium /bă-**sid**-ee-ŭm/ (*pl.* **basidia**) The cell of the basidiomycete fungus that produces the haploid basidiospores. The basidia develop in the basidiocarp from hyphal tips in the hymenium and may be club-shaped or cylindrical. A basidium usually produces four basidiospores, which develop externally on short stalks called *sterigmata*.

basifixed /**bay**-să-fixt/ Stamens in which the anther lobes are attached at their base to the filament, and are not capable of independent movement. *Compare* dorsifixed; versatile.

basipetal /bay-**sip**-ĕ-t'l/ Developing or moving from the apex downward. Basipetal differentiation is seen in the formation of proto- and metaxylem in the stem. *Compare* acropetal. *See also* centrifugal; centripetal.

basophil /**bay**-sŏ-fïl/ A white blood cell (LEUKOCYTE) containing granules that stain with basic dyes. It has a lobed nucleus. Basophils comprise only about 0.5% of all leukocytes. They move about in an ameboid fashion and when activated release the contents of their granules – histamine and other substances that promote inflammation, bringing more phagocytic cells to the infection site. They are equivalent to wandering MAST CELLS, similar to those found in the lining of blood vessels.

bats *See* Chiroptera.

B-cell (**B-lymphocyte**) A type of lymphocyte (*see* white blood cell) that originates in the bone marrow and serves as the main instrument of the specific immune response, particularly the production of antibodies. (In birds, where antibodies were first studied, the antibody-producing lymphocytes mature in part of the cloaca called the bursa of Fabricius. (The bursa was named for the Italian anatomist and embryologist Hiernonymus Fabricius ab Aquapendente (1537–1619).) Hence they were termed bursal cells, or B-cells. This organ has no known counterpart in mammals, and mammalian B-cells are thought to mature in the bone marrow.) Lymphoid tissue contains vast numbers of B-cells, each with a different type of immunoglobulin receptor on its surface. When an antigen binds to the one or several B-cells carrying its specific receptor, it triggers them to undergo repeated division forming a large clone of cells dedicated to producing antibody specific to that antigen. Hence, according to the *clonal selection theory* of antibody specificity, the antigen selects the few appropriate B-cells from among the huge number present in the body. These B-cells then enter the bloodstream as antibody-secreting *plasma cells*.

Another type of lymphocyte, known as a T-helper cell, is required to trigger clonal expansion and antibody secretion by B-cells. This recognizes the antigen bound to the surface of the B-cell in association with class II MHC proteins (*see* major histocompatibility complex). The T-helper cell binds to the antigen-MHC complex and releases substances (lymphokines) that act as the stimulus for B-cell growth.

After an infection has been dealt with and all antigens removed by the antibodies, a small set of B-cells from the clone, called *memory cells*, remain in the circulation. These carry receptors that will bind avidly to the same antigen in any subsequent infection, prompting a much more rapid response by the immune system. *See* immunity; immunoglobulin; T-cell.

bees *See* Hymenoptera.

beetles *See* Coleoptera.

behavior, animal A general term applied to any observable activity of a whole animal. Behavior includes all the processes by which an animal senses its external surroundings and the internal state of its body and responds to any changes it perceives. An animal behaves continuously in order to survive – to feed, drink, reproduce, and avoid being eaten. Some behavior is *innate* and some is learnt through experience. *See also* instinct.

Benson–Calvin–Bassham cycle *See* photosynthesis.

benthic /ben-thik/ Describing organisms that live on or in the sea bed. Benthic epifauna live upon the seafloor or upon bottom objects and benthic infauna live within the surface sediments. *Compare* pelagic.

benzene /ben-zeen, ben-zeen/ (C_6H_6) A colorless liquid hydrocarbon with a characteristic odor. Benzene is a highly toxic compound and continued inhalation of the vapor is harmful. It was originally isolated from coal tar and for many years this was the principal source of the compound. Benzene is a typical example of an *aromatic compound*.

berry A succulent fruit, usually containing more than one seed, that does not burst open when ripe. The ovary wall remains fleshy after fertilization of the ovule except for the development of a thin outer skin. A berry with a hard rind, such as a cucumber, is called a *pepo*. A berry with a leathery rind in which the seeds are separated by segments, as in the citrus fruits, is called a *hesperidium*.

beta-oxidation /bay-tă-oks-ă-**day**-shŏn/ The series of biochemical reactions by which living cells metabolize fatty acids to produce energy. This metabolic pathway is essential for utilizing dietary fats, but is especially significant in animals mobilizing body reserves of fat during periods of starvation, and in germinating seeds. In animal cells beta-oxidation takes place in the matrix of mitochondria, whereas in plant cells it occurs in organelles called MICROBODIES, particularly glyoxysomes. The fatty acid's hydrocarbon chain is repeatedly cleaved from its end into two-carbon fragments (i.e. at the beta carbon atom each time). Each resulting fragment combines with coenzyme A to form acetyl CoA, which enters the KREBS CYCLE to undergo further oxidation, or to take part in the biosynthesis of other molecules.

beta sheet A type of protein structure in which polypeptide chains run close to each other and are held together by hydrogen bonds at right angles to the main chain. The structure is folded in regular 'pleats'. Fibers having this type of structure are usually composed of amino acids with short side chains. The chains may run in the same direction (parallel) or opposite directions (antiparallel).

biceps /bÿ-seps/ The muscle that in humans runs through the upper arm connecting the shoulder blade (scapula) with the radius bone of the forearm. When it contracts it flexes (bends) the elbow joint and raises the forearm. Its antagonist muscle is the TRICEPS, which extends the arm.

bicollateral bundle /bȳ-kŏ-**lat**-er-ăl/ The arrangement of tissues in the vascular bundles in which the phloem is situated on both sides of the xylem. This is thought to aid transport of nutrients in plants with long trailing stems, such as the marrow. *Compare* centric bundle; collateral bundle.

bicuspid valve /bȳ-**kus**-pid/ *See* mitral valve.

biennial A plant that completes its life cycle within two years. In the first year it produces foliage only and photosynthesizes. The food is stored during the winter in a swollen underground root or stem. In the second year, the stored food is used to produce flowers, fruits, and seeds. Many important crops, such as carrot and parsnip, are biennials. Some biennials can be induced to act as annuals and flower in the first year by appropriate cold or hormone treatments. *Compare* annual; ephemeral; perennial.

bilateral symmetry The arrangement of parts in an organism in such a way that the structure can only be divided into similar halves (mirror images) along one plane. Bilateral symmetry is characteristic of most free-moving animals, where one end constantly leads during movement. However some secondary asymmetry of internal organs has occurred in humans and other vertebrates. In plants, bilateral symmetry is seen particularly in flowers (e.g. snapdragon), the condition commonly being termed *zygomorphy*. *See also* radial symmetry.

bile A secretion of the liver that enters the duodenum via the bile duct. It is a mixture of bile salts, bile pigments (bilirubin and biliverdin), cholesterol, and traces of other substances. The bile salts aid digestion by facilitating the emulsification of fats, while the bile pigments and cholesterol are merely excretory products, playing no part in digestion.

bile duct A duct in vertebrates that transports bile from the liver to the duodenum. *See also* gall bladder. *See illustration at* alimentary canal.

bile pigments Pigments excreted in the bile as the products of the degradation of hemoglobin. When hemoglobin is destroyed in the body the protein portion, globin, is degraded to amino acids, while the porphyrin or heme portion gives rise to the bile pigments. The green pigment *biliverdin* is the first of the bile pigments; it is easily reduced to the red-brown pigment *bilirubin*. Bilirubin is the major pigment in human bile, there being only slight traces of biliverdin, which is the chief pigment in the bile of birds. A specific enzyme, biliverdin reductase, catalyzes the reduction of biliverdin to bilirubin. The formation of biliverdin and bilirubin from heme takes place in the reticuloendothelial cells of the liver, spleen, and bone marrow. The pigments are then transported to the liver where they are excreted in the bile into the duodenum. In the intestine, bilirubin undergoes further chemical modification and is secreted in the feces.

bile salts Components of bile that aid digestion in the duodenal region of the gut. Sodium taurocholate and sodium glycocholate emulsify fat droplets by lowering their surface tension, causing them to break up into numerous smaller droplets and facilitating the digestive action of the enzyme lipase.

bilirubin /bil-ă-**roo**-bin/ *See* bile pigments.

biliverdin /bil-ă-**ver**-din/ *See* bile pigments.

binary fission *See* fission.

binding site A region on the surface of a large molecule that specifically combines with another molecule, chemical group, or ion. Molecules in living cells often combine in a precise manner via binding sites. For example, cell receptor proteins have binding sites that will only combine with particular signal molecules approaching the cell.

binocular vision A type of vision in which the eyes point forwards so that the image of a single object can be focused onto the fovea of both eyes at once. This allows perception of depth and distance. Binocular vision is found in primates and other vertebrates, especially active predators, such as owls.

binomial nomenclature A system of classification introduced by Linnaeus, the Swedish botanist, in which each species is given two names. The first is the generic name, written with a capital letter, which designates the genus to which the species belongs. The second is the specific name, indicating the species. The generic and specific names are in Latin and are printed in italic type. For example, humans belong to the species *Homo sapiens*. *Homo* is the generic name and *sapiens* is the specific name.

bioassay /bÿ-oh-**ass**-ay, bÿ-oh-ă-**say**/ An experimental technique for measuring quantitatively the strength of a biologically active chemical by its effect on a living organism. For example, the vitamin activity of certain substances can be measured using bacterial cultures. The increase in bacterial numbers is compared against that achieved with known standards for vitamins. Plant growth hormones can be estimated by their effect in causing curvature of oat coleoptiles.

Biochemical Oxygen Demand (BOD) The standard measurement for determining the level of organic pollution in a sample of water. It is the amount of oxygen used by microorganisms feeding on the organic material over a given period of time. Sewage effluent must be diluted to comply with the statutory BOD before it can be disposed of into clean rivers.

biochemical taxonomy The use of chemical characteristics to help classify organisms; for example, the Asteroideae and Cichorioideae, which are the two main divisions of the plant family Compositae, are separated by the presence or absence of latex. This area of taxonomy has increased in importance with the development of chromatography, electrophoresis, serology, and other analytical techniques. *See also* molecular systematics.

biochemistry The study of chemical reactions occurring in living organisms.

biodegradable /bÿ-oh-di-**gray**-dă-băl/ Describing organic compounds that are able to be decomposed by bacteria and other microorganisms, such as the constituents of sewage, as compared with non-biodegradable compounds, such as most plastics. *See also* pollution.

biodiversity /bÿ-oh-di-**ver**-să-tee/ The number and variety of organisms in a given locality, community, or ecosystem. High biodiversity is typical of complex and highly productive ecosystems, such as tropical rainforests, where a small area can contain many different species of animals, plants, and other organisms. Biodiversity is often used as an indicator of the health of such ecosystems.

bioengineering 1. The design, manufacture, and use of replacements or aids for body parts or organs that have been removed or are defective, e.g. artificial limbs, hearing aids, etc.
2. The design, manufacture, and use of equipment for industrial biological processes, such as fermentation.

biofeedback A technique whereby a person can consciously control physiological responses that are normally controlled unconsciously by the autonomic nervous system. The technique is learnt by measuring the function, e.g. heart rate using an electrocardiogram, and indicating the rate to the subject. When the subject has learnt to discriminate between different rates he can then learn to control his heart rate consciously. Biofeedback has been used to control heart rate, blood pressure, and migraine and to relax spastic muscles.

biogenesis /bȳ-oh-**jen**-ĕ-sis/ The theory that living things originate only from other living things as opposed to nonliving matter. The theory became accepted as a result of the work of Redi and Pasteur, who showed that dirt, for example, does not itself produce bacteria or maggots, but that bacteria and maggots only come from spores or eggs already existing in the dirt. This theory satisfactorily explains the occurrence of existing organisms, but does not explain the origins of the first organisms. *See* abiogenesis.

bioinformatics The use of computerized systems for the collection, storage, and analysis of DNA and protein sequence data. Such data are held in several large databases, such as SwissProt, maintained by the University of Geneva, and GenBank at the National Institutes of Health, Bethesda, Maryland. The data are made available to researchers worldwide via the Internet. Newly discovered sequences can therefore be readily compared with existing data and any similarities identified. Such homologous sequences often point to functional similarities in three-dimensional protein structures, and can provide clues about the function of a newly sequenced protein. Moreover, the comparative analysis of sequence data from the genomes of different organisms provides information about the array of proteins any particular organism (e.g. a bacterium or protist) is likely to synthesize from its genome. This sheds light on its biology, even where this has not been investigated directly, and on its taxonomic relationships to other organisms. *See also* genomics.

biological clock The internal mechanism of an organism that regulates CIRCADIAN RHYTHMS and various other periodic cycles. The mechanism of the biological clock has long proved elusive. However, a molecular basis for such a clock in the fruit fly *Drosophila* has been discovered, and similar mechanisms may well apply in other organisms. It is based on the cyclical rise and fall in the concentrations of certain proteins, notably PER and TIM, which form part of a negative feedback loop that con-

trols transcription of their own genes. The clock is entrained to the day–night cycle because some of the proteins are light sensitive, being degraded in daylight. *See also* photoperiodism.

biological control The use of natural predators or parasites, instead of chemicals, to control pests. The most famous successful example was the introduction of the gray moth, *Cactoblastis cactorum*, into Australia to control the prickly pear, *Opuntia inermis*, which was overrunning vast tracts of land. The moth's caterpillars eat the shoots of the plant. Another example is the introduction of parasitic wasps that lay their eggs inside the eggs of pest insects, such as corn borers (which attack maize). Populations of insect pests may also be reduced by releasing sterile males to mate with the females, or by using sex-attractant chemicals (pheromones) to trap males or females.

biology The study of living organisms, including their structure, function, evolution, interrelationships, behavior, and distribution.

bioluminescence The production of light by living organisms. Bioluminescence is found in many marine organisms, especially deep-sea organisms. It is also a property of some insects, for example fireflies, and certain bacteria. The light is produced as a result of a chemical reaction whereby the compound LUCIFERIN is oxidized and gives off photons. An enzyme, luciferase, catalyzes the reaction, and there are several quite different types of luciferin. Most luminescent animals have a light-producing organ (*photophore*). This may contain their own specialized light-producing cells, or house symbiotic luminescent bacteria.

biomass /bȳ-oh-mass/ The weight or volume of living organisms of one animal or plant species per unit area (*species biomass*), or of all the species in the community (*community biomass*). *See also* pyramid of biomass.

biome /bȳ-ohm/ A major regional community of plants and animals with similar life forms and environmental conditions. It is the largest geographical biotic unit, and is named after the dominant type of vegetation, such as tropical rain forest, grassland, or coral reef.

biomolecule /bÿ-oh-**mol**-ĕ-kyool/ Any molecule that forms part of or is produced by living organisms.

biophysics /bÿ-oh-**fiz**-iks/ The branch of science that studies biological systems and living organisms in terms of physical laws or by applying the techniques of physics. For example, the principles of electricity have proved instructive in understanding the functions of nervous systems, while the laws of mechanics help to elucidate the workings of the musculoskeletal system. Various physical techniques have advanced our knowledge of biologically important molecules (e.g. proteins and nucleic acids), notably X-ray crystallography, nuclear magnetic resonance imaging, and spectroscopy.

biopoiesis /bÿ-oh-poi-**ee**-sis/ The origin of organisms from replicating molecules. Biopoiesis is a cornerstone of ABIOGENESIS. Deoxyribonucleic acid (DNA) is the best example of a self-replicating molecule, and is found in the chromosomes of all higher organisms. In some bacterial viruses (bacteriophages) ribonucleic acid (RNA) is self-replicating. Various chemical and physical conditions must be met before either DNA or RNA is able to replicate.

biorhythm /**bÿ**-oh-ri*th*-<m/ A periodic physiological or behavioral change that is controlled by a biological clock. CIRCADIAN RHYTHMS, hibernation, and migration are examples.

biosphere /**bÿ**-ŏ-sfeer/ The part of the earth and its atmosphere that is inhabited by living organisms. The earth's surface and the top layer of the hydrosphere (water layer) have the greatest density of living organisms. The geosphere, or nonliving world, is made up of the lithosphere (solid earth), hydrosphere, and atmosphere.

biosynthesis /bÿ-oh-**sin**-th'ĕ-sis/ Chemical reactions in which a living cell builds up its necessary molecules from other molecules present. *See* anabolism.

biosystematics /bÿ-oh-sis-tĕ-**mat**-iks/ The area of systematics in which experimental taxonomic techniques are applied to investigate the relationships between taxa. Such techniques include serological methods, biochemical analysis, breeding experiments, and cytological examination, in addition to the more established procedures of comparative anatomy. Evidence from ecological studies may also be brought to bear. *See also* molecular systematics.

biotechnology /bÿ-oh-tek-**nol**-ŏ-jee/ The application of technology to biological processes for industrial, agricultural, and medical purposes. For example, bacteria such as *Penicillium* and *Streptomycin* are used to produce antibiotics and fermenting yeasts produce alcohol in beer and wine manufacture. Genetic engineering now enables the large-scale production of hormones, blood serum proteins, and other medically important products. Genetic modification of farm crops offers improved protection against pests, or products with novel characteristics, such as new flavors, colors, or extended storage properties. *See also* enzyme technology; genetic engineering.

biotic environment /bÿ-ot-ik/ The biological factors acting on an organism, which arise from the activities of other living organisms, as distinct from physical factors.

biotic potential (intrinsic rate of natural increase) Symbol: *r* The maximum possible number of offspring produced by an individual under ideal conditions. Environmental factors such as limitation of resources, predation, and disease mean that the biotic potential is seldom realized.

biotin /**bÿ**-ŏ-tin/ A water-soluble vitamin generally found, together with vitamins in the B group, in the VITAMIN B COMPLEX. It is widely distributed in natural foods, egg yolk, kidney, liver, and yeast being good sources. Biotin is required as a coenzyme for carboxylation reactions in cellular metabolism. It functions in the synthesis of proteins and fatty acids, and in transamination reactions.

biotype /bÿ-ŏ-tÿp/ **1.** A naturally occurring group of individuals all with the same genetic composition, i.e. a clone of a pure line. *Compare* ecotype. **2.** A physiological race or form within a species that is morphologically identical with it, but differs in genetic, physiological, biochemical, or pathogenic characteristics.

bipinnaria /bÿ-pi-**nair**-ee-ă/ A form of dipleurula larva characteristic of starfish. It bears lobes that carry ciliated bands, used for feeding and locomotion. *Compare* pluteus. *See* dipleurula.

biramous appendage /bÿ-**ray**-mŭs/ A two-branched appendage found in crustaceans. It consists of three sections, the *protopodite* (basal section), which is attached at one end to the body and at the other to the two branches, the *expodite* (outer branch) and *endopodite* (inner branch). Each section may consist of several segments.

birds *See* Aves.

bisexual /bÿ-**sek**-shoo-ăl/ *See* hermaphrodite.

biuret test A standard colorimetric test for proteins and their derivatives, e.g. peptides and peptones. Sodium hydroxide is first mixed with the test solution and then copper(II) sulfate solution is added drop by drop. A violet color indicates a positive result. The reaction is due to the presence in the molecule of –NH–CO–, the color not appearing with free amino acids. The compound biuret also readily forms from urea around 150°–170°C and thus the biuret test can be used to identify urea.

bivalent /bÿ-**vay**-lĕnt, biv-ă-/ A term used for any pair of homologous chromosomes when they pair up during meiosis. Pairing

of homologous chromosomes (*synapsis*) commences at one or several points on the chromosome and is clearly seen during pachytene of meiosis I.

bivalve mollusks /bÿ-valv/ *See* Pelecypoda.

bladder 1. A modified leaf, found on the stems of members of the bladderwort family, that develops into a distended structure for trapping small invertebrates. The bladder is thin-walled and closed by a one-way valve. **2.** An air-filled sac found in large numbers in the thalli of certain seaweeds (e.g. bladderwrack). **3.** (**urinary bladder**) A thin-walled muscular sac used as a temporary store for urine in most vertebrates (except birds and some reptiles). In mammals, the urine enters the bladder directly from the ureters and is discharged to the urethra, under the control of a sphincter muscle. It develops as an enlargement of the Wolffian duct or cloaca. **4.** *See* gall bladder. **5.** *See* swim bladder.

bladderworm The intermediate larval stage in the life cycle of certain cestodes (tapeworms) formed when the hexacanth encysts in the muscle of the intermediate host (the pig in the case of the pork tapeworm). It consists of a fluid-filled cyst containing the inverted head of the worm. When raw or insufficiently cooked meat is eaten and digested, the head everts (turns inside-out) and attaches itself to the lining of the gut of the final host (humans), becoming an adult worm and producing proglottids. *See* cysticercus; hydatid cyst.

blastocoel /blast-ŏ-seel/ (**segmentation cavity**) The internal fluid-filled cavity of a blastula, which first appears during cleavage of the egg. The cavity seen in sections of chick and fish eggs, between the blastoderm and yolk, is not a blastocoel but an artifact of histological processing.

blastocyst /blast-ŏ-sist/ A mammalian egg in the later stages of cleavage, before implantation. It consists of a hollow fluid-

44

filled ball of cells and the inner cell mass, from which the embryo develops. *See also* trophoblast.

blastoderm /blast-ŏ-derm/ The cellular mass that results from cleavage of the cytoplasm (blastodisc) of very yolky eggs, such as those of birds, sharks, and cephalopods. The term is also used for the cellular coat of cleaved insect eggs.

blastomere /blast-ŏ-meer/ One of the cleavage products (sometimes called cells) of an animal zygote. The zygote usually divides into two, then four, then eight blastomeres, and so on until the normal nuclear/cytoplasmic ratio of the embryo's cells is achieved. *See* growth; mosaic.

blastopore /blast-ŏ-por/ The opening in a gastrula between the archenteron and the outside, through which invagination occurs at gastrulation. In amniotes it is represented by the primitive streak. The dorsal lip of the amphibian gastrula is the primary organizer of the axial structure of the embryo and corresponds to Hensen's node in a chick or mammal; the future notochord is invaginated over this lip.

blastula /blass-tyŭ-lă/ The stage in an animal embryo following cleavage. It is a hollow fluid-filled ball of cells.

blind spot A point on the retina of a vertebrate eye that is not sensitive to light. It contains no rods or cones, but only nerve fibers passing to the optic nerve, which joins the eyeball at this point. *See illustration at eye.*

blood The transport medium of an animal's body. It is a fluid tissue that circulates by muscular contractions of the heart (in vertebrates) or other blood vessels (in invertebrates). It usually carries oxygen and food to the tissues and carbon dioxide and nitrogenous waste from the tissues, to be excreted. It also conveys hormones and circulates heat throughout the body. Blood consists of liquid plasma in which are suspended white cells (LEUKOCYTES), which devour bacteria and produce antibodies. In most animals (except insects) the blood carries oxygen combined with a pigment (hemoglobin in vertebrates; hemocyanin in some invertebrates). In some invertebrates the pigment is dissolved in the plasma, but in vertebrates it is contained in the red cells (ERYTHROCYTES). *See also* blood plasma; platelet.

blood–brain barrier The anatomical arrangement of epithelial cells in blood capillaries of the brain that prevents large molecules and solid particles passing from blood into brain tissue. This helps to ensure a stable environment for brain cells. The barrier consists of tight junctions between the epithelial cells, coupled with a covering layer of cells (astrocytes). Substances such as water, glucose, oxygen, and carbon dioxide can easily cross the barrier, as do small lipid-soluble molecules (e.g. ethanol). However, the blood–brain barrier prevents many potentially useful drugs from entering the brain.

blood capillary *See* capillary.

blood cell (**blood corpuscle**) Any of the cells contained within the fluid plasma of blood. 45% of blood volume is made up of red cells (ERYTHROCYTES) and 1% of white cells (LEUKOCYTES).

blood clotting (**blood coagulation**) The conversion of blood from a liquid to a solid state, which occurs when an injury to the blood vessels exposes blood to air. The clot closes the wound and prevents further blood loss. Blood clotting is brought about in a series of changes that will occur only when at least 14 different *clotting factors* are all present. When platelets in the blood encounter a damaged blood vessel they adhere to the site, forming a plug. They also activate various clotting factors, culminating in the formation of the enzyme thrombokinase (Factor X). This converts the precursor molecule prothrombin (Factor II) into the enzyme thrombin. This then converts fibrinogen, a soluble protein in the plasma, into insoluble fibrin, which forms a network of fibers in which the blood cells become entangled to form a blood clot. *See also* platelet.

blood corpuscle *See* blood cell.

blood groups Types into which the blood is classified. The *ABO system*, discovered in 1900, assigns human blood to four groups, A (42% of the population), B (9%), AB (3%), and O (46%), based on the presence or absence of certain molecular groups (antigens), called A and B, on the surface of the red blood cells. In group AB, for example, both antigens are present, while group O has neither antigen. Knowledge of a patient's blood group is essential when a blood transfusion is to be given. If blood from a group A donor is given to a group B recipient, the recipient's anti-A antibodies will attack the donor's A antigens, causing the red cells to clump together. Group O blood, having no antigens, can be given to patients of any blood group since it will not provoke an antibody reaction. Group AB, having both antigens and therefore neither antibody, can receive blood from any group. There are various other blood group systems, including that based on the presence or absence of the RHESUS FACTOR.

blood plasma The straw-colored liquid that remains when all the cells are removed from blood. It consists of 91% water and 7% proteins, which are albumins, globulins (mainly antibodies), and prothrombin and fibrinogen (concerned with clotting). Plasma also contains the ions of dissolved salts, especially sodium, potassium, chloride, bicarbonate, sulfate, and phosphate. Plasma is slightly alkaline (pH 7.3) and the proteins and bicarbonate act as buffers to keep this constant. It transports dissolved food (as glucose, amino acids, fat, and fatty acids), excretory products, (urea and uric acid), dissolved gases (about 40 mm^3 oxygen, 19 mm^3 carbon dioxide, and 1 mm^3 nitrogen in 100 mm^3 plasma), hormones, and vitamins. Most of the body's physiological activities are concerned with maintaining the correct concentration and pH of all these solutes (this is homeostasis),

since plasma supplies the extracellular fluid that is the environment of all the cells.

blood platelet *See* platelet.

blood pressure The pressure of the blood against the walls of the blood vessels, in particular the main arteries, that results from the pumping action of the heart and the elasticity of the arterial walls. In a resting human, it normally oscillates between 120 mmHg at systole and 80 mmHg at diastole, although it varies with changes in physiological state of the body. Blood pressure drops considerably by the time it reaches the veins. Abnormally raised blood pressure (*hypertension*) is linked to increased risk of stroke, kidney failure, heart disease, and other disorders.

blood serum The pale fluid that remains after blood has clotted. It consists of plasma without any of the substances involved in clotting. *See also* blood plasma.

blood vascular system In mammals, a continuous system of vessels containing blood, which transports food materials, excretory products, hormones, respiratory gases, etc., from one part of the body to another. The blood is circulated by muscular contractions of the heart: it is first pumped to the lungs but returns to the heart to be pumped around the body.

Certain invertebrates (mollusks and arthropods) have an open system in which blood flows in blood spaces (e.g. the hemocoel of Crustacea). Vertebrates and most invertebrates have a closed system with the blood contained in blood vessels and generally circulated by muscular contractions of the vessels or a heart. In fish, the blood flows only once through the heart before circulating around the body (single circulation) but in other vertebrates blood returning to the heart is circulated to the lungs, where it is oxygenated, before being repumped around the body (double circulation).

blood vessel A tubular structure found in vertebrates and some invertebrates that transports blood throughout the body.

Blood vessels vary in diameter and hence may help to regulate the blood flow to different parts of the body; for example to the surface layers to regulate heat loss. *See* artery; capillary; vein.

blue-green algae *See* Cyanobacteria.

blue-green bacteria *See* Cyanobacteria.

BMR *See* basal metabolic rate.

BOD *See* Biochemical Oxygen Demand.

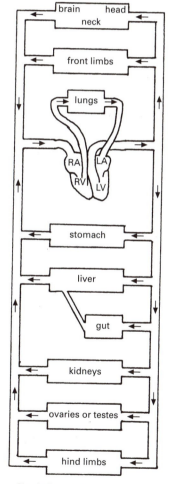

Circulation of blood in mammals

body cavity In most metazoan animals, the cavity bounded by the body wall, which contains the heart, viscera, and many other organs. The body cavity of many triploblastic animals is the COELOM; the body cavity of arthropods is called the hemocoel.

Bohr effect /bor, bohr/ The phenomenon whereby the affinity of the respiratory pigment of the blood (hemoglobin in vertebrates) for oxygen is reduced as the level of carbon dioxide is increased. This facilitates gaseous exchange, because more oxygen is released in the tissues where the amount of carbon dioxide is rising due to metabolic activity. At the same time, more oxygen is taken up at the lungs or gills where the amount of carbon dioxide is low.

bone A hard connective tissue that makes up most of the skeleton of vertebrates. It consists of fine-branched cells (osteoblasts) embedded in a matrix, which they have secreted. The matrix is 30% protein (collagen) and 70% inorganic matter, mainly calcium phosphate.

In *compact bone* the matrix is laid down in concentric cylinders called *lamellae*, which surround *Haversian canals* containing blood vessels; the osteoblasts lie in spaces (*lacunae*) between the lamellae and are linked by fine canals (*canaliculi*), which contain the cells' branches. *Spongy bone* has lamellae that form an interlacing network with red marrow in the spaces. Most bones are first formed in cartilage, which is then replaced by bone. A few, called *dermal* or *membrane bones*, are formed by the ossification of connective tissue. *See also* cartilage bone; membrane bone; ossification; osteoblast; osteoclast.

bony fishes *See* Osteichthyes.

boron /bor-on, boh-ron/ *See* trace element.

botany The scientific study of plants.

bovine spongiform encephalopathy /boh-vȳn spun-jă-form en-seff-ă-lop-ă-th'ee/ (BSE; 'mad cow disease') A progressive degenerative brain disease of cattle that is thought to be caused by an infectious abnormal protein called a PRION. There is a build-up of protein fibrils in the brain, which causes nervous signs such as abnormal posture or aggressiveness, and ultimately death. BSE is one of several related diseases, including scrapie in sheep and chronic wasting disease in deer and elk. It was first diagnosed in 1986 in British cattle, and the subsequent large-scale outbreak was attributed largely to feeding meat-and-bone meal derived from scrapie-infected sheep or BSE-infected cattle. It is thought that some people contracted a variant form of Creutzfeldt–Jakob disease (vCJD) by eating meat products derived from BSE-infected cattle. There is currently no vaccine or treatment, and all infected animals are destroyed.

Bowman's capsule /**boh**-mănz/ The cup-shaped end of a uriniferous tubule of the vertebrate kidney. It surrounds a knot of blood capillaries (glomerulus) and together they form a MALPIGHIAN BODY. *See* nephron.

bp Symbol for base pair, used as a unit for measuring the length of double-stranded nucleotides, such as DNA. A length of 1000 bp is equivalent to 1 kilobase (1 kb) of a single-stranded nucleotide. *Compare* kilobase; megabase.

Brachiopoda /brak-ee-**op**-ŏ-dă, bray-kee-/ A small phylum of marine invertebrates, the lamp shells (e.g. *Lingula*), living attached to a firm substratum in shallow waters. Lamp shells superficially resemble bivalve mollusks in the possession of a bivalve shell, but the valves are placed dorsally and ventrally (as opposed to laterally, as in bivalves). The internal lophophore, a filter-feeding organ composed of ciliated tentacles, protrudes from the shell. Excretion is carried out by nephridia. The sexes are separate and there is a free-swimming larva. The phylum was very much larger in Paleozoic times; *Lingula* is identical to its Cambrian fossil ancestors.

brachysclereid /brak-ă-**skleer**-ee-id/ *See* stone cell.

bract /brakt/ A modified leaf that develops below a flower or an inflorescence. It may be reduced or highly colored, as in the scarlet bracts of *Poinsettia*. In the daisy family the numerous bracts which subtend the inflorescence are known as an *involucre*. In the grasses bracts are borne in pairs below the spikelets and are called *glumes*. Each individual floret of the spikelet is also subtended by two bracts, the lower being called the *lemma* and the upper the *palea*.

bracteole /**brak**-tee-ohl/ A small bract, sometimes a secondary bract, on the pedicel of a flower, as in the bluebell.

bract scale The structure found in large numbers in the female cone of gymnosperms, each bearing an ovuliferous scale in its axil.

bradykinin /brad-ă-**kŷ**-nin/ A peptide formed from a blood protein (kininogen) in certain inflammatory conditions. It has a strong dilating action on blood vessels and causes a fall in blood pressure. *See also* kinin.

brain The most highly developed part of the nervous system, which is located at the anterior end of the body in close association with the major sense organs and is the main site of nervous control within the animal.

The human brain has two greatly folded CEREBRAL HEMISPHERES, which cover most of its surface. The outer layer of these forms the CEREBRAL CORTEX, which is the principal site of integration for the entire nervous system and is also concerned with memory and learning. The CEREBELLUM, lying beneath the cerebral hemisphere at the rear of the brain, makes fine adjustments to muscle actions initiated by the cortex. Involuntary muscle actions, such as those involved in breathing and swallowing, are governed by the MEDULLA OBLONGATA, located where the spinal cord enters the brain. The HYPOTHALAMUS, which lies deep within the brain, controls various metabolic functions and also influences the activity of the pituitary gland. The brain has four interconnected internal cavities, the VENTRICLES, which are filled with cerebrospinal fluid. External protection is provided by three membranes, the MENINGES.

In fishes and birds, the major sensory and motor centers occur in the greatly enlarged deeper regions of the cerebral hemispheres – the corpus striatum. This is thought to reflect the predominance of instinct in bird behavior whereas in mammals, in which learning and memory are paramount, the cerebral cortex (neopallium) is the dominant region of the brain.

brain stem The parts of the adult brain that constitute its basic structure as seen in the developing embryo. It comprises the medulla oblongata, the pons, the midbrain, and part of the forebrain. The brain stem of all vertebrates has a similar basic structure, derived from an expanded and folded neural tube, which subsequently undergoes development depending on the vertebrate group involved.

branchial arch /brank-ee-ăl/ One of a number of visceral arches in fish that support the gills. There are usually five, each typically consisting of nine elements – a mid-ventral basibranchial, two hypobranchials, two ceratobranchials, two epibranchials, and two pharyngobranchials.

Branchiopoda /brank-ee-op-ŏ-dă/ The most primitive class of the Crustacea. Most branchiopods live in fresh water (except *Artemia*, the brine shrimp), and have flat fringed appendages for filter feeding, respiration, and locomotion. Parthenogenesis is common. The class includes *Daphnia* (water flea).

Branchiostoma /brank-ee-os-tŏ-mă/ (*Amphioxus*) A genus of small marine burrowing cephalochordates, the lancelets. *Branchiostoma* has a fish-shaped body with a dorsal and caudal fin and segmentally arranged muscle blocks (myotomes). The pharynx and gill slits are modified for food collection as well as respiration. Excretion is by nephridia, which is unique among chordates. *See* Cephalochordata.

breast *See* mammary gland.

breathing *See* respiration; ventilation.

breed A domesticated VARIETY of plants or animals, for example German shepherd dogs or Holstein cattle. Members of a breed share certain defining traits, which have been developed and are maintained by artificial selection, and can (in theory at least) be mated with other breeds belonging to the same species to produce viable offspring (crossbreeds).

bristle worms *See* Polychaeta.

brittle stars *See* Ophiuroidea.

bronchiole /bronk-ee-ohl/ One of the smaller air passages in the lungs of mammals. A branching system of bronchioles carries air from the bronchi to all regions of the lung. The finest bronchioles end in alveoli. The walls of bronchioles are lined with cells that secrete mucus, which traps dirt and bacteria, and cells with cilia, which beat to carry foreign matter out of the lungs.

The walls of the larger bronchioles are stiffened by incomplete rings of cartilage, to prevent kinking. The walls of the finer bronchioles are thin and allow limited gaseous exchange with capillaries surrounding them.

bronchus /bronk-ŭs/ (*pl.* bronchi) An air passage leading from the trachea and entering the lung in a tetrapod. Each main bronchus is a wide tube whose walls are stiffened by thick incomplete rings of cartilage and contain mucus-secreting glands. This allows flexibility yet prevents collapse due to excess external pressure and kinking due to bending of the tubes. The main bronchi branch to form a number of smaller bronchi, which in turn lead to the bronchioles.

brown algae *See* Phaeophyta.

brown earth The type of soil found under deciduous forests in temperate climates. It is slightly acid and when cleared provides good fertile agricultural soil.

brown fat *See* adipose tissue; thermogenesis.

Brownian motion /**brown**-ee-ăn/ The random motion of microscopic particles due to their continuous bombardment by the much smaller and invisible molecules in the surrounding liquid or gas. Particles in Brownian motion can often be seen in colloids under special conditions of illumination. *See* colloid.

brush border The outer surface of columnar epithelial cells lining the intestine, kidney tubules, etc. With a light microscope it appears as a narrow layer with vertical striations, but the electron microscope shows it to be made of fine hairlike processes called *microvilli*. These greatly increase the surface area of the cell for absorption of dissolved substances. There may be as many as 3000 microvilli on a single epithelial cell.

Bryophyta /brÿ-**off**-ă-tă/ A phylum of simple, mainly terrestrial, erect leafy plants commonly found in moist habitats and comprising the mosses. They show a heteromorphic alternation of generations, the gametophyte being the dominant generation. When mature the gametophyte shows differentiation into stem and leaves but there are no roots or vascular tissues. The sporophyte, which is wholly dependent on the gametophyte, is simply a spore capsule borne on a stalk. Mosses are far more widely distributed than the liverworts (*see* Hepatophyta), and have greater differentiation of the gametophyte. Also, the mechanism of capsule dehiscence (splitting) is more complex. There are three classes: Sphagnopsida (peat mosses, e.g. *Sphagnum*); Andreaopsida (granite mosses); and Bryopsida ('true mosses', e.g. *Polytrichum*). The liverworts were formerly also included in the Bryophyta, but are now more commonly placed in their own phylum (Hepatophyta). *See also* Hepaticae; Musci.

Bryozoa /brÿ-ŏ-**zoh**-ă/ A phylum of mainly marine sessile invertebrates, the moss animals and sea mats, which live in colonies resembling seaweed or form an encrusting sheet on rocks and shells. The oval or tubular individuals live in self-secreted horny, chalky, or gelatinous protective cases, into which they retreat when disturbed. They have a coelom, a ciliated tentacular food-catching organ (lophophore) surrounding the mouth, and a U-shaped gut. There are two classes, often regarded as separate phyla. In the Ectoprocta, the anus lies outside the tentacles and there are no special excretory organs. The Entoprocta have the anus within the tentacles, no true coelom, and protonephridia as excretory organs.

BSE *See* bovine spongiform encephalopathy.

buccal cavity /**buk**-ăl/ The mouth cavity, by which food is taken into the alimentary canal. In mammals it is surrounded by the lips and cheeks, which enclose the food while it is being chewed by the teeth and tasted by the tongue, which also helps to move the food back for swallowing. Ducts from salivary glands open into the mouth so that food is moistened and lubricated and, in some cases, digested. Other vertebrates have buccal cavities without cheeks and cannot chew their food, but must swallow it whole. Invertebrates have a variety of buccal cavities depending on the type of food and method of ingestion.

The buccal cavity is lined with stratified squamous epithelium derived from the embryonic ectoderm and is part of the stomodaeum.

bud 1. (*Botany*) A compacted undeveloped shoot consisting of a shortened stem and immature leaves or floral parts. The young leaves are folded about the growing tip, and the outermost leaves may be scaly and reduced to protect the growing point. A bud has the potential to develop into a new shoot or flower. *Terminal buds* are formed at the stem or branch tip. *Axillary* or *lateral buds* develop in the leaf axils. Buds can develop adventitiously on other parts of a plant and are sometimes a means of asexual reproduction.

2. (*Zoology*) An outgrowth of an animal that is capable of vegetative reproduction. In lower animals the production of buds that grow into new individuals and then break away from the parent is a common form of asexual reproduction.

budding 1. The production of buds on plants.

2. A type of grafting in which the grafted part is a bud.

3. A type of asexual reproduction in which a new individual is produced as an outgrowth (bud) of the parent organism. It is common in certain animal groups, such as cnidarians, sponges, and urochordates, where it is also termed *gemmation*. It also occurs in the unicellular fungi, especially the yeasts.

buffer A solution that resists any change in acidity or alkalinity. Buffers are important in living organisms because they guard against sudden changes in pH. They involve a chemical equilibrium between a weak acid and its salt or a weak base and its salt. In biochemistry, the main buffer systems are the phosphate ($H_2PO_4^-/HPO_4^{2-}$) and the carbonate (H_2CO_3/HCO_3^-) systems.

Bufo /boo-foh/ (**toad**) *See* Anura.

bugs *See* Hemiptera.

bulb A modified shoot that acts as an organ of perennation and often vegetative reproduction. The stem is reduced to a disclike structure, bearing concentric layers of fleshy leaf bases that comprise the food store. Each leaf has a bud at its base that is able to develop into a subsidiary bulb. The whole bulb is protected by scale leaves and adventitious roots arise from the base of the stem. In spring or summer one or more buds grow and produce leaves and flowers, exhausting the food supply. The new leaves photosynthesize and food is stored in their bases thus giving rise to a new bulb. If more than one bud develops, eventually more than one bulb develops.

bulbil /bul-băl/ A small bulblike organ of vegetative reproduction that may form in a leaf axil, an inflorescence, or at a stem base.

bulla /bûl-ă, bul-/ A small projection of the mammalian skull that encases and protects the middle ear.

bundle of His /hiss/ (**atrioventricular bundle**) A bundle of specialized muscle fibers in the mammalian heart that conducts electrical impulses from the atrioventricular node, between the left and right atria, to the PURKINJE FIBERS in the midline between the ventricles. It thus triggers the synchronized wave of excitation that results in simultaneous contraction of both ventricles. It is named for the German physician Wilhelm His Jr (1863–1934).

bundle sheath The ring of parenchymatous or sclerenchymatous tissue, usually one cell thick, that surrounds the vascular bundle in an angiosperm leaf. The individual cells are closely packed with no apparent intercellular spaces, and conduct water and solutes from the vascular bundle to the surrounding tissues. Chloroplasts may be present in the bundle sheath, and are thought to be connected with starch storage in the tropical grasses.

Burgess shale /ber-jis/ A deposit of shale in British Columbia, Canada, dating from the middle of the Cambrian period (about 530 million years ago) and noted for the remarkable preservation of fossils of soft-bodied marine animals. These include not only many unique arthropods, such as the elegant spined *Marella*, but other organisms with no known relatives, either living or fossil. Such an example is the strange segmented *Opabina*, with its prominent frontal nozzle and five eyes. Study of the Burgess fauna has led to a dramatic reassessment of life in the early Cambrian, indicating a huge diversity of marine life forms, both soft-bodied and hard-bodied (e.g. trilobites), and prompting speculation about why some forms survived and so many others became extinct.

cis (maleic)

trans (fumaric)

Butenedioic acids

butenedioic acid /byoo-teen-dÿ-**oh**-ik/ Either of two isomers. *Transbutenedioic acid* (fumaric acid) is a crystalline compound found in certain plants. *Cisbutenedioic acid* (maleic acid) is used in the manufacture of synthetic resins. It can be converted into the *trans* isomer by heating at 120°C.

butterflies *See* Lepidoptera.

buttress root An asymmetrically thickened prop root, found at the base of certain trees.

C

Caenozoic /see-nŏ-**zoh**-ik, sen-ŏ-/ *See* Cenozoic.

Cainozoic /see-nŏ-**zoh**-ik, sen-ŏ-/ *See* Cenozoic.

calcicole /**kal**-să-kohl/ Describing plants that thrive on neutral to alkaline chalk, carboniferous, or limestone soils, such as marls. *Compare* calcifuge.

calciferol /kal-**sif**-ĕr-ol, ohl/ *See* vitamin D.

calcifuge /**kal**-să-fyooj/ Describing plants that grow on soils containing very little calcium carbonate, such as loams. Calcifuge plants are generally absent from chalky soils. *Compare* calcicole.

calcitonin /kal-să-**toh**-nin/ A polypeptide hormone, produced by the C-cells of the thyroid gland, that lowers the concentration of calcium in the blood. It acts by preventing calcium and phosphorus from leaving bone but not from entering it, thus antagonizing the effect of parathyroid hormone (*see* parathyroid glands).

calcium /**kal**-see-ŭm/ An essential mineral salt for animal and plant growth. It is present between plant cell walls as pectate, and is found in the bones and teeth of animals. Calcium ions, Ca^{2+}, are important in triggering muscle contraction where their rapid release from the cisternae of the sarcoplasmic reticulum triggers the reaction between ATP and the myofilaments. Calcium is important in resting muscles in maintaining the relative impermeability of the cell membranes. If the calcium concentration falls, the potential difference across the membrane also falls so that muscles may spontaneously contract without activation by acetylcholine, giving twitching and spasms. The concentration of calcium ions is also important in influencing the breakdown of glycogen in muscles. Calcium is important in the clotting of blood in the conversion of prothrombin to thrombin. In mammalian stomachs it is also important in precipitating the protein casein from milk.

callose /**kal**-ohs/ An insoluble carbohydrate that is laid down around the perforations in sieve plates. As the sieve tube ages the callose layers become thicker, eventually blocking the sieve element. Such blocking may be seasonal or permanent.

callus /**kal**-ŭs/ A mass of undifferentiated parenchyma cells formed by the cambium in response to wounding of the vascular tissue. If parenchyma is injured, then the surrounding uninjured parenchyma cells form a cork cambium that produces a layer of suberized cells sealing off the wound. In tissue cultures, callus can be induced to form by various hormone treatments. Adventitious shoots and roots often differentiate from calluses, a phenomenon exploited in the rooting of cuttings. *See also* graft.

calmodulin A small calcium-binding protein found in the cytosol of cells that regulates many cellular activities. Each calmodulin molecule binds four calcium ions (Ca^{2+}) and in so doing undergoes a change in its three-dimensional shape, enabling it to further bind to and activate various enzymes.

calorie /**kal**-ŏ-ree/ (**gram calorie; small calorie**) Symbol: cal Any of several obsolete units of energy in the c.g.s. system based on the amount of heat needed to raise the temperature of 1 g of water by 1°C from a specified starting temperature, and all approximately equal to 4.185 J.

Calorie (kilocalorie; kilogram calorie; large calorie) Symbol: Cal A unit equal to 1000 calories and formerly used in measuring food energy. Equal to 4186.8 joules, it should no longer be used.

calorimetry /kal-ŏ-**rim**-ĕ-tree/ The measurement of thermal changes involved in chemical or physical reactions, in either an *in vitro* system or an intact organism. For example, a *bomb calorimeter* is used to determine the calorific value of foods. This is a steel chamber in which the foodstuff is placed and ignited, in the presence of oxygen. The increase in temperature of water in a jacket surrounding the calorimeter indicates the amount of heat produced by the oxidized food.

Calvin cycle /**kal**-vin/ *See* photosynthesis.

calyptra /kă-**lip**-tră/ A layer of cells derived from the venter of the archegonium that covers the developing sporophyte. In bryophytes it ruptures as the seta elongates, being taken up as a hood over the capsule in mosses, and forming the sheath of tissue at the base of the seta in liverworts. The presence of the calyptra is necessary for the proper development of the capsule in mosses and the embryo in ferns.

calyptrogen /kă-**lip**-trŏ-jĕn/ A layer of meristematic cells covering the root apical meristem in some plants (e.g. grasses) that gives rise to the root cap.

calyx /**kay**-liks, **kal**-iks/ (*pl.* **calyces**) The outermost part of a flower, enclosing the other floral parts during the bud stage. It consists of leaflike sepals, which are normally green. The symbol *K* denotes the calyx in the floral formula. If the sepals are joined at their lateral margins a *calyx tube* is formed, the mouth of which may be extended into lobes or teeth.

cambium /**kam**-bee-ŭm/ The ring of dividing cells responsible for lateral growth in plants. The primary cambium is found in the stem and root between the phloem and xylem cells, and by division gives rise to the secondary phloem and xylem in woody dicotyledonous plants. *See* wood. *See also* annual ring; meristem; phellogen; interfascicular cambium; intrafascicular cambium.

Cambrian /**kam**-bree-ăn/ The earliest period of the Paleozoic era, about 590–510 million years ago. It is characterized by the appearance of algae and a proliferation of marine invertebrate animal forms, including ancestors of most modern animals – the so-called Cambrian explosion. In Britian, Cambrian rocks are found in Wales and North-West Scotland. *See* Burgess shale. *See also* geological time scale.

campylotropous /kam-pă-**lot**-rŏ-pŭs/ (**orthocampylotropous**) Describing the position of the ovule in the ovary, when the funicle appears to be attached half way between the chalaza and the micropyle. The micropylar end is turned through 90° relative to the orthotropous condition so that the ovule is horizontal, as in mallow. *Compare* anatropous; orthotropous. *See illustration at* ovule.

Canada balsam /**bawl**-săm/ A clear resin similar to glass in its optical properties, used for mounting microscope specimens.

cancer A malignant tumor, or disease caused by it. Malignant tumors are distinguished from benign ones in that they are not encapsulated, their cells show uncontrolled reproduction and lack differentiation of structure, and they are capable of producing secondary growths (*metastases*) in a part of the body distant from the original tumor. Cancers are classified into two main groups according to the tissue in which they arise: *carcinomas* arise in epithelial tissue; *sarcomas* in connective tissue. *See also* oncogene; oncogenic; tumor suppressor gene.

cane sugar *See* sucrose.

canine tooth (eye tooth) A mammalian

tooth with a single pointed crown, occurring on either side of the jaw between the incisors in front and the premolars behind. There is one canine on each side of each jaw, making a total of four. They are typically conical and pointed, and in carnivorous animals, such as the dog, they are long and fanglike. These animals use the canine teeth for killing their prey, by piercing and tearing the flesh. In rodents, such as mice and rats, the canine teeth are absent. *See also* teeth.

capillary /**kap**-ă-lair-ee/ One of numerous tiny blood vessels (5–20 μm diameter) that branches out from an arteriole to form a dense network (*capillary bed*) amongst the tissues and reunites into a venule. They have thin walls of endothelium through which oxygen, carbon dioxide, inorganic ions, dissolved food, excretory products, etc., are exchanged between the blood and the cells via the tissue fluid.

capitulum /kă-**pich**-ŭ-lŭm/ An inflorescence typical of the family Compositae (e.g. daisy). It is made up of a large number of unstalked florets inserted on the flattened disclike end of the peduncle and surrounded by a ring of sterile bracts, the involucre. Each floret may or may not be borne in the axil of a bract on the disk.

capsid /**kap**-sid/ The protein coat of a VIRUS, surrounding the nucleic acid. A capsid is present only in the inert extracellular stage of the life cycle. Capsids are composed of subunits called *capsomeres*.

capsomere /**kap**-sŏ-meer/ *See* capsid.

capsule /**kap**-sewl/ 1. A dry dehiscent fruit that is formed from several fused carpels. The numerous seeds may be released through pores (e.g. snapdragon), a lid (e.g. poppy), or by complete splitting of the capsule (e.g. iris). The carcerulus, pyxidium, regma, silicula, and siliqua are all forms of capsule.
2. The mucilaginous covering often found around the cell membrane in bacteria.
3. The structure within which spores of the sporophyte generation of mosses and liverworts are formed. It is borne at the end of a long stalk, the seta, and ruptures to release the spores.
4. (*Zoology*) A protective or supportive sheath or envelope that surrounds an organ or part of the body. It is usually composed of connective tissue, as in the capsule of a joint or capsule of a kidney. *See also* auditory capsule.

carapace 1. In some arthropods (e.g. crustaceans, spiders, and king crabs), a shield-like protective covering of the dorsal and lateral surfaces of the cephalothorax.
2. In reptiles of the order Chelonia (turtles, tortoises, etc.), the domed dorsal part of the shell, consisting of platelike bones covered on the outside by horny plates (*scuta*). The thoracic vertebrae and ribs are incorporated in the carapace, but the limb girdles are separate and located inside the carapace. The flatter ventral part of the shell is the *plastron*.

carbohydrates /kar-boh-**hy**-dayts/ A class of compounds occurring widely in nature and having the general formula type $C_x(H_2O)_y$. (Note that although the name suggests a hydrate of carbon these compounds are in no way hydrates and have no similarities to classes of hydrates.) Carbohydrates are generally divided into two main classes: SUGARS and POLYSACCHARIDES.

Carbohydrates are both stores of energy and structural elements in living systems; plants having typically 15% carbohydrate and animals about 1% carbohydrate. The body is able to build up polysaccharides from simple units (anabolism) or break the larger units down to more simple units for releasing energy (catabolism). Carbohydrates require neutral or slightly alkaline conditions for the operation of enzymes such as maltase and amylase. Thus, carbohydrate digestion is an intestinal rather than a stomach process.

carbon /**kar**-bŏn/ An essential element in plant and animal nutrition that occurs in all organic compounds and thus forms the basis of all living matter. It enters plants as carbon dioxide and is assimilated into car-

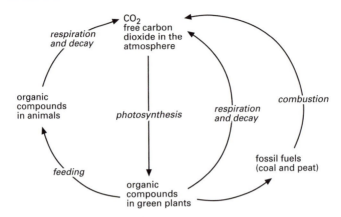

Some of the main stages in the carbon cycle

bohydrates, proteins, and fats, forming the backbones of such molecules. The element carbon is particularly suited to such a role as it can form stable covalent bonds with other carbon atoms, and with hydrogen, oxygen, nitrogen, and sulfur atoms. It is also capable of forming double and triple bonds as well as single bonds and is thus a particularly versatile building block. Carbon, like hydrogen and nitrogen, is far more abundant in living materials than in the Earth's crust, indicating that it must be particularly suitable to fulfill the requirements of various living processes. *See also* carbon cycle.

carbon cycle The circulation of carbon between living organisms and the environment. The carbon dioxide in the atmosphere is taken up by autotrophic organisms (mainly green plants) and incorporated into carbohydrates. The carbohydrates so produced are the food source of the heterotrophs (mainly animals). All organisms return carbon dioxide to the air as a product of respiration and of decay. The burning of fossil fuels also releases CO_2. In water, carbon combined as carbonates and bicarbonates is the source of carbon for photosynthesis.

carbon-14 dating *See* radioactive dating.

carbon dioxide /dÿ-**oks**-ÿd/ A colorless odorless gas, CO_2, that is present in the atmosphere at a low concentration (currently about 380 parts per million) and occurs dissolved in water as carbonic acid. It is taken from the atmosphere by photosynthesizing organisms (e.g. plants, algae, and some bacteria) for the manufacture of organic compounds, and returned by organisms through the chemical reactions of respiration (*see* carbon cycle). Carbon dioxide, along with methane and water vapor, is responsible for the GREENHOUSE EFFECT, which maintains the earth's surface temperature within tolerable limits. Since 1800 the level of CO_2 in the atmosphere has risen by about a third, due largely to human activities, including the burning of fossil fuels and large-scale deforestation. This rise in CO_2 is intensifying the greenhouse effect, causing average surface temperatures to rise, resulting in the phenomenon of global warming.

Carboniferous /kar-bŏ-**nif**-er-ŭs/ The second most recent period of the Paleozoic era, some 355–280 million years ago. It is characterized by the evolution on swampy land of amphibians, a few primitive early reptiles, and giant ferns. Aquatic life included sharks and coelacanths. The period is named after the extensive coal deposits that formed from the remains of vast forests of swamp plants. *See also* geological time scale.

carboxylase /kar-**boks**-ă-layss/ An enzyme that catalyzes the decarboxylation of ketonic acids. Carboxylases are found in yeasts, bacteria, plants, and animal tissues. Pyruvic carboxylase brings about the decarboxylation of pyruvic acid whilst oxaloacetic carboxylase helps the breakdown of oxaloacetic acid into pyruvic acid and carbon dioxide. Carboxylases are thus involved in the transfer of carbon dioxide in respiration.

Carboxylic acid: R is an organic group

carboxylic acid /kar-boks-**il**-ik/ An organic compound of general formula RCOOH, where R is an organic group and –COOH is the carboxylate group. Many carboxylic acids are of biochemical importance. Those of particular significance are:
1. The lower carboxylic acids (such as citric, succinic, fumaric, and malic acids), which participate in the Krebs cycle.
2. The higher acids, which are bound in lipids. These are also called *fatty acids*, although the term 'fatty acid' is often used to describe any carboxylic acid of moderate-to-long chain length. The fatty acids contain long hydrocarbon chains, which may be saturated (no double bonds) or unsaturated (C=C double bonds). Animal fatty acids are usually saturated, the most common being stearic acid and palmitic acid. Plant fatty acids are often unsaturated: oleic acid is the commonest example. *See also* lipid; triglyceride.

carcerulus /kar-**ser**-yŭ-lŭs/ A type of capsular fruit that breaks up at maturity into one-seeded segments or nutlets. The carcerulus is typical of the Labiatae family, e.g. deadnettle.

carcinogen /kar-**sin**-ŏ-jĕn/ Any substance that causes living tissues to become cancerous. Chemical carcinogens include many organic compounds, e.g. hydrocarbons in tobacco smoke, as well as inorganic ones, e.g. asbestos. Carcinogenic physical agents include ultraviolet light, x-rays, and radioactive materials. Many carcinogens are mutagenic, i.e. they cause changes in the DNA; dimethylnitrosamine, for example, methylates the bases in DNA. A potential carcinogen may therefore be identified by determining whether it causes mutations, as by the Ames test, which uses bacteria.

cardia /**kar**-dee-ă/ (**cardiac sphincter**) The opening from the esophagus to the stomach in vertebrates, which functions as a sphincter but is not adapted anatomically.

cardiac cycle /**kar**-dee-ak/ The sequence of events within the mammalian heart that is repeated during each heart beat. Each cycle consists of a phase when the heart chambers are contracting (systole) followed by a relaxation phase (diastole). During diastole, blood enters the right and left atria from the venae cavae and pulmonary veins, respectively, and flows into the ventricles through the open atrioventricular valves (tricuspid and bicuspid valves). Contraction of the atria (*atrial systole*) forces blood into the ventricles, which immediately begin to contract (*ventricular systole*). As pressure rises inside the ventricles, the atrioventricular valves close, preventing backflow into the atria. For a brief period the ventricles are sealed chambers, then the pulmonary and aortic valves open, and blood is forced into the pulmonary arteries and the aorta. The ventricles then relax (*ventricular diastole*), the pulmonary and aortic valves close, and the atria begin to fill with blood for another cycle. In a human the complete cycle takes about 0.8 s at a normal resting heart rate.

cardiac muscle /**kar**-dee-ak/ The muscle of the vertebrate heart. It consists of short cylindrical fibers with branching ends that connect with each other, forming a network, so that waves of contraction can travel through the muscle from the sinoatrial node. Each fiber has one or more nuclei and contains mitochondria and striated fibrils similar to those of skeletal muscle. The rhythmic contractions of the heart arise within the muscle itself, without nervous stimulation; the contraction is said to be *myogenic*. *See also* heart; pacemaker.

cardiac sphincter *See* cardia.

cardinal vein Either of two pairs of veins found in fish that carries deoxygenated blood back to the heart. The anterior cardinal veins serve the head while the posterior cardinal veins serve the rest of the body. They unite to form the common cardinal vein (*Cuvierian duct*), which enters the sinus venosus of the heart. In tetrapods the anterior cardinal veins are replaced by the jugular veins and the anterior venae cavae, and the posterior cardinal veins by the posterior venae cavae.

carina /kă-rỹ-nă/ *See* keel.

carnassial teeth /kar-**nass**-ee-ăl/ Specialized teeth found in dogs and other carnivorous mammals. They comprise the last premolar in the upper jaw and the first molar in the lower jaw, which are very large and are used to cut up meat and shear it from bones. These teeth have a single line of sharp-edged points (cusps) parallel to the jaw. The smooth flat inner surface of the upper carnassial slides over the outer smooth surface of the lower one, like the two blades of shears. When dogs are eating bones or meat they move the food back in the mouth so that they can gnaw it with these teeth.

Carnivora /kar-**niv**-ŏ-ră/ The order that contains the flesh-eating mammals, including *Canis* (wolf, dog), *Felis* (cat), *Meles* (badger), and *Lutra* (otter). The teeth of carnivores are specialized for biting and tearing flesh. The long pointed canines are used for killing prey, two pairs of carnassials (modified cheek teeth) shear the flesh and the sharp molars and premolars crush and grind it. The claws are well developed and sometimes retractile. Carnivores are typically intelligent mammals with keen senses. Most eat flesh, but the bears are omnivorous and the panda is a herbivore.

carnivore /**kar**-nă-vor, -vohr/ A flesh-eating animal, especially a mammal of the order Carnivora (e.g. cats, wolves, seals, etc.). Carnivores generally have powerful jaws, teeth modified for tearing flesh and cracking bones, and well-developed claws. *Compare* herbivore; omnivore.

carotene /**ka**-rŏ-teen/ A carotenoid pigment, examples being lycopene, and α- and β-carotene. The latter compounds are important in animal diets as a precursor of vitamin A. *See* carotenoids; photosynthetic pigments.

carotenoids /kă-**rot**-ĕ-noidz/ A group of yellow, orange, or red pigments comprising the carotenes and xanthophylls. They are found in all photosynthetic organisms, where they function mainly as accessory pigments in photosynthesis, and in some animal structures, e.g. feathers. They contribute, with anthocyanins, to the autumn colors of leaves since the green pigment chlorophyll, which normally masks the carotenoids, breaks down first. They are also found in some flowers and fruits, e.g. tomato. Carotenoids have three absorption peaks in the blue-violet region of the spectrum.

Carotenes are hydrocarbons. The most widespread is β-carotene. This is the orange pigment of carrots whose molecule is split into two identical portions to yield vitamin A during digestion in vertebrates. *Xanthophylls* resemble carotenes but contain oxygen. *See* absorption spectrum; photosynthetic pigments.

carotid arch /kă-**rot**-id/ *See* carotid artery.

carotid artery One of a pair of blood vessels that supplies oxygenated blood to the head and neck. They are derived from the third aortic arch, which in tetrapods forms the *carotid arch* that arises from the aorta as the common carotid artery. This branches in the neck region into an internal and external carotid artery. *See illustration at* heart.

carotid body A vascular structure at the base of the external carotid artery that contains chemoreceptors, which monitor carbon dioxide and oxygen concentrations and pH of the blood. It responds to a change in any of these factors by firing nervous impulses that bring about reflex changes in respiratory and heart rates.

carotid sinus A swollen portion of the internal carotid artery, near its origin in the neck, containing sensory receptors that monitor changes in blood pressure. *See* baroreceptor.

carpal bones /**kar**-păl/ Bones in the distal region of the forelimb of tetrapods; in humans they constitute the wrist (carpus).

In the typical pentadactyl limb there are 12 carpal bones, arranged in three rows. However, there are various modifications and reductions to this basic pattern; in humans there are only eight. They articulate with each other, and with the metacarpal bones distally. Three carpal bones form the wrist joint with the radius of the forearm. *Compare* tarsal bones. *See illustration at* pentadactyl limb.

carpel /**kar**-pĕl/ The female reproductive organ of a flowering plant. It usually consists of an ovary, containing one or more ovules, a stalk, or style, and a terminal receptive surface, the stigma. Each flower may have one or more carpels that may be borne singly giving an apocarpous gynoecium or fused together giving a syncarpous gynoecium. The carpel is homologous with the megasporophylls of certain pteridophytes and the ovuliferous scales of gymnosperms. It has evolved by fusion of the two edges of the megasporophyll. This development can be most clearly seen in simple carpels such as those of the Leguminosae.

carpus /**kar**-pŭs/ (*pl.* **carpi**) The collection of carpal bones that forms the wrist in humans. *Compare* tarsus.

carrier 1. An organism that carries a recessive, often harmful, gene masked in the phenotype by a normal dominant gene. Deleterious sex-linked genes may be carried by the homogametic sex, e.g. women can carry color blindness and hemophilia. There is a 50% chance that the sons of a human carrier will show the effects of the harmful gene, and each daughter of a carrier stands a 50% chance of being a carrier. **2.** An individual infected with pathogenic microorganisms without showing symptoms of disease. Such carriers can transmit the infection to others.

carrying capacity Symbol: K The maximum number of organisms of any particular species that can be supported indefinitely by a given ecosystem or habitat. When the number of individuals (N) in a population approaches the K value, there is increased competition between population members for resources (e.g. space, food, light), and the population growth rate declines to zero, at which point the birth rate equals the death rate.

cartilage /**kar**-t'l-ij/ Gristle: a firm but flexible skeletal material that makes up the entire skeleton of the cartilaginous fishes (sharks, etc.). In more advanced vertebrates the skeleton is first formed as cartilage in the embryo and then changed into bone; in adults cartilage persists in a few places, such as the end of the nose, the pinna of the ear, the disks between vertebrae, over the ends of bones, and in joints. Cartilage is a connective tissue containing cells (*chondroblasts*) embedded in a matrix of solid protein (*chondrin*), which may have elastic or tough white fibers in it. When the cells divide they cannot move apart, so they generally remain in groups of two or four.

cartilage bone (**replacing bone**) The type of bone that is formed from cartilage in the embryo. The cartilage is invaded by bone-forming cells (OSTEOBLASTS), which convert it into bone in the process of ossification. *Compare* membrane bone.

cartilaginous fishes /kar-tă-**laj**-ă-nŭs/ *See* Chondrichthyes.

caruncle /**ka**-runk-ăl, kă-**runk**-ăl/ A fleshy outgrowth from a seed, similar to, but smaller than, an aril. Caruncles may arise from the placenta, micropyle, or funicle and are seen in castor oil seeds at the micropyle and in violet seeds at the hilum.

caryopsis /ka-ree-**op**-sis/ (**grain**) A dry indehiscent fruit, typical of the grasses. It is similar to an achene except that the ovary wall is fused with the seed coat.

casein /**kay**-seen, -see-in/ A phosphoprotein that is present in milk (as calcium caseinate). It belongs to a group of proteins whose main function it is to store amino acids as both nutrients and as building blocks for growing animals.

Casparian strip /kass-**par**-ee-ăn/ An impervious band of thickening on the radial and transverse walls of the endodermis. It consists of deposits of suberin and cutin and insures that all the water and solutes entering the stele pass through the cytoplasm of the endodermal cells.

caste One of the several specialized groups of individuals that exist in a community of social insects, especially ants, bees, wasps, or termites. They are distinguished by structural and functional differences. For example, honeybees have three castes: the queen (a fertile female) reproduces; workers (sterile females) gather food; and drones (males) mate with the queen. The caste system is an example of polymorphism in animals.

catabolism /kă-**tab**-ŏ-liz-ăm/ Metabolic reactions involved in the breakdown of complex molecules to simpler compounds, in a series of stages. The main function of catabolic reactions is to provide energy, which is used in the synthesis of new structures, for work (e.g. contraction of muscles), for transmission of nerve impulses, and for the maintenance of functional efficiency. *See also* metabolism.

catalase /**kat**-ă-layss/ An enzyme present in both plant and animal tissues that cat-alyzes the breakdown of hydrogen peroxide, a toxic compound produced during metabolism, into oxygen and water.

catalyst /**kat**-ă-list/ A substance that increases the rate of a chemical reaction without being used up in the reaction. Enzymes are highly efficient and specific biochemical catalysts.

catecholamines /kat-ĕ-**koh**-lă-meenz/ A group of chemicals (amine derivatives of catechol) that occur in animals, especially vertebrates, and act as neurohormones or neurotransmitters. Examples are norepinephrine, epinephrine, and dopamine.

cathepsin /kă-**th'ep**-sin/ One of a group of enzymes that break down proteins to amino acids within the various mammalian tissues. Several cathepsins have been isolated and are differentiated by their different activities.

cation /**kat**-ÿ-ŏn, -on/ A positively charged ion, formed by removal of electrons from atoms or molecules. In electrolysis, cations are attracted to the negatively charged electrode (the cathode). *Compare* anion.

catkin An inflorescence, often pendulous, that is a modified spike adapted for wind pollination. It is made up of numerous reduced flowers, which are usually unisexual. Male flowers produce very large quantities of pollen, as in the hazel, and are often protected from dew and rain by rooflike bracts above them. When the pollen or seeds have been shed the catkin falls as a unit.

caudal vertebrae /**kaw**-d'l/ Bones of the vertebral column that protect the spinal cord in the tail region. As the tip of the tail is approached they lose the general features of vertebrae until the terminal ones consist solely of a cylindrical centrum. In humans (and other primates) the caudal vertebrae are much reduced and fuse to form the coccyx. *See also* urostyle.

cdl *See* critical day length.

CD marker Cluster of differentiation marker: any of a group of antigenic molecules occurring on the surface of white blood cells and other cells that are used to distinguish subsets of very similar cell populations, e.g. in immunology. For example, T-helper cells possess CD4 antigens, whereas CD8 antigens occur on cytotoxic T-cells. They are identified using specific monclonal antibodies. *See also* T-cell.

cDNA *See* complementary DNA.

cecum /see-kŭm/ (*pl.* ceca) A blind pouch at the junction of the ileum and colon. It ends in the vermiform appendix and in most mammals is very short; in herbivores (e.g. the rabbit) it is large and has an important function. These animals eat only plant food consisting mainly of cellulose, which mammals are normally unable to digest. Herbivores rely on the activity of millions of symbiotic bacteria living in the cecum. These produce enzymes capable of digesting cellulose and releasing simple soluble substances that can be absorbed into the bloodstream. The cecum has no villi and its only secretion is mucus. *See illustration at* alimentary canal.

In insects a number of ceca are found projecting from the midgut, increasing the surface area for absorption.

cell The basic unit of structure of all living organisms, excluding viruses. Cells were discovered by Robert Hooke in 1665, but Schleiden and Schwann in 1839 were the first to put forward a clear *cell theory*, stating that all living things were cellular. Prokaryotic cells (typical diameter 1 μm) are significantly smaller than eukaryotic cells (typical diameter 20 μm). The largest cells are egg cells (e.g. ostrich, 5 cm diameter); the smallest are mycoplasmas (about 0.1 μm diameter). All cells contain genetic material in the form of DNA, which controls the cell's activities; in eukaryotes this is enclosed in the nucleus. All contain cytoplasm, containing various organelles (see diagrams and relevant headwords), and are surrounded by a plasma membrane. This controls entry and exit of substances. Plant cells and most prokaryotic cells are surrounded by rigid cell walls. Differences between animal and plant cells can be seen in the diagrams; differences between prokaryotic and eukaryotic cells can be seen in the table. In multicellular organisms cells become specialized for different functions; this is called division of labor. Within the cell, further division of labor occurs between the organelles.

cell body The part of a nerve cell (NEURON) that contains the nucleus and most of its organelles. It has a swollen appearance and contains Nissl granules. The cell body is a center of synthesis, supplying materials to the rest of the neurone.

cell coat *See* glycocalyx, def. 1.

cell cycle The ordered sequence of phases through which a cell passes leading up to and including cell division (mitosis). It is divided into four phases G_1, S, and G_2 (collectively representing interphase), and M-phase, during which mitosis takes place. Synthesis of messenger RNA, transfer RNA, and ribosomes occurs in G_1, and replication of DNA occurs during the S phase. The materials required for spindle formation are formed in G_2. The time taken to complete the cell cycle varies in different tissues. For example, epithelial cells of the intestine wall may divide every 8–10 hours.

The events of the cell cycle are tightly controlled, principally by various protein kinases. Each consists of a regulatory subunit, called a cyclin, and a kinase subunit, called a cyclin-dependent kinase (Cdk). The Cdk activates other proteins by the addition of a phosphate group (phosphorylation), as directed by the cyclin subunit. The concentrations of these cyclin-Cdk components rise and fall during successive phases of the cell cycle, thereby causing sequential activation of the enzymes, transcription factors, and other components required for each phase. Progress through the cycle is monitored at several checkpoints, where the cell ensures that the chromosomes and DNA are undamaged, and that all opera-

COMPARISON OF PROKARYOTIC AND EUKARYOTIC CELLS		
	Prokaryote	*Eukaryote*
Occurrence	bacteria	animals, plants, and fungi
Average diameter	1 μm	20 μm
Nuclear material	not separated from cytoplasm by membrane	bounded by nuclear membrane
DNA	circular and forming only one linkage group	linear and divided into a number of chromosomes
Nucleolus	–	+
Cell division	amitotic	usually by mitosis or meiosis
Cytoplasmic streaming	–	+
Vacuoles	–	+
Plastids	–	+
Ribosomes	smaller (70S)	larger (80S)
Endoplasmic reticulum	–	+
Golgi apparatus	–	+
+ indicates presence; – indicates absence		

tions have been completed, before entering the next phase.

cell division The process by which a cell divides into daughter cells. In unicellular organisms it is a method of reproduction. Multicellular organisms grow by cell division and expansion, and division may be very rapid in young tissues. Mature tissues may also divide rapidly when continuous replacement of cells is necessary, as in the epithelial layer of the intestine. In plants certain growth regulators (e.g. cytokinins) stimulate renewed cell division. *See* meiosis; mitosis; amitosis.

cell fractionation The separation of the different constituents of the cell into homogenous fractions. This is achieved by breaking up the cells in a mincer or grinder and then centrifuging the resultant liquid. The various components settle out at different rates in a centrifuge and are thus separated by appropriately altering the speed and/or time of centrifugation.

cell junction (junctional complex) Any of the various types of connections between individual cells. In animal tissues the main types are the beltlike adherens junctions, GAP JUNCTIONS, DESMOSOMES, and TIGHT

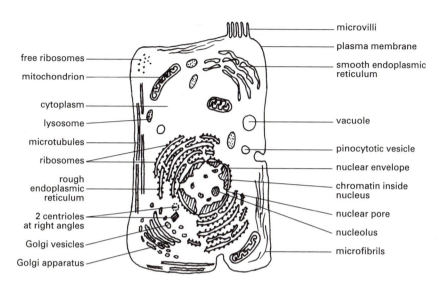

free ribosomes
mitochondrion

cytoplasm
lysosome
microtubules
ribosomes
rough endoplasmic reticulum
2 centrioles at right angles
Golgi vesicles
Golgi apparatus

microvilli
plasma membrane
smooth endoplasmic reticulum

vacuole

pinocytotic vesicle
nuclear envelope
chromatin inside nucleus
nuclear pore
nucleolus
microfibrils

Generalized animal cell as seen under the electron microscope

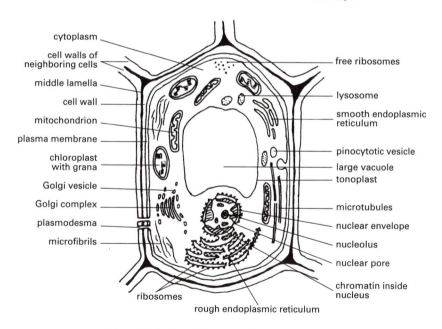

cytoplasm
cell walls of neighboring cells
middle lamella
cell wall
mitochondrion
plasma membrane
chloroplast with grana
Golgi vesicle
Golgi complex
plasmodesma
microfibrils

free ribosomes

lysosome

smooth endoplasmic reticulum

pinocytotic vesicle
large vacuole
tonoplast

microtubules
nuclear envelope
nucleolus
nuclear pore
chromatin inside nucleus

ribosomes
rough endoplasmic reticulum

Generalized plant cell as seen under the microscope

JUNCTIONS. Plant cells can communicate via a junction called a PLASMODESMA.

cell lineage The theory stating that cells arise only from pre-existing cells. The cell lineage of a structure traces the successive stages that the cells pass through from the time of their formation in the zygote to their appearance in the mature functional structure.

cell membrane *See* plasma membrane.

cell plate A structure that appears in late anaphase in dividing plant cells and is involved in formation of a new cell wall. It is formed by fusion of vesicles from the Golgi apparatus; resulting in a flattened membrane-bounded sac spreading across and effectively dividing the cell. Cell wall polysaccharides contained in the vesicles contribute to growth of the new wall inside the sac. The process begins at the *phragmoplast*, a barrel-shaped region at the former site of the spindle equator, where microtubules remain and the vesicles cluster, possibly oriented by the microtubules. Endoplasmic reticulum and ribosomes are also present. The cell plate membranes eventually form the two new plasma membranes of the daughter cells. *See* cytokinesis.

cell theory The theory that all organisms are composed of cells and cell products and that growth and development results from the division and differentiation of cells. This idea resulted from numerous investigations that started at the beginning of the 19th century, and it was finally given form by Schleiden and Schwann in 1839.

cellulase /sel-yŭ-layss/ An enzyme that hydrolyzes 1,4-glycosidic linkages in cellulose, yielding cellobiose (a disaccharide) and glucose. It is important in the degradation of plant cell walls in living plants (e.g. in leaf abscission), while the cellulase enzymes of gut bacteria are essential for digestion in animals, such as ruminants, that consume plant material.

cellulose /sel-yŭ-lohs/ A polysaccharide forming the framework of the CELL WALLS of all plants, many algae, and some fungi. Cellulose molecules are unbranched chains that together form a rigid structure of high tensile strength. Bundles of molecules form microfibrils, which may be aligned in the primary cell wall either transversely or longitudinally. Cellulose forms an important source of carbohydrate in the diets of herbivores, and is a major constituent of dietary fiber in human diets. The individual units are β-1,4 linked D-glucose molecules.

cell wall A rigid wall surrounding the cells of plants, fungi, bacteria, and algae. Plant cell walls are made of cellulose fibers in a cementing matrix of other polysaccharides. Fungi differ, with their walls usually containing chitin. The walls of some algae also differ, e.g. the silica boxes enclosing diatoms. Bacterial walls are more complex, containing peptidoglycans – complex polymers of amino acids and polysaccharides. Cell walls are freely permeable to gases, water, and solutes. They have a mechanical function, allowing the cell to become turgid by osmosis, but preventing bursting. This contributes to the support of herbaceous plants. Plant cell walls can be strengthened for extra support by addition of lignin (as in xylem and sclerenchyma) or extra cellulose (as in collenchyma). Plant cell walls are an important route for movement of water and mineral salts. Other modifications include the uneven thickening of guard cells, the sieve plates in phloem, and the waterproof coverings of epidermal and cork cells.

At cell division in plants the *primary wall* is laid down on the middle lamella of the cell plate as a loose mesh of cellulose fibers. This gives an elastic structure that allows cell expansion during growth. Later the *secondary wall* grows and acquires greater rigidity and tensile strength. New cellulose fibers are laid down in layers, parallel within each layer, but orientated differently in different layers. The Golgi apparatus provides polysaccharide-filled vesicles that deposit wall material by exocytosis, guided by microtubules.

cement (**cementum**) A bonelike substance that covers the root of a mammalian tooth and helps to fix it in the socket (alveolus) of the jaw bone. It is very compact and hard, having a higher mineral content than bone. Cement normally lacks a system of Haversian canals. *See illustration at* teeth.

cementum /sĕ-**men**-tŭm/ *See* cement.

Cenozoic /see-nŏ-**zoh**-ik, sen-ŏ-/ (**Caeno-zoic**; **Cainozoic**) The present geological era, beginning some 65 million years ago, and divided into two periods, the Tertiary and the Quaternary. It is characterized by the rise of modern organisms, especially mammals and flowering plants. *See also* geological time scale.

centipedes *See* Chilopoda.

central nervous system (**CNS**) The part of the nervous system that receives sensory information from all parts of the body and, through the many interconnections that are possible, causes the appropriate messages to be sent out to muscles and other organs. In vertebrates the CNS consists of the brain and spinal cord. The CNS of invertebrates consists of a connected pair of ganglia in each body segment and a pair of ventral nerve cords running the length of the body.

The development of a CNS is associated with the increasing sensory awareness and complex actions that are involved in locomotion, feeding, reproduction, etc., and the need for central integration of all sensory input and motor output. This compares with the simple localized integration found in the nerve net of coelenterates. *See also* autonomic nervous system; peripheral nervous system.

centrarch /**sen**-trark/ Denoting a protostele in which protoxylem is at the center of the axis. *Compare* endarch; exarch; mesarch.

center A cluster of nerve cells that are concerned with a common function in the nervous system. Spinal cord centers deal with relatively simple reflex actions of the body while centers in the brain regulate such functions as breathing, thirst, hunger, pain, pleasure, etc.

centric bundle /**sen**-trik/ A vascular bundle in which the xylem and phloem are arranged in rings, one completely surrounding the other. *Compare* bicollateral bundle; collateral bundle. *See* amphiphloic; amphixylic.

centrifugal /sen-**trif**-ŭ-găl, -yŭ-găl/ Developing from the center outward so that the youngest structures are at the outer edge. Centrifugal xylem differentiates from the center outwards, and a centrifugal inflorescence (e.g. the dichasial cyme) is one in which the progression of flower opening is from the center to the periphery. *Compare* centripetal. *See also* acropetal; basipetal.

centrifuge /**sen**-tră-fyooj/ An apparatus in which suspensions may be rotated at very high speeds in order to separate the component solids by centrifugal force. If different components have different sedimentation coefficients then they may be separated by removing pellets of sediment at given intervals. *See also* ultracentrifuge.

centriole /**sen**-tree-ohl/ A cell organelle consisting of two short tubular structures orientated at right angles to each other. It lies within the CENTROSOME, just outside the nucleus of animal and protoctist cells, but is absent in cells of most higher plants. Prior to cell division it replicates, and the sister centrioles move to opposite ends of the cell to lie within the spindle-organizing structure, the centrosome. However, the centriole is not essential for spindle formation, although an analogous structure, the basal body, is responsible for organizing the microtubules of undulipodia (cilia and flagella). Under the electron microscope, each 'barrel' of the centriole is seen to consist of a cylinder of nine triplets of microtubules surrounding two central ones. *See illustration at* cell.

centripetal /sen-**trip**-ă-t'l/ Developing from the outside inwards so the youngest structures are at the center. Centripetal xylem differentiates from the outside towards the center, and a centripetal inflorescence (e.g. the capitulum) is one in which the outer flowers open before those in the center. *Compare* centrifugal. *See also* acropetal; basipetal.

centromere /**sen**-trŏ-meer/ The region of the chromosome that becomes attached to the nuclear spindle during mitosis and meiosis. Following the replication of chromosomes, resultant chromatids remain attached at the centromere. The centromere attaches to the spindle microtubules via a platelike structure called the *kinetochore*. This contains proteins that bind to DNA sequences in the centromere region. The centromere is a specific genetic locus and remains relatively uncoiled during prophase, appearing as a primary constriction. It does not stain with basic dyes.

centrosome /**sen**-trŏ-sohm/ A structure found in all eukaryotic cells, except fungi, that forms the spindle during cell division. It lies close to the nucleus in nondividing cells, but at the commencement of cell division it divides, and the sister centrosomes move to opposite ends of the cell, trailing the microtubules of the spindle behind them. In animal and protoctist cells the centrosome contains two short barrel-shaped structures, the CENTRIOLE, but this is not necessary for spindle formation. In fungi, the function of the centrosome is served instead by the spindle pole body. *See* centriole; mitosis; spindle.

centrum /**sen**-trŭm/ (*pl.* **centra**) The main weight-bearing body or center of a vertebra, present in all except the atlas and axis and situated ventrally to the spinal cord. It is approximately cylindrical in shape, with flat concave anterior and posterior surfaces, which are attached to fibrocartilaginous intervertebral disks.

cephalization /sef-ă-li-**zay**-shŏn/ The development of a head – in which sense organs, brain, and feeding organs are concentrated – in animals.

Cephalochordata /sef-ă-loh-kor-**day**-tă/ A marine subphylum of chordates, comprising the lancelets, in which the characteristics of metameric segmentation, notochord, dorsal nerve cord, and gill slits are retained in the adult. The best known member is *Branchiostoma (Amphioxus)*. *Compare* Urochordata.

Cephalopoda /sef-ă-**lop**-ŏ-dă/ The most advanced class of mollusks, containing the cuttlefishes (e.g. *Sepia*), squids (e.g. *Loligo*), and the octopus (*Octopus*). All are marine and typically have a ring of prehensile suckered tentacles around the mouth for food capture, a well-developed nervous system, and eyes. Some can learn to distinguish various shapes

Part of the foot is modified as a siphon through which water is forced by contraction of the muscular mantle during swimming. *Nautilus* has a large coiled shell, which acts as a buoyancy chamber, and numerous unsuckered tentacles. Squids and cuttlefishes have an internal shell and ten tentacles; in the octopus, which has only eight tentacles, the shell is absent.

cephalothorax /sef-ă-loh-**thor**-aks, -**thoh**-raks/ In crustaceans and arachnids, the anterior part of the body, formed by the fusion of the head and thorax. It is connected to the abdomen and bears the mouthparts and walking legs. In arachnids the cephalothorax is called the *prosoma*.

cercaria /ser-**kair**-ee-ă/ (*pl.* **cercariae**) A free-swimming larval stage of trematode worms (flukes). A *sporocyst* in the body of the intermediate host (often a snail) produces *redia* larvae by internal budding. Rediae are simple forms that bud again to produce several cercarias each, while remaining inside the body of the host. The cercaria, which has a tail and two suckers, bores its way out of the body of the intermediate host. In some species it then encysts on plants and is later eaten by the final host (e.g. a sheep for the liver fluke); in others it may swim to and penetrate the body of the final host (e.g. the human blood fluke); in some it may enter a second intermediate host, encyst in its body, and then be eaten by the final host (e.g. the Chinese liver fluke, which encysts in fish and is eaten by humans).

cerci /**ser**-kÿ, -sÿ/ (*sing*. **cercus**) Appendages at the hind end of the abdomen of some insects (e.g. mayflies, earwigs, and cockroaches).

cerebellum /se-rĕ-**bel**-ŭm/ A part of the brain consisting of a pair of grayish deeply-folded hemispheres lying dorsal to the medulla oblongata and partially hidden by the overlying cerebral hemispheres. It monitors the position of limbs and the tension of their muscles and makes any necessary adjustments to the messages sent out to voluntary muscles by the cerebral cortex. It is thus important in maintaining balance, in locomotion, etc. *See also* brain.

cerebral cortex /**se**-rĕ-brăl, sĕ-**ree**-/ The surface layer of the CEREBRUM of the brain. It contains billions of nerve cell bodies, collectively called gray matter, and is responsible for the senses of vision, hearing, smell, and touch, for stimulating the contraction of voluntary muscles, and for higher brain activities, such as language and memory. Many of these activities occur in specific regions of the cortex. *See also* cerebral hemispheres; neopallium.

cerebral hemispheres A pair of structures, originating from the forebrain, that contain the centers concerned with the major senses, voluntary muscle activities, and higher brain functions, such as language and memory.

In humans, each hemisphere has a greatly enlarged and much folded outer area of gray matter – the CEREBRAL CORTEX – and overlies other parts of the brain. Beneath the gray matter is the white matter, comprising nerve fibers connecting to other regions of the brain. Each hemisphere controls that side of the body opposite to it, although one hemisphere is dominant in such functions as speech. *See also* cerebrum.

cerebral nuclei /**new**-klee-ÿ/ (**basal ganglia**) Masses of gray matter lying deep within the white matter near the base of each cerebral hemisphere in the brain. In humans they include the amygdaloid nucleus, and the caudate and lenticular nuclei. The latter two form part of the CORPUS STRIATUM. Cerebral nuclei relay motor impulses from the cerebrum to the midbrain, and help subconsciously to regulate voluntary muscle movements.

cerebrospinal fluid /se-rĕ-broh-**spÿ**-năl/ A clear watery fluid containing glucose, salts, and a few white blood cells, that is found in the internal cavities and between the surrounding membranes of the central nervous system. It is filtered from blood by the choroid plexuses in the brain and eventually returns via lymph vessels or in venous blood. It cushions and protects nerve tissues.

cerebrum /**se**-rĕ-brŭm, sĕ-**ree**-/ The most prominent region of the brain in humans, consisting of two CEREBRAL HEMISPHERES joined by a band of nerve fibers (the *corpus callosum*). It is the site of such functions as vision, hearing, touch, smell, voluntary muscle activity, speech, and memory. *See also* cerebral cortex.

cervical vertebrae /**ser**-vă-kăl/ The small VERTEBRAE of the neck, which form a very flexible portion of the vertebral column. In mammals there are usually seven: the first and second are modified to form the atlas and axis. They are all characterized by relatively large neural canals.

cervix /**ser**-viks/ The neck of the uterus (womb) in mammals, a narrow cylindrical passage situated at the posterior end of the uterus leading into the vagina. It contains numerous glands producing mucus, the viscosity of which alters throughout the menstrual cycle.

The term is also used for other necklike anatomical parts, such as the *cervix cornu*, part of the gray matter of the spinal cord.

Cestoda /sess-**toh**-dă/ A class of the phylum Platyhelminthes containing the tapeworms (cestodes), which are all internal parasites of other animals. They have a complex life cycle involving one or more intermediate hosts; for example, *Taenia* uses the pig as an intermediate host and humans as the final host. Undercooked pork is a source of infection. Tapeworms live in the gut of vertebrates, absorbing predigested food through their body wall. The body, up to 10 m long, is covered by a tough cuticle to prevent digestion by the host and has a small head (*scolex*), with hooks and suckers for attachment; the rest of the body is divided into segments (*proglottids*). Each proglottis contains the complete hermaphrodite reproductive system. The eggs develop into six-hooked embryos, which are excreted and eaten by the intermediate host, in which they develop into cysticercus larvae (bladderworms), which eventually become sexually mature in the final host.

Cetacea /sĕ-**tay**-shă/ The order that contains the only completely marine mammals – the whales, dolphins, and porpoises. They have a hairless streamlined body, no hind limbs, forelimbs modified as flippers, and a tail with horizontal flukes used for propulsion. An insulating layer of blubber beneath the skin helps to conserve heat, there are no external ears, and the respiratory outlet is the dorsal blowhole. The toothed whales, e.g. *Delphinus* (dolphin) and *Orcinus* (killer whale), feed on fish and other animals and have many peglike teeth. The whalebone whales, e.g. *Balaenoptera musculus* (blue whale), feed on plankton filtered from the sea by baleen plates hanging from the roof of the mouth.

CFCs *See* chlorofluorocarbons.

chaeta /**kee**-tă/ (*pl.* **chaetae**) (**seta**) A stiff bristle-like structure in some animals. Chaetae are found on the outside of the body of an earthworm, for example, in which they are used for gripping the soil during locomotion.

Chaetognatha /ki-**tog**-nă-th'ă/ A phylum of marine invertebrates called arrow worms, characterized by long slender arrow-shaped bodies. There are about 90 known species, found widely distributed but especially abundant in warm seas. Arrow worms are members of the plankton and feed on smaller plankton organisms, such as copepods and fish larvae, which are grasped using moveable spines located near the mouth. They swim by flicking their tail fin, and have two pairs of lateral fins for stability. Many have transparent bodies, and some are brightly colored. Arrow worms are hermaphrodite, and the adults lie side by side to exchange eggs and sperm. There are no larval stages, the fertilized eggs hatching into miniature arrow worms.

chalaza /kă-**lay**-ză/ **1.** (*Botany*) The region of an angiosperm ovule where the nucellus and integuments merge. When ovule orientation is orthotropous the chalaza corresponds to the point where the funicle is attached but in anatropous and campylotropous ovules the chalaza is some distance from the funicle. *See illustration at* ovule.
2. (*Zoology*) A twisted cord of albumen that joins the shell membrane to the egg membrane in the egg of a bird. There is a chalaza at each end of the egg, supporting the yolk sac centrally in the shell.

chalazogamy /kal-ă-**zog**-ă-mee/ A method of fertilization in angiosperms in which the pollen tube enters the ovule by the chalaza instead of through the micropyle. Chalazogamy is seen in certain trees and shrubs, e.g. beech. *Compare* porogamy.

chamaephyte /**kam**-ee-fÿt/ A perennial plant that is able to produce new growth from resting buds near the soil surface. Chamaephytes are usually small bushes (e.g. heather). *See also* Raunkiaer's plant classification.

chasmogamy /kaz-**mog**-ă-mee/ The production of flowers that open their petals so that cross pollination is possible. *Compare* cleistogamy. *See also* anemophily; entomophily.

chela /**kee**-lă/ (*pl.* **chelae**) **1.** The large pinching claw on the ninth pair of appendages in such crustaceans as crabs and lobsters. It is used for attack and defense. 2. The large claw on the second pair of appendages in scorpions, used for catching prey and tearing it apart.

chelicera /kĕ-**liss**-er-ă/ (*pl.* **chelicerae**) A paired appendage on the third segment of the head of arachnids, in the form of a sharp claw or pincer. Ducts lead from poison glands to these appendages, which the arachnid uses to inject paralyzing poisons into the bodies of prey.

Chelicerata /kĕ-liss-ĕ-**ray**-tă/ In some classifications, a phylum of arthropods containing the spiders, mites and ticks (class ARACHNIDA); the sea spiders (class Pycnogonida); and the horseshoe crabs (class Merostomata). They lack antennae, and are characterized by the anterior pair of clawed jointed feeding appendages (chelicerae). The head and thorax are fused to form a single unit, the cephalothorax, which is distinct from the abdomen (opisthosoma).

chemical fossils Particularly resistant organic chemicals present in geological strata that are thought to indicate the existence of life in the period when the rocks were formed. Chemical fossils (e.g. alkanes and porphyrins) are often the only evidence for life in rocks of Precambrian age. *See* Precambrian.

chemiosmotic theory /kem-ee-oz-**mot**-ik/ *See* electron-transport chain.

chemoautotrophism /kee-moh-aw-tŏ-**troff**-ik, kem-oh-/ (**chemosynthesis**) *See* autotrophism; chemotrophism.

chemoheterotrophism /kee-moh-het-ĕ-roh-**troff**-iz-ăm, kem-oh-/ *See* chemotrophism; heterotrophism.

chemoreceptor /kee-moh-ri-**sep**-ter, kem-oh-/ A receptor, e.g. a taste bud, that responds to chemical compounds.

chemosynthesis /kee-moh-**sin**-th'ĕ-sis, kem-oh-/ *See* autotrophism; chemotrophism.

chemotaxis /kee-moh-**taks**-is, kem-oh-/ (**chemotactic movement**) A TAXIS in response to a chemical concentration gradient. The spermatozoids of primitive plants are often positively chemotactic, swimming towards the female organs in response to a chemical secreted by the latter. For example, the archegonium (female organ) of the moss *Funaria* secretes sucrose.

chemotrophism /kee-moh-**troff**-iz-ăm, kem-oh-/ A type of nutrition in which the source of energy for the synthesis of organic requirements is chemical. Most chemotrophic organisms are heterotrophic (i.e. *chemoheterotrophic*) and their energy source is always an organic compound; animals, fungi, and most bacteria are chemoheterotrophs. If autotrophic (i.e. *chemoautotrophic* or *chemosynthetic*) the energy is obtained by oxidation of an inorganic compound; for example, by oxidation of ammonia to nitrite or a nitrite to nitrate (by nitrifying bacteria), or oxidation of hydrogen sulfide to sulfur (by colorless sulfur bacteria). Only a few specialized bacteria are chemoautotrophic. *Compare* phototrophism. *See* autotrophism; heterotrophism.

chemotropism /kee-moh-**trop**-iz-ăm, kem-oh-/ (**chemotropic movement**) A TROPISM in which the stimulus is chemical. The hyphae of certain fungi (e.g. *Mucor*) are positively chemotropic, growing towards a particular source of food. Pollen tube growth down the style is chemotropic.

chiasma /kÿ-**az**-mă/ (*pl.* **chiasmata**) A connection between homologous chromosomes seen during the prophase stage of meiosis. Chiasmata represent a mutual exchange of material between homologous, non-sister chromatids (crossing over) and provide one mechanism by which RECOMBINATION occurs, through the splitting of linkage groups.

Chilopoda /kÿ-**lop**-ŏ-dă/ The class of arthropods that contains the centipedes (e.g. *Lithobius*), characterized by a flat body divided into numerous segments, each bearing one pair of walking legs. They are terrestrial and breathe air through tracheae. Excretion is by Malpighian tubules. Centipedes are carnivorous, with poison claws on the first body segment. They are sometimes placed with the millipedes (DIPLOPODA) in the group MYRIAPODA.

chimaera /kă-**meer**-ă, kÿ-/ An individual or part of an individual in which the tissues are a mixture of two genetically different tissues. It may arise naturally due to mutation in a cell of a developing embryo, producing a line of cells with the mutant gene, and hence different characteristics compared to surrounding cells. It may also be induced experimentally. For example, two mouse embryos at the eight-cell stage from different parents can be fused and develop into a mouse of normal size. Analysis of the genotypes of the tissues and organs of such a mouse reveals that there is a random mixture of the two original genotypes.

In plants, chimaeras produced from two different species are known as *graft hybrids*. For example, a bud may develop at the junction between the scion and stock with a mixture of tissues from both. Many variegated plants are examples of *periclinal chimaeras*, in which a mutation has occurred in a sector of tissue derived from the tunica or corpus, resulting in subsequent chlorophyll deficiency. For example, in a white-edged form of *Pelargonium*, the outermost layer is colorless, indicating a lack of chlorophyll, and is the result of a mutation. There is no genetic mixture throughout the plant.

chirality /kÿr-**al**-ă-tee/ *See* optical activity.

Chiroptera /kÿr-**op**-tĕ-ră/ The order that contains the bats, the only flying mammals. Bats have a thin elastic hairless flight membrane (*patagium*) extending from the elongated forearm and four of the elongated fingers to the hind limbs and, usually, the tail. The first finger and the toes are smaller, free, and clawed. Bats are nocturnal. They have specialized ears and use echolocation to avoid objects and to locate their prey.

chitin /**kÿ**-tin/ A nitrogen-containing heterosaccharide found in some animals and the cell walls of most fungi. The outer covering of arthropods, the cuticle, is impregnated in its outer layers with chitin, which makes the exoskeleton more rigid. It is associated with protein to give a uniquely tough yet flexible and light skeleton, which also has the advantage of being waterproof. The chitinous plates are thinner for bending and flexibility or thicker for stiffness as required. The plates cannot grow once laid down and are broken down at each molt. Chitin is also found in the hard parts of several other groups of animals. Chitin is a polymer of N-acetylglucosamine. It consists of many glucose units, in each of which one of the hydroxyl groups has been replaced by an acetylamine group (CH_3CONH).

chlorenchyma /klor-**enk**-ă-mă, kloh-**renk**-/ A form of parenchyma in which the cells contain many chloroplasts, as in the palisade layer of the leaf.

chlorine /**klor**-een, -in, **kloh**-reen, -rin/ An element found in trace amounts in plants and one of the essential nutrients in animal diets. Common table salt, a very important item of the diet, is made up of crystals of sodium chloride. The chlorine ion is important in buffering body fluids, and, because it can pass easily through cell membranes, it is also important in the absorption and excretion of various cations. The hydrogen chloride secreted in gastric juices is important in lowering the pH of the stomach so that the enzyme pepsin is able to act.

chlorofluorocarbons (CFCs) A group of hydrocarbon-based compounds in which some of the hydrogen atoms are replaced by chlorine or fluorine atoms. They are chemically inert and were formerly widely used as refrigerant liquids (freons) in fridges, aerosol propellants, and blowing agents in the manufacture of plastics. However, there is strong evidence that when they escape to the upper atmosphere they cause depletion of the ozone layer, hence their use has been phased out since the late 1980s.

chlorophylls /**klor**-ŏ-filz, **kloh**-rŏ-/

A group of photosynthetic pigments. They absorb blue-violet and red light and hence reflect green light, imparting the green color to green plants. The molecule consists of a hydrophilic (water-loving) head, containing magnesium at the center of a porphyrin ring, and a long hydrophobic (water-hating) hydrocarbon tail (the phytol chain), which anchors the molecule in the lipid of the membrane. Different chlorophylls have different chemical groups attached to the head. *See* absorption spectrum; photosynthetic pigments. *See also* bacteriochlorophyll.

Chlorophyta /klor-**roff**-ă-tă, kloh-roff-/ (green algae)

A phylum of protoctists comprising mainly freshwater algae with some marine and terrestrial forms. They contain the pigments chlorophyll *a* and *b* together with carotenes and xanthophylls. The Chlorophyta store food as starch and fat and have cell walls containing cellulose and hemicellulose. They are of interest because their pigments, metabolism, and ultrastructure resemble those of the lower plants more closely than do any other algal phyla. Some of the commoner orders include: Volvocales, unicellular and colonial plants, e.g. *Chlamydomonas*, *Volvox*; Chlorococcales, including unicellular and coenobic plants, e.g. *Chlorella*, *Pediastrum*; Ulotrichales, filamentous and thallose plants, e.g. *Ulothrix*, *Ulva*; Oedogoniales, e.g. the filamentous *Oedogonium*; and the Conjugales (Zygnematales), e.g. *Spirogyra*.

chloroplast /**klor**-ŏ-plast, **kloh**-rŏ/

A photosynthetic PLASTID containing chlorophyll and other photosynthetic pigments. It is found in all photosynthetic cells of plants and protoctists but not in photosynthetic prokaryotes. It has a membrane system containing the pigments and on which the light-dependent reactions of PHOTOSYNTHESIS occur. The surrounding gel-like ground substance, or *stroma*, is where the light-independent reactions occur. The typ-

ical higher plant chloroplast is lens-shaped and about 5 μm in length. Various other forms exist in the algae, e.g. spiral in *Spirogyra*, stellate in *Zygnema*, and cup-shaped in *Chlamydomonas*. The number per cell varies, e.g. one in *Chlorella* and *Chlamydomonas*, two in *Zygnema*, and about one hundred in palisade mesophyll cells of leaves.

Chloroplast membranes form elongated flattened fluid-filled sacs called *thylakoids*. The sheetlike layers of the thylakoids are called *lamellae*. In all plants except algae, the thylakoids overlap at intervals to form stacks, like piles of coins, called *grana*. In this way the efficiency of the light reactions seem to be improved.

The stroma may contain storage products of photosynthesis, e.g. starch grains. The chloroplasts of most algae contain one or more *pyrenoids*. These are dense protein bodies associated with polysaccharide storage. In green algae, for example, starch is deposited in layers around pyrenoids during development.

The stroma also typically contains *plastoglobuli*, spherical droplets of lipid staining intensely black with osmium tetroxide. They become larger and more numerous as the chloroplast senesces, when carotenoid pigments accumulate in them. Apart from photosynthetic enzymes, the stroma also contains typical prokaryotic protein-synthesizing machinery including circular DNA and smaller ribosomes. There is now strong evidence that chloroplasts and other cell organelles, such as mitochondria, represent prokaryotic organisms that invaded heterotrophic eukaryotic cells early in evolution and are now part of an indispensable symbiotic union (*see* endosymbiont theory). Chloroplast DNA codes for some chloroplast proteins but most are encoded by nuclear DNA.

In C_4 plants there are two types of chloroplast. *See illustration at* cell.

chlorosis /klor-**oh**-sis, kloh-**roh**-/

The loss of chlorophyll from plants resulting in yellow (*chlorotic*) leaves. It may be the result of the normal process of senescence, lack of key minerals for chlorophyll synthesis (particularly iron and magnesium), or be caused by disease.

choanocyte /koh-**an**-ŏ-sÿt/ (**collar cell**) In sponges, a cell bearing a flagellum (undulipodium) and surrounded at the base by a raised cylindrical collar. Choanocytes line chambers of the sponge and the beating flagella circulate water through the chambers and canals.

cholecalciferol /kol-ĕ-kal-**sif**-er-ol, -ohl/ *See* Vitamin D.

cholecystokinin /kol-ĕ-sis-tŏ-**kÿ**-nin/ (**pancreozymin**) A peptide hormone secreted by cells in the intestinal mucosa. It stimulates contraction of the gall bladder and hence discharge of bile into the DUODENUM. It also triggers the release of digestive enzyme precursors in pancreatic tissue.

cholesterol /kŏ-**les**-tĕ-rol, -rohl/ A sterol (fat derivative) found in animal cells. It occurs in bile, blood cells, cell membranes, blood plasma, and egg yolk. It can accumulate in the gall bladder as gallstones. An elevated level of cholesterol in the blood – especially when transported as a constituent of low-density lipoproteins ('bad cholesterol') – is thought to be a contributory cause of hardening and narrowing of the arteries (atherosclerosis).

choline /**koh**-leen, -lin, **kol**-een, -in/ An amino alcohol, sometimes classified as a member of the vitamin B complex. Although humans can synthesize choline in small amounts, it must be supplied in the diet to maintain health. It occurs in the body chiefly as a component of certain phospholids, notably phosphatidylcholine (lecithin) and sphingomyelins. These are important structural components of cell membranes, and also precursors of the intracellular signal molecules diacylglycerol and ceramide. Phosphatidylcholine is also required for the formation of the lipoproteins in which fats and cholesterol are packaged by the liver for transport in the blood. Choline is, moreover, a constituent of the neurotransmitter acetylcholine, and is oxidized to betaine, which is an impor-

tant source of methyl groups used for methylation reactions. A normal balanced diet is thought to supply adequate amounts of choline, mainly as phosphatidylcholine – milk, eggs, liver, and peanuts are rich sources.

cholinergic /koh-lă-**ner**-jik, kol-ă/ Designating the type of nerve fiber that releases acetylcholine from its ending when stimulated by a nerve impulse. In vertebrates, motor fibers to striated muscle, parasympathetic fibers to smooth muscle, and preganglionic sympathetic fibers are cholinergic. *Compare* adrenergic.

Chondrichthyes /kon-drik-**th'ÿ**-eez/ (**Elasmobranchii**) The class of vertebrates that contains the cartilaginous fishes – the sharks, skates, and rays, e.g. *Scyliorhinus*, (dogfish). They are predominantly marine predators characterized by a cartilaginous skeleton, a skin covering of denticles (placoid scales) that are modified in the mouth as rows of teeth, pectoral and pelvic fins, and a heterocercal tail; lungs and a swim bladder are absent, therefore the fish sink when they stop swimming. Most have separate gill openings, not covered by an operculum, and a small spiracle. The pelvic fins in the male often bear claspers through which sperm are transmitted in internal fertilization. *Compare* Osteichthyes. *See* Selachii; Scyliorhinus.

chondroblasts /**kon**-drŏ-blast/ (**chondrocytes**) The cells that are embedded in the matrix of CARTILAGE.

chondrocranium /kon-drŏ-**kray**-nee-ŭm/ The first part of the skull to form in vertebrate embryos. It consists of cartilaginous structures – plates and capsules – that protect and support the brain, olfactory organs, eyes, and the inner ear. It usually becomes ossified in adults, although it remains cartilaginous in Chondrichthyes (cartilaginous fish). *See also* neurocranium.

chondrocytes /**kon**-drŏ-sÿts/ *See* chondroblasts.

chordae tendineae /**kor**-dee ten-dǎ-**nay**-ee/ *See* tendinous cords.

Chordata /kor-**day**-ta/ A major phylum of bilaterally symmetrical metamerically segmented coelomate animals characterized by the possession at some or all stages in the life history of a dorsal supporting rod, the notochord. The dorsal tubular nerve cord lies immediately above the notochord and a number of visceral clefts (gill slits) are present in the pharynx at some stage of the life history. The post-anal flexible tail is the main propulsive organ in aquatic chordates. The phylum includes the subphylum Craniata (Vertebrata; *see* craniate), in which the notochord is replaced by a vertebral column (backbone).

There are two other subphyla, the URO-CHORDATA and CEPHALOCHORDATA (sometimes known collectively as the Acrania or Protochordata). These invertebrate chordates are marine.

chorion /**kor**-ee-on, koh-ree-/ **1.** One of the three embryonic membranes of reptiles, birds, and mammals. It arises from the outer layer of the amniotic hood and encloses the amnion (with embryo or fetus inside), yolk sac, and allantois. The trophoblast of mammals is part of the chorion. The allantois is usually fused with it, forming the chorio-allantoic membrane; this forms a 'lung' attached to the inside of the egg shell in birds and the embryonic part of the placenta in mammals.
2. The tough outer membrane of some eggs, notably those of insects. There is usually a pore, the *micropyle*, to admit spermatozoa.

chorionic gonadotropin /kor-ee-**on**-ik/ (**choriogonadotropin**) A glycoprotein hormone secreted in higher mammals by the chorionic villi of the placenta (fingerlike projections of the chorion into the uterus). It prevents the regression of the corpus luteum in the earlier stages of pregnancy. The detection of human chorionic gonadotropin (HCG) in the urine is often used as a pregnancy test.

choroid /**kor**-oid, koh-roid/ The middle layer of the vertebrate eye, between the sclera and the retina. It is rich in blood vessels and contains pigment that absorbs light and stops it going through the eye past the retina. Anteriorly, it projects into the ciliary body and the iris. *See also* tapetum. *See illustration at* eye.

choroid plexus One of two thin-walled greatly folded highly vascular regions in the inner wall of the brain through which exchange of materials between blood and cerebrospinal fluid takes place. Located in the roof of the fourth ventricle and at the junction of the first and second ventricles with the third ventricle, the choroid plexuses (which form the *blood–brain barrier*) bar the entry of blood cells and large molecules (including potentially harmful foreign materials and certain drugs) into the cerebrospinal fluid.

chromatid /**kroh**-mǎ-tid/ One of a pair of replicated chromosomes found during the prophase and metaphase stages of mitosis and meiosis. During mitosis, sister chromatids remain joined by their centromere until anaphase. In meiosis it is not until anaphase II that the centromere divides, the chromatids being termed daughter chromosomes after separation.

chromatin /**kroh**-mǎ-tin/ The material of eukaryotic chromosomes, consisting of DNA and proteins (mainly HISTONES). In nondividing cells the DNA molecule is highly condensed, being wound around the histones to form a series of globular structures called NUCLEOSOMES. Under the microscope these resemble beads on a string. This form is usually condensed further into a spiral or solenoid form, seen as thicker fibers, 30 nm in diameter, in which the genes are inaccessible. To permit transcription, some unfolding to the beads-on-a-string form is necessary. During mitosis or meiosis the chromatin becomes yet more condensed, with the 30 nm fiber itself coiling (supercoiling) to produce a solenoid some ten times thicker.

Chromatin can be classified as *euchromatin* or *heterochromatin* on the basis of its staining properties, the latter staining much more intensely with basic stains because it is more condensed. Euchromatin is partially unfolded and actively involved in transcription and therefore protein synthesis, while heterochromatin is inactive. Euchromatin stains more intensively than heterochromatin during nuclear division. *See also* chromosome.

chromatography /kroh-mă-**tog**-ră-fee/ A method of analyzing materials involving the separation by selective absorption of the various compounds as identifiable bands. For instance, a mixture of substances in solution is passed slowly down a long column packed with alumina. The different compounds move at different rates and separate into bands.

In general chromatography involves a test material being carried by a moving phase (liquid or gas) through a stationary phase (solid or liquid). Different substances move at different rates (depending on their absorption-desorption) and are therefore spatially separable, the least readily absorbed being carried the farthest. Colorless materials can be used if some means of detecting them is used (electronic detection, radioactive labeling, or ninhydrin developer). *See also* paper chromatography; thin-layer chromatography; gas–liquid chromatography; gel filtration.

chromatophore /kroh-**mat**-ŏ-for, -fohr/ A name generally applied to a pigment-bearing structure. **1.** (in prokaryotes) The pigment-bearing membranes of photosynthetic bacteria, being an invagination of the plasma membrane.
2. (*Botany*) *See* chromoplast.
3. (*Zoology*) A type of effector. It is a pigment-containing cell, usually in the skin, whose color can be changed by expansion or contraction in response to various stimuli, e.g. light intensity, temperature, fright, or the opposite sex in courtship. They often result in camouflage, as in the chameleon and some fish.

chromomere /**kroh**-mŏ-meer/ A region of a chromosome where the chromosomal material is relatively condensed, and consequently stains darker. Clusters of chromomeres produce distinct bands, the pattern of which is characteristic for a particular chromosome and is used to distinguish the chromosomes of a particular organism.

chromoplast /**kroh**-mŏ-plast/ (**chromatophore**) A colored plastid, i.e. one containing pigment. They include chloroplasts, which contain the green pigment chlorophyll and are therefore photosynthetic, and non-photosynthetic chromoplasts. The term is sometimes confined to the latter, which are best known in flower petals, fruits (e.g. tomato) and carrot roots. They are yellow, orange, or red owing to the presence of carotenoid pigments. *See* carotenoids.

chromosome /**kroh**-mŏ-sohm/ A thread-like structure containing nucleic acid and various proteins that carries the genes of a cell. Organisms differ in the number, length, and shape of the chromosomes in their cells. Bacteria generally have a single circular chromosome, usually aggregated into a region of the cell called the nucleoid. Eukaryotes typically have several or many linear chromosomes contained within the nucleus of each cell. These consist of CHROMATIN – a single molecule of DNA and proteins (mostly histones) with associated RNA. During nuclear division the chromosomes are tightly coiled and are easily visible through the light microscope. After division, they uncoil and become difficult to see. The number of chromosomes per nucleus is characteristic of the species, for example, humans have 46. Normally one set (haploid) or two sets (diploid) of chromosomes are present in the nucleus. In early prophase of mitosis and later prophase of meiosis, the chromosomes split lengthwise into two identical chromatids held together by the CENTROMERE. In diploid cells, there is a pair of sex chromosomes; the remainder are termed autosomes. *See* cell division; chromomere; chromosome map; gene.

chromosome map (**genetic map**) A diagram showing the order of genes or other landmarks along a chromosome. Such maps have traditionally been constructed from information gained by linkage studies (to give a *linkage map*) or by observations made on the polytene (giant) salivary-gland chromosomes of certain insects, e.g. *Drosophila*, to give a *cytological map*. The techniques employed differ according to the type of organism being studied. For example, many plants and animals can be crossed experimentally to study inheritance patterns of particular genes, but this is not possible in humans, where family pedigrees were, until recently, often the only available evidence. However, the advent of new molecular techniques has dramatically changed the nature of chromosome mapping, making it feasible to construct detailed *physical maps* that depict the base sequence of all genes along any chromosome. The HUMAN GENOME PROJECT is an international project that has succeeded in sequencing all the estimated 30 000 or so genes on human chromosomes. The results will transform the detection and treatment of a wide range of diseases.

Mapping very large genomes, such as those of humans and other vertebrates, involves several steps. The first is to assign each gene to a particular chromosome. This can be achieved by, for example, SO-MATIC-CELL HYBRIDIZATION or using a GENE PROBE. The next step is to determine the relative positions of the genes on a particular chromosome. This involves comparing RE-STRICTION FRAGMENT LENGTH POLYMOR-PHISMS between individuals and constructing a linkage map of all restriction sites, i.e. sites that are cleaved by restriction enzymes. These restriction sites can then be used as markers for closely neighboring genes. The last step is to construct a physical map of the base sequence of the chromosomal DNA. One approach uses cloned DNA segments obtained from a GENE LIBRARY of the chromosome. These clones can be fitted together to form a series of overlapping segments (contig) that corresponds to a particular region of the chromosome. The base sequence of the contig is then determined, and hence the base sequence of the chromosomal DNA. Genetic mapping data for many species of plants, animals, and microorganisms are now held on databases that are freely available to geneticists and others via the Internet. *See* restriction map.

chromosome mutation A change in the number or arrangement of genes in a chromosome. If chromosome segments break away during nuclear division they may rejoin the chromosome the wrong way round, giving an *inversion*. Alternatively, they may rejoin a different part of the same chromosome, or another chromosome, giving a *translocation*. If the segment becomes lost, this is termed a *deficiency* or *deletion*; it is often fatal. A part of a chromosome may be duplicated and occur either twice on the same chromosome or on two different nonhomologous chromosomes: this is a *duplication*. Chromosome mutations can occur naturally but their frequency is increased by the effect of x-rays and chemical mutagens.

chrysalis /kris-ă-lis/ The pupa of insects of the order Lepidoptera (butterflies and moths). *See* pupa.

Chrysomonada /kris-ŏ-**mon**-ă-dă/ A phylum of protoctists comprising the golden-brown algae, characterized by having golden-yellow pigmented plastids. Chrysomonads are mainly single-celled, but some form multicellular colonies, which can be large and branching or filamentous. Most occur in temperate lakes and ponds, although some live in the sea, assimilating silica dissolved in seawater to form complex external shells (tests) with intricate patterns. At some stage in their life cycle all chrysomonads produce swarmer cells each with two unequal undulipodia (flagella), enabling the cell to swim away to produce a new colony. In some classifications, the golden-brown algae belong to a class within a phylum (Chrysophyta) that also contains classes for the diatoms and the yellow-green algae (*see* Xanthophyta).

chyle /kÿl/ A milky fluid consisting of lymph containing globules of emulsified fat, which is found in the lymphatic vessels of the small intestine (the lacteals) and in the thoracic duct of vertebrates during the digestion and absorption of fat.

chylomicron /kÿ-loh-**mÿ**-kron/ *See* lipoprotein.

chyme /kÿm/ The creamy semi-fluid contents of the stomach as they are introduced from the stomach into the duodenum through the pyloric sphincter.

chymosin /kÿ-**moh**-sin/ (**rennin**) An enzyme found in gastric juices and responsible for the coagulation of milk. It acts by hydrolyzing peptide links. At 37°C, rennin can coagulate 10^7 times its own weight of milk in ten minutes. It has been crystallized out but is commercially manufactured from the stomach of animals and sold under the name *rennet*. It is used in the manufacture of cheese and junkets.

chymotrypsin /kÿ-mŏ-**trip**-sin/ An enzyme used to carry out the partial hydrolysis of peptide chains. It is found in pancreatic tissue and in pancreatic juices as an inactive form, chymotrypsinogen. Chymotrypsin will catalyze the hydrolysis of the peptide bonds formed by the carbonyl groups from only certain amino acids, notably phenylalanine, tryosine, and tryptophan residues.

ciliary body /**sil**-ee-air-ee/ A ring of muscular tissue at the junction of the choroid with the iris in the vertebrate eye. It surrounds the lens, to which it is attached by the suspensory ligament. It is important in altering the curvature of the lens, to accommodate for near or distant vision. It contains the ciliary glands, which secrete the aqueous humor. *See also* accommodation. *See illustration at* eye.

ciliary feeding A method of feeding in some invertebrates, in which cilia, after creating a current of water towards the body, filter out food particles and transport them into the mouth. Most bivalve mollusks and the urochordates and cephalochordates are ciliary feeders. *See also* filter feeding.

Ciliata /sil-ee-**ay**-tă/ *See* Ciliophora.

ciliated epithelium /**sil**-ee-ay-tid/ A single layer of tightly packed columnar or cubical epithelial cells with numerous cilia projecting from the free surface. The cilia beat in metachronal rhythm (each moves a fraction of a second after the one in front), causing the movement of surrounding fluid or particles. *See also* cilium.

Ciliophora /sil-ee-**off**-ŏ-ră/ A phylum of protoctists containing some of the best known protozoans. All have cilia for locomotion, a contractile vacuole, and a mouth. Most have two types of nuclei, the meganucleus controlling normal cell metabolism, and the smaller micronucleus controlling sexual reproduction (conjugation). Binary fission also takes place. Some (e.g. *Paramecium*) are covered with cilia. Others (e.g. *Vorticella*) have cilia only round the mouth, and in some (e.g. *Stentor*) these cilia are specialized for feeding. *See also* Paramecium.

cilium /**sil**-ee-ŭm/ (*pl.* **cilia**) A whiplike extension of certain eukaryotic cells that beats rapidly, thereby causing locomotion or movement of fluid over the cell. Cilia and flagella represent the two types of eukaryotic undulipodium. Cilia are identical in structure to flagella, though shorter, typically 2–10 μm long and 0.5 μm in diameter and usually arranged in groups. Each cilium has a basal body at its base. In ciliated protoctists, sperm, and some marine larvae they allow locomotion. In multicellular animals they may function in respiration and nutrition, wafting water containing respiratory gases and food over cell surfaces, e.g. filter-feeding mollusks. In mammals the respiratory tract is lined with ciliated cells, which waft mucus, containing trapped dust, bacteria, etc., towards the throat.

Cilia and flagella have a '9 + 2' structure (the axoneme), consisting of 9 outer pairs of microtubules with 2 single central microtubules enclosed in an extension of the plasma membrane. The beat of each cilium comprises an effective downward stroke followed by a gradual straightening (limp recovery). Cilia beat in such a way that each is slightly out of phase with its neighbor (*metachronal rhythm*), thus producing a constant rather than a jerky flow of fluid. The basal bodies of cilia are connected by threadlike strands (neuronemes), which coordinate the beating of neighboring cilia. *Compare* flagellum. *See* basal body; undulipodium.

circadian rhythm /ser-**kay**-dee-ăn/ (**diurnal rhythm**) A daily rhythm of various metabolic activities in animals and plants. Such rhythms persist even when the organism is not exposed to 24-hour cycles of light and dark, and are controlled by an endogenous BIOLOGICAL CLOCK. Circadian rhythms are found in the most primitive and the most advanced of organisms. Thus *Euglena* shows a diurnal rhythm in the speed at which it moves to a light source, while humans are believed to have at least 40 daily rhythms. *See also* biorhythm.

circulatory system A continuous series of vessels or spaces in nearly all animals that transports materials around the body. The system is best developed in mammals. Circulatory systems developed in association with the differentiation of specific organs and tissues in multicellular animals. They enable all parts of the body to receive a constant supply of oxygen, food, etc., and to have waste products removed promptly. *See also* blood vascular system; lymphatic system.

circumnutation /ser-kŭm-new-**tay**-shŏn/ *See* nutation.

cisterna /sis-**ter**-nă/ (*pl.* **cisternae**) A flattened membrane-bounded sac of the endoplasmic reticulum or the Golgi apparatus, being the basic structural unit of these organelles.

***cis-trans* effect** /sis-**tranz**/ The phenomenon resulting from recombination within a gene (cistron), in which a mutation is only expressed in the phenotype if the mutant pseudoalleles are on different homologous chromosomes. *See* cis-trans test.

***cis-trans* test** A test that determines whether two mutations that have the same effect occur in the same gene or in different genes. The mutations may be in either the *cis* position (i.e. on the same chromosome) or the *trans* position (one on each homolog). If the mutations are in different genes then a normal phenotype results whether the mutations are in the *cis* or *trans* position, since they are masked by corresponding dominant genes on the other homolog. However, if the mutations are in the same gene then a normal phenotype will result only if the mutations are in the *cis* position. In the *trans* position the mutant phenotype is expressed, since both alleles of the locus are mutants.

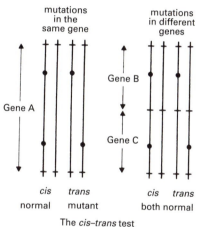

The *cis–trans* test

cistron /**sis**-tron/ A unit of function, i.e. a segment of DNA that determines a single polypeptide chain of a protein molecule. Its extent may be defined by the *cis-trans* test. *See* gene.

citric acid /**sit**-rik/ A carboxylic acid, occurring in the juice of citrus fruits, particularly lemons, and present in many other fruits. Citric acid is biologically important because it participates in the KREBS CYCLE.

citrulline /**sit**-rŭ-leen/ *See* amino acids.

cladistic /klă-**dis**-tik/ Describing the evo-

CLASSIFICATION
(An example of one plant and one animal classification,
to show the hierarchical nature of the classification schemes)

Taxon	*Beech tree*	*Dogfish*
kingdom	Plantae	Animalia
phylum	Angiospermophyta	Chordata
subphylum		Craniata
class	Dicotyledonae	Chondrichthyes
order	Fagales	Selachi
family	Fagaceae	Scyliorhinidae
genus	*Fagus*	*Scyliorhinus*
species	*F. sylvatica*	*S. caniculus*

---

lutionary pathways of taxonomic groups, particularly the relationships between organisms due to recent common ancestry. *See* cladistics.

cladistics A method of classification in which the relationships between organisms are based on selected shared characteristics. These are generally assumed to have been derived from a common ancestor, in the evolutionary process of cladogenesis; i.e. they are homologous derived traits (SYNAPOMORPHIES). However, the 'transformed cladists' believe that shared characteristics alone provide a logical basis for classification without postulating evolutionary relationships. The patterns of these shared characteristics are demonstrated in a branching diagram called a *cladogram*. The branching points of the cladogram may be regarded either as an ancestral species (as in an evolutionary tree) or solely as representing shared characteristics.

Cladistics assumes the closeness of relationship depends on the recentness of common ancestry, indicated by the number and distribution of shared characteristics that can be traced back to a recent common ancestor. Cladistics also regards the only true natural groups as monophyletic, ones containing *all* the descendants of a common ancestor.

cladode /klad-ohd/ A modified internode of the stem that functions as a leaf, being flattened and highly photosynthetic. It is a

xerophytic adaption and is seen in butcher's broom and asparagus.

cladogenesis /klad-oh-jen-ĕ-sis/ The branching of an evolutionary line into two or more separate lineages. *See* cladistics.

cladogram /klad-ŏ-gram/ *See* cladistics.

claspers The copulatory organs of male cartilaginous fish (e.g. dogfish). They consist of a pair of grooved rodlike projections, one on each side of the cloaca between the pelvic fins.

The term may also refer to a pair of tubular outgrowths on the hind end of the abdomen of male insects, which are used to grasp the female during mating.

class A taxonomic rank that is subordinate to a phylum (or sometimes a division in plant taxonomy) and superior to an order. Classes may be divided into subclasses.

classification 1. The grouping and arrangement of organisms into a hierarchical order.
2. The arrangement of organisms resulting from classification procedures. An important aspect of classifications is their predictive value. For example if a characteristic is found in one member of a group of plants, then it is also likely to be found in the other members of that group even though the characteristic in question was not used in the initial construction of the classification.

---

clavicle /**klav**-ă-kăl/ One of a pair of bones that lie anteriorly, on each side of the base of the neck, in some vertebrates; they form the collar bones in humans. In humans and many other mammals they form the ventral side of the pectoral girdle, extending from the scapulae to the sternum, and serve as props for the shoulders. They are membrane bones. *Compare* coracoid.

clay Extremely fine-textured SOIL made up of small mineral particles, less than 0.002 mm in diameter, formed mainly from aluminum and magnesium silicates. Clay soils become very sticky and difficult to work when wet and can easily become waterlogged. Nutrient availability to plants can be a problem as the nutrients may become chemically bound to the surfaces of the particles.

clearing In the preparation of permanent microscope slides, the stage between dehydration and embedding, the purpose of which is to remove the dehydrating agent from the sample and replace it by a substance that is miscible with the embedding substance. Clearing also renders the tissues transparent. Clearing agents include benzene and xylene.

cleavage The mitotic divisions that divide the fertilized egg into smaller cells (*blastomeres*) with equivalent nuclei. The egg cytoplasm is not usually homogeneous, being divided into special regions that foreshadow the major parts of the future embryo and affect the nuclei of the blastomeres. An alternative old name for cleavage is *segmentation. See* holoblastic; meroblastic.

cleidoic egg /klÿ-**doh**-ik/ An egg with a tough shell, which permits gaseous exchange but restricts water loss (although it may take up water). Characteristic of reptiles, birds, and insects, it usually has a large food store (yolk and, additionally in birds, albumen).

cleistogamy /klÿ-**stog**-ă-mee/ The production of closed flowers. It is a method of ensuring self pollination and occurs to-

wards the end of the flowering season in certain plants (e.g. wood sorrel) when no seed has been set by the cross-pollinating flowers. *Compare* chasmogamy.

cleistothecium /klÿ-stoh-**th'ee**-see-ŭm/ (*pl.* **cleistothecia**) *See* ascocarp.

climacteric /klÿ-**mak**-tĕ-rik, klÿ-măk-**te**-rik/ 1. The rise in respiration rate, found in some species, associated with fruit ripening and senescence.
2. The cessation of ovarian function in women, or the decline in sexual activity of men, usually occurring from middle age.

climax The final or stable community in a succession of natural plant communities in one area under a particular set of conditions. Oakwood is the natural climax vegetation in much of Britain, but on chalk it is beechwood. A climax is self-perpetuating and in equilibrium with the physical and biotic environment. It can be held at any stage by biotic factors, such as grazing on chalk downland.

cline A graded series of characters exhibited by a species or other related group of organisms, usually along a line of environmental or geographical transition. The populations at each end of the cline may be substantially different from each other.

clinostat /**klÿ**-nŏ-stat/ (**klinostat**) An apparatus used in tropism experiments to remove the unidirectional influence of a stimulus on a plant organ. Typically it consists of a clockwork or electric motor with an arm to which a chamber containing growing seedlings can be attached. Slow rotation of the arm results in all parts of the seedlings receiving an identical stimulus. *See* tropism.

clisere /**klis**-eer/ A succession of CLIMAX communities in a given area, each giving way to the next as a result of climatic changes. *See also* sere.

clitellum /kli-**tel**-ŭm, klÿ-/ (*pl.* **clitella**) (**saddle**) A glandular and vascular swelling of the epidermis of some Annelida (earthworms and leeches), involved in reproduction. In the earthworm, it occurs over segments 32 to 37 and forms the site of attachment for segments 9 to 15 of a mate at copulation. It secretes a mucus tube that binds the two worms together. After copulation it secretes a cocoon, which is later deposited containing fertilized eggs.

clitoris /**klit**-ŏ-ris/ A small erectile rod of tissue in female mammals that lies anterior to the vagina and urethra. It is homologous with the male penis.

cloaca /kloh-**ay**-kă/ (*pl.* **cloacae**) In most vertebrates, the single chamber into which the alimentary canal, kidney ducts, and genital ducts discharge their contents. In placental mammals there are two (or in females three) separate openings for feces, urine, and gametes.

clonal selection theory /**kloh**-năl/ *See* B-cell.

clone 1. A group of organisms or cells that are genetically identical. In nature, clones are derived from a single parental organism or cell by asexual reproduction or parthenogenesis. Clones of sexually reproducing higher animals have been produced experimentally by new embryo-splitting techniques, and even from single adult body cells. This was first accomplished in 1997 when DNA from a sheep's udder cell was transferred to a fertilized egg cell from which the DNA had been removed. *See* nuclear transfer.
2. In genetic engineering, an identical copy of a gene or other DNA segment produced in cloning vectors such as plasmids and phages (*see* gene cloning).

clotting factors A group of substances that are activated when blood leaves the circulatory system, usually by injury, and cause BLOOD CLOTTING. They include 14 or so proteins, calcium ions, and platelets.

club moss *See* Lycopodophyta.

Cnidaria /nŷ-**dair**-ee-ă/ A large and successful phylum of aquatic, mostly marine, invertebrates and the most primitive of the truly multicellular (metazoan) animals. Cnidarians are typically radially symmetrical and are diploblastic, the body wall having two layers separated by a layer of jelly (*mesoglea*) and enclosing the body cavity (*coelenteron*). The single opening (*mouth*) is surrounded by a circle of tentacles, which are used for food capture and defense and may bear stinging cells (*cnidoblasts*). Two structural forms occur, the sedentary POLYP (e.g. the solitary *Hydra* and the sea anemones and the colonial corals) and the mobile MEDUSA (jellyfish); either or both forms occur in the life cycle. *See also* Anthozoa; Hydra; Hydrozoa; Obelia; Scyphozoa.

cnidoblast /**nŷ**-dŏ-blast/ A specialized stinging cell in the ectoderm of cnidarians, mainly on the tentacles. Each contains a *nematocyst*, which consists of a threadlike structure enclosed in a cavity. A fine trigger-like hair (*cnidocil*) projects from the outer end of the thread cell; this is sensitive to substances dissolved in the water; for example, from nearby prey. When the cnidocil is stimulated the thread is discharged from the nematocyst. It penetrates the prey, injecting poisonous substances that paralyze it. As well as this *penetrant* type of thread, there is the *volvent* type that coils around hairs or bristles on the prey and two kinds of *glutinant* thread that produce sticky substances to prevent the prey from escaping.

cnidocil /**nŷ**-dŏ-sil/ *See* cnidoblast.

CNS *See* central nervous system.

CoA *See* coenzyme A.

coarctate /koh-**ark**-tayt/ Designating pupae (of certain Diptera) in which the last larval cuticle is retained and forms a hardened shell or puparium around the body. *Compare* exarate; obtect.

cobalt /**koh**-bawlt/ *See* trace element.

coccus /**kok**-ŭs/ (*pl.* **cocci**) A spherical-shaped bacterium. Cocci may be found singly, in pairs (e.g. *Diplococcus*), or chains (e.g. *Streptococcus*), or in regularly or irregularly packed clusters. Different species are characteristically found in certain conformations.

coccyx /**kok**-siks/ (*pl.* **coccyges**) A small triangular bone at the end of the vertebral column in humans and other primates. It is formed from the fusion of three to five (usually four) caudal vertebrae.

cochlea /**kok**-lee-ă/ In mammals, the spirally coiled membranous tube in the inner ear that is concerned with sound reception. The sense organ (organ of Corti), with its rows of supporting and sensory cells, lies on the lower (basilar) membrane. Sound waves reaching the inner ear via the oval window (fenestra ovalis) are transmitted through the fluid (perilymph) surrounding the cochlea and cause the basilar membrane to vibrate, which in turn stimulates corresponding sensory cells to send nervous impulses via the auditory nerve to the brain for interpretation. Sounds of different frequencies are thought to affect different regions of the membrane – the apex responding to low frequency and the base to high frequency – and enable the distinction between sounds of different pitch to be made. A cochlea is also present in crocodiles and birds.

cockroach *See* Dictyoptera.

cocoon A protective covering for eggs and/or larvae produced by many invertebrates, e.g. spiders and earthworms.

The larvae of many insects also spin cocoons for the pupal stage.

coding strand *See* nontemplate strand.

co-dominance The situation in which two different alleles are equally dominant. If they occur together the resulting phenotype is intermediate between the two respective homozygotes. For example, if white antirrhinums (AA) are crossed with red antirrhinums (A'A') the progeny (AA') will be pink. Sometimes one allele may be slightly more dominant than the other

(*partial* or *incomplete dominance*) in which case the offspring, though still intermediate, will resemble one parent more than the other.

codon /**koh**-don/ A group of three nucleotide bases (i.e. a nucleotide triplet) in a messenger RNA (mRNA) molecule that codes for a specific amino acid or signals the beginning or end of the message (start and stop codons). Since four different bases are found in nucleic acids there are 64 ($4 \times 4 \times 4$) possible triplet combinations. The arrangement of codons along the mRNA molecule constitutes the *genetic code*. Gene expression involves TRANSCRIPTION of the coding regions of the gene, so that the base sequence of the DNA directs the formation of an mRNA molecule with a complementary base sequence. When the mRNA migrates to the ribosomes, its string of codons is paired with the anticodons of TRANSFER RNA molecules, each of which is carrying one of the amino acids necessary to make up the protein. *See* translation.

coelacanth /**see**-lŏ-kanth/ A fish thought to have become extinct in the Cretaceous until the first live specimen of modern times was caught off South Africa in 1938. It belongs to an order (Coelacanthiformes) containing just a single genus, *Latimeria*. This is a large deep-sea fish with blue scales and long strong lobed fins, supported by bony skeletons, which are used to stir up the mud of the sea floor in search of prey. *Latimeria* has a swim bladder although a lung is present in fossils. There are no internal nostrils.

coelenterate /si-**len**-ter-ayt, -it/ Any invertebrate belonging to either of the phyla Cnidaria (jellyfish, sea anemones, etc.) or Ctenophora (comb jellies), which were formerly united in the single phylum Coelenterata. Both groups possess a gastrovascular cavity (coelenteron).

coelenteron /si-**len**-ter-on/ (**gastrovascular cavity**) The body cavity of cnidarians and ctenophores. It has a single opening (the mouth) through which food is ingested and waste products egested.

coelom /see-lŏm/ A fluid-filled cavity arising in the mesoderm of the more advanced animals. It functions as a hydrostatic skeleton in some worms (e.g. earthworm) providing an incompressible barrier for the muscles to act against. The cavity separates an internal splanchnic mesoderm from an external somatic mesoderm and is lined with the coelomic epithelium. It separates the body wall from the gut wall, allowing independent muscular movement of these structures. It allows for, and demands, greater body complexity, notably development of a blood vascular system.

In annelids, mollusks, echinoderms, and chordates it is the main body cavity, containing the viscera; in mammals it is divided into separate cavities enclosing the heart (pericardial cavity), lungs (pleural cavity), and gut (peritoneal cavity). In arthropods the coelom is reduced to cavities surrounding the gonads and excretory organs, the main body cavity being a *hemocoel*, a blood-filled cavity. *See also* coelomoduct.

coelomoduct /si-loh-mŏ-dukt/ A ciliated duct connecting the coelom with the external environment. Coelomoducts provide a means of exit for gametes and waste products; in higher animals they are specialized as oviducts, etc.

coenocyte /see-nŏ-sÿt/ An area of cytoplasm containing many nuclei, typically found in certain fungi and algae. *Compare* plasmodium; syncytium.

coenzyme /koh-en-zÿm/ A non-protein group without which certain enzymes are inactive or incomplete. The protein part of an enzyme is known as the apoenzyme and when united with a coenzyme, either permanently or temporarily, the two form an active enzyme known as a holoenzyme.

coenzyme A (CoA) A coenzyme that is important in the synthesis and reactions of fatty acids. In the Krebs cycle it combines with pyruvic acid, leading to loss of carbon dioxide. It is a complex nucleotide containing an active –SH group. The compound is readily acetylated to CoAS–COCH$_3$ (acetyl CoA).

coenzyme Q *See* ubiquinone.

coevolution /koh-ev-ŏ-loo-shŏn/ The evolution in separate species of adaptations that are complementary or counteractive in some way, due to the selection pressure that each species applies on the other. The classic example is the coevolution of certain flowers and their insect pollinators. Here, mutations that make the flowers more recognizable (e.g. changes in shape, color, or scent) or more attractive to insect pollinators (e.g. richer nectar or more nutritious pollen) are of selective advantage to the plant because they increase the likelihood of insects visiting the plant. Similarly, mutations that enhance an insect's ability to detect or exploit a particular flower (e.g. a longer proboscis to reach the nectar) will benefit the insect and its offspring. Over time, many plant and insect species have formed tight-knit mutually beneficial associations that interlock by virtue of the special features each species has evolved to assist its partner. Another form of coevolution is seen between many predators and their prey. This involves a biological 'arms race', in which the prey species evolves defensive features such as spines, poisons, camouflage, or speedy locomotion, in its efforts to avoid capture, while the predator evolves countermeasures, such as better senses, more powerful jaws, or faster gait.

cofactor /koh-fak-ter/ A non-protein substance that helps an enzyme to carry out its activity. Cofactors may be cations or organic molecules, known as coenzymes. Unlike enzymes they are, in general, stable to heat. When a catalytically active enzyme forms a complex with a cofactor a *holoenzyme* is produced. An enzyme without its cofactor is termed an *apoenzyme*.

cohesion /koh-hee-zhŏn/ An attraction between the molecules of a substance. Water molecules exhibit relatively strong cohesion, a property that permits water to be transported through plants in the TRAN-

SPIRATION stream. According to the *cohesion theory*, a continuous column of water is drawn from the roots to the leaves, over vertical distances sometimes exceeding 30 m in the tallest trees. The driving force is the evaporation of water from the leaves, which creates a tension that is transmitted through the column of water.

colchicine /kol-chă-seen, -sin, **kol**-kă-/ A drug obtained from the autumn crocus *Colchicum autumnale* that is used to prevent spindle formation in mitosis or meiosis. It has the effect of halting cell division at metaphase, the stage at which the chromosomes have duplicated to give four homologs for each chromosome. If a resting nucleus forms after colchicine treatment it is thus likely to be tetraploid. Colchicine is also used to double the chromosome number of haploid plants derived from cultured pollen grains (*see* anther culture).

cold-blooded *See* poikilothermy.

Coleoptera /koh-lee-**op**-ter-ă, kol-ee-/ The largest order of insects and possibly the largest order in the animal kingdom, containing the beetles and weevils. Their forewings are modified to form hard leathery *elytra*, which protect the membranous hind wings and soft abdomen when resting. The head, projected into a snout in weevils, has biting mouthparts. Beetles are found universally in a variety of terrestrial and freshwater habitats. The larvae vary between legless grubs, caterpillar-like forms, and predators. Metamorphosis is complete. Many larvae and adults are serious pests, e.g. *Anobium* larvae (woodworm), and *Elater* larvae (wire-worm). Others are beneficial, e.g. *Coccinella* (ladybird), which eats aphids.

coleoptile /koh-lee-**op**-t'l, kol-ee-/ A sheathlike structure that protects the developing plumule in grasses. Some scientists think that, together with the scutellum, it represents the cotyledon, while others believe it is the first plumular leaf. The coleoptile contains very little chlorophyll and is usually light-sensitive. Research into plant tropisms using the oat coleoptile led

to the isolation and characterization of IAA. *See* plant hormone.

coleorhiza /koh-lee-ŏ-rÿ-ză, kol-ee-/ The protective sheath surrounding the radicle in grasses.

coliform bacteria /**kol**-ă-form, **koh**-lă-/ Gram-negative rod-shaped bacteria able to obtain energy aerobically or by fermenting sugars to produce acid or acid and gas. Most are found in the vertebrate gut, (e.g. *Escherichia coli*), but some are present in soil, water, or as plant pathogens. Many are pathogenic to humans (e.g. *Salmonella* species).

collagen /**kol**-ă-jěn/ The protein of fibrous connective tissues, present in bone, skin, and cartilage. It is the most abundant of all the proteins in the higher vertebrates. Collagen contains about 35% glycine, 11% alanine, 12% proline and small percentages of other amino acids. Collagen is chemically inert which suggests that its reactive side groups are immobilized by ionic bonding. Collagen fibrils are highly complex and have a variety of orientations depending on the biological function of the particular type of connective tissue. All have the same basic molecular structure of three polypeptide chains wound together in a triple helix. These rodlike molecules pack together side by side to form *collagen fibrils*. Parallel bundles of fibrils form the *collagen fibers*.

collar cell *See* choanocyte.

collateral bundle A kind of vascular bundle in which the phloem is external to the xylem and on the same radius. *Compare* bicollateral bundle; centric bundle.

collenchyma /kŏ-**lenk**-ă-mă/ A specialized type of parenchyma, usually located just beneath the epidermis, that functions as supporting tissue. The cell walls are irregularly thickened with cellulose and pectin, the thickening giving distinct patterns to the cells in cross-section. Collenchyma is the first strengthening tissue to be formed in young plants and is able to expand as the young tissues continue development. *Compare* sclerenchyma.

colloid /kol-oid/ A heterogeneous system in which the interfaces between phases, though not visibly apparent, are important factors in determining the system properties. The three important attributes of colloids are:
1. They contain particles, commonly made up of large numbers of molecules, forming the distinctive unit or *disperse phase*.
2. The particles are distributed in a continuous medium (the *continuous phase*).
3. There is a stabilizing agent, which has an affinity for both the particle and the medium; in many cases the stabilizer is a polar group.

Particles in the disperse phase typically have diameters in the range 10^{-6}–10^{-4} mm. Milk, rubber, and emulsion paints are typical examples of colloids. *See also* sol.

colon /koh-lŏn/ The first part of the large intestine, between the ileum and the rectum. It is thin-walled and of wide diameter; in man it has ascending, transverse, and descending limbs. Its lining has no villi and its only secretion is mucus. It contains the indigestible residue from the food, mostly cellulose, together with digestive juices and millions of bacteria, which are usually harmless and frequently beneficial. In the colon most of the water in ingested food is absorbed back into the blood, together with vitamins manufactured by the bacteria. The contents are thus converted into solid masses of feces, which are moved on into the rectum. *See illustration at* alimentary canal.

colony A group of organisms of the same species, generally attached to each other and dependent on each other to some degree. Colonial organization occurs in some hydrozoans (e.g. *Obelia* and *Physalia*), some anthozoans (corals and sea fans), and bryozoans.

colony-stimulating factor (CSF) Any of a group of cytokines that control the growth and differentiation of blood cells. Various ones are responsible for different types of cells. For example, *erythropoietin*, produced in the kidney and liver, promotes the formation of red blood cells (erythrocytes); interleukin-3 controls the production of certain white blood cells, namely granulocytes and monocytes/macrophages. *See* cytokine; interleukin.

color blindness Imperfect perception of color thought to be caused by a malfunction or absence of one of the three pigments in the light-sensitive cells (cones) of the retina of the eye. Although it can occasionally be acquired by disease or injury, the defect is usually inherited as a sex-linked recessive character on the X chromosome and is therefore more common in men (about 8% of the population) than in women (about 0.5%). However, women can be carriers of the gene. Complete color blindness is extremely rare, the most common form (*Daltonism*, named for the British chemist and physicist John Dalton (1766–1844)) being the inability to distinguish between reds and greens.

colostrum /kŏ-**los**-trŭm/ Liquid secreted by the mammary glands immediately and for the first few days after parturition, preceding the secretion of milk. It is rich in nutrients and antibodies and contains enzymes to clear mucus from the digestive tract of the newborn.

columella /kol-yŭ-**mel**-ă/ (*pl.* **columellae**)
1. The structure present in sporangia of many zygomycete fungi (e.g. *Mucor*) produced by formation of a dome-shaped septum cutting off the sporangium from the sporangiophore.
2. The central column of sterile tissue in the sporangium of liverworts and mosses.

columellar auris /kol-yŭ-**mel**-er **aw**-ris/ A rod of bone or cartilage that forms the only ear ossicle in amphibians, birds, and reptiles. It is homologous with the hyomandibular of fishes.

commensalism /kŏ-**men**-să-liz-ăm/ An association between two organisms in which one, the *commensal*, benefits and the other remains unaffected either way, e.g. the saprophytic bacteria in animal guts. *Compare* mutualism; parasitism; symbiosis.

commissure /kom-ă-**shoor**/ A cord of nerve fibers connecting symmetrical parts of the central nervous system. A commissure is usually transverse, such as the corpus callosum, which joins the two sides of the vertebrate brain, and the short commissures connecting the segmental ganglia in arthropods and annelids.

community A general term covering any naturally occurring group of different organisms living together in a certain environment, and interacting with each other. *See* association; consociation.

community biomass *See* biomass.

companion cell An elongated thin-walled cell cut off longitudinally from the same meristematic cell as the sieve element with which it is closely associated. It has a nucleus and dense cytoplasm and is thought to provide, via plasmodesmata, the needs of the less metabolically active, enucleated sieve element.

compass plant A plant with its leaf edges permanently aligned due north and south. Such plants thus avoid receiving the strong midday rays of the sun directly on the leaf blades, but are positioned to use fully the weaker rays of the morning and evening sun from the east and west. The best known example is the compass plant of the prairies *Silphium laciniatum.*

compensation point The light intensity at which the rate of photosynthesis is exactly balanced by the combined rates of respiration and photorespiration, so that net exchange of oxygen and carbon dioxide is zero. At normal daylight intensities the rate of photosynthesis exceeds respiration. Shade plants tend to reach their compensation points faster than sun plants but are unable to utilize high light intensities to the same extent. The point at which photosynthesis does not increase with increased light intensity is termed the *light saturation point.* This point occurs at much higher light intensities in C_4 plants than C_3 plants. *See* C_4 plant; photorespiration.

compensatory hypertrophy Replacement of a lost or damaged part of an organ by an increase in size of the remainder. *Compare* regeneration.

competent Describing embryonic tissue that is able to respond to natural (induction) or experimental (evocation) stimuli by becoming or making a specialized tissue. For example, the ectoderm over the optic cup of vertebrate embryos is competent to produce lens tissue.

competition The utilization of the same resources by one or more organisms of the same or of different species living together in a community, when the resources are not sufficient to fill the needs of all the organisms. The closer the requirements of two species, then the less likely it is that they can live in the same community, unless they differ in behavioral ways, such as periods of activity or feeding patterns.

complement A group of about 20 proteins normally found in vertebrate blood that react in an ordered sequence when an antigen-antibody complex has formed. The reaction causes lysis of the foreign cells or bacteria, attracts phagocytic cells to the reaction site, and promotes ingestion of antigen-bearing cells by phagocytes. Complement plays a role in INFLAMMATION, and is also involved in the tissue damage associated with certain autoimmune disorders.

complemental males Small and usually degenerate males of certain animals (e.g. some barnacles and angler fish) that live on or in the body of the female. They are dependent on the female for nutrition and are sometimes reduced to little more than an attached testis. Their attachment insures that fertilization (cross-fertilization in hermaphrodite barnacles) occurs.

complementary DNA (cDNA) A form of DNA synthesized by genetic engineering techniques from a messenger RNA template using a reverse transcriptase. It is used in cloning to obtain gene sequences from mRNA isolated from the tissue to be cloned. It differs from the original DNA sequence in that it lacks intron and promoter sequences. Labeled single-stranded cDNA is used as a gene probe to identify common gene sequences in different tissues and species, for example, as the reagent in DNA microarrays. *See* gene cloning.

complementary genes Genes that can only be expressed in the presence of other genes; for example if one gene controls the formation of a pigment precursor and another gene controls the transformation of that precursor into the pigment, then both genes must be present for the color to develop in the phenotype. Such interactions between genes lead to apparent deviations from the 9:3:3:1 dihybrid ratio in the F_2. For example, if two complementary genes control a certain character and dominant alleles of each of the two genes must be present for the character to appear then a 9:7 ratio is seen (9:(3 + 3 + 1)). *Compare* complementation; epistasis.

complementation /kom-plĕ-men-**tay**-shŏn/ The production of a normal phenotype from a genotype that apparently has two mutant alleles at a given locus. It is assumed that, although the mutants have the same effect, they actually occur in different cistrons of the same gene. In a diploid cell the two different mutated regions are prevented from expression by corresponding dominant alleles on the other homologous chromosome. *Compare* cis–trans effect.

complement fixation The combination of complement with antibody–antigen complexes. It is a property used to test for the presence of a specific antigen or antibody.

composite fruit /kŏm-**poz**-it/ A type of pseudocarpic ('false') fruit that incorporates the inflorescence. *See also* strobilus; sorosis; syconus.

compound eye The type of eye found in crustaceans and insects. It consists of several thousands of units (*ommatidia*). The spots of light focused by these give a mosaic image, which lacks visual acuity. However, the compound eye is very efficient at detecting the slightest movement over a wide area. *See also* ommatidium.

conceptacle /kŏn-**sep**-tă-kăl/ A flask-shaped reproductive cavity that develops on the swollen tips (receptacles) of the thalli of certain brown algae, (e.g. bladder wrack). Female conceptacles are lined with unbranched sterile hairs (paraphyses) and the oogonia develop on short stalks projecting from the chamber wall. Male conceptacles contain branched paraphyses that bear the antheridia. Both female and male conceptacles open to the exterior via a pore, the *ostiole*.

condensation /kon-den-**say**-shŏn/ A type of chemical reaction in which two molecules join together to form a larger molecule, with the associated production of a small molecule such as water (H_2O).

conditioned reflex A reflex action by an animal that is modified by experience (or conditioning) so that instead of occurring in response to the original stimulus it follows a different 'learned' stimulus. However, it is not permanent and requires periodic reinforcement with the original stimulus. Conditioned reflexes were first demonstrated in Pavlov's experiment with dogs, which received food at the ring of a bell. Eventually the bell alone was able to evoke salivation.

conditioning A form of learning in which a stimulus or signal becomes increasingly effective or a piece of behavior occurs with increasing regularity as a result of reinforcement each time it is exhibited. In *classical conditioning* behavior is altered by pairing two stimuli so that eventually the second stimulus alone elicits a response that initially was only produced by the first stimulus. For example, in Pavlov's dog experiments, the animals were conditioned to salivate at the sound of a bell by giving food and ringing a bell at the same time until the two stimuli became associated, after which the dogs salivated whenever a bell rang. In *instrumental conditioning* reinforcement only occurs after the animal performs a set piece of behavior; there is no initial stimulus (e.g. food) to initiate the behavior.

condyle /**kon**-d'l/ The curved surface at the end of a bone that articulates with another bone and forms a joint, which allows movement in definite planes. The occipital condyles of the skull are an example.

cone 1. One of the two types of light-sensitive cells in the retina of the vertebrate eye. They are concerned with vision in bright light and color vision and they have a high visual acuity. Cones have a different shape and pigment to rods. The greatest concentration of cones occurs at the fovea, but they are also found in the rest of the retina except at the periphery.

There are three groups of cones, each type containing a different type of photopigment (iodopsin) sensitive to red, green, and blue light. Iodopsins each consist of a glycoprotein (opsin) combined with retinal, derived from vitamin A. Cones also connect with a number of ganglion cells, but in the fovea the ratio of cones to ganglion cells is about 1:1. *See also* retina.
2. (*Botany*) *See* strobilus.

congenital /kŏn-**jen**-ă-tăl/ Present at birth. The term describes all deformities and other conditions that are present at birth, whether they are inherited or caused by environmental factors. Some congenital deformities, e.g. cleft palate and harelip, undoubtedly run in families but their occurrence is determined by environmental as well as hereditary factors.

conidiophore /kŏ-**nid**-ee-ŏ-for, -fohr/ *See* conidium.

conidiospore /kŏ-**nid**-ee-ŏ-spor, -spohr/ *See* conidium.

conidium /kŏ-**nid**-ee-ŭm/ (*pl.* **conidia**) (**conidiospore**) An asexual spore of certain fungi, e.g. *Pythium* and *Albugo*. They are

cut off externally in chains at the apex of a specialized hypha, the *conidiophore*.

Coniferophyta /koh-nă-fĕ-**roff**-ă-tă, kon-ă-/ The largest phylum of gymnosperms, comprising evergreen trees and shrubs, with many important species, e.g. *Pinus* (pine), *Picea* (spruce), *Taxus* (yew), and *Abies* (fir). They generally show a pyramidal growth form, bear simple leaves, and have the male and female reproductive structures contained in cones or strobili.

conjugated protein A protein that on hydrolysis yields not only amino acids but also other organic and inorganic substances. They are simple proteins combined with non-protein groups (prosthetic groups). *See also* glycoprotein; lipoprotein; phosphoprotein.

conjugation /kon-jŭ-**gay**-shŏn/ 1. The sexual fusion of gametes, particularly isogametes. *See* isogamy.
2. A type of sexual reproduction found in some bacteria, most ciliates, and certain algae, involving the union of two individuals for the purpose of transferring genetic material. In bacteria (*Escherichia* and related genera), two individuals join by a conjugation bridge and part of the genetic material of one, the donor (or male) cell, is transferred to the recipient (or female) cell. In ciliates (e.g. *Paramecium*) the two individuals unite by a bridge; their macronuclei disintegrate and their micronuclei divide by meiosis to form two gamete nuclei, one of which moves to the other cell and fuses with the remaining gamete nucleus to form a zygote. Each zygote divides and eventually forms four daughter cells. In algae such as *Spirogyra*, which are normally haploid, a conjugation tube forms between cells of two individuals and the gamete formed in one cell (the male gamete) moves through the tube and fuses with the gamete of the other cell (the female gamete).

conjunctiva /kon-junk-**tÿ**-vă/ A thin transparent layer of epidermis and connective tissue covering the cornea of the eye and inner eyelid in vertebrates. It is kept moist by secretions from the tear glands. *See illustration at* eye.

connective The tissue that joins the two lobes of the anther and contains the vascular strand.

connective tissue A type of tissue in which the cells are isolated from each other by a matrix. It supports, binds, connects, and holds in position the organs of the body and arises from the mesoderm germ layer of the embryo. The matrix is gelatinous glycoprotein in the loose *areolar connective tissue* of the mesenteries and the dermis of the skin. Connective tissue usually contains varying amounts of branching yellow elastic and tough white collagen fibers. The cells that secrete the matrix are called mast cells; those that produce the fibers are fibroblasts. Macrophages and lymphocytes are also present. *See also* macrophage; mast cell.

consensus sequence The sequence of nucleotides that is most likely to occur in a particular segment of a nucleic acid (i.e. DNA or RNA), or the sequence of amino acids that is most probable in a given part of a protein molecule.

conserved sequence A sequence of nucleotides in a nucleic acid (DNA or RNA), or a sequence of amino acids in a protein, that shows a high degree of similarity among comparable molecules obtained from different species of organisms.

consociation /kŏn-soh-see-**ay**-shŏn/ A climax of natural vegetation dominated by one particular species, such as oakwood, dominated by the oak tree, or *Calluna* heathland dominated by the heather, *Calluna vulgaris*. Many consociations together may form an association, for example oakwood, beechwood, and ashwood consociations together make up a deciduous forest association. *See also* association.

constitutive enzyme /kon-stă-tew-tiv/ One of a group of enzymes that are always present in nearly constant amounts in a given cell. These enzymes are formed at constant rates and in constant amounts regardless of the metabolic state of the organism.

consumer An organism that feeds upon another organism, e.g. all animals and parasitic and insectivorous plants. *Compare* producer. *See also* trophic level.

consummatory act /kŏn-**sun**-ă-tor-ee, -toh-ree/ A behavioral pattern, often stereotyped, that occurs once a goal has been achieved following a phase of appetitive behavior. For example, eating is the consummatory act of feeding behavior, copulation is the consummatory act of sexual behavior.

continental drift The theory that present-day continents have arisen by the breaking up and drifting apart of a previously existing ancient land mass (Pangaea). There is much evidence to support the theory, and it serves to explain the distribution of contemporary and fossil plants and animals. Continental drift is now believed to reflect the movement over geological time of underlying plates in the earth's crust – the theory of *plate tectonics*.

continuous variation *See* quantitative variation.

contour feathers Feathers that give the bird a streamlined shape and provide the flight surfaces of the wings and tail. The longer contour feathers are the *flight* or *quill feathers*; almost all their barbs are interlocked, except for a small tuft of separate barbs (the *aftershaft*) at the base of the vane. The shorter contour feathers (*coverts*) have a greater region with separate barbs and the aftershaft is a small feather with separate barbs.

contractile roots Specialized roots developed by certain bulb and corm forming plants that serve to pull the bulb or corm down to the appropriate depth in the soil. This counteracts the tendency for each new year's growth to be raised above the growth of the previous years.

contractile vacuole One or more membrane-bound cavities in many protoctists that act as osmoregulators. They periodically expand as they fill with water by osmosis and contract to discharge their contents to the exterior.

control The part of an experiment that provides a baseline or standard against which the effects of the experiment can be judged.

conus arteriosus /koh-nŭs ar-teer-ee-**oh**-sŭs/ The fourth chamber of the heart in primitive vertebrates, which leads forward from the anterior base of the ventricle and into the ventral aorta. It is tubular in shape, with thick muscular walls and two rows of three semilunar valves to prevent the backflow of blood. In lungfish and amphibians there is also a thin twisted fold (the *spiral valve*), which aids the separation of oxygenated and deoxygenated blood.

convergent evolution (**convergence**) The development of similar structures in unrelated organisms as a result of living in similar ecological conditions. The wings of vertebrates and insects are an example of convergence, in which quite distinct groups of animals have independently adapted in a similar way to life in the air. *See also* analogous; homoplasy.

convoluted tubule /kon-vŏ-**loo**-tid/ *See* kidney tubule.

Copepoda /koh-**pep**-ŏ-dă/ A large class of minute Crustacea whose members lack a carapace and compound eyes and have the first thoracic appendages modified for feeding. The remaining thoracic appendages are used for swimming. Many are important members of the plankton, e.g. the marine *Celanus* and the freshwater *Cyclops*.

copper /**kopp**-er/ *See* trace element.

coracoid /**kor**-ă-koid, koh-ră-/ One of a pair of cartilage bones forming the ventral side of the pectoral girdle in birds, reptiles, and bony fish (Osteichthyes). They contribute, with the scapulae, to the formation of the glenoid cavities – the articular surfaces for the forelimbs – and in birds act as a wing brace. In mammals, they have been functionally replaced by the clavicles and are reduced to small processes on the scapulae.

coral *See* Anthozoa.

cork (**phellem**) A protective layer of radially arranged cells produced to the outside of the cork cambium. The cork replaces the epidermis in certain woody plants, forming an impervious layer broken only by lenticels. The older cork cells are dead, suberized, and frequently only air-filled, although lignin, fatty acids, and tannins often accumulate. The cork oak, *Quercus suber*, forms a very thick layer used commercially.

cork cambium (**periderm cambium; phellogen**) A subepidermal layer of cells forming a lateral meristem that arises following the onset of secondary growth. The cells of the cork cambium give rise externally to the cork of phellem and internally to the phelloderm.

corm An organ of perennation and vegetative reproduction, consisting of a short erect fleshy swollen underground stem, usually broader than high and covered with membranous scales. It stores food material in the stem and bears buds in the axils of the scalelike leaf remains of the previous years' growth. Examples of corms are crocus and gladiolus. *Compare* bulb.

cornea /**kor**-nee-ă/ The firm transparent front part of the outer coat of the vertebrate eye, covering the iris and the pupil. It bulges slightly (having a smaller radius than the eye as a whole) and its curved surface bends the light rays passing through it. It consists of white fibrous connective tissue covered by stratified epithelium (CONJUNCTIVA) with free nerve endings, which, when the cornea is damaged, cause intense pain. It is lined by an elastic membrane and an inner endothelial layer. The endothelium regulates the hydration of the cornea, which keeps it transparent. *See illustration at* eye.

cornification /kor-nă-fă-**kay**-shŏn/ *See* keratinization.

corn rule *See* optical activity.

corolla /kŏ-**rol**-ă/ A collective term for the petals of a flower. The corolla is denoted in the floral formula by the symbol C.

corona /kŏ-**roh**-nă/ **1.** In flowering plants, any type of outgrowth from the petals or sepals, such as the trumpet of the daffodil flower.
2. A group of cells at the tip of the oogonium in *Chara* species of algae.

coronary vessels /**kor**-ŏ-nair-ee, ko-rŏ-/ Either of two pairs of blood vessels present in vertebrates that serve the heart. In mammals, the coronary arteries arise from the aorta and carry oxygenated blood into the muscle of the ventricles. The coronary veins return the deoxygenated blood to the right atrium. 30–60% of the coronary blood flow may be returned to any of the four chambers by luminal vessels. 7–10% of the aortic output goes to the coronary circulation and any stoppage to this supply leads to coronary heart disease. Blocked coronary arteries may be treated surgically by a coronary bypass operation.

corpora allata *and* **corpora cardiaca** /**kor**-pŏ-ra ă-**lay**-tă *and* kar-**dy̆**-ă-kă/ Glands found in insects. They produce hormones, including juvenile hormone, that control such factors as gamete production, metamorphosis, growth, and molting. The *corpora cardiaca* lie posterior to the brain and are usually blue in color. The *corpora allata* lie behind the corpora cardiaca and are yellow. General body growth is controlled by the corpora cardiaca and molting and egg production by the corpora allata.

corpus callosum /**kor**-pŭs kă-**loh**-sŭm/ (*pl.* **corpora callosa**) A thick band of nerve fibers that connects the two cerebral hemispheres in the brain of placental mammals. It enables coordination of the functions of the two hemispheres. *See* cerebral hemispheres.

corpus luteum /loo-**tee**-ŭm/ (*pl.* **corpora lutea**) A yellow mass of glandular tissue formed temporarily within the Graafian follicle after ovulation. It secretes progesterone. *See* estrous cycle.

corpus meristem The central region of the meristem below the tunica where cell divisions are in all directions giving both increased width and length to the apex. The tissues of the stele and cortex are derived from the corpus. *See* tunica–corpus theory.

corpus striatum /strȳ-**ay**-tŭm/ (*pl.* **corpora striata**) A complex mass of nerve cell bodies and fibers lying deep within each cerebral hemisphere of the brain. It is highly developed in birds, in which it is the site of the highest brain functions, but is much less important in mammalian brains.

cortex (*pl.* **cortices**) **1.** (*Botany*) A primary tissue in roots and stems of vascular plants derived from the corpus meristem, that extends inwards from the epidermis to the phloem. It usually consists of parenchyma cells but other tissues (e.g. collenchyma) may be present. Some algae, fungi, mosses, and lichens have a well defined region that is termed the cortex although this is different in origin and composition from the cortex of vascular plants. *See* parenchyma.
2. (*Zoology*) The outermost layer of an organ or part. For example, the outer region of the kidney is called the *renal cortex* and the surface layer of gray matter in the cerebral hemispheres of the brain is the *cerebral cortex*. *Compare* medulla.

cortical granules Membrane-bound vesicles in the cortex of many animal eggs, whose contents are extruded at fertilization. The contents turn the vitelline membrane into the fertilization membrane and effect the zona reaction in mammals, preventing further spermatozoa penetrating the egg.

corticosteroid /kor-tă-koh-**steer**-oid/ Any steroid hormone produced by the adrenal cortex. The release of corticosteroids is controlled by corticotropin. They are classified into two groups: mineralocorticoids and glucocorticoids. Natural and synthetic corticosteroids have widespread use in the treatment of adrenal insufficiency, allergy, and skin and inflammatory diseases. *See* glucocorticoid; mineralocorticoid.

corticosterone /kor-tă-**kos**-tĕ-rohn/ A steroid hormone produced by the adrenal cortex and having glucocorticoid activity. *See* glucocorticoid.

corticotropin /kor-tă-koh-**trop**-in/ (**ACTH; adrenocorticotropic hormone**) A polypeptide hormone secreted by the anterior pituitary gland. It acts on the adrenal cortex, stimulating the secretion of corticosteroid hormones. Its release is controlled by the hypothalamus and by circulating corticosteroids, whose production it stimulates. Stress also stimulates its secretion. It is used in the diagnosis of disorders of the anterior pituitary gland and adrenal cortex and may be used therapeutically, for example to stimulate corticosteroid production in children.

cortisol /kor-tă-sohl, -zohl, kor-tă-zol/ *See* hydrocortisone.

cortisone /**kor**-tă-zohn/ A steroid hormone, produced by the adrenal cortex, that is mainly inactive until converted into hydrocortisone.

corymb /**kor**-imb, -im, **koh**-rimb, -im/ An inflorescence with flower stalks of different lengths, the lowest being the longest. This gives a flat-topped cluster of flowers at the same level that is characteristic of many crucifers (e.g. candytuft). *See* inflorescence.

cosmid /**koz**-mid/ A hybrid cloning vector formed from a virus and a plasmid. It contains the COS site from the lambda phage genome, hence the name. Cosmids are used to infect *E. coli* bacterial host cells, in which they behave not as viruses but as plasmids, being replicated by their hosts.

They are used to clone foreign DNA fragments of up to about 45 kb.

cosmoid scales /**koz**-moid/ Spiny SCALES in the skin of cartilaginous fish. The outer part is formed from a dentine-like substance (*cosmine*), covered with an enamel-like substance (*vitrodentine*). *Compare* ganoid scales.

cotyledon /ko-tă-**lee**-dŏn/ (**seed leaf**) The first leaf of the embryo of seed plants, which is usually simpler in structure than later formed leaves. Cotyledons play an important part in the early stages of seedling development. For example they act as storage organs in seeds without an endosperm, such as peas and beans, and they form the first photosynthetic organ in seeds showing epigeal germination (e.g. sunflower). Monocotyledons and dicotyledons are so termed because they contain one and two cotyledons respectively. Gymnosperms may contain two cotyledons (e.g. *Taxus* and *Cycas*) or a varied number (e.g. *Pinus*). The first two leaves of the *Selaginella* embryo are also termed cotyledons. In some seeds the cotyledon may have a haustorial function remaining within the seed to absorb the endosperm, as in the onion. *See* endosperm.

courtship The specialized patterns of behavior that are preliminary to mating and reproduction. It may consist of a few simple stimuli but is often a long elaborate and ritualized series of actions. Its function is to synchronize precisely the activities of male and female so that copulation can occur, to reduce the female's fear of the male and the male's aggressiveness towards the female, and to arouse sexual responses while suppressing any other tendencies (such as feeding). It is also important in enabling the participants to identify each other as potential mates.

covert /**koh**-vert, koh-**vert**/ *See* contour feathers.

coxa /**koks**-ă/ (*pl.* **coxae**) The first (basal) segment of an insect leg, which articulates with the thorax and second leg segment (trochanter).

coxal gland /koks-ăl/ A type of excretory organ present in Arachnida. Coxal glands occur in one or two pairs on the floor of the cephalothorax and are drained by ducts that open to the exterior near the base of the legs. *See also* Malpighian tubules.

coxopodite /koks-**op**-ŏ-dÿt/ The joint of the protopodite nearest to the body. *See* biramous appendage.

C₃ plant A plant in which the first product of photosynthesis is a 3-carbon acid, phosphoglyceric acid. Most temperate and many other plants are C_3 plants. They are characterized by high carbon dioxide compensation points owing to photo-respiration, and are not as efficient photosynthetically as C_4 PLANTS.

C₄ plant A plant in which the first product of photosynthesis is a 4-carbon dicarboxylic acid. C_4 plants have evolved from C_3 PLANTS by a modification in carbon dioxide fixation, leading to more efficient photosynthesis. The modified pathway is called the *Hatch–Slack pathway* or the *C_4 dicarboxylic acid pathway*. C_4 plants are mainly tropical or subtropical, including many tropical grasses (e.g. maize and sorghum). In the leaves, the mesophyll cells surrounding the vascular bundles (bundle sheath cells), contain the carbon dioxide fixing enzyme phosphoenolpyruvate carboxylase (PEP carboxylase) in their cytoplasm. This has a higher affinity for carbon dioxide than ribulose bisphosphate carboxylase (RUBP carboxylase). The product of carbon dioxide fixation is oxaloacetate, which is rapidly converted to the C_4 acids malate and aspartate. The decarboxylation of C_4 acids releases CO_2, which is then re-fixed as in C_3 plants.

C_4 plants are more efficient than C_3 plants: they are capable of utilizing much higher light intensities and temperatures; have up to double the maximum rate of photosynthesis; and lose less water by tran-

spiration because, with more efficient CO_2 uptake, smaller stomatal apertures are needed to obtain sufficient CO_2 for photosynthesis. The Hatch–Slack pathway is named for the Australian plant physiologist M. D. Hatch and the English plant physiologist C. R. Slack.

cranial nerves The paired nerves that originate directly from the brain of vertebrates. Most supply the sense organs and muscles of the head although some, such as the vagus nerve, supply other parts of the body also. In humans and other mammals there are 12 cranial nerves: olfactory (I), optic (II), oculomotor (III), trochlear (IV), trigeminal (V), abducens (VI), facial (VII), auditory (VIII), glossopharyngeal (IX), vagus (X), accessory (XI), and the hypoglossal (XII).

craniate /kray-nee-ayt/ (vertebrate) Any chordate animal in which the notochord is replaced by a dorsal vertebral column (the backbone) and the brain is housed in the cranium (skull). Hence craniates include the cyclostomes (Agnatha), cartilaginous fishes (Chondrichthyes), bony fishes (Osteichthyes), amphibians (Amphibia), reptiles (Reptilia), birds (Aves), and mammals (Mammalia). Some authorities group all these classes together as the Craniata, which is variously given the status of subphylum or phylum. Craniates have an internal skeleton of cartilage or bone and the backbone encloses the tubular nerve cord (the spinal cord). There is a complex nervous system with a well-developed brain. The circulatory and digestive systems are located ventral to the vertebral column. All but the most primitive vertebrates have jaws formed from the anterior pair of visceral arches.

cranium /**kray**-nee-ŭm/ The skull of vertebrates.

crawfish *See* Astacus.

creatine /**kree**-ă-tin, -teen/ A constituent of vertebrate muscle, averaging 0.3 to 0.4 per cent of the tissue. The greatest concentration is found in voluntary muscle and the least in involuntary muscle. Vertebrate animals, especially carnivores, obtain some creatine in the diet, but all, and particularly herbivores, can synthesize it. In the resting muscle, creatine is combined with phosphoric acid to form phosphocreatine (creatine phosphate). When the muscle contracts, phosphocreatine is split back into creatine and phosphoric acid, with the release of energy. This is energy responsible for the actual muscle contraction through the medium of ATP. The energy necessary for synthesis of the creatine phosphate is provided by carbohydrate breakdown (glycolysis).

creatinine /kree-**at**-ă-nin, -neen/ A characteristic constituent of the urine of all mammals. It is a waste product produced by catabolism of creatine.

cremocarp /**krem**-ŏ-karp, **kree**-mŏ-/ A dry fruit splitting into two one-seeded portions. The portions are termed *mericarps*. The mericarps are indehiscent and remain attached to the plant for some time before being dispersed. The cremocarp is characteristic of certain members of the Umbelliferae (e.g. hogweed).

Cretaceous /kri-**tay**-shŭs/ The most recent period of the Mesozoic era, 145–66 million years ago. It is marked by continued domination of land and sea by dinosaurs until a rapid extinction towards the end of the period. The marine ammonites and aquatic reptiles also became extinct in this period. Primitive mammals were present, but were relatively insignificant in number, size, and variety until the Cenozoic era, which followed. Birds and fishes evolved into structurally modern forms during the Cretaceous. The flowering plants replaced the gymnosperms as the dominant terrestrial vegetation. The Cretaceous is named after the large amounts of chalk (fossilized plankton) found in rocks of the period. *See also* geological time scale.

Crinoidea /kri-**noid**-ee-ă/ The most primitive class of the Echinodermata and the only echinoderms with the mouth on the upper surface of the body. The mouth is surrounded by feathery arms bearing tube feet and ciliary grooves used in feeding. The larvae are always sessile, attached to the substratum by a stalk. The deep-sea *Metacrinus* (sea lily) remains stalked as an adult but the coastal *Antedon* (feather star) is free-swimming.

crista /**kris**-tă/ (*pl.* **cristae**) The structure formed by folding of the inner mitochondrial membrane. The extent and nature of the folding varies, active cells having complex and closely packed cristae, less active cells having fewer and less complex cristae. The surface of cristae is covered with stalked particles (respiratory granules) that contain the oxidative enzymes (e.g. ATPase and the cytochromes).

critical day length (cdl) The amount of light per day that is the maximum a short-day plant may receive and still flower, and conversely, the minimum a long-day plant needs to flower. Cocklebur, a short-day plant, will not flower if given more than 15½ hours light per day; i.e. its cdl is 15½ hours. Henbane, a long-day plant, will only flower if given more than 11 hours light per day; i.e. the cdl is 11 hours. *See* photoperiodism.

Cro-Magnon man /kroh-**mag**-non, -**man**-yon/ A group of early representatives of the species *Homo sapiens* that lived in Europe about 40 000–13 000 years ago. They are direct ancestors of modern humans.

crop A part of the alimentary canal, present in such animals as earthworms, insects, and birds, that is modified for the storage and partial digestion of food. Food passes from the esophagus into the crop before going, a little at a time, into the gizzard.

The crop in birds is large and thin-walled, projecting from the ventral region of the esophagus. In female pigeons crop milk is produced by glands to feed the nestlings.

cross The act of cross-fertilization or the organism resulting from cross-fertilization.

crossing over The exchange of material between homologous chromatids by the formation of chiasmata. *See also* chiasma; mitosis.

cross-over value (COV) The percentage of recombinations between two linked genes in the progeny of a given cross. It is a measure of the strength of linkage between two genes and can be used in the construction of chromosome maps. However, if two genes are very far apart on the chromosome more than one crossover may occur between them, in which case misleading values are obtained.

crown The part of a tooth outside the gum, covered by enamel. *See* teeth.

Crustacea /krus-**tay**-shă/ A large group of arthropods containing the mostly aquatic gill-breathing crawfish, crabs, lobsters, barnacles, water fleas, etc., and the terrestrial woodlice. The body is divided into a head, thorax, and abdomen. The head bears compound eyes, two pairs of antennae, and mouthparts composed of a pair of mandibles and two pairs of maxillae. The thorax is often covered with a dorsal carapace. The appendages are typically forked and specialized for different functions. The sexes are usually separate and development is indirect, via a nauplius larva. The taxon named Crustacea is variously given the rank of class, superclass, or phylum by different authorities. *See also* Branchiopoda; Copepoda; Decapoda.

cryptic coloration A coloring that conceals an animal in its usual surroundings, protecting it from predators. Cryptic coloration is very diverse; examples are the strong contrasting patterns that break up the outline of the body, making it difficult to identify (e.g. angel fish or ring plover), countershading of the underside of the body to counteract shadows cast (e.g. many caterpillars), and imitation of inanimate objects, such as petals, leaves, and bird droppings.

cryptobiosis /krip-toh-bÿ-**oh**-sis/ A state of suspended activity in which certain minute invertebrates, including rotifers, nematodes, and collembolans (springtails), survive desiccation or other adverse conditions. The animal typically reduces its surface area to a minimum by curling into a ball, and then shrivels as it loses all but a bare minimum of body water. Biochemical changes protect vital cell components, and the cytoplasm becomes glasslike. The animal absorbs moisture when this returns, causing the body to swell and activity to resume in a matter of hours.

crypt of Lieberkühn /**lee**-ber-koon/ (**intestinal crypt**) A circular cleft that surrounds each villus in the small intestine. Cells in the intestinal epithelium of each crypt divide rapidly and migrate up the sides of neighboring villi to replace cells shed from the villi into the lumen of the intestine. Among them are large numbers of mucus-secreting goblet cells. The crypts are named for the German anatomist Johann N. Lieberkühn (1711–56).

cryptogam /**krip**-tŏ-gam/ In early classifications, any plant that reproduces by spores or gametes rather than by seeds. Cryptogams were thus named because early botanists considered their method of reproduction to be cryptic. They included the algae, fungi, bryophytes, and pteridophytes (ferns), the latter group often being termed vascular cryptogams. *Compare* phanerogam.

cryptophyte /**krip**-tŏ-fÿt/ (**geophyte**) A plant in which the resting buds are below the soil surface. *See also* Raunkiaer's plant classification.

crystallography /kris-tă-**log**-ră-fee/ The study of the geometric structure and internal arrangement of crystals. It is often used in the identification of macromolecules, because each type of macromolecular crystal has a characteristic refractive index, i.e. a light ray passing through the crystal will change direction at a constant angle. *See also* x-ray crystallography.

CSF *See* cerebrospinal fluid.

Ctenophora /ten-**off**-ŏ-ră/ A small phylum of marine invertebrates, the sea gooseberries or comb jellies, which show similarities to the cnidarians. The transparent, often globular, saclike body bears eight rows of fused cilia (*combs* or *ctenes*) used in locomotion and the enteron forms a canal system in the body. Most have tentacles armed with adhesive cells for food capture on each side of the body. *Compare* Cnidaria.

cultivar /**kul**-tă-var/ Any agricultural or horticultural 'variety'. The term is derived from the words *culti*vated *var*iety.

culture A population of microorganisms or dissociated cells of a tissue grown on or within a solid or liquid medium for experimental purposes. This is done by inoculation and incubation of the nutrient medium. *See also* tissue culture.

culture medium A mixture of nutrients used, in liquid form or solidfied with agar, to cultivate microorganisms, such as bacteria or fungi, or to support tissue cultures.

cumulus cells /**kyoo**-myŭ-lŭs/ Cells from the Graafian follicle that surround the ovulated mammalian egg. They disperse quickly if sperms are present (30 minutes) but more slowly (2 hours or more) if they are not. *See* zona pellucida.

cupule /**kyoo**-pyool/ 1. The cup-shaped structure in which certain fruits (e.g. acorn and hazelnut) are borne.
2. A protective cup made up of six modified leaves surrounding the young gemma of *Lycopodium*.
3. The bright red tissue surrounding the ovule of *Taxus*.
4. The ovule-bearing structure terminating the pinna in the extinct gymnosperms Caytoniales.

cusp A conical point on the surface of the crown of the molar and premolar teeth of carnivorous and omnivorous mammals, such as dogs and humans. Small premolars have only two cusps, and the larger molars

have three or four. They are used for crushing, cutting, and chewing.

cuticle /**kyoo**-tă-kăl/ A protective layer secreted by an epidermis.
In plants, the cuticle is a waterproof layer of waxy cutin covering the epidermis, mainly of aerial parts and some seeds. It may be covered by a layer of wax, e.g. apple, and sometimes resin, e.g. horse chestnut buds. Its thickness varies with the species and environment. Xerophytes tend to have thick cuticles, and dry conditions often induce cuticular thickening. On average, only about 5% of the water lost from a plant is via the cuticle (cuticular transpiration). *See illustration at* leaf.
Cuticles are found in a variety of animals, e.g. the thick cuticles of endoparasites such as tapeworms and flukes; the thin collagen cuticle of earthworms; the chitin-containing cuticle of arthropods, which is calcium-impregnated in crustaceans for extra hardness; and the calcium-impregnated shells of mollusks. The arthropod cuticle contributes greatly to the success of the group. Apart from its protective role, it acts as an exoskeleton, serving as an attachment for muscles and being flexible at the joints. The insect cuticle is extremely waterproof and is covered by a thin waxy layer. Arthropods must, however, periodically molt their cuticles to allow growth during which time vulnerability to predators is high. *See* chitin.

cuticularization /kyoo-tik-yŭ-lă-ri-**zay**-shŏn/ The formation of cuticle by the secretion of fluid materials, which subsequently harden.

cutin /**kyoo**-tin/ A group of substances chemically related to fatty acids forming a continuous layer called the cuticle on the epidermis of plants, interrupted only by stomata or lenticels. Being fatty in nature, cutin is water-repellent, therefore helping to reduce transpiration. It is also protective, for example preventing invasion by parasites. *See* suberin.

cutinization /kyoo-tin-i-**zay**-shŏn/ The

impregnation of a plant cell wall with cutin.

cutis /**kyoo**-tis/ *See* dermis.

cutting A part of a plant that is removed from the parent and encouraged to grow into another plant. Cuttings are a means of asexual propagation and may vary in size from buds or root segments to large shoots. *See also* graft; vegetative propagation.

Cuvierian duct /kyoo-vee-**air**-ee-ăn, kyoo-**veer**-ee-ăn/ (**ductus Cuvieri**) *See* cardinal vein.

Cyanobacteria /sÿ-an-oh-bak-**teer**-ee-ă/ A phylum of bacteria containing the blue-green bacteria (formerly called blue-green algae) and the green bacteria (chloroxybacteria). Both groups convert carbon dioxide into organic compounds using photosynthesis, generally using water as a hydrogen donor to yield oxygen, like green plants. However, under certain circumstances they use hydrogen sulfide instead of water, yielding sulfur. Cyanobacteria are an ancient group, and their fossils (stromatolites) have been dated at up to 2500 million years old. Today, most species are found in soil and freshwater. They are spherical (coccoid) or form long microscopic filaments of individual cells. Many species, e.g. *Nostoc* and *Oscillatoria*, are nitrogen fixers. They reproduce asexually by binary fission, or by releasing sporelike propagules or filament fragments. It is thought that certain ancient cyanobacteria became permanent symbionts of ancestral algae and green plants, taking up residence in their cells as photosynthetic organelles (plastids). This theory would account for the striking similarities between cyanobacteria and plastids. *See* stromatolite.

cyanocobalamin /sÿ-an-oh-kŏ-**bal**-ă-min/ (**cobalamin; vitamin B$_{12}$**) One of the water-soluble B-group of vitamins. It has a complex organic ring structure at the center of which is a single cobalt atom. Foods of animal origin are the only important dietary source. A deficiency in humans leads to the development of pernicious anemia since the vitamin is required for the development of red blood cells. It is also needed for the synthesis of proteins and nucleic acids. Deficiency in humans is most commonly due to lack of INTRINSIC FACTOR, which is required for absorption of the vitamin by the intestine. *See also* vitamin B complex.

Cycas /sÿ-kas/ A genus of the primitive gymnosperm phylum Cycadophyta. Cycads may grow to a height of 15m and generally have an unbranched stem with a rosette of large pinnate leaves at the apex, giving a palmlike growth habit. The male and female strobili and the micro- and megaspores are larger than any found in other plant groups. The male spermatozoids are motile, bearing a spiral band of flagella, and swim towards the megaspore to effect fertilization.

cyclic AMP (**cAMP; adenosine-3′,5′-monophosphate**) A form of adenosine monophosphate (*see* AMP) formed from ATP in a reaction catalyzed by the enzyme ADENYLATE CYCLASE. It has many functions, acting as an enzyme activator, genetic regulator, chemical attractant, second messenger, and as a mediator in the activity of many hormones, including epinephrine, norepinephrine, vasopressin, ACTH, and the prostaglandins.

Cyclops /sÿ-klops/ A genus of minute crustaceans found universally in fresh water and an important member of the freshwater plankton. *Cyclops*, so called because of its one central simple eye, has an oval body divided into an anterior region covered in a thin chitinous shield and a segmented posterior region that ends in a forked tail. The first thoracic appendages are modified for feeding; the remaining appendages are used in rapid swimming. The female has two oval egg sacs hanging from the last thoracic segment. *See also* Copepoda; Crustacea.

cyclosis /sÿ-**kloh**-sis/ 1. The streaming of cytoplasm in a circular motion around the cell observed in some plants, particularly young sieve tube elements. It is an example

of the more generalized and widespread phenomenon of *cytoplasmic streaming* in which movement of parts of the cytoplasm occur relative to other parts, sometimes in fixed channels. It is an energy-consuming process, thought to be produced by interaction between myosin-like motor proteins and actin microfilaments arrayed around the cell perimeter. The motor proteins are attached to cell organelles, and as these are propelled along, the cytosol is swept along with them, resulting in bulk flow of cytoplasm.
2. The circulation of cell organelles through the cytoplasm, e.g. the food vacuoles of *Paramecium*.

Cyclostomata /sỹ-kloh-**stom**-ă-tă/ The order of agnathans that contains the most primitive living vertebrates – the lampreys (e.g. *Petromyzon*) and the hagfish (e.g. *Myxine*). They are fishlike animals with no scales or paired fins. Jaws are functionally replaced by a round suctorial mouth with horny teeth and a protrusible tongue. Lampreys live in the sea or rivers, attaching to fish by means of the mouth and feeding on their blood and flesh. They spawn in fresh water. Hagfishes are marine scavengers on the sea bottom.

The skeleton is secondarily cartilaginous and there is a single nasal opening and a row of spherical gill pouches. The notochord persists throughout life. In lampreys, development is via an ammocoete larva; hagfish have no larval stage.

cyme /sỹm/ *See* cymose inflorescence.

cymose inflorescence /sỹ-mohs, sỹ-**mohs**/ (**cyme; definite inflorescence**) An inflorescence in which apical growth is terminated by the formation of a flower at the apex. Subsequent growth is then from lateral buds below the apex, which themselves form flowers and more lateral shoots. If one shoot develops behind each axis a *monochasial cyme* is formed, which may be scorpioid or helicoid in shape. If two shoots develop below each axis this gives a *dichasial cyme*. *Compare* racemose inflorescence. *See also* sympodial.

cypsela /**sip**-sĕ-lă/ A small dry indehiscent fruit characteristic of the family Compositae, (e.g. sunflower and daisy). It is formed from a bicarpellary inferior ovary in which only one of the ovules continues to develop to maturity.

cysteine /**sis**-tee-een/ *See* amino acids.

cysticercus /sis-tă-**ser**-kŭs/ (*pl.* **cysticerci**) A larval stage of tapeworms belonging to the genus *Taenia*, which infest dogs, cats, humans, and other mammals. It consists of a fluid-filled sac containing the head and neck of a single tapeworm, which develops into an adult tapeworm when eaten by the final host. *See* bladderworm.

cystine /**sis**-teen, -tin/ A compound formed by the joining of two cysteine amino acids through a –S–S– linkage (a

helicoid   scorpioid
monochasial cymes

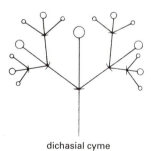

dichasial cyme

Types of cymose inflorescence

cystine link). Bonds of this type are important in forming the structure of proteins.

cystolith /sis-tŏ-lith/ A deposit of calcium carbonate arising internally on a stalk from the cell walls of large modified epidermal cells in some flowering plants (e.g. stinging nettle).

cytidine /sȳ-tă-din, sit-, -deen/ (**cytosine nucleoside**) A nucleoside formed when cytosine is linked to D-ribose via a β-glycosidic bond.

cytochromes /sȳ-tŏ-krohmz/ Conjugated proteins containing heme, that act as intermediates in the ELECTRON-TRANSPORT CHAIN. There are four main classes, designated *a*, *b*, *c*, and *d*.

cytogenetics /sȳ-toh-ji-**net**-iks/ The area of study that links the structure and behavior of chromosomes with inheritance.

cytokine /sȳ-toh-kȳn/ Any of a large group of proteins released by mammalian cells that act as highly potent chemical messengers for other cells. They may cause a wide variety of responses in their target cells, for example, triggering differentiation or stimulating secretion. Cytokines released by lymphocytes are called *lymphokines*; these regulate and coordinate the activities of the different types of lymphocytes that participate in immune responses. Other common cytokines include the interferons and interleukins. *See* colony-stimulating factor; insulin-like growth factor; interferon; interleukin; lymphokine.

cytokinesis /sȳ-toh-kin-**ee**-sis, -kȳ-**nee**-/ The division of the cytoplasm after nuclear division (mitosis or meiosis). In animal cells cytokinesis involves constriction of cytoplasm between daughter nuclei; in plant cells it involves formation of a new plant cell wall.

cytokinin /sȳ-toh-**kȳ**-nin/ One of a class of plant hormones concerned with the stimulation of cell division, nucleic acid metabolism, and root-shoot interactions.

Cytokinins are often purine derivatives: e.g. *kinetin* (6-furfuryl aminopurine), an artificial cytokinin commonly used in experiments; and zeatin, found in maize cobs.

Cytokinins are produced in roots, where they stimulate cell division. They are also transported from roots to shoots in the transpiration stream, where they are essential for healthy leaf growth. Subsequent movement from the leaves to younger leaves, buds, and other parts may occur in the phloem and be important in sequential leaf senescence up the stem. Senescence of detached leaves can be delayed by adding kinins, which mobilize food from other leaf parts and preserve green tissue in their vicinity.

Cytokinins promote bud growth, working antagonistically to auxins in causing bud regeneration in tobacco callus tissue and in releasing lateral buds from apical dominance. They work synergistically with auxins and gibberellins in stimulating cambial activity. The richest sources of cytokinins have been fruit and endosperm tissues, notably coconut milk.

cytology /sȳ-**tol**-ŏ-jee/ The study of cells; cell biology.

cytolysis /sȳ-**tol**-ă-sis/ The destruction of cells, usually by the breakdown of their cell membranes.

cytoplasm /sȳ-tŏ-**plaz**-ăm/ The living contents of a cell, excluding the nucleus and large vacuoles, in which many metabolic activities occur. It is contained within the plasma membrane and comprises a colorless substance (*hyaloplasm*) containing organelles and various inclusions (e.g. crystals and insoluble food reserves). The cytoplasm is about 90% water. It is a true solution of ions (e.g. potassium, sodium, and chloride) and small molecules (e.g. sugars, amino acids, and ATP); and a colloidal solution of large molecules (e.g. proteins, lipids, and nucleic acids). It can be gel-like, usually in its outer regions, or sol-like. *See* organelle; protoplasm.

cytoplasmic inheritance /sȳ-tŏ-**plaz**-mik/ The determination of certain characters by genetic material contained in plasmids or organelles other than the nucleus, e.g. mitochondria and chloroplasts. Characters controlled by the DNA of extranuclear organelles are not inherited according to Mendelian laws and are transmitted only through the female line, since only the female gametes have an appreciable amount of cytoplasm. Cytoplasmic inheritance is known in a wide variety of animals, plants, and unicellular organisms, e.g. *Paramecium*.

cytoplasmic streaming *See* cyclosis.

cytosine /sÿ-tŏ-seen, -sin/ A nitrogenous base found in DNA and RNA. Cytosine has the pyrimidine ring structure.

cytoskeleton /sÿ-tŏ-**skel**-ĕ-tŏn/ A network of fibers within the cytoplasm of a cell that maintains its shape, enables movement of the cell, and provides anchorage and movement of its organelles. It comprises various elements, including MICROTUBULES, MICROFILAMENTS, and INTERMEDIATE FILAMENTS.

cytosol /sÿ-tŏ-sol, -sohl, -zol, -zohl/ The soluble fraction of cytoplasm remaining after all particles have been removed by centrifugation.

cytotaxonomy /sÿ-tŏ-taks-**on**-ŏ-mee/ The use of chromosome number, size, and shape in the classification of organisms. *See also* taxonomy.

![D]

2,4-D (2,4-dichlorophenoxyacetic acid) A synthetic auxin used as a potent selective weedkiller. Monocotyledenous species with narrow erect leaves (e.g. cereals and grasses) are generally resistant to 2,4-D while dicotyledenous plants are often very susceptible. The compound is thus particularly effective in controlling weeds in cereal crops and lawns. *See* auxin.

Daltonism /**dawl**-tŏn-i-zăm/ *See* color blindness.

dance of the bees A precise pattern of movements performed by returning forager (worker) honeybees in order to indicate the direction and distance of a food source to other workers in the hive. The dance is performed on the vertical surface of the comb and two types are recognized: the round dance, performed when food is nearby, and the waggle dance, when food is further away. In the waggle dance the bee moves in a figure 8; the angle between the 8 and the vertical indicates direction.

Daphnia /**daf**-nee-ă/ A genus of tiny crustaceans – water fleas – common universally in ponds and ditches. *Daphnia* has a laterally compressed body (often reddish due to the presence of hemoglobin). The head bears a single compound eye and large forked antennae, by means of which *Daphnia* swims in a characteristic jerky manner. The thorax is entirely encased in a large transparent carapace. *Daphnia* is omnivorous and the fringed thoracic limbs are used in filter feeding. The eggs are fertilized in a thoracic brood pouch, which is shed at the next molt. Parthenogenesis occurs in favorable conditions. *See also* Branchiopoda.

dark reactions The traditional name for a group of reactions in PHOTOSYNTHESIS that follow the 'light reactions' and form glucose and other reduced products from carbon dioxide. The name is misleading since several key enzymes are inactive without light, and reduction of carbon dioxide does not occur in darkness.

Darwinism /**dar**-win-i-zăm/ The theory named for the British naturalist Charles Robert Darwin (1809–82). Darwin's explanation of the mechanism of evolutionary change, namely, that in any varied population of organisms only the best adapted to that environment will tend to survive and reproduce. Individuals that are less well adapted will tend to perish without reproducing. Hence the unfavorable characteristics, possessed by the less well-adapted individuals, will tend to disappear from a species, and the favorable characteristics will become more common. Over time the characteristics of a species will therefore change, eventually resulting in the formation of new species. The main weakness of Darwin's theory was that he could not explain how the variation, which natural selection acts upon, is generated, since at the time it was believed that the characteristics of the parents become blended in the offspring. This weakness was overcome with the discovery of Mendel's work and its description of particulate inheritance. *See also* neo-Darwinism; pangenesis.

Darwin's finches The fourteen types of finches, first described by Charles Darwin, unique to the Galápagos Islands of the equatorial Pacific, but related to finches on the South American continent from which they evolved. Of particular interest is the amount of adaptive radiation they display, and the fact that they occupy ecological niches that, in other parts of the world, are occupied by different birds, e.g. woodpeckers.

dating techniques Methods used in determining the age of rocks, fossils, or archaeological remains. There are two main methods. *Relative dating* assesses the age of a specimen in comparison to other specimens. *Absolute dating* involves assessing the actual age of a specimen by using some reliable measure of time. *See also* dendrochronology; radioactive dating.

day-neutral plant A plant that requires no particular photoperiod to flower. *See* photoperiodism.

DDT /dee-dee-**tee**/ (**dichlorodiphenyl-trichloroethane**) An organochlorine insecticide introduced in the late 1930s and subsequently widely used to control insect carriers of diseases, such as malaria, typhus, and yellow fever. However, its persistence in the environment led to widespread poisoning of certain wild predators, especially hawks and other birds of prey, which accumulated high concentrations of DDT in their body tissues. It has now been banned in many countries. *See* pollution.

deamination /dee-an-ă-**may**-shŭn/ A type of chemical reaction in which an amino group (NH_2) is removed. It occurs in animals when excess amino acids are to be excreted, by the action of deaminating enzymes (in the liver and kidneys of mammals). Depending on the type of organism, ammonia produced by the reaction may be excreted directly or first converted to urea or to uric acid.

Decapoda /de-**kap**-ŏ-dă/ The order that contains the most specialized members of Crustacea. The prawns (e.g. *Palaemon*), shrimps (e.g. *Crangon*), lobsters (e.g. *Homarus*), and crawfish (e.g. *Astacus*) all have a long abdomen ending in a tail for swimming backwards. The crabs (e.g. *Cancer*) have a much reduced abdomen. The

head and thorax are characteristically fused and covered with a carapace. There are five pairs of walking legs, the first and second pairs often having pincers (chelae) used in feeding and defense.

decarboxylase /dee-kar-**boks**-ă-layss/ An enzyme that catalyzes the decarboxylation of carboxylic acids, including the conversion of amino acids to amines.

decidua /di-**sij**-oo-ă/ The thickened and modified endometrium that lines the uterus during pregnancy in most mammals. It is expelled shortly after birth of young. *See* afterbirth.

deciduous /di-**sij**-oo-ŭs/ Denoting plants that seasonally shed all their leaves, for example before the winter, or dry season. It is an adaptation to prevent excessive water loss by transpiration when water is scarce. *Compare* evergreen.

deciduous teeth (**milk teeth**) The first set of teeth of a mammal. They are temporary and soon fall out, to be replaced by the permanent set. The milk teeth are smaller than the permanent teeth, and are fewer in number since there are no molars. *See also* diphyodont.

decomposer An organism that feeds upon dead organisms breaking them down into simpler substances. Decomposers recycle nutrients making them available to producer organisms. Bacteria and fungi belong to the decomposer group of an ecosystem. *See also* trophic level.

deficiency disease A disease caused by deficiency of a particular essential nutrient (such as a vitamin or a trace element), usually with a characteristic set of symptoms. Green plants, being autotrophic, are only likely to suffer mineral deficiency diseases, whereas animals, which are heterotrophic, are susceptible to a wider range of such diseases due to dietary deficiencies. *See* chlorosis; micronutrient; vitamins.

definite inflorescence *See* cymose inflorescence.

definitive nucleus *See* polar nuclei.

degeneration 1. Evolution to an apparently simpler structural form. It is seen in the wings of the flightless birds (e.g. emus). Such organs are said to be vestigial. It is also common in parasites; for example, parasitic protozoa possess few organelles compared with their free-living relatives. **2.** Death and deterioration of cells, nerve fibers, etc.

deglutition /dee-gloo-**tish**-ŏn/ (**swallowing**) The action of passing food through the mouth and pharynx into the esophagus. It is initiated voluntarily by movements of the tongue, which pass food backwards into the pharynx. This then triggers a complex reflex in which breathing is inhibited – the soft palate is raised against the internal nasal openings and the epiglottis is lowered over the tracheal entrance – and the muscles of the pharynx are contracted to force the food into the esophagus, where peristalsis begins.

dehiscent /di-**hiss**-ĕnt/ Describing a fruit or fruiting body that opens at maturity to release the seeds or spores. Dehiscence is often violent to aid seed dispersal. *Compare* indehiscent.

dehydration /dee-hȳ-**dray**-shŏn/ (*Microscopy*) A process followed when tissues are prepared for permanent microscope slides. Water is removed by immersing the tissue in increasing strengths of ethyl alcohol. The alcohol concentration must be increased gradually as otherwise the cells would dehydrate too quickly and shrink. Dehydration is necessary because water does not mix with the chemicals used in cleaning and mounting sections.

dehydrogenase /dee-hȳ-**droj**-ĕ-nayss/ An enzyme that catalyzes the removal of certain hydrogen atoms from specific substances in biological systems. Hydrogenases are usually called after the name of their substrate, e.g. lactate dehydrogenase. Some dehydrogenases are highly specific, both with respect to their substrate and coenzyme, whilst others catalyze the oxidation of a wide range of substrates. Many require the presence of a coenzyme, which is often involved as a hydrogen acceptor. Dehydrogenases catalyze the transfer of two hydrogen atoms from substrates to NAD and NADP.

deletion /di-**lee**-shŏn/ *See* chromosome mutation.

denaturation /dee-nay-chŭ-**ray**-shŏn/ A process that causes unfolding of the peptide chain of proteins or of the double helix of DNA. These changes may be brought about by a variety of physical factors: change in pH, temperature, violent shaking, and radiation. The primary structure remains intact. Denatured proteins and nucleic acids show changes in physical and biological properties; proteins, for example, are often insoluble in solvents in which they were originally soluble.

dendrite /**den**-drȳt/ One of several slender branching projections that arise from or near the cell body of a nerve cell (NEURON) to make contact with other nerve cells. The entire array of dendrites is sometimes called the *dendritic tree.*

dendrochronology /den-droh-krŏ-**nol**-ŏ-jee/ A method of archaeological dating by the annual rings of trees, used when the lifespans of living and fossil trees in an area overlap. Exact dates for sites can be calculated and the method is more accurate than radioactive dating techniques. Bristlecone pines, which can live for up to 5000 years, have been used in such work.

dendrogram /**den**-drŏ-gram/ A branched diagram used in taxonomy to demonstrate relationships between species, families, etc., or used to show the relationships between individuals, as in family trees. A cladogram (*see* cladistics) is a type of dendrogram.

denitrification /dee-nȳ-tră-fă-**kay**-shŏn/ The chemical reduction of nitrate by soil bacteria. The process is important in terms of soil fertility since the products of denitrification (e.g. nitrites and ammonia) cannot be used by plants as a nitrogen source. *Compare* nitrification. *See* nitrogen cycle.

dental formula A formula that shows the number of teeth of each type on the upper and lower jaws of a mammal. The initial letter of each type of tooth – i for incisor, c for canine, pm (or p) for premolar, and m for molar – is followed by the number of teeth of that type on one side of the jaw. The number on the upper jaw is put above a horizontal line, with the number on the lower jaw below it. The total number of teeth is found by adding all the numbers together and multiplying by two.

dentary /den-tă-ree/ A tooth-bearing membrane bone. In mammals it is the lower jaw bone, consisting of a single membrane bone on each side, fused together in front.

denticle /den-tă-kăl/ (**placoid scale**) A small toothlike structure in the skin of cartilaginous fish, e.g. dogfish or shark. Each has a spine that projects backwards through the skin and consists of an outer layer of enamel covering dentine round a central pulp cavity. The spine is attached to a flat base of bonelike substance beneath the skin. The teeth of these fishes are strong denticles with five pointed spines. Denticles are similar in structure and origin to the teeth of higher vertebrates and are homologous with them.

dentine /den-teen, -tin/ A hard substance, closely resembling bone, that makes up the bulk of a tooth. It has a higher mineral content than bone and is perforated by fine canals (*canaliculae*) containing cytoplasmic processes of the cells (*odontoblasts*) that line the pulp cavity. Unlike bone, dentine has no system of Haversian canals, lacunae, or cells within its substance. *See illustration at* teeth.

dentition /den-tish-ŏn/ The number, type, arrangement, and physiology of teeth in any given species. *See also* dental formula.

deoxycorticosterone /dee-oks-ă-kor-tă-kos-tĕ-rohn/ (**deoxycortone**) A steroid hormone, produced by the adrenal cortex, having mineralocorticoid activity. *See* mineralocorticoid.

deoxyribonuclease /dee-oks-ă-rÿ-boh-new-klee-ayss/ *See* DNase.

deoxyribonucleic acid /dee-oks-ă-rÿ-boh-new-klee-ik/ *See* DNA.

deoxy sugar /dee-oks-ee/ A sugar in which oxygen has been lost by replacement of a hydroxyl group (OH) with hydrogen (H). The most important example is deoxyribose, the sugar component of DNA.

depolarization /dee-poh-lă-ri-zay-shŏn/ A reduction in the potential difference that exists across the membrane of a nerve or muscle cell; i.e. a reduction in the resting potential. It occurs during the passage of an impulse when the membrane becomes more permeable to ions, which previously have accumulated on one side and caused the difference. The ions diffuse through ion channels in the membrane, tending to equalize their concentration on both sides.

derived trait *See* apomorphy.

dermal bone /der-măl/ *See* membrane bone.

Dermaptera /der-map-tĕ-ră/ A small order of nocturnal insects – the earwigs (e.g. *Forficula*). Earwigs have a long body covered with a hard shiny exoskeleton and are omnivorous, with biting and sucking mouthparts. The thin transparent hindwings are folded in a complicated way and covered by the short scaly fore wings (*elytra*) when resting. The forceps-like cerci at the end of the abdomen aid in folding the wings as well as being used in attack and defense.

dermatogen /der-mat-ŏ-jĕn/ (**protoderm**) *See* histogen theory.

dermatome /der-mă-tohm/ That part of the mesodermal somites of vertebrate embryos that underlies the epidermis and develops into dermis. *See also* mesoderm.

dermis /**der**-mis/ (**cutis**) The inner layer of the skin of a vertebrate animal, beneath the epidermis. It consists of connective tissue in which are embedded small blood vessels, nerve fibers, nerve endings (sensitive to heat, cold, pain, and pressure), and sweat glands (in mammals only). Beneath the dermis is the subcutaneous tissue. *See illustration at* skin.

desmosomes /**dess**-mŏ-sohm/ Patchlike CELL JUNCTIONS found in vertebrate tissues that serve to attach neighboring cells and may help distribute mechanical stresses through tissues. *Spot desmosomes* act like 'spot welds' between cells. These consist of linker proteins spanning the intercellular space and joining together protein plaques on the inner faces of adjoining plasma membranes. Inside the cell the plaques appear to connect with keratin intermediate filaments of the cytoskeleton. *Hemidesmosomes* are structurally similar to spot desmosomes but anchor the cell to the extracellular matrix rather than to other neighboring cells.

determinate growth /di-**ter**-mă-nit/ *See* growth.

determined Describing embryonic tissue whose developmental possibilities are restricted. For example the neural plate is determined to form nervous tissue and can no longer make epidermis.

detritivore /di-**trÿ**-tă-vor, -vohr/ An animal that feeds on detritus – particles of organic matter derived from dead and decomposing organisms. Examples of detritivores include earthworms, some beetles, pillbugs, and gastropod mollusks. Detritivores are thus extremely important DECOMPOSERS.

deuterated compound A compound in which one or more ^1H atoms have been replaced by deuterium (^2H) atoms.

deuterium /dew-**teer**-ee-ŭm/ Symbol: D, ^2H A naturally occurring stable isotope of hydrogen in which the nucleus contains one proton and one neutron. The atomic mass is thus approximately twice that of ^1H; deuterium is known as 'heavy hydrogen'. Chemically it behaves almost identically to hydrogen, forming analogous compounds, although reactions of deuterium compounds are often slower than those of the corresponding ^1H compounds. This is made use of in kinetic studies where the rate of a reaction may depend on transfer of a hydrogen atom (i.e. a kinetic isotope effect).

Deuteromycota /dew-ter-oh-mÿ-**koh**-tă/ (**Fungi Imperfecti**) A phylum of so-called 'imperfect' fungi in which sexual reproduction is unknown or has been lost during evolution. Its members, for example the *Penicillium* molds, are assigned as a taxonomic convenience, rather than by any true criteria, and some authorities prefer to allocate the deuteromycotes to either the ascomycetes (the majority) or the basidiomycetes, on the basis of available evidence. Indeed, *Penicillium* is now known to have a sexual stage, formerly regarded as a quite distinct ascomycete fungus (*Talaromyces*).

deutoplasm /**dew**-tŏ-plaz-ăm/ Yolk, or the yolk-laden cytoplasm at the vegetal pole of many eggs.

Devonian /di-**voh**-nee-ăn/ The geological period known as the 'Age of Fish', some 405–355 million years ago, between the Silurian and the Carboniferous periods of the Paleozoic era. It was characterized by an enormous number and variety of fish, most of which have become extinct without leaving any modern relatives. During the late Devonian, primitive amphibians were evolving from crossopterygians (lobe-finned fish). Vascular land plants appeared, such as the psilophytes and pteridophytes, while terrestrial fauna included insects and spiders. The period is named after rocks found in Devon. *See also* geological time scale.

dextrin /**deks**-trin/ Any of a class of intermediates produced by the hydrolysis of starch. Further hydrolysis eventually produces the monosaccharide glucose.

dextrorotatory /deks-troh-**roh**-tă-tor-ee, -toh-ree/ Describing compounds that rotate the plane of polarized light to the right (clockwise as viewed facing the oncoming light). *Compare* levorotatory. *See* optical activity.

dextrose /deks-trohs/ (**grape sugar**) Naturally occurring glucose belongs to the stereochemical series D and is dextrorotatory, indicated by the symbol (+). Thus the term 'dextrose' has been traditionally used to indicate D-(+)-glucose. As other stereochemical forms of glucose have no significance in biological systems the term 'glucose' is often used interchangeably with dextrose. *See also* glucose.

diabetes /dy-ă-**bee**-tis, -teez/ (**diabetes mellitus**) A condition caused by deficiency of the hormone insulin and characterized by large quantities of glucose in the blood and urine. The volume of urine also increases. Diabetes that starts early in life is usually more severe (*insulin-dependent diabetes*); such patients require regular injections of insulin. A mild form of diabetes (*noninsulin-dependent diabetes*) is also common in middle-aged to elderly overweight people. In such patients, insulin is not usually required and the condition may be treated by weight reduction, dietary control, and (sometimes) the administration of drugs to lower the blood-glucose level.

diageotropism /dy-ă-jee-**ot**-rŏ-piz-ăm/ A geotropic response in which the direction of growth is horizontal. *See* geotropism; tropism.

diakinesis /dy-ă-ki-**nee**-sis, -ky-/ The last stage of the prophase in the first division of meiosis. Chiasmata are seen during this stage, and by the end of diakinesis the nucleoli and nuclear membrane have disappeared.

dialysis /dy-**al**-ă-sis/ A technique for separating compounds with small molecules from compounds with large molecules by selective diffusion through a semipermeable membrane. For example, a mixed solution of starch (large molecules) and

glucose (small molecules) is placed in a bag or piece of tubing made of thin cellophane or other suitable material. If the container is put in water, the glucose molecules diffuse out, leaving the starch behind.

Dialysis is performed naturally by the kidneys to extract wastes from the blood, or, in the case of lost or damaged kidneys, artificially by machine.

diapause /dy-ă-pawz/ A period of dormancy in the life cycle of some insects during which growth and development cease and metabolism is greatly decreased. It is often seasonal, e.g. hibernation, or in some cases may last for several years. It results from changes in hormone production and enables the insect to survive unfavorable environmental conditions.

diaphragm /dy-ă-fram/ A dome-shaped sheet of muscle and tendon that completely divides the body cavity of a mammal into two parts, thorax and abdomen. The muscles are at the rim of the diaphragm and when they contract the diaphragm is made flatter. This increases the volume of the thorax, reducing pressure there and causing air to be drawn into the lungs. When the muscles relax, the diaphragm recovers its domed shape and air is expired. The esophagus, dorsal aorta, and posterior vena cava pass through the diaphragm. Contractions of the diaphragm are controlled by the respiratory center in the medulla of the brain, via the phrenic nerves. *See illustration at* alimentary canal.

diaphysis /dy-**af**-ă-sis/ The shaft of a long bone in mammals. It contains the primary center from which the majority of a bone is ossified: the bone ends (epiphyses) are formed separately. *See also* epiphyses.

diastole /dy-**as**-tŏ-lee/ The phase of the CARDIAC CYCLE when the heart muscle is relaxed and the chambers fill with blood.

diatom /dy-ă-tom, -tŏm/ A unicellular alga of the phylum Bacillariophyta, found in freshwater, the sea, and soil. Much of plankton is composed of diatoms and they are thus important in food chains. They have silica cell walls (frustules) composed of two valves ornamented with perforations, which are arranged differently in each species. Diatoms are typically pill-box shaped (*centric*) or coffin-shaped (*pennate*). The chloroplasts contain chlorophyll *a* and *c*, carotenes, and xanthophylls. The principal storage product is the oil chrysolaminarin. Diatoms reproduce asexually by binary fission producing successively smaller generations until size is restored through sexual reproduction by auxospores.

dicaryon /dȳ-**ka**-ree-on/ *See* dikaryon.

dichasial cyme /dȳ-**kay**-zhee-ăl, -zee-/ (**dichasium**) A cyme in which each branch gives rise to two other branches. *See* cymose inflorescence.

dichogamy /dȳ-**kog**-ă-mee/ The condition in which the anthers and stigmas mature at different times thus helping prevent self pollination. *Compare* homogamy. *See* protandry; protogyny.

dichotomy /dȳ-**kot**-ŏ-mee/ Forked branching produced by division of the growing point into two equal parts, seen for example in *Fucus*.

Dicotyledoneae /dȳ-ko-tă-lee-**doh**-nee-ee/ The former name for the MAGNOLIOPSIDA.

Dictyoptera /dik-tee-**op**-ter-ă/ The order of insects that contains the cockroaches (e.g. *Periplaneta*). Cockroaches are cosmopolitan nocturnal omnivorous insects; they are pests and can spread disease in dirty places. They have a flattened body enabling them to hide in crevices and the hardened fore wings protect the larger delicate hind wings. They seldom fly. The eggs are laid in capsules (*oothecae*), which may be carried about by the female. The young resemble adults but are wingless.

dictyosome /**dik**-tee-ŏ-sohm/ A cell organelle characteristic of plant cells and consisting of a short stack of flattened membranous vesicles. Dictyosomes receive proteins from the cell's endoplasmic reticulum and modify and package these in vesicles for despatch to destinations inside the cell, including the plasma membrane for secretion from the cell surface. Dictyosomes may also polymerize sugars to polysaccharides. The GOLGI APPARATUS found in animal cells is an aggregation of numerous dictyosomes, but this rarely occurs in plant cells.

dictyostele /**dik**-tee-ŏ-steel, -stee-lee/ A modified solenostele that is broken up by large leaf gaps so crowded together that they overlap. The tube of a stelar tissue is thus broken up into a mesh, each small segment of remaining vascular tissue being called a *meristele*. It is found in certain fern stems (e.g. *Dryopteris*). *See* solenostele.

diestrus /dȳ-**es**-trŭs/ *See* estrous cycle.

differentially permeable membrane *See* osmosis.

differentiation /dif-ĕ-ren-shee-**ay**-shŏn/ A process of change during which cells with generalized form become morphologically and functionally specialized to produce the different cell types that make up the various tissues and organs of the organism. Differentiation has been best studied in experimental organisms, such as the fruit fly *Drosophila*. Here, proteins called *morphogens*, encoded by maternal genes of follicle cells, diffuse into the developing early embryo where they lay the foundations of the general body plan. Gradients of concentration of the various morphogens cause genes in different zones of the embryo to be activated to different extents, creating a rudimentary pattern of body segments. This pattern is reinforced and refined by the embryo's own genes – the so-called *segment genes*. Within each segment the differentiation of limbs and other appendages is controlled by a class of master genes, called *homeotic genes*. Mutations of these in *Drosophila* can result in, for example, legs developing on the head instead of antennae. These homeotic genes show remarkable similarities in base sequence across a wide range of species, from plants to humans, and produce a protein that binds to DNA, acting as a switch for various other genes. *See also* totipotency.

diffuse-porous /di-**fyooss**/ Describing wood in which vessels of approximately equal diameter tend to be evenly distributed so there is no obvious growth ring. Diffuse-porous wood is seen for example in birch. *Compare* ring-porous.

diffusion pressure deficit (DPD) *See* osmosis.

digestion The breakdown of complex organic foodstuffs by enzymes into simpler soluble substances, which can be absorbed and assimilated by the tissues. In most animals (e.g. vertebrates and arthropods) it is extracellular, occurring in an alimentary canal or gut into which the enzymes are secreted. In simpler animals (e.g. protoctists, cnidarians, and some other invertebrates) it is intracellular, with solid particles being engulfed and digested by ameboid cells. *See also* endocytosis; phagocyte.

digit A finger or toe. There are typically five digits terminating the limbs of tetrapods, each made up of a series of small bones (phalanges). However, there are reductions and modifications to this general plan. Some species have digits bearing nails, claws, or hooves distally. *See also* pentadactyl limb; hallux; pollex.

digitigrade /**dij**-ă-tă-grayd/ Describing the mode of progression in some mammals in which only digits (fingers or toes) are in contact with the ground. It is seen in dogs, cats, and most fast-running animals. *Compare* plantigrade; unguligrade.

dihybrid /dÿ-**hÿ**-brid/ A hybrid heterozygous at two loci and obtained by crossing homozygous parents with different alleles at two given loci: for example, Mendel's cross between yellow round (YYRR) and green wrinkled (yyrr) garden peas to give a yellow round dihybrid (YyRr). When a dihybrid is selfed a characteristic *dihybrid ratio* of 9:3:3:1 is obtained in the offspring. Nine plants exhibit both dominant characters, six plants show one dominant and one

recessive character, and one plant exhibits both recessives. *Compare* monohybrid.

dikaryon /dÿ-ka-ree-on/ (**dicaryon**) A cell containing two different nuclei, arising from the fusion of two compatible cells, each with one nucleus. The nuclei do not fuse immediately, instead dividing independently, but simultaneously. The term is usually applied to fungal mycelia, notably of ascomycetes and basidiomycetes.

dimorphism /dÿ-**mor**-fiz-ăm/ The existence of two different forms of an organism. An example is sexual dimorphism in animals. *See* polymorphism.

dinitrogenase *See* nitrogenase.

dinoflagellate /dÿ-noh-**flaj**-ĕ-lit, -layt, dÿ-noh-flă-**jel**-ayt/ A marine or freshwater protoctist of the phylum Dinoflagellata (Dinomastigota) that swims in a twirling manner by means of two undulipodia (flagella). These lie at right angles to each other in two grooves within the organism's rigid body wall (test). Many possess stinging organelles that they discharge to catch prey; and some produce potent toxins that are capable of killing fish. Roughly half of all known dinoflagellates are capable of photosynthesis.

dinosaur /**dÿ**-nŏ-sor/ Any of the large extinct terrestrial reptiles that existed during the Mesozoic era. Dinosaurs of the order Ornithischia were typically herbivorous quadripedal reptiles, such as *Stegosaurus* and *Triceratops*; the order Saurischia included bipedal carnivores, such as *Tyrannosaurus*, as well as quadripedal herbivores, such as *Apatosaurus* (*Brontosaurus*). Dinosaurs were a very successful and diverse group, dominating the terrestrial environments of the earth for 140 million years. There are many theories to explain their extinction at the end of the Cretaceous period, the most popular being major climatic changes induced by continental shift or by the impact of a giant meteorite on the earth's surface.

dinucleotide /dȳ-**new**-klee-ŏ-tȳd/ A compound of two nucleotides linked by their phosphate groups. Important examples are the coenzymes NAD and FAD.

dioecious /dȳ-**ee**-shŭs/ Denoting a plant species in which male and female flowers are borne on separate individuals. *Compare* hermaphrodite; monoecious.

diphycercal tail /**dif**-ă-ser-kăl/ The type of tail found in lampreys, hagfish, the young stages of all fish, and the larvae of such amphibians as frogs and toads. The vertebral column extends into the tail, which has caudal fins of equal size above and below it. *Compare* heterocercal tail; homocercal tail.

diphyodont /**dif**-ee-ŏ-dont/ A type of dentition (found in most mammals) in which an animal has two sets of teeth in succession. The milk (or deciduous) teeth of the young animal fall out and are replaced by the permanent teeth, which are larger and more numerous since there are molars as well as premolars. *Compare* monophyodont; polyphyodont.

dipleurula /dȳ-**ploor**-yŭ-lă/ The hypothetical free-swimming larva of early echinoderms. It is believed to have been flattened and bilaterally symmetrical, with the digestive canal opening to the exterior by a mouth and anus and the coelom opening to the dorsal surface by a narrow tube. It had large lobes on its body, fringed with bands of cilia. These were used for locomotion and for feeding. This basic form has been modified in existing echinoderms. *Compare* trochophore. *See* bipinnaria; pluteus; tornaria.

diplobiontic /dip-lŏ-bȳ-**on**-tik/ Describing life cycles showing a typical alternation of generations with haploid and diploid somatic bodies. Ferns are diplobiontic organisms. *Compare* haplobiontic.

diploblastic /dip-lŏ-**blas**-tik/ Describing animals in which the body wall consists of two layers, an ectoderm and an endoderm, separated by mesoglea. The Cnidaria are diploblastic animals.

diploid /**dip**-loid/ A cell or organism containing twice the haploid number of chromosomes (i.e. 2n). In animals the diploid condition is generally found in all but the reproductive cells and the chromosomes exist as homologous pairs, which separate at meiosis, one of each pair going into each gamete. In plants exhibiting an alternation of generations the sporophyte is diploid, while higher plants are normally always diploid. Exceptions are those species in which polyploidy occurs.

diplont /**dip**-lont/ A diploid organism that represents the vegetative stage in life cycles in which haploidy is restricted to the gametes. *Diplontic* life cycles are found in most organisms. *Compare* haplont.

diplontic /dă-**plon**-tik/ *See* diplont.

Diplopoda /dă-**plop**-ŏ-dă/ The class of arthropods that contains the millipedes (e.g. *Julus*), characterized by a cylindrical body divided into numerous segments each bearing two pairs of walking legs. Millipedes are terrestrial herbivores, breathing air through tracheae. Excretion is by Malpighian tubules. They are sometimes placed with the centipedes (CHILOPODA) in the group MYRIAPODA.

diplotene /**dip**-lŏ-teen/ In meiosis, the stage in late prophase I when the pairs of chromatids begin to separate from the tetrad formed by the association of homologous chromosomes. Chiasmata can often be seen at this stage.

Dipnoi /**dip**-noi/ The order of bony fish that contains the lungfishes. Found in fresh water in areas subject to seasonal drought, they are characterized by a functional lung and internal as well as external nostrils, allowing air-breathing at the water surface without opening the mouth. The young breathe by temporary external gills. *Neoceratodus* of Queensland rivers is the most primitive and cannot live out of water. It has paddle-shaped rayed fins, a single lung, and very large scales. *Protopterus* of the Nile and *Lepidopterus* of South America have rayless whiplike fins, paired lungs, smaller scales, and survive when rivers dry up completely by living in the mud, leaving a small opening for breathing. The heart and blood system are adapted for pulmonary respiration and resemble those of amphibians.

Diptera /**dip**-ter-ă/ The order of insects that contains the flies, characterized by only one pair of wings (balancing organs (*halteres*) replace the hind wings). The adults have sucking or piercing mouthparts and feed on plant juices, decaying organic matter, or blood. The legless larvae (*maggots*) feed on plants, decaying organic matter, or are carnivorous. Metamorphosis is complete and the pupa is often protected by a barrel-shaped *puparium*. Many flies are medically important as the carriers of various diseases; for example the *Anopheles* mosquito transmits malaria.

disaccharide /dӯ-**sak**-ă-rӯd, -rid/ A sugar with molecules composed of two monosaccharide units. Sucrose and maltose are examples. These are linked by a –O– linkage (*glycosidic link*). *See also* sugar.

discontinuous replication *See* replication.

discontinuous variation *See* qualitative variation.

displacement activity A type of behavior that appears irrelevant to the situation in which it is performed. It occurs when an animal is torn between equal and opposing tendencies, such as threat and escape, and often consists of comfort actions, e.g. grooming, eating, or scratching. For example, fighting cocks may stop and peck at the ground as if feeding, and fighting herring gulls may stop and pluck at nest material for a few moments. Displacement activity occurs when there is conflict between antagonistic drives in animals fighting at a boundary between their territories, where the drive to escape and the fighting drive are both aroused. It may also occur when the animal has a strong urge that cannot be fulfilled.

display behavior Activities such as movements, postures, sounds, etc., that are used by an animal to communicate specific information to another, especially one of the same species. Display behavior is most frequently seen in courtship and aggression; for example, a male bird may puff out its feathers, bow or turn its head, and sing to attract a female at the beginning of the breeding season.

distal /**dis**-tăl/ Denoting the part of an organ, limb, etc., that is furthest from the origin or point of attachment. *Compare* proximal.

distely /**dis**-tĕ-lee/ Having two steles, e.g. the stem of *Selaginella kraussiana*. The steles are joined only at branches. *See also* polystely.

disulfide bridge /dӯ-**sul**-fӯd/ (**disulfide bond**) A covalent bond between two sulfur atoms, typically formed by oxidation between two sulfhydryl (–SH) groups. Such bonds can link neighboring cysteine residues within protein molecules, helping to stabilize the three-dimensional arrangement (i.e. tertiary structure) of the polypeptide chains.

diuresis /dӯ-yû-**ree**-sis/ The increased flow of urine. It may be due to excessive water in the body, or the administration of certain drugs (diuretics), for example. Diuresis is counteracted by secretion of the hormone VASOPRESSIN.

diurnal rhythm /dӯ-**er**-năl/ *See* circadian rhythm.

diverticulum /dӯ-ver-**tik**-yŭ-lŭm/ (*pl.* **diverticula**) A blind tubular or saclike outgrowth from a tube or cavity. For instance, the appendix and cecum of a rabbit form a diverticulum. The primitive chordate animal, *Branchiostoma*, has such an outgrowth at the point where the esophagus meets the intestine. It projects forwards beside the pharynx and may be homologous with the liver of vertebrates.

division One of the major groups into which the plant kingdom is classified: a taxonomic rank used instead of phylum in some plant classifications. Division names (and hence phylum names too) end in -*phyta* (e.g. Bryophyta, Filicinophyta). Divisions may be divided into subdivisions.

dizygotic twins /dÿ-zÿ-**got**-ik/ *See* fraternal twins.

DNA (deoxyribonucleic acid) A nucleic acid, mainly found in the CHROMOSOMES, that contains the hereditary information of organisms. The molecule is made up of two helical polynucleotide chains coiled around each other to give a *double helix*. Phosphate molecules alternate with deoxyribose sugar molecules along both chains and each sugar molecule is also joined to one of four nitrogenous bases – adenine, guanine, cytosine, or thymine. The two chains are joined to each other by bonding between bases. The sequence of bases along the chain makes up a code – the genetic code – that determines the precise sequence of amino acids in proteins (*see* messenger RNA; protein synthesis; transcription).

The shape of the DNA molecule is shown in the illustration. The two purine bases (adenine and guanine) always bond with the pyrimidine bases (thymine and cytosine), and the pairing is quite specific: adenine with thymine and guanine with cytosine.

DNA is the hereditary material of all organisms with the exception of RNA viruses. Together with RNA and histones it makes up the chromosomes of eukaryotic cells. *See* junk DNA; replication; selfish DNA. *See also* RNA.

DNA fingerprinting *See* genetic fingerprinting.

DNA hybridization *See* nucleic acid hybridization.

S – P   sugar-phosphate chain
≡   hydrogen bonds linking bases

The double helix of the DNA molecule

DNA polymerase *See* polymerase.

DNA probe (gene probe) A nucleic acid consisting of a single strand of nucleotides whose base sequence is complementary to that of a particular DNA fragment being sought, for example a gene on a chromosome or a restriction fragment in a DNA digest. The probe is labeled (e.g. with a ra-

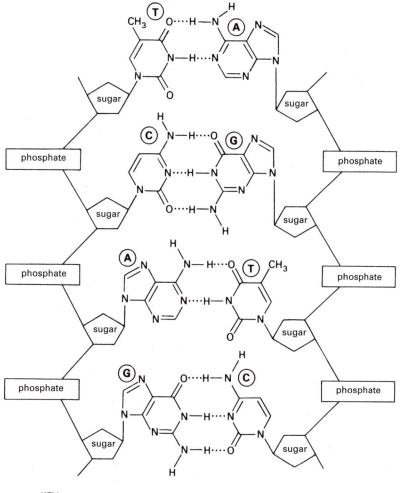

KEY

(A) adenine

(C) cytosine

(T) thymine

(G) guanine

Part of the structure of DNA showing
hydrogen bonding (dotted lines)
between complementary bases

DNA structure

dioisotope or a fluorescent compound) so that when it binds to the target sequence, both it and the target can be identified (by autoradiography or fluorescence microscopy).

DNA replication *See* replication.

DNase /dee-en-**ay**-ayss/ (**deoxyribonuclease**) Any enzyme that hydrolyzes the phosphodiester bonds of DNA. DNases are classified into two groups, according to their site of action in the DNA molecule (*see* endonuclease; exonuclease).

DNase /dee-en-**ay**-ayss/ (**deoxyribonuclease**) Any enzyme that catalyzes hydrolytic cleavage of the phosphodiester bonds joining the nucleotides along a strand of a DNA molecule. Some DNases cleave terminal nucleotides only, whereas others cleave at sites within the strand.

dogfish *See* Scyliorhinus.

domain 1. (*Taxonomy*) A category used in some classification systems to contain one or more related KINGDOMS of organisms. Studies in molecular systematics have led some authorities to organize the living world into three domains: Eubacteria (sometimes called simply Bacteria); Archaebacteria (or Archaea); and Eukaryota (or Eukarya). It is argued that this better reflects the very ancient evolutionary 'three-way split' between these groups of organisms, two of which are prokaryotes (i.e. the two groups of bacteria) and the other consisting of eukaryotes (plants, animals, etc.).
2. (*Biochemistry*) A region of a protein molecule where different parts of the polypeptide chain are juxtaposed and folded to form a particular three-dimensional pattern with certain physical and chemical properties. These properties may relate to some function, such as providing a site for catalytic activity in an enzyme, or for DNA binding in a transcription factor. The nature of the domain is determined by the amino acid sequence of the polypeptide, and hence by the gene encoding the polypeptide.

dominance hierarchy (peck order) A strict hierarchy existing in many vertebrates that live in social groups in which each individual occupies a particular position that is recognized by others in the group. There is much fighting in the initial establishment of the hierarchy and the dominant animal emerges as the one that cannot be dominated by any other. After establishment a hierarchy is usually stable as subordinates avoid threatening dominant animals and perform submissive actions to avoid being threatened themselves.

dominant An allele that, in a heterozygote, prevents the expression of another (recessive) allele at the same locus. Organisms with one dominant and one recessive allele thus appear identical to those with two dominant alleles, the difference in their genotypes only becoming apparent on examination of their progenies. The dominant allele usually controls the normal form of the gene, while mutations are generally recessive.

donor A person or animal that donates blood, tissues, or organs for use by another person or animal (the *recipient*).

dopamine /**doh**-pă-meen/ A catecholamine precursor of epinephrine and norepinephrine. In mammals it is found in highest concentration in the corpus striatum of the brain, where it functions as an inhibitory neurotransmitter. High levels of dopamine are associated with Parkinson's disease in humans.

dormancy A period of minimal metabolic activity of an organism or reproductive body. It is a means of surviving a period of adverse environmental conditions, e.g. cold or drought. Seeds, spores, cysts, and perennating organs of plants are potentially dormant structures. Biennial and perennial plants often lose their leaves and produce dormant buds, either in underground perennating organs or, in the case of woody plants, above ground. The onset and breaking of dormancy are normally controlled environmentally, factors such as day-length (*photoperiod*) and temperature being involved. Dormancy may be promoted by certain hormones (e.g. abscisic acid) and broken by other hormones (e.g. gibberellins). Animal dormancy may take several forms. *See* estivation; diapause; hibernation. *See also* after-ripening; photoperiodism.

dorsal 1. Designating the side of an animal furthest from the substrate, i.e. the upper surface. However, in bipedal animals, such as humans, the dorsal side is directed backwards corresponding to the posterior side of other animals.
2. (In the lateral organs of plants, e.g. leaves) Designating the lower or abaxial surface.
Compare ventral.

dorsal aorta In mammals, the descending portion of the aorta, which carries oxygenated blood to the trunk and hind limbs. In fish it is formed from the fusion of the six aortic arches. In tetrapods it arises from the systemic arch.

dorsal root *See* spinal nerve.

dorsifixed /**dor**-să-fikst/ A stamen in which the filament is fused to the back of the anther. *Compare* basifixed; versatile.

double circulation *See* blood vascular system.

double fertilization The fusion of one pollen nucleus with the egg to form the zygote and of the other pollen nucleus with a polar nucleus to form the triploid endosperm nucleus. The process is restricted to certain angiosperms.

double helix *See* DNA.

double recessive An organism containing both recessive alleles of a particular gene and thus expressing the recessive form of the gene in its phenotype. Double recessives, being of known genotype, are often used in test crosses to establish whether the organism to which it is crossed is heterozygous or homozygous for the same gene. *See* back cross.

Down's syndrome A condition seen in humans, characterized by short stature and a rounded head with obliquely slanted eyes. Affected children may have learning difficulties. It is caused by the presence in all body cells of an extra chromosome 21, due to its nondisjunction at meiosis – a state termed *trisomy 21*. The syndrome is named for the English physician John Langdon Haydn Down (1826–96). *See also* nondisjunction.

DPD Diffusion pressure deficit. *See* osmosis.

dragonflies *See* Odonata.

drive A basic urge or motivation towards a particular goal. Drive usually arises as a result of some deficiency (e.g. hunger or thirst) or to satisfy some instinctive urge (e.g. reproduction), which impels the animal to act towards achievement of its goal.

Drosophila A genus of fruit flies containing the species *D. melanogaster*, which is widely used for research into genetics, development, and other areas of biology. Fruit flies are well suited to such work because they have a short lifespan and produce large numbers of offspring, and also their salivary glands contain giant (POLYTENE) chromosomes, which are amenable to microscopic study.

drupe (**pyrenocarp**) A fleshy fruit containing one or more seeds each surrounded by a hard stony wall, the endocarp. Drupes with one seed include plums and cherries while many-seeded drupes include holly and elder fruits. Blackberries and raspberries are collections of small drupes or drupelets.

ductus arteriosus /**duk**-tŭs ar-teer-ee-**oh**-sŭs/ A blood vessel present in embryo tetrapods that links the pulmonary artery with the aorta. It therefore enables blood to be shunted from the right ventricle into the systemic circulation and bypass the lungs. It is derived from the outer portion of the sixth aortic arch, closes at birth, and remains as a solid strand in the adult.

duodenum /dew-ŏ-**dee**-nŭm/ The first part of the small intestine into which the food passes when it leaves the stomach. It forms a loop (30 cm long in man) into which the duct from the pancreas and the bile duct from the liver open. The lining is covered with villi, between which are glands secreting intestinal juice (*succus entericus*) containing enzymes. When acid chyme from the stomach enters the duodenum, the lining cells secrete a hormone (cholecystokinin) that stimulates the pancreas to release pancreatic juice containing enzymes. Cholecystokinin also causes contraction of the gall bladder, resulting in the passage of bile into the duodenum. These alkaline secretions neutralize the acid from the stomach and continue the process of digestion. *See illustration at* alimentary canal.

duplex Double, or having two distinct parts. The term is particularly used to describe the double helix of the Watson–Crick DNA model.

duplication The occurrence of extra genes or segments of a chromosome in the genome. *See* chromosome mutation.

dura mater /**dewr**-ă **may**-ter/ The tough thick outer membrane that surrounds and protects the brain and spinal cord in vertebrates. *See* meninges.

duramen /dewr-**ay**-mĕn/ *See* heartwood.

dynein /**dy**-neen, -nee-in/ A large protein that uses energy from the hydrolysis of ATP to act as a motor and propel itself along a protein fibril. Some dyneins occur in the cell cytoplasm and serve to propel transport vesicles along microtubules within cells, or help in orienting and moving chromosomes on the spindle microtubules during mitosis. Other dyneins are involved in the beating of undulipodia (eukaryotic cilia and flagella), generating the sliding forces between the microtubules that run through the undulipodium shaft.

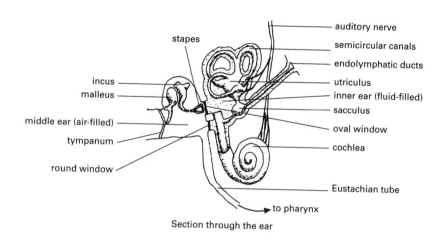

incus —
malleus —

middle ear (air-filled) —

tympanum —

round window —

stapes

auditory nerve
semicircular canals
endolymphatic ducts
utriculus
inner ear (fluid-filled)
sacculus
oval window
cochlea

Eustachian tube

to pharynx

Section through the ear

ear One of a pair of sense organs, located either side of the head in vertebrates, that are concerned with hearing and balance. In mammals the ear consists of an outer ear, separated by the tympanum (eardrum) from the middle ear, which communicates, via the oval window (fenestra ovalis) and round window (fenestra rotunda), with the inner ear. In fish only the inner ear is concerned solely with balance. *See* inner ear; middle ear; outer ear.

ear ossicles /oss-ă-kălz/ In mammals, three small bones in the middle ear – the malleus, incus, and stapes. They form a series of levers whereby vibrations induced by sound waves falling on the eardrum (tympanum) are transmitted to the oval window (fenestra ovalis) and so to the inner ear. They are homologous with certain jaw bones of lower vertebrates. *See also* columella auris; Weberian ossicles.

earthworms *See* Lumbricus.

earwigs *See* Dermaptera.

ecdysis /ek-dă-sis/ (**molting**) **1.** The periodic shedding of the rigid cuticle of arthropods, especially insects and crustaceans, to enable growth to occur. Some useful materials are reabsorbed from the old cuticle, which is then split along lines of weakness, revealing a soft new cuticle underneath. The animal then enlarges its body, by taking in air or water, so the new cuticle hardens a size larger than the old. Ecdysis is controlled by the hormone ecdysterone (β-ecdysone).
2. The periodic shedding of the outer epidermal layer of reptiles (except crocodiles). It is shed in a single piece by snakes, but in small patches by lizards.

A similar process occurs continuously in mammals, including humans, in which very small flakes of epidermis are shed.

ecdysterone /ek-**dis**-tĕ-rohn/ (**β-ecdysone**) A steroid hormone, produced by arthro-

115

pods (e.g. insects, spiders, and scorpions), that induces molting and metamorphosis. It acts on DNA to initiate the synthesis of new proteins and enzymes involved in the process of molting and cuticle formation. Ecdysterone is formed from its inactive precursor, ecdysone, which is secreted by the prothoracic glands.

ecesis /i-**see**-sis/ The germination and establishment of colonizing plants in an area, this being the first stage in a succession. *See* succession.

ECG *See* electrocardiogram.

Echinodermata /i-kỹ-nŏ-**der**-mă-tă/ A phylum of marine invertebrates containing the starfishes, sea urchins, feather stars, sea cucumbers, and brittle stars. Most echinoderms exhibit radial symmetry, with typically five rays extending from a central disk. All have calcareous skeletal plates and most have spines. Part of the coelom is modified as the WATER VASCULAR SYSTEM, which extends into hydraulic *tube feet*, used typically in locomotion. The water vascular system has an external opening, the *madreporite*. The nervous system is simple and there are no excretory organs. The development of the bilaterally symmetrical larvae shows affinities to the Chordata. The phylum is divided into five classes. *See* Asteroidea; Crinoidea; Echinoidea; Holothuroidea; Ophiuroidea. *See also* dipleurula.

Echinoidea /i-kỹ-**noi**-dee-ă/ The class of the Echinodermata that contains the sea urchins (e.g. *Echinus*), found on the sea bed or buried in sand. The spherical, heart-shaped, or flattened body lacks arms and is covered by a rigid shell (*test*) bearing movable spines used in locomotion and defense. The mouth, with a complicated jaw apparatus (*Aristotle's lantern*), is on the ventral surface.

echolocation /ek-oh-loh-**kay**-shŏn/ A method used by some animals (e.g. bats and dolphins) to locate objects. They emit high-pitched sounds, often inaudible to humans, which are reflected back off the object and detected by the ear or other sensory receptor.

eclosion hormone /I-**kloh**-zhŏn/ A peptide hormone secreted by neurosecretory cells in the brain of insects that stimulates the emergence of the adult from the pupa. It triggers events that lead to shedding of the cuticle in the final molt of the pupa.

ECM *See* extracellular matrix.

E. coli /ee koh-lÿ/ *See* Escherichia coli.

ecological niche The functional role of an organism in a community. If two species occupy the same niche then competition occurs until one has replaced the other. A similar niche may be occupied by different species in different areas, for example the fallow deer of Africa occupies the same niche as the red deer of Eurasia. Conversely one type of organism may evolve by adaptive radiation to fill several different niches, such as the finches of the Galápagos Islands. *See* Darwin's finches.

ecology The study of the relationships of organisms to one another and to their environment.

ecospecies /ee-kŏ-spee-sheez, -seez/ A group of ecotypes within a species, between which crossing is possible without loss of fertility in the offspring.

ecosystem (ecological system) A unit made up of all the living and nonliving components of a particular area that interact and exchange materials with each other. The concept of the ecosystem differs from that of the community in that more emphasis is placed on abiotic factors. Various studies have been made to attempt to itemize the energy flow of an entire ecosystem, taking into account factors such as incoming radiation, photosynthetic efficiency, etc.

ecotype /ee-koh-tÿp/ A group of organisms within a species adapted genetically to the combination of environmental factors in their habitat, but able to reproduce with other ecotypes belonging to the same species. Differences between ecotypes may be physiological or morphological. *Compare* biotype. *See also* adaptive radiation; speciation.

ectoderm /**ek**-tŏ-derm/ The germ layer of metazoans (including vertebrates) that remains on the outside of the embryo and develops into the epidermis and its derivatives (e.g. feathers, hairs, various glands, enamel) and the lining of mouth and cloaca. *See* germ layers.

ectoparasite /ek-tŏ-**pa**-ră-sÿt/ An organism that lives on the outside of another organism (the host) and benefits from the relationship at the expense of the host, such as the flea. *Compare* endoparasite. *See* parasitism.

ectoplasm /**ek**-tŏ-plaz-ăm/ *See* plasmagel.

ectotherm /**ek**-tŏ-th'erm/ An animal that derives most of its body heat from its surroundings. Ectotherms are typically 'cold-blooded', and their body temperature tends to fluctuate with the temperature of their environment (*see* poikilothermy). Yet many show considerable ability to regulate their temperature, whether by physiological or behavioral means. For example, lizards bask in the sun to warm themselves early in the day, then seek shade to prevent overheating during the hottest periods. *Compare* endotherm; heterotherm.

ectotrophic mycorrhiza /ek-toh-**troff**-ik/ An association between a basidiomycete fungus and a woody plant (e.g. pine). The fungal hyphae form an external covering around the root and also grow between the cells of the host root cortex. *Compare* endotrophic mycorrhiza. *See* mycorrhiza.

edaphic factors /i-**daf**-ik/ The physical, chemical, and biological characteristics of the soil that together form an important component of the habitat because of their influence on plant distribution. The main edaphic factors are water content, pH, organic matter, and soil texture.

EEG *See* electroencephalogram.

effector /i-**fek**-ter/ An organ or cell that responds in a particular way to a nervous impulse. Effectors include muscles and glands.

efferent /**ef**-ĕ-rĕnt/ Carrying or conducting away from an organ or from the center. An efferent blood vessel collects blood from a capillary network, e.g. the arteries of the aortic arches of fish that lead from the gills to the dorsal aorta. An efferent nerve or neurone conducts impulses from the central nervous system to the periphery, e.g. to a muscle. *Compare* afferent.

egg 1. A nonmotile female gamete, or egg cell. *See* ovum.
2. A fertilized ovum (zygote) or developing embryo enclosed in a protective shell or capsule, usually with food reserves. Examples include the eggs of birds, reptiles, and insects.

egg apparatus The three haploid nuclei that are situated at the micropylar end of the embryo sac. The central nucleus is the female gamete and those to either side of it are called the synergids. *See* embryo sac. *See also* antipodal cells.

egg cell *See* ovum.

egg membrane 1. The vitelline or fertilization membrane: a thin membrane that surrounds the egg cell and is secreted by the oocyte and follicle cells.
2. The tough membrane beneath the shell of a bird's egg, which is secreted by the oviduct wall before it secretes the shell.

elaioplast /i-**lay**-oh-plast/ (**lipidoplast**; **oleoplast**) A plastid storing lipids (fats or oils).

elasmobranch /i-**las**-mŏ-brank/ *See* Chondrichthyes.

elastin A protein component of connective tissue, particularly elastic fibers. It consists of chainlike polymers linked together to form an extensible network,

elater

and is found at sites such as the lung and aorta.

elater /el-ă-ter/ One of many elongated spirally thickened cells that are formed within the capsule of certain liverworts (e.g. *Marchantia* and *Pellia*). When the elaters are exposed to the air they dry out unevenly because of the differential thickening. The resulting twisting movements of the cells help in dispersing the spores.

In *Equisetum* the outer layer of the spore coat forms an X-shaped structure, the arms of which are wrapped around the spore. On drying out, the arms, or *haptera*, act in similar manner to the liverwort elaters, and are thus also termed elaters.

electric organ An organ located in the tail of certain fishes that generates electric currents. In weakly electric fishes, the organ produces weak electrical fields (typically less than 1 V) that are used for communication and navigating, whereas strongly electric fishes, such as the electric eel and electric rays, can generate powerful discharges (up to 150 V) capable of stunning prey. The organ consists of modified muscle cells, called electroplax cells, which are arranged in series like the cells of a battery. *See* electroreceptor.

electrocardiogram /i-lek-troh-**kar**-dee-ŏ-gram/ (ECG) A recording of the changes in electrical potential that occur during a cycle of heart muscle activity. These changes are detected by electrodes attached to the skin, usually on the arms or legs, and are recorded by an oscilloscope or similar device (an *electrocardiograph*) as a characteristic series of waves. A deviation from this pattern may be an indication of a heart rate or rhythm disorder or of heart muscle injury, caused, for example, by coronary thrombosis.

electroencephalogram /i-lek-troh-en-**sef**-ă-lŏ-gram/ (EEG) A recording of the changes in electrical potential that are produced by activity of nerve cells in the brain, mainly the cerebral cortex. These changes are detected by electrodes placed against the scalp and are recorded by an oscillo-

scope or similar device (an *electroencephalograph*) as a series of waves ('brain waves'). Variations in wave pattern can be correlated with different mental states and neurological conditions, such as epilepsy.

electrolyte /i-**lek**-trŏ-lÿt/ A liquid containing positive and negative ions that conducts electricity by the flow of those charges. Electrolytes can be solutions of acids or metal salts ('ionic compounds'), usually in water. Alternatively they may be molten ionic compounds – again the ions can move freely through the substance. Liquid metals (in which conduction is by free electrons rather than ions) are not classified as electrolytes. *See also* electrolysis.

electron micrograph *See* micrograph.

electron microscope *See* microscope.

electron-transport chain (respiratory chain) A chain of chemical reactions involving the stepwise transfer of electrons through a series of electron carriers (enzyme complexes) and resulting ultimately in the formation of ATP and the transfer of hydrogen atoms to oxygen to form water. Such reaction sequences play a crucial role in respiration in virtually all living cells; a similar set of reactions forms part of the light-dependent reactions of PHOTOSYNTHESIS. The enzymes and other components are, in eukaryotic cells, located in the inner membrane of the mitochondria. During aerobic respiration the reduced coenzyme NADH, produced by the Krebs cycle in the mitochondrial matrix, gives up two electrons to the first component in the chain, NADH dehydrogenase, and four hydrogen ions (H^+) are discharged from the matrix of the mitochondrion into the intermembrane space. The electrons are transferred along the chain to a carrier molecule (ubiquinone), and then in sequence to a series of cytochromes, finally acting with the enzyme cytochrome oxidase to reduce an oxygen atom, which combines with two H^+ ions to form water. During this electron transfer, a further six H^+ ions are pumped into the intermembrane space, making a total of 10 per molecule of NADH. If

$FADH_2$ is the electron donor, only six H^+ ions are pumped across.

The function of electron transport in the mitochondrion is to phosphorylate ADP to ATP. According to the *chemiosmotic theory*, the H^+ ions in the intermembrane space diffuse back to the matrix through the inner mitochondrial membrane down a concentration gradient. As they do so they drive the synthesis of ATP from ADP by the enzyme ATP synthase, which consists of a proton channel and an ATP-synthesizing site, and is embedded in the inner membrane. The current consensus of opinion is that four protons must pass through the ATP synthase proton channel for every molecule of ATP that the enzyme synthesizes. Hence, for every molecule of NADH donating electrons to the chain, 2.5 molecules of ATP are produced; and for each molecule of $FADH_2$, 1.5 molecules of ATP are produced.

electrophoresis /i-lek-troh-fõ-**ree**-sis/ The migration of electrically charged particles towards oppositely charged electrodes in solution under an electric field – the positive particles to the cathode and negative particles to the anode. The rate of migration varies with molecular size and shape. The technique can be used to separate or analyze mixtures (e.g. of proteins or nucleic acids). Wetted filter paper, starch gel, or a similar inert porous medium is used in the technique. It can be carried out in conjunction with paper chromatography. In *immunoelectrophoresis*, components are firstly separated by electrophoresis on a gel. Then a solution of antibodies is added to a trough in the gel. The antibodies diffuse through the gel, and where they encounter their specific antigen they form a precipitate. This allows specific antigenic components to be identified.

electroreceptor /i-lek-troh-ri-**sep**-ter/ A sensory organ that is specialized for detecting electric current. Certain fishes have electroreceptors in their skin, which they use in navigating or for detecting objects in their vicinity, including prey. Each electroreceptor consists of a pore, approximately 0.1 mm in diameter, in the skin at the base of which is a cluster of electrically sensitive cells. These receptor cells act like tiny voltmeters, responding to changes in the electrical field surrounding the fish. Each electroreceptor sends impulses to the brain via an afferent nerve fiber. A single fish may have thousands of such receptors, usually concentrated on the head and dorsal surfaces. Some electroreceptive fish generate their own electric fields using ELECTRIC ORGANS, whereas others, such as sharks and catfish, simply detect the fields generated by the muscular activity of other nearby fish.

ELISA /i-**lee**-să/ (**enzyme-linked immunosorbent assay**) A sensitive and convenient type of IMMUNOASSAY used for determining the concentration of proteins or other (potentially) antigenic substances in biological samples. Specific antibodies against the substance being tested are adsorbed onto an insoluble carrier surface, such as a PVC sheet. Then a known amount of the sample is added, so that molecules of the test substance are bound by the antibodies. The carrier is rinsed and a second antibody, specific to a second site on the test substance, is added. Molecules of this also carry an enzyme, which causes a color change in a fourth reagent. The intensity of the color change can then be measured photometrically and compared against known standard solutions of the test substance. ELISA is widely used in medical and veterinary diagnostics, and in research.

elution /i-**loo**-shŏn/ The removal of an adsorbed substance in a chromatography column or ion-exchange column using a solvent (*eluent*), giving a solution called the *eluate*. The chromatography column can selectively adsorb one or more components from the mixture. To ensure efficient recovery of these components graded elution is used. The eluent is changed in a regular manner starting with a non-polar solvent and gradually replacing it by a more polar one. This will wash the strongly polar components from the column. *See* column chromatography.

elytra /el-ă-tră/ (*sing.* **elytron**) The hardened fore wings found in beetles and in some other insects. They protect the delicate hind wings when these are not in use.

emasculation The removal of the stamens of a plant to prevent self-pollination or unwanted crosses to nearby plants.

Embden–Meyerhoff pathway /em-děn mÿ-er-hoff/ *See* glycolysis.

embedding The sealing of tissue prepared for permanent microscope slides in a solid block of paraffin wax prior to sectioning. After clearing, tissues are placed in two or three baths of molten paraffin wax. When the tissue is completely infiltrated by the wax it is allowed to harden. As the wax is opaque the block must be marked to insure correct orientation when sectioning. Tissue prepared for electron microscopy may be embedded in Araldite and cut with a diamond knife. *See also* clearing; fixing.

embryo /em-bree-oh/ **1.** (*Zoology*) The organism formed after cleavage of the fertilized ovum and before hatching or birth. In mammals the embryo in its later well-differentiated stages is called a fetus.
2. (*Botany*) The organism that develops from the zygote of bryophytes, pteridophytes, and seed plants before germination. *See* embryo sac.

embryology /em-bree-ol-ŏ-jee/ The study of the development of organisms, especially animals, usually restricted to the period from fertilization to hatching or birth.

embryo sac A large oval cell in the nucellus of flowering plants in which egg fertilization and subsequent embryo development occurs. It corresponds to the female GAMETOPHYTE of lower plants and contains a number of nuclei derived by division of the MEGASPORE nucleus.
The number of nuclei in the sac may vary. It commonly consists of the egg apparatus at the micropylar end, made up of an egg nucleus and two synergid nuclei. There are three antipodal cells at the opposite chalazal end that probably aid embryo nourishment, and two polar nuclei in the center that fuse to form the primary ENDOSPERM nucleus. At fertilization one male nucleus fuses with the egg nucleus to form the zygote, while the second male nucleus fuses with the primary endosperm nucleus to form a triploid cell that later gives rise to the endosperm. In the gymnosperms the megaspore gives rise to a cell that is termed the embryo sac because of its similarity to the angiosperm structure. *See* nucellus; micropyle. *See illustration at* ovule.

emulsion A colloid in which a liquid phase (small droplets with a diameter range 10^{-5}–10^{-7} cm) is dispersed or suspended in a liquid medium. Emulsions are classed as lyophobic (solvent-repelling and generally unstable) or lyophilic (solvent-attracting and generally stable).

enamel The hard white outer coating of the teeth of vertebrates. It is a protective layer, epidermal in origin. Enamel is constructed from hexagonal crystals of calcium phosphate, calcium carbonate, and calcium fluoride bound together by keratin fibers. *See illustration at* teeth.

enantiomer /en-an-tee-ŏ-mer/ (**enantiomorph**) A compound whose structure is not superimposable on its mirror image: one of any pair of optical isomers. *See also* isomers; optical activity.

encephalin /en-sef-ă-lin/ *See* endorphin.

endarch /en-dark/ Denoting a stele in which metaxylem develops to the outside of protoxylem. *Compare* centrarch; exarch; mesarch.

endemic /en-dem-ik/ Describing a population or species that is restricted to a particular region.

endocarp /en-doh-karp/ *See* pericarp.

endocrine gland /en-doh-krin, -krÿn/ (**ductless gland**) A gland that has no duct or opening to the exterior. It produces hormones, which pass directly into the bloodstream. The circulatory system then transmits them to other body tissues or organs, where activity is modified. Thus an alternative method of communication to the nervous system is provided. *Compare* exocrine gland.

endocrinology /en-doh-kri-**nol**-ŏ-jee/ The study of the endocrine glands and their secretions (hormones).

endocytosis /en-doh-sÿ-**toh**-sis/ The entry into cells of materials by infolding of a small area of the plasma membrane to form a membranous sac, or transport vesicle. This vesicle then pinches off and moves to the cell's interior, where the contents may be sorted and transferred to other vesicles, or to cell organelles, such as lysosomes.

One form is *pinocytosis* ('cell drinking'), when a cell indiscriminately takes in a droplet of extracellular fluid, and any dissolved substances it might contain. More commonly, cells select materials from their environment by using *receptor-mediated endocytosis*, whereby particular substances bind to specific receptors on the cell surface. The receptors with the bound substance (e.g. a hormone, low-density lipoprotein) gather in clusters at pits in the membrane. The pit expands and deepens, eventually forming a vesicle. The receptors are subsequently separated and recycled back to the plasma membrane. *Compare* exocytosis; phagocytosis.

endoderm /**en**-doh-derm/ (**entoderm**) The innermost germ layer of most metazoans (including vertebrates) that develops into the gut lining and its derivatives (e.g. liver, pancreas). It also forms the yolk sac and allantois in birds and mammals. *See* germ layers.

endodermis /en-doh-**der**-mis/ The innermost part of the cortex of plant tissue, consisting of a single layer of cells that controls the passage of water and solutes between the cortex and the stele. A clearly defined endodermis is seen in all roots and in the stems of the pteridophytes and in some of the dicotyledons. *See* Casparian strip; passage cells.

endogamy /en-**dog**-ă-mee/ The fusion of gametes produced by closely related organisms. *See* inbreeding.

endogenous /en-**doj**-ĕ-nŭs/ Produced or originating within an organism. *Compare* exogenous.

endolymph /**en**-doh-limf/ A fluid filling the structures of the inner ear of vertebrates. *See* labyrinth.

endometrium /en-doh-**mee**-tree-ŭm/ The glandular mucous membrane lining the internal surface of the uterus in mammals. It passes through cyclical periods of growth and development and regression or degeneration, in association with the estrous cycle. It develops to receive and nourish any developing embryo present. If fertilization does not occur it either returns to its natural state or, in the case of humans and some primates, breaks down and is discharged at menstruation.

endomitosis /en-doh-mÿ-**toh**-sis/ The duplication of chromosomes without division of the nucleus. Endomitosis may take two forms: the chromatids may separate causing endopolyploidy, e.g. in the macronucleus of ciliates, or the chromatids may remain joined leading to multistranded chromosomes or *polyteny*, e.g. during larval development of dipteran flies. Both processes lead to an increase in nuclear and cytoplasmic volume. *Compare* amitosis; mitosis.

endonuclease /en-do-**new**-klee-ayss/ An enzyme that catalyzes the hydrolysis of internal bonds of polynucleotides such as DNA and RNA, producing short segments of linked nucleotides (oligonucleotides). *See also* DNase; restriction endonuclease.

endoparasite /en-doh-**pa**-ră-sÿt/ An organism that lives inside the body of another organism (the host) and benefits from the relationship at the expense of the host, such as the malarial parasite. *Compare* ectoparasite. *See* parasitism.

endoplasm /**en**-doh-plaz-ăm/ *See* plasmasol.

endoplasmic reticulum /en-doh-**plaz**-mik/ (ER) A system of membranes forming tubular channels and flattened sacs (cisternae), running through the cytoplasm of all eukaryotic cells and continuous with the nuclear envelope. Although often extensive, it was only discovered with the advent of electron microscopy. Its surface is often covered with ribosomes, forming *rough ER*. The polypeptides they make pass into the cisternae of the ER, where they undergo various modifications, including folding into their correct conformation, addition of carbohydrate groups, and assembly with other polypeptides into multimeric proteins. These are then transported to other parts of the cell or for secretion via the GOLGI APPARATUS. ER lacking ribosomes is called *smooth ER* and is involved with lipid synthesis, including phospholipids and steroids.

In muscle cells a specialized form of ER called SARCOPLASMIC RETICULUM is present. *See illustration at* cell.

endopodite /en-**dop**-ŏ-dȳt/ *See* biramous appendage.

endorphin /en-**dor**-fin/ (encephalin; enkephalin) One of a group of peptides produced in the brain and other tissues that are released after injury and have pain-relieving effects similar to those of opiate alkaloids, such as morphine. They include the *enkephalins*, which consist of just five amino acids. Other larger endorphins occur in the pituitary, while some are polypeptides, found mainly in pancreas, adrenal gland, and other tissues. Pain relief from acupuncture may be due to stimulated production of endorphins.

endoscopic /en-dŏ-**skop**-ik/ Describing the type of development of a plant embryo in which the inner cell formed by the first division of the zygote develops into the embryo, while the outer cell develops into the suspensor. It is seen in many ferns and in all seed plants. *Compare* exoscopic.

endoskeleton /en-doh **skel**-ĕ-tŏn/ A skeleton that occurs inside the animal body, such as the bony or cartilaginous skeleton of vertebrates. It comprises a skull, vertebral column, pectoral and pelvic girdles, ribs, and limb or fin elements. It gives shape and support to the body, protects vital organs, and provides a system of rigid levers to which muscles can attach and produce movement. An endoskeleton also allows the steady growth in size of the animal. An endoskeleton is also present in echinoderms and certain other invertebrates. *Compare* exoskeleton.

endosperm /en-doh-sperm/ The nutritive tissue that surrounds the embryo in angiosperms. In non-endospermic seeds most of the endosperm is absorbed by the developing embryo and the food stored in the cotyledons. In endospermic seeds the endosperm replaces the nucellus and is often a rich source of growth regulating substances. Many endospermic seeds (e.g. cereals and oil seeds) are cultivated for their food reserves. The endosperm develops from the primary endosperm nucleus and is therefore triploid.

In the gymnosperms the female prothalial tissue is sometimes termed the endosperm. However this tissue develops before fertilization and in fact is the haploid female gametophyte. It is therefore not homologous with the angiosperm endosperm although it does have a nutritive function.

endospore /**en**-doh-spor, -spohr/ A resting stage produced by certain bacteria under unfavorable conditions. Endospores are formed within the cell and are surrounded by a thick coat containing dipicolinic acid. On germination the wall is lysed and one vegetative cell is produced. Endospores can remain viable for several centuries and are resistant to heat, desiccation, and x-rays.

endosporic /en-doh-**spor**-ik, -**spoh**-rik/ Describing the formation of spores within the spore-producing body; for example, the production of ascospores inside the ascus in Ascomycete fungi. *Compare* exosporic.

endosporium /en-doh-**spor**-ee-ŭm, -**spoh**-ree-/ *See* intine.

endostyle /**en**-doh-stÿl/ A shallow groove along the ventral wall of the pharynx of *Branchiostoma* and other primitive chordates. It consists of four tracts of mucus-producing cells and five tracts of ciliated cells. Food particles are trapped in the mucus and moved forwards by the cilia. It is homologous with the thyroid gland of vertebrates.

endosymbiont theory /en-doh-**sim**-bÿ-ont/ The theory that eukaryotic organisms evolved from symbiotic associations between bacteria. It proposes that integration of photosynthetic bacteria, for example purple bacteria and cyanobacteria, into larger bacterial cells led to their permanent incorporation as forerunners of the mitochondria and plastids (e.g. chloroplasts) seen in modern eukaryotes. There is compelling supporting evidence for the theory, particularly from studies of mitochondrial and plastid DNA and ribosomes, which demonstrate remarkable similarities with those of bacteria.

endothelium /en-doh-**th'ee**-lee-ŭm/ (*pl.* endothelia) The tissue lining the blood vessels and heart. It consists of a single layer of thin flat cells fitting very close together. In capillaries it is the only layer, providing the barrier between the blood and the fluid bathing the cells. Water and all dissolved substances with small molecules pass through the cells. White blood cells pass between the endothelial cells by a movement known as *diapedesis*.

endotherm /**en**-doh-th'erm/ An animal that generates all or most of its body heat from its own metabolism. The birds and mammals (i.e. 'warm-blooded' vertebrates) make up the majority of endotherms, but many terrestrial reptiles and some insects also fit this description. Endotherms are generally able to maintain their body temperature within a narrow range that is optimal for their metabolic reactions (*see* homoiothermy). To achieve this they have efficient means of insulation (e.g. hair,

feathers, subcutaneous fat) coupled with effective cooling mechanisms (e.g. sweating, panting). *Compare* ectotherm; heterotherm.

endotoxins /en-doh-**toks**-inz/ Toxic substances formed inside the cells of Gram-negative bacteria (e.g. *Salmonella*) and released on disintegration of the cell. They are heat-stable polysaccharide-protein complexes causing nonspecific effects in their hosts, e.g. fevers. *Compare* exotoxins. *See* toxin.

endotrophic mycorrhiza /en-doh-**troff**-ik/ An association between a fungus and a woody or herbaceous plant, e.g. orchid. The fungal hyphae grow both between and within the cells of the host root cortex. *Compare* ectotrophic mycorrhiza. *See* mycorrhiza.

end plate A flattened nerve ending that occurs at the junction of a motor axon and a muscle cell, i.e. a NEUROMUSCULAR JUNCTION. It transmits nerve impulses in a way similar to that of other SYNAPSES.

end-plate potential (EPP) A brief localized depolarization or potential change across the membrane in the motor end-plate region of a muscle fiber, at a neuromuscular junction. A neurotransmitter substance is released from the presynaptic nerve endings on stimulation by an impulse and increases the permeability of the post-synaptic (muscle) membrane to ions, causing the EPP. The size of the EPP depends on the amount of neurotransmitter released but normally it is sufficiently large to cross the threshold level for response and set off an action potential, which is propagated along the length of the muscle fiber.

enkephalin /en-**kef**-ă-lin/ *See* endorphin.

enterokinase /en-tĕ-roh-**kÿ**-nayss/ *See* enteropeptidase.

enteropeptidase /en-tĕ-roh-**pep**-ti-dayss/ (**enterokinase**) A peptidase enzyme that converts trypsinogen to trypsin.

enthalpy /en-**thal**-pee, en-thal-/ Symbol: H A thermodynamic property of a system defined as $U + pV$, where U is the internal energy, p the pressure, and V the volume. Changes in enthalpy are important for chemical reactions, for which the heat absorbed (or evolved) is equal to the change of internal energy (ΔU) plus the external work done during the change (ΔpV). In many biochemical systems no work is done and the enthalpy change is equal to the change in internal energy.

entoderm /en-tŏ-derm/ *See* endoderm.

entomology /en-tŏ-**mol**-ŏ-jee/ The scientific study of insects.

entomophily /en-tŏ-**moff**-ă-lee/ Pollination by insects. Various structures and mechanisms have evolved to attract insects (e.g. showy petals and nectar) and to insure that they carry pollen away on their bodies.

Entoprocta /en-toh-**prok**-tă/ A phylum of tiny transparent aquatic animals containing about 150 known species, chiefly marine and widely distributed in coastal waters. Entoprocts have a cup-shaped body bearing a ring of ciliated tentacles around the mouth and attached to a substrate via a contractile stalk, which can produce flicking movements of the body. Entoprocts form large colonies on rocks, shells, seaweeds, and other surfaces, and use their tentacles to filter food particles from the water. They have a U-shaped gut, with the anus near the mouth, a simple nervous system, and paired protonephridia for excretion. There are no circulatory or gas-exchange systems. Entoprocts are hermaphrodites, and sexual reproduction produces free-swimming larvae, which settle and metamorphose into adults. Asexual reproduction by budding is also common.

entropy /en-trŏ-pee/ Symbol: S In any system that undergoes a reversible change, the change of entropy is defined as the heat absorbed divided by the thermodynamic temperature:

$$\delta S = \delta Q/T$$

A given system is said to have a certain entropy, although absolute entropies are seldom used: it is change in entropy that is important. The entropy of a system measures the availability of energy to do work.

In any real (irreversible) change in a closed system the entropy always increases. Although the total energy of the system has not changed (first law of thermodynamics) the available energy is less – a consequence of the second law of thermodynamics.

The concept of entropy has been widened to take in the general idea of disorder – the higher the entropy, the more disordered the system. For instance, a chemical reaction involving polymerization may well have a decrease in entropy because there is a change to a more ordered system. The 'thermal' definition of entropy is a special case of this idea of disorder – here the entropy measures how the energy transferred is distributed among the particles of matter.

environment The complete range of external conditions under which an organism lives, including physical, chemical, and biological factors, such as temperature, light, and the availability of food and water.

enzyme /en-zÿm/ A compound that catalyzes biochemical reactions. Enzymes are proteins, which act with a given compound (the substrate) to produce a complex, which then forms the products of the reaction. The enzyme itself is unchanged in the reaction; its presence allows the reaction to take place. The names of most enzymes end in -ase, added to the substrate (e.g. lactase) or the reaction (e.g. hydrogenase).

Enzymes are extremely efficient catalysts for chemical reactions, and very specific to particular reactions. They may have a non-protein part (cofactor), which may be an inorganic ion or an organic constituent (coenzyme). The mechanism of action of most enzymes appears to be by *active sites* on the enzyme molecule. The substrate acting with the enzyme changes shape to fit the active site, and the reaction proceeds. Enzymes are very sensitive to their environment – e.g. temperature, pH, and the presence of other substances. *See also* ribozyme.

enzyme-linked immunosorbent assay /im-yŭ-noh-**sor**-bĕnt, -zor-/ *See* ELISA.

enzyme technology (enzyme engineering) A branch of biotechnology that utilizes enzymes for industrial purposes. For example rennet (impure rennin) is manufactured on a large scale to make cheese and junkets. Enzymes are also used to determine the concentration of reactants or products in specific reactions catalyzed by them.

Eocene /ee-ŏ-seen/ The second oldest epoch of the Tertiary period, 55–38 million years ago, represented in Britain by clay deposits in the London basin. It is characterized by predominance of early hoofed mammals, the ungulates (perissodactyls and artiodactyls). Many other mammals (e.g. carnivores, bats, and whales) and birds are also present. *See also* geological time scale.

eosin /ee-ŏ-sin/ *See* staining.

eosinophil /ee-ŏ-**sin**-ŏ-fil/ A white blood cell (LEUKOCYTE) with a lobed nucleus and cytoplasmic granules that stain with acidic dyes. Eosinophils comprise 2–5% of all leukocytes, but the number increases in allergic conditions, such as asthma and hay fever, as they have antihistamine properties. They can also destroy bacteria and larger pathogens (e.g. protozoa and parasitic worms) by phagocytosis and by releasing hydrogen peroxide.

ephemeral /i-**fem**-ĕ-răl/ A plant that has a very short life cycle, often completing many life cycles within a year. Examples are shepherd's purse and certain desert plants that grow, flower, and set seed in brief periods of rain. *Compare* annual; biennial; perennial.

ephyra /**ef**-ă-ră/ (*pl.* **ephyrae**) A small medusa-like stage in the life cycle of scyphozoans (jellyfish). After breaking off from the scyphistoma, it swims freely and grows into an adult jellyfish. *See* scyphistoma.

epiboly /i-**pib**-ŏ-lee/ During embryonic development, the spreading out of blastomeres, or later cleavage products, of the animal pole of a blastula to engulf the vegetal part of the egg, including its yolky cells. Epiboly occurs in the eggs of some amphibians and most teleosts.

epicalyx /ep-ă-**kay**-liks/ (*pl.* **epicalyces**) A ring of bracts or bracteoles below the calyx forming a structure that resembles the calyx found, for example, in the sweet william flower.

epicarp /**ep**-ă-karp/ *See* pericarp.

epicotyl /i-**pik**-ŏ-til/ The part of the plumule above the cotyledons. *Compare* hypocotyl.

epidemiology /ep-ă-dee-mee-**ol**-ŏ-jee/ The study of diseases affecting large numbers within a population. These include both epidemics of infectious diseases and also diseases associated with environmental factors and dietary habits (e.g. lung cancer, some forms of heart disease, etc.).

epidermis /ep-ă-**der**-mis/ **1.** (*Botany*) The outer protective layer of cells in plants. In aerial parts of the plant the outer wall of the epidermis is usually covered by a waxy cuticle that prevents desiccation, protects the underlying cells from mechanical damage, and increases protection against fungi, bacteria, etc. The cells are typically platelike and closely packed together except where they are modified for a particular function, as are guard cells. The epidermis arises from the tunica meristem. When it is damaged it is replaced by a secondary layer, the periderm. The specialized epidermal area of the roots from which the root hairs arise is termed the *piliferous layer*.
2. (*Zoology*) The outer layer of cells or outer tissue of an animal that generally protects the tissues beneath and insures that the body is waterproof. In vertebrates, the epidermis consists of several layers of cells and forms the outer layer of skin. As

it wears away at the surface it is renewed continuously by growth of new cells in the Malpighian layer, which is immediately beneath. The harder cornified cells of the stratum corneum are the chief protective cells. Products of the epidermis of vertebrates include hair, claws, nails, hooves, horns of cattle and sheep, feathers and beaks of birds, and the scales on the legs of birds and on the shells of tortoises. The epidermis of invertebrates is a single layer of cells, often secreting a protective cuticle. In arthropods, this cuticle forms the exoskeleton. *See also* dermis. *See illustration at* skin.

epididymis /ep-ă-**did**-ă-miss/ (*pl.* epididymides) A long narrow coiled tube attached to the surface of the testis in reptiles, birds, and mammals. It acts as a temporary storage organ for spermatozoa received from the seminiferous tubules, until their release to the vas deferens and the exterior during mating. The epididymis is derived from the embryonic mesonephros.

epigamic /ep-ă-**gam**-ik/ Designating an animal feature that is attractive to the opposite sex during courtship. Examples of epigamic features are the color of feathers and bird song.

epigeal germination /ep-ă-**jee**-ăl/ Seed germination in which the cotyledons form the first photosynthetic organs above the ground (e.g. sunflower). *Compare* hypogeal germination.

epigenesis /ep-ă-**jen**-ĕ-sis/ The process by which a developing organism increases in complexity, which is brought about by interaction between parts of the nuclear genetic program, the organized cytoplasm of the developing egg, and the egg's environment. *Compare* preformation.

epiglottis /ep-ă-**glot**-is/ A flap of tissue in mammals that closes the glottis during swallowing in order to prevent food entering the windpipe. It is covered with mucous membrane and stiffened by elastic cartilage. During swallowing, the muscles of the pharynx contract and the larynx at the top of the windpipe moves upwards to meet the epiglottis, so closing the opening (the glottis). As a result, the bolus of food passes over the top of the epiglottis and into the esophagus. *See illustration at* alimentary canal.

epigyny /i-**pij**-ă-nee/ The type of flower structure in which the perianth and androecium are inserted above the gynoecium, giving an inferior ovary, fused with the receptacle. It is seen in the Compositae and Rosaceae. *Compare* hypogyny; perigyny. *See illustration at* receptacle.

epinasty /ep-ă-nas-tee/ (**epinastic movements**) The curving of a plant organ away from the axis caused by greater growth on the upper surface. *See* nastic movements.

epinephrine /ep-ă-**nef**-rin/ (**adrenaline**) A hormone produced by the medulla of the adrenal glands, and a neurotransmitter in certain parts of the central nervous system and autonomic nervous system. Epinephrine secretion into the bloodstream in stress causes acceleration of the heart, constriction of arterioles, and dilation of pupils. In addition, epinephrine produces a marked increase in metabolic rate thus preparing the body for emergency. It is chemically related to the neurotransmitter norepinephrine (*see* catecholamine).

epiphyses /i-**pif**-ă-seez/ (*sing.* **epiphysis**) The ends of a long bone in mammals. They are formed separately from the shaft (diaphysis) by secondary centers of ossification. A narrow disk of cartilage (epiphysial plate) persisting between each epiphysis and the diaphysis lays down new bone tissue and provides for growth in length of a bone. When maximum growth is reached it becomes ossified and the epiphyses and diaphysis are fused together.

epiphyte /**ep**-ă-fÿt/ Any plant growing upon or attached to another plant or object merely for physical support. They are often known as air plants because they are not attached to the ground, but obtain water and minerals from the rain and from debris that falls on the support. Examples of tropical epiphytes are ferns and orchids. Lichens, mosses, liverworts, and algae are epiphytes of temperate regions.

episome /ep-ă-sohm/ A genetic element that exists inside a cell, especially a bacterium, and can replicate either as part of the host cell's chromosome or independently. Homology with the bacterial chromosome is required for integration, therefore a plasmid may behave as an episome in one cell but not in another. Examples of episomes are temperate phages. *See* plasmid.

epistasis /i-**pis**-tă-sis/ The action of one gene (the *epistatic gene*) in preventing the expression of another, nonallelic, gene (the *hypostatic gene*). Epistatic and hypostatic genes are analogous to dominant and recessive alleles. *Compare* complementary genes.

epithelium /ep-ă-**th'ee**-lee-ŭm/ (*pl.* **epithelia**) A tissue consisting of a sheet (or sheets) of cells that covers a surface or lines a cavity. The cells are close together, and may be described as *cubical, columnar, ciliated,* or *squamous,* depending on their shape. In some cases there are several layers, as in the epidermis of the skin. Epithelial cells are sealed together by tight junctions just below their apical surface (i.e. the surface facing outwards into the lumen or cavity), which block the passage of materials between the cells. Additional junctions, called DESMOSOMES, give strength and integrity to the epithelial cell layer. Materials are forced to move through the epithelial cells, for example, via transport proteins in their plasma membranes. Hence the cells can regulate the flow of materials across the epithelium, between the apical and the basal surfaces. The latter forms junctions with the underlying extracellular matrix or basal lamina. *See also* ciliated epithelium.

epitope /ep-ă-tohp/ The region of an antigen molecule that is unique to the antigen and therefore responsible for its specificity in an antigen–antibody reaction. The epitope combines with the complementary region on the antibody molecule.

EPP *See* end-plate potential.

EPSP *See* excitatory postsynaptic potential.

equatorial plate The equator of the nuclear spindle upon which the centromeres of the chromosomes become aligned during metaphase of mitosis and meiosis.

Equisetum /ek-wă-**see**-tŭm/ (**horsetail**) *See* Sphenophyta.

ergosterol /er-**gos**-tĕ-rol, -rohl/ A sterol present in plants. It is converted, in animals, to vitamin D_2 by ultraviolet radiation, and is the most important of vitamin D's provitamins.

ergot /**er**-got/ *See* sclerotium.

erythroblast /i-**rith**-roh-blast/ One of the cells in the red bone marrow from which ERYTHROCYTES develop. At first they are colorless, but by the time they are released into the blood the cytoplasm is full of hemoglobin and, in mammals, the nucleus has disappeared. In humans, 200 000 million new erythrocytes are made each day to replace those that are worn out.

erythrocyte /i-**rith**-roh-sÿt/ (**red blood cell**) A type of blood cell that contains HEMOGLOBIN and is responsible for the transport of oxygen in the BLOOD. Mammalian red cells are circular biconcave disks without nuclei (the red cells of other vertebrates are oval and nucleated). Human blood contains 5 million red cells per cubic millimeter; each cell lives for about 120 days, after which it is destroyed in the liver and replaced by a new cell from the red bone marrow. The number of red cells increases in regions of oxygen shortage, such as high altitudes. In addition to hemoglobin, erythrocytes also contain an enzyme, carbonic anhydrase, and therefore have an important role in transporting carbon dioxide and maintaining a constant pH.

Escherichia coli /esh-ĕ-**rik**-ee-ă **koh**-lÿ/ A bacterium widely used in genetic research, occurring naturally in the intestinal tract of animals and in soil and water. It is Gram-negative and the cells are typically straight round-ended rods, usually occurring singly or in pairs. Some strains are pathogenic, causing diarrhea or more serious gastrointestinal infections. The strains can be distinguished serologically on the basis of their antigens. *E. coli* is killed by pasteurization and many common disinfectants.

esophagus /ee-**sof**-ă-gŭs/ (*pl.* **esophagi**) The tube that connects the buccal cavity or pharynx to the stomach. The lining of mucous membrane is folded to allow expansion during the passage of food. In vertebrates, layers of circular and longitudinal muscle surround the tube; their rhythmic contractions cause peristaltic waves to move the food down to the stomach. In birds and insects the esophagus may include the crop. *See illustration at* alimentary canal.

essential amino acid *See* amino acids.

essential element An element that is indispensable for the normal growth, development and maintenance of a living organism. Some, the *major elements*, are required in relatively large quantities and may be involved in several different metabolic reactions (*see* carbon; hydrogen; oxygen; nitrogen; sulfur; phosphorus; potassium; magnesium; calcium). Others are required in only small or minute amounts, such as iron, manganese, molybdenum, boron, zinc, copper, cobalt, iodine, and selenium (*see* trace element).

essential fatty acids Fatty acids (*see* carboxylic acid) required for growth and health that cannot be synthesized by the body and therefore must be included in the diet. Linoleic acid and possibly (9,12,15)-linolenic acid are the only essential fatty acids in humans, being required for cell membrane synthesis and fat metabolism. Arachidonic acid is essential in some animals, such as the cat, but in humans it is synthesized from linoleic acid. Essential fatty acids occur mainly in vegetable-seed oils, e.g. safflower-seed and linseed oils.

ester A compound formed by reaction of a carboxylic acid with an alcohol:

$$RCOOH + HOR_1 \rightarrow RCOOR_1 + H_2O$$

Glycerides are esters of long-chain fatty acids and glycerol.

estivation /es-tă-**vay**-shŏn/ 1. The way in which sepals and petals are folded in the flower bud before expansion. *Compare* ptyxis; vernation.
2. A period of inactivity seen in some animals during the summer or dry hot season. For example, lungfish respond to the drying up of water by burying themselves in the mud bottom. They re-emerge at the start of the rainy season. *See also* hibernation.

estradiol /es-tră-**dÿ**-ol, -ohl/ The most active estrogen produced by the body. It promotes proliferation of the endometrium of the womb during the first half of the estrous cycle to prepare the womb for ovulation, and is also important in female development. It is metabolized to estrone and then estriol. *See* estrogen.

estriol /es-**tree**-ol, -ohl/ A metabolite of estradiol and estrone, produced mainly by the placenta during pregnancy, when it occurs in the urine in high concentrations. It has only weak estrogenic activity. *See* estrogen.

estrogen /**es**-trŏ-jĕn/ Any of various female sex hormones (steroids) involved in the development and maintenance of accessory sex organs and secondary sex characteristics (e.g. growth of the breasts). During the menstrual cycle estrogens act on the female sex organs to produce an environment suitable for fertilization of the egg cell and implantation and growth of the embryo. They are used therapeutically to correct estrogen deficiency, for example at the menopause, and to treat some forms of breast cancer. In oral contraceptives, synthetic estrogens and progesterone act to prevent ovulation and the release of gonadotropins, e.g. follicle-stimulating hormone. Excessive blood estrogen levels lead to sickness, as in pregnancy morning sickness. The ovary produces mainly two estrogens: estradiol, and estrone. These hormones are produced in smaller amounts by the adrenal cortex, testis, and placenta. *See also* progesterone.

estrone /es-trohn/ An estrogen hormone produced by the ovary and by peripheral tissues, with actions similar to estradiol. *See* estradiol; estrogen.

estrous cycle /es-trŭs/ A rhythmic cycle, varying from 5 to 60 days, occurring in sexually mature females of most mammal species in the absence of pregnancy. It can occur either throughout the year or only during a breeding season. The central event, *estrus*, is a brief period during which the female is both receptive and attractive to males and during which both ovulation and copulation normally occur. *Polyestrous* species have estrous cycles following in quick succession, the males being continuously sexually active. *Monestrous* species, which are typically larger and require good environmental conditions for raising offspring, have only one estrus each year, the cycles of females throughout the population being synchronized. Males of monestrous species also exhibit a cycle, being sexually active only during the females' estrus. The estrous cycle is controlled hormonally, with a complex of pituitary and ovarian hormones under the overall regulation of the hypothalamus.

The estrous cycle can be separated into phases.
1. *Follicular phase.* Graafian follicles grow in the ovary and secrete the hormone estradiol, which causes the uterine lining to proliferate and thicken.
2. *Estrus.* The period of 'heat'. Ovulation occurs, stimulated by a rise in the secretion of pituitary luteinizing hormone, and copulation also takes place. In some mammals, such as the cat and the rabbit, ovulation is delayed until copulation has occurred.
3. *Luteal phase* (*metestrus*). A corpus luteum forms from the follicle and secretes progesterone. Estradiol secretion de-

creases. If fertilization follows estrus, the cycle is suspended in this phase for the duration of the pregnancy.
4. *Regression phase* (*diestrus*). In the absence of fertilization, the corpus luteum and endometrium diminish, ovarian hormone levels fall, and a new follicle begins growth. In some animals this phase is lengthened if pregnancy has not occurred, giving rise to symptoms of pregnancy (known as pseudopregnancy). The completion of an estrous cycle or a pregnancy is followed by *pro-estrus*, in which there is a slow increase of pituitary follicle-stimulating hormone and estradiol leading up to the next full cycle. *See also* menstrual cycle.

estrus /es-trŭs/ *See* estrous cycle.

etaerio /i-teer-ee-oh/ A compound fruit that has developed from a single flower with an apocarpous gynoecium. It may be composed of achenes (e.g. buttercup), follicles (e.g. larkspur), or drupes (e.g. blackberry).

ethanedioic acid /eth-ayn-dÿ-oh-ik/ *See* oxalic acid.

ethanoic acid /eth-ă-noh-ik/ *See* acetic acid.

ethanol /eth-ă-nol, -nohl/ (**ethyl alcohol**) A simple water-soluble ALCOHOL, C_2H_5OH, that is produced by yeasts in alcoholic FERMENTATION and is the active ingredient in alcoholic beverages.

Ethiopian One of the six zoogeographical regions of the earth. It encompasses Africa south of the Sahara Desert, the southern half of Arabia, and (according to certain authorities) Madagascar. The animals characteristic of this region are the gorilla, chimpanzee, African elephant, rhinoceros, lion, hippopotamus, giraffe, certain antelopes, ostrich, guinea fowl, secretary bird, and, in Madagascar, the lemur.

ethology /i-thol-ŏ-jee/ The study of the behavior of animals in their natural environments.

ethylene /eth-ă-leen/ (ethene) A gaseous hydrocarbon (C_2H_4), produced in varying amounts by many plants, that functions as a plant hormone. It is synthesized from the amino acid methionine, and its production is usually stimulated by auxins. Ethylene has wide-ranging effects on plant growth and development, but is known particularly to stimulate senescence and fruit ripening, and is made in response to stress.

etiolation /ee-tee-ŏ-**lay**-shŏn/ The type of growth exhibited by plants grown in darkness, usually from seed. They lack chlorophyll and therefore appear white or yellow. They show less differentiation and contain reduced amounts of supporting material, such as lignin, concentrating resources on elongation of internodes. They are therefore more fragile. Dicotyledons have very small unexpanded leaves and a hooked plumule (for protection if growing through soil). Monocotyledons tend to have normal-length leaves, but these are thinner and rolled. Under natural conditions such growth maximizes the chances of the shoot reaching light, on which it depends for photosynthesis and hence food. Light restores normal growth.

etioplast /ee-tee-ŏ-plast/ A modified chloroplast formed from proplastids in leaves grown in total darkness. Instead of the normal chloroplast membrane system, etioplasts contain a highly organized semicrystalline array of tubular membranes (called a *prolamellar body*) showing a hexagonal symmetry. Radiating from the prolamellar body are single thylakoids. A normal membrane system develops from this body once the plant is exposed to light. Etioplasts contain few if any polyribosomes and rarely contain starch.

eubacteria /yoo-bak-**teer**-ee-ă/ A large and diverse group of BACTERIA, principally distinguishable from the other major group of bacteria, the ARCHAEBACTERIA, by the presence of a peptidoglycan layer in their cell walls. There are also differences in the base sequences of RNA subunits of the ribosomes which are thought to reflect a very ancient evolutionary divergence between the two groups. Most are unicells that divide by binary fission. The cells can be spherical, rod-shaped, or helical, and some form assemblages of cells, such as branching filaments. Most are immotile, but some possess flagella. They are a ubiquitous group, and are often assigned to the DOMAIN Eubacteria.

eucaryote /yoo-**ka**-ree-oht/ *See* eukaryote.

euchromatin /yoo-**kroh**-mă-tin/ *See* chromatin.

eudicot /**yoo**-dă-kot/ Any dicotyledonous flowering plant whose pollen grains have three apertures (i.e. triaperturate pollen). The majority of dicots (*see* Magnoliopsida) are eudicots, and these form an evolutionarily more derived group than the remaining dicots, represented by the magnolias and their relatives. The latter have pollen with only a single germination pore (i.e. uniaperturate pollen), which is regarded as a relict feature inherited from a common ancestral species. Evidence from molecular comparisons now indicates that magnolias may be more closely related to the monocots than to the eudicots.

eugenics /yoo-**jen**-iks/ The theory that the human race could (or should) be improved by controlled selective breeding between individuals with 'desirable' characteristics – health, physique, intelligence, etc. It is a controversial subject, partly because of difficulties in judging the relative importance of genetic and environmental factors. More fundamentally, it is thought by many to be a moral issue involving the freedom of the individual and the danger of its use for sinister political ends.

Euglenophyta /yoo-glee-**noff**-ă-tă/ A phylum of aquatic single-celled protoctists that swim using one or more undulipodia (flagella). It contains both photosynthesizing and non-photosynthesizing members. For example members of the genus *Euglena* possess a flexible pellicle surrounding the cell, and a single undulipodium. Many species contain chloroplasts, with pigments similar to those found in plants, although these organisms may also consume dissolved or particulate food from their surroundings. Reproduction is asexual, by binary fission. *E. gracilis* is a popular laboratory organism, in which the chloroplasts regress and are effectively turned off in sustained darkness, a process reversed when light is reintroduced. Some individuals may lose their chloroplasts permanently.

eukaryote /yoo-**ka**-ree-oht/ Any member of a group (sometimes called Eukaryota) comprising all organisms except eubacteria and archaebacteria (which comprise the PROKARYOTES). There are several key features that distinguish eukaryotes from prokaryotes. Firstly, their genetic material is packaged in chromosomes within a membrane-bound nucleus. Also, unlike prokaryotes, they possess mitochondria and (in plants and other photosynthetic eukaryotes) chloroplasts in the cell cytoplasm. Thirdly, they have complex cilia and flagella (undulipodia) composed of arrays of microtubules, wheres the prokaryotic flagellum is a simple shaft of protein.

euploidy /yoo-ploid-ee/ The normal state in which an organism's chromosome number is an exact multiple of the haploid number characteristic of the species. For example, if the haploid number is 7, the euploid number would be 7, 14, 21, 28, etc., and there would be equal numbers of each different chromosome. *Compare* aneuploidy.

euryhaline /yoo-**rȳ**-hă-lin, -lȳn/ Describing organisms that are able to tolerate wide variations of salt concentrations (and hence osmotic pressure) in the environment, for example the eel can live in both fresh and salt water. *Compare* stenohaline.

eusporangiate /yoo-spŏ-**ran**-jee-ayt/ Describing the condition, found in certain club mosses, in which the sporangia develop from a group of initial cells. *Compare* leptosporangiate.

Eustachian tube /yoo-**stay**-shăn, -shee-ăn, -kee-ăn/ A tube connecting the middle ear with the back of the throat in tetrapods. It maintains atmospheric pressure on both sides of the eardrum (tympanum): any external change in pressure is equalized by swallowing or yawning, which opens the tube and admits air to, or releases air from, the middle ear. The tube is named for the Italian anatomist Bartolommeo Eustachio (*c.* 1524–74). *See illustration at ear.*

eustele /yoo-steel, -stee-lee/ The stele arrangement found in most gymnosperms and dicotyledons in which the vascular tissue is arranged into a ring of discrete bundles all contained within a single ring of endodermis and pericycle. *See* stele.

Eutheria /yoo-**theer**-ee-ă/ The infraclass that contains the placental mammals – the most advanced and the majority of living mammals, including humans. Placental mammals have a complex and well-developed brain. Their young are born at a comparatively advanced stage of development after a long gestation period in the maternal uterus (womb), where they are nourished by the placenta. The group shows great adaptive radiation, being found in terrestrial, aquatic, and aerial habitats. *See* Artiodactyla; Carnivora; Cetacea; Chiroptera; Insectivora; Lagomorpha; Perissodactyla; Primates; Proboscidea; Rodentia. *See also* Mammalia.

eutrophic /yoo-**troff**-ik/ Describing lakes or ponds that are rich in nutrients and consequently are able to support a dense population of plankton and littoral vegetation. *Eutrophication* is the process that results when an excess of nutrients enters a lake, for example as sewage or from water draining off land treated with fertilizers. The nutrients stimulate the growth of the algal population giving a great concentration or 'bloom' of such plants. When these die they are decomposed by bacteria, which use up the oxygen dissolved in the water, so that aquatic animals such as fish are deprived of oxygen and die from suffocation. *Compare* oligotrophic.

evergreen Describing plants that retain their leaves through the winter and into the following summer or through several years. Many tropical species of broad-leaved flowering plants are evergreen and their leaves are thicker and more leathery than DECIDUOUS trees. In polar and cold temperate regions the evergreens are often cone-bearing shrubs or trees with needle-like or scalelike leaves.

evolution The gradual process of change that occurs in populations of organisms over a long period of time. It manifests itself as new characteristics in a species, and the formation of new species. *See* Darwinism; Lamarckism; natural selection.

evolutionary clock *See* molecular clock.

exarate /eks-ă-rayt/ Designating pupae in which the wings and legs are free and which are therefore capable of limited movement. *Compare* coarctate; obtect.

exarch /eks-ark/ Denoting a stele in which the metaxylem develops to the inside of the protoxylem. *Compare* centrarch; endarch; mesarch.

excitation–contraction coupling A process by which excitation of the muscle fiber membrane at a neuromuscular junction results in contraction of a muscle. The resulting depolarization spreads along infoldings of the membrane, the TRANSVERSE TUBULE (or T) system, which activates the sarcoplasmic reticulum to release calcium ions into the sarcoplasm. The calcium ions act by removing the effect of an inhibitory protein so that the filaments which make up a myofibril become linked by crossbridges. Each crossbridge changes its configuration in rapid succession to result in the filaments sliding over one another (contraction). The energy for this process is derived from the breakdown of ATP. On cessation of stimulation (when the T system is no longer depolarized) calcium ions are resorbed into the sarcoplasmic reticulum and relaxation occurs.

excitatory postsynaptic potential /eks-ÿ-tă-tor-ee, -toh-ree- pohst-să-**nap**-tik/ (EPSP) A localized depolarization at an excitatory synapse, due to the release of neurotransmitter from the presynaptic membrane, on stimulation by an impulse. The neurotransmitter acts by increasing the membrane permeability to certain ions. The size of the EPSP depends on the amount of neurotransmitter released; if it is sufficiently large it will set off an action potential in the postsynaptic nerve fiber. It can be raised either by several impulses arriving in quick succession, at one synapse, or by simultaneous impulses arriving at different synapses. *Compare* inhibitory postsynaptic potential.

excretion /eks-**kree**-shŏn/ The process by which excess, waste, or harmful materials, resulting from the chemical reactions that occur within the cells of living organisms, are eliminated from the body. The main excretory products in animals are water, carbon dioxide, salts, and nitrogenous compounds: in unicellular or simple multicellular animals these substances are excreted by diffusion through the cell or body surface, but in more complex animals excretion occurs largely from special organs. In humans and other vertebrates the main excretory organs are the kidneys: they eliminate excess water, salts, and nitrogenous compounds as urine. In addition, the lungs excrete carbon dioxide and water from respiration; the liver excretes bile pigments derived from the breakdown of hemoglobin, and small amounts of water, sodium chloride, and urea are lost from the skin in sweat.

In invertebrates, the excretory organs include Malpighian tubules (of arthropods) and nephridia (of many invertebrates).

exine /eks-in, -ÿn/ (**exosporium**) The outermost wall of a spore or pollen grain. The patterning of the exine may be characteristic of the species and has become an important taxonomic character since the advent of the scanning electron microscope. *See also* pollen analysis.

exocarp /**eks**-oh-karp/ (**epicarp**) *See* pericarp.

exocrine gland /**eks**-oh-krin, -krÿn/ A gland that produces a secretion that passes along a duct to an epithelial surface. The ducts may pass to the body surface (e.g. sweat, lacrimal, and mammary glands), or they may be internal (e.g. in the mouth, stomach, and intestines). *Compare* endocrine gland.

exocytosis /**eks**-oh-sÿ-**toh**-sis/ The bulk transport of materials out of the cell across the plasma membrane. It involves fusion of vesicles or vacuoles with the plasma membrane in a reversal of endocytosis. The materials thus lost may be secretory, excretory (e.g. from autophagic vacuoles), or may be the undigested remains of materials in food vacuoles. Typical secretions are enzymes and hormones from gland cells, often brought to the plasma membrane by Golgi vesicles. *See also* lysosome.

exodermis /**eks**-oh-**der**-mis/ An outermost layer of thickened or suberized cortical cells that sometimes replaces the epidermal layer in the older parts of roots if the epidermal cells have died.

exogamy /**eks**-**og**-ă-mee/ The fusion of gametes produced by organisms that are not closely related. *See* outbreeding.

exogenous /**eks**-**oj**-ĕ-nŭs/ Produced or originating outside an organism. *Compare* endogenous.

exo-intine /**eks**-oh-**in**-tin, -**in**-tÿn/ (**mesosporium**) The middle layer of the wall of a spore or pollen grain between the exine and intine.

exon /**eks**-on/ A segment of a gene that is both transcribed and translated and hence carries part of the code for the gene product. Most eukaryotic genes consist of exons interrupted by noncoding sequences (*see* intron). Both exons and introns are transcribed to heterogeneous nuclear RNA (hnRNA), an intermediary form of messenger RNA (mRNA); the introns are then re-

moved leaving mRNA, which has only the essential sequences and is translated into the protein.

exonuclease /**eks**-oh-**new**-klee-ayss/ An enzyme that catalyzes the hydrolysis of the terminal linkages of polynucleotides such as DNA and RNA, thereby removing terminal nucleotides. *See also* DNase.

exopodite /**eks**-**op**-ŏ-dÿt/ *See* biramous appendage.

exoscopic /**eks**-ŏ-**skop**-ik/ Describing the type of development of a plant embryo in which the apex of the sporophyte develops from the outer cell formed by the first division of the zygote. It is seen in bryophytes and certain pteridophytes, such as the whisk ferns (Psilophyta). *Compare* endoscopic.

exoskeleton /**eks**-oh-**skel**-ĕ-tŏn/ The hard outer covering of the body of certain animals, such as the thick cuticle of arthropods (e.g. insects and crustaceans). It forms a rigid skeleton, which protects and supports the body and its internal organs and provides attachment for muscles. Growth of the body may only occur in stages, by a series of molts (ecdyses) of the cuticle. The term is also applied to other hard external protective structures, including a mollusk shell and the shell of a tortoise. *Compare* endoskeleton.

exosporic /**eks**-oh-**spor**-ik. -**spoh**-rik/ The formation of spores outside the spore-producing organ. In the basidiomycete fungi, for example, the basidiospores are borne on the tips of the sterigmata, which are outgrowths of the basidium. Spore production is more commonly endosporic. *Compare* endosporic.

exosporium /**eks**-oh-**spor**-ee-ŭm, -**spoh**-ree-/ *See* exine.

exotoxins /**eks**-oh-**toks**-inz/ A group of heat-labile proteinaceous toxic substances that are produced by bacteria and secreted into the tissues of their host. Exotoxins may act by interfering with a vital biochemical pathway or the molecular structure of the host cell. Others are neurotoxins. Diseases caused by exotoxins include tetanus, diphtheria, and botulism. *Compare* endotoxins. *See* toxin.

explantation The culture of isolated tissues of adults or embryos in an artificial environment, usually *in vitro*, for maintenance, growth, and/or differentiation. *Compare* implantation; transplantation.

exploratory behavior The early activities of young animals during play, interactions with other individuals, and exploration, through which they learn about their immediate environment and acquire behavioral skills.

exponential growth A type of growth in which the rate of increase in numbers at a given time is proportional to the number of individuals present. Thus, when the population is small multiplication is slow, but as the population gets larger, the rate of multiplication also increases. An exponential growth curve starts off slowly and increases faster and faster as time goes by. However, at some point factors such as lack of nutrients, accumulated wastes, etc., limit further increase, when the curve of number against time begins to level off. The total curve is thus sigmoid (S-shaped).

extinction 1. The loss of an entire species (or other taxon) of organisms, marked by the death of its last living member.
2. The loss of a learned behavior by an animal when that behavior is no longer appropriate or falls into disuse.

extracellular /eks-tră-**sell**-yŭ-ler/ Occurring or situated outside a cell.

extracellular matrix (**ECM**) The viscous watery mixture of proteins, polysaccharides, and other substances that surrounds cells in the tissues of animals. It provides a framework that helps bind cells together, giving strength and integrity to tissues. It is also the environment in which tissue cells migrate and develop in the embryo, and through which they communicate and their

activities are regulated by means of hormones and other signal molecules. The components of the ECM are secreted by the tissue cells themselves, and include three classes of proteins: COLLAGENS, which form various chains, bundles, and other structures; PROTEOGLYCANS, which form the bulk of the matrix; and multiadhesive matrix proteins, which bind to adjacent cells. Connective tissue, such as bone, cartilage, and adipose tissue, contains a relatively large proportion of ECM, sometimes called 'ground substance'.

extrachromosomal DNA /eks-tră-krohmŏ-**soh**-măl/ In eukaryotes, DNA found outside the nucleus of the cell and replicating independently of the chromosomal DNA. It is contained within self-perpetuating organelles in the cytoplasm, e.g. mitochondria, chloroplasts, and plastids, and is responsible for cytoplasmic inheritance.

extraembryonic membranes /eks-trăem-bree-**on**-ik/ (**embryonic membranes**) Structures and membranes developed by embryos for purposes of nutrition or protection and not directly involved in development of embryonic structures. The membranes of later mammalian embryos, particularly human embryos, are known as *fetal membranes*. *See* allantois; amnion; chorion; yolk sac.

extremophile /ik-stree-mŏ-fÿl, -fil/ A prokaryote organism that thrives under extreme environmental conditions, such as very salty pools, glaciers, sulfur-rich hot acid pools, or hot thermal springs. Their enzymes and other cellular components are adapted to function under such exotic conditions, and are often extracted for use in industrial processes. Many extremophiles are archaebacteria.

extrorse /eks-**trors**/ Denoting anthers in which dehiscence lines are to the outside of the flower, promoting cross pollination. *Compare* introrse.

eye An organ of sight, or light perception. Invertebrates usually have eyes that are simple photoreceptors (ocelli), sensitive to

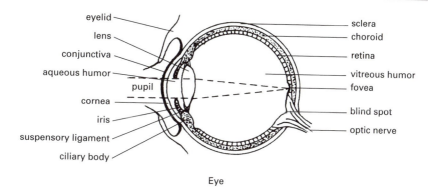

Eye

the direction and intensity of light. The higher mollusks and arthropods have compound eyes that form images. The vertebrate eye is a complicated spherical structure, connected to the brain by the optic nerve. It has an outer white sclerotic coat with a transparent front, called the cornea. This is lined by the vascular pigmented choroid, continuous with the ciliary body and the iris in front. In the center of the iris is a hole, the pupil, through which light enters, to be focused by the lens onto the retina. This is the innermost layer and contains light-sensitive cells (rods and cones).

eyespot (**stigma**) A light-sensitive structure of certain protoctists and invertebrate animals. The eyespot of unicellular and colonial algae and their gametes and zoospores contains globules of orange or red carotenoid pigments. It controls locomotion, ensuring optimum light conditions for photosynthesis. Its location varies. In *Chlamydomonas* it is just inside the chloroplast; in *Euglena* it is near the base of the flagellum. A light-sensitive pigmented spot is also found in the cells of some jellyfish and flatworms, e.g. the miracidium larva of liver fluke.

eye tooth *See* canine tooth.

F_1 The first filial generation; i.e. the first generation that results from a particular hybrid cross.

F_2 The second filial generation, obtained by crossing within the F_1 generation. It is in the F_2 that the characteristic monohybrid and dihybrid ratios become apparent.

facial nerve (cranial nerve VII) One of the pair of nerves that arises from the anterior hindbrain in vertebrates to supply the muscles of the face. It also carries sensory nerve fibers from taste buds and autonomic nerve fibers to salivary glands and the lacrimal gland. *See* cranial nerves.

facilitated diffusion A passive transport of molecules across a cell membrane along a concentration gradient, mediated by carrier molecules or complexes. No energy is expended in this process, but it greatly increases the rate of diffusion through the membrane of molecules that otherwise could not pass or would pass only slowly.

facilitation The phenomenon in which passage of an impulse across a synapse renders the synapse more sensitive to successive impulses so increasing the postsynaptic response. Eventually one stimulus will evoke a response large enough to trigger an impulse. *Compare* summation.

facultative /**fak**-ŭl-tay-tiv/ Occurring under one set of conditions but not another. For example, a *facultative aerobe* is an organism that grows best in the presence of oxygen but can grow without it, i.e. anaerobically.

FAD (flavin adenine dinucleotide) A de-rivative of riboflavin that is a coenzyme in electron-transfer reactions. *See also* flavoprotein.

Fallopian tube /fă-**loh**-pee-ăn/ One of a pair of ducts in female mammals, also called oviducts, that conveys ova from the ovary to the womb (uterus) with the aid of muscular and ciliary action. It opens anteriorly into a ciliated funnel, which lies close to the ovary. The Fallopian tubes each represent the upper region of a Müllerian duct. The tubes are named for the Italian anatomist Gabriel Fallopius (1523–62) and the duct for the German physiologist Johannes Peter Müller (1801–58).

false fruit *See* pseudocarp.

family A collection of similar genera. Families may be subdivided into subfamilies, tribes, and subtribes. Plant family names generally end in *aceae* whereas animal family names usually end in *idae*. Similar families are grouped into orders.

fascia /**fash**-ee-ă/ (*pl.* **fasciae**) A sheet of connective tissue. For example, the layer of adipose tissue under the human dermis and the sheets of tough connective tissue around muscles are types of fasciae.

fascicular cambium /fă-**sik**-yŭ-ler/ *See* intrafascicular cambium.

Fasciola /fas-ee-**oh**-lă/ (**liverfluke**) *See* Trematoda.

fast green *See* staining.

fat Triglycerides of long-chain carboxylic acids (fatty acids) that are solid below 20°C. They commonly serve as energy storage material in higher animals and some plants. *See also* adipose tissue; lipid.

fat body A mass of fatty (adipose) tissue forming a definite structure within the body cavity of some animals. In amphibians and reptiles, a pair of solid fat bodies are attached to the kidneys or near the rectum and act as a food store for use during hibernation and breeding. In insects the fat forms a more diffuse tissue around the gut and reproductive organs and stores protein and glycogen as well as fat.

fate map A map of the probable fates of the various regions of the blastula surface of animals with indeterminate development, usually shown as colors: colors are assigned to the various tissues and structures of the formed embryo (or of any later stage) and, by imagining the developmental process reversed, these colored tissues are followed back to the blastula. The three major germ layers occupy large discrete areas of the fate maps of vertebrates. Mosaic eggs are assigned cell lineages instead of fate maps.

fatty acid *See* carboxylic acid.

faunal regions /**faw**-năl/ *See* zoogeography.

feathers The body covering of birds. They provide heat insulation, help to streamline the body, and are used in flight. A feather has a basal *quill*, which is attached to the feather follicle in the skin and continues into the vane of the feather as the *rachis* (or shaft), which carries two rows of stiff paired *barbs*. Each bears two rows of *barbules*. The barbules of adjacent barbs overlap and hook together to give a firm structure to the vane. The *distal barbules* (on the side of the barb furthest from the quill) bear hooks that engage with the curved edges of the *proximal barbules*.

feces /**fee**-sees/ Solid or semisolid material, consisting of undigested food, bacteria, mucus, bile, and other secretions, that is expelled from the alimentary canal through the anus.

fecundity The capacity of an organism to produce offspring. Most organisms show such enormous fecundity that the size of their population would rapidly increase if all of them survived. In practice most offspring do not survive. *See also* natural selection.

feedback inhibition The inhibition of the activity of an enzyme (often the first) in a reaction sequence by the product of that sequence. When the product accumulates beyond an optimal amount it binds to a site (allosteric site) on the enzyme, changing the shape so that it can no longer react with its substrate. However, once the product is utilized and its concentration drops again, the enzyme is no longer inhibited and further formation of product results. The mechanism is used to regulate the concentration of certain substances within a cell.

Fehling's solution /**fay**-lingz/ A freshly mixed solution used for testing for the presence of reducing sugars (e.g. glucose) and aldehydes in solution. When boiled with equal amounts of Fehling's A (copper(II) sulfate solution) and Fehling's B (sodium potassium tartrate and sodium hydroxide solution) reducing sugars and aldehydes produce a brick-red precipitate of copper(I) oxide.

femur 1. The long bone forming the upper bone or thigh bone of the hindleg in tetrapods, extending from the hip to the knee. Its upper end bears a projecting round head, which articulates with the pelvic girdle at the acetabulum in a ball-and-socket joint. The lower end consists of two articular surfaces (condyles) for the tibia, to form the hinge joint of the knee. 2. The third segment of an insect leg, between the trochanter and tibia.

fenestra ovalis /fi-**nes**-tră oh-**val**-is/ *See* oval window.

fenestra rotunda /roh-**tun**-dă/ *See* round window.

fermentation The breakdown of organic substances, particularly carbohydrates, under anaerobic conditions. It is a form of anaerobic respiration and is seen in certain bacteria, yeasts, some invertebrates, and in skeletal muscle cells during prolonged contraction. The incompletely oxidized products of alcoholic fermentation – ethanol and carbon dioxide – are important in the brewing and baking industries. *See also* glycolysis; lactic acid bacteria.

ferns *See* Filicinophyta.

ferredoxins /fe-ri-**doks**-inz/ A group of red-brown proteins found in green plants, many bacteria and certain animal tissues. They contain non-haem iron in association with sulfur at the active site. They are strong reducing agents (very negative redox potentials) and function as electron carriers, for example in photosynthesis and nitrogen fixation. They have also been isolated from mitochondria.

fertilization (**syngamy**) The fusion of a male gamete with a female gamete to form a zygote; the essential process of sexual reproduction. In animals, a fertilization membrane forms around the egg after the penetration of the sperm, preventing the entry of additional sperm. *External fertilization* occurs when gametes are expelled from the parental bodies before fusion; it is typical of aquatic animals and lower plants. *Internal fertilization* takes place within the body of the female and complex mechanisms exist to place the male gametes into position. Internal fertilization is usually an adaptation to life in a terrestrial environment, although it is retained in secondarily aquatic organisms, such as pondweeds or sea turtles.

Internal fertilization is necessary for terrestrial animals because the male gametes are typically very small and require external water for swimming towards the female gametes. In addition, the propagules produced on land require waterproof integuments, which would be impenetrable to male gametes, so they must be fertilized before being discharged from the female's body. Internal fertilization also allows a considerable degree of nutrition and protection of the early embryo, which is seen in both mammals and seed plants. Because plants are relatively immotile, they are dependent on other agents such as wind or insects to carry the male gamete to the female plant.

fetal membranes /**fee**-tăl/ *See* extraembryonic membranes.

fetus /**fee**-tŭs/ (**foetus**) The embryo of a mammal, especially a human embryo, when its external features resemble those of the mammal after birth, i.e. after it has developed limbs, eyelids, etc. Technically, the term should be restricted to those embryos with an umbilical cord (not a short stalk). In humans it usually refers to the unborn child from after the seventh week of the pregnancy.

Feulgen's stain /**foil**-gĕn/ The stain is named for the German chemist Robert Joachim Feulgen (1884–1955). *See* staining.

fiber 1. (*Botany*) A form of sclerenchyma cell that is often found associated with vascular tissue. Fibres are long narrow cells, with thickened walls and finely tapered ends. Their function is more as supporting tissue than conducting tissue. Where they occur interspersed with the xylem they may be distinguished from tracheids by their narrower lumen. The fibers of many plants (e.g. flax) are economically important.
2. (*Zoology*) A narrow thread of material, usually flexible and having high tensile strength. Examples include the fibers in such tissues as skin, cartilage, and tendons, which are strengthened by the protein collagen; the silk of the web of a spider; the fibroin fibers of the horny sponges; and the fibrin fibers formed from fibrinogen at the site of a wound. The elongated cells of muscles and the axons of neurones are also called fibers. *See also* cytoskeleton.
3. (*Nutrition*) The indigestible fraction of the diet, consisting of various plant cell-wall materials, that passes through the body largely unchanged. Adequate dietary fiber (over 30 g per day in humans) is considered important in the prevention of cer-

tain disorders of the digestive system common in Western societies, e.g. diverticulosis, bowel cancer. Foods high in fiber include cereals, fruit, and vegetables.

fiber-tracheid An elongated cell found in wood, intermediate in form between a fiber and a tracheid.

fibrin /**fy**-brin/ An insoluble protein material that aids blood clotting. It is not present as such in any quantity in blood but is formed from fibrinogen, which is normally present in blood plasma. The conversion from soluble fibrinogen to insoluble fibrin is brought about by the enzyme thrombin. If fresh blood is rapidly whipped a stringy mass of fibrin is obtained.

fibrinogen /fy-**brin**-ŏ-jĕn/ (**Factor I**) A protein present in blood; it is the precursor of fibrin, the structural element of blood clots.

fibrinolysis /fy-bră-**nol**-ă-sis/ The destruction of blood clots as a result of dissolution of fibrin by the enzyme plasmin (fibrinolysin).

fibroblast /**fy**-brŏ-blast/ A cell that produces fibers in connective tissue. Usually they are long flat cells found alongside the fibers. *See also* connective tissue.

fibula /**fib**-yŭ-lă/ One of two long bones of the lower hindlimb of tetrapods. In humans, it is a slender bone bearing little weight. Its upper end articulates with the tibia just behind and below the outer side of the knee and it extends down, lateral to the tibia, to its lower end (the lateral malleolus) – seen as a prominence on the lateral side of the ankle joint. In some species it is reduced and partly or wholly fused with the tibia.

field capacity The point at which the soil contains all the water it can hold by capillary and chemical attraction. Any more water added to soil at field capacity would drain away by gravity.

filament 1. The stalk of the stamen bearing the anther in angiosperms.
2. The vegetative body of the filamentous algae (e.g. *Spirogyra*), composed of a line of similar cells joined by their end walls.
3. (*Zoology*) A narrow threadlike structure. Examples include any of the fine processes on the gills of fish and the shaft of a down feather.

filamentous bacteria *See* actinomycetes.

Filicinophyta /fil-ă-sin-**off**-ă-tă/ (**Pteridophyta**) A phylum of seedless vascular plants comprising the ferns. It is divided into the orders Ophioglossales, Marattiales, Filicales (e.g. *Dryopteris*), Marsileales, and Salviniales, the latter two orders sometimes being included in the Filicales. There are also a number of fossil genera. Ferns have large spirally arranged leaflike fronds (megaphylls) bearing sporangia on their margins or undersurfaces. The sporangia develop from one cell (leptosporangiate) or a number of cells (eusporangiate) and develop into thin- or thick-walled sporangia respectively.

filoplumes /**fil**-ŏ-ploomz, **fy**-lŏ-/ Hairlike feathers consisting of a bare shaft (*rachis*) tipped with a few barbs that are not held together by barbules. *See* feathers.

filter feeding The method of feeding of some aquatic animals, especially invertebrates, in which small suspended food particles are strained from the surrounding water. Many animals simply allow water to flow over or through them, but others actively produce a current, often with cilia. The type of filter used varies from minute hairlike cilia to the large horny plates of certain whales. *See also* ciliary feeding.

fimbriae /**fim**-bree-ee/ *See* pili.

fine structure *See* ultrastructure.

fins Flattened organs, usually supported by fin rays, used in swimming. Most fish have paired *pectoral* and *pelvic fins*, homologous to the forelimbs and hindlimbs, respectively, of terrestrial vertebrates and used for controlling the angle of ascent or descent. In some species the pelvic fins are small and modified for copulation. Some fish also have one or more *dorsal fins* and one or more *ventral fins* (*anal fins*), used for preventing sideways movement and rolling, and a *caudal fin* (*tail fin*), used for propelling the fish in continuous forward motion.

fish *See* Chondrichthyes; Osteichthyes.

fission /**fish**-ŏn/ A type of asexual reproduction in which a parent cell divides into two (binary fission) or more (multiple fission) similar daughter cells. Binary fission occurs in many unicellular organisms (protoctists, bacteria); multiple fission occurs in apicomplexans, such as the malaria parasite *Plasmodium*. Fission begins with division of the nucleus by mitosis, followed by cytoplasmic division and sometimes sporulation.

fitness In an evolutionary context, the ability of an organism to produce a large number of offspring that survive to a reproductive age. 'Fit' in this sense has nothing to do with being healthy although healthy animals and plants are more likely to leave more offspring than weak individuals. In human populations fitness is affected more by social conditions and traditions than by health and indeed large families are more often found in the poorer Third World countries in which people are generally less healthy. The phrase 'survival of the fittest' summarizes the principles of the theory of NATURAL SELECTION.

fixation (*Genetics*) The situation in a usually small population when an allele is either completely lost or reaches a 100% frequency.

fixing In the preparation of microscope slides, the process by which tissues are rapidly killed by chemicals to preserve their morphology. As well as preventing deterioration, fixing agents (fixatives) should also render cell organelles and inclusions more visible and harden the tissue to prevent shrinkage and distortion during dehydration, embedding, and sectioning. Examples of fixatives are formaldehyde and acetic acid. *See also* dehydration; embedding.

flaccid /**flaks**-id/ Lacking turgor. *See* plasmolysis.

Flagellata /flă-**jel**-ă-tă/ *See* Mastigophora.

flagellum /flă-**jel**-ŭm/ (*pl.* **flagella**) A whiplike extension of prokaryote cells with a BASAL BODY at its base, whose beat causes locomotion of the cell. Strictly the term is now reserved for the bacterial flagellum. The flagella and cilia of eukaryote cells have a quite different structure and are called undulipodia. Bacterial flagella are much simpler than undulipodia, being hollow cylinders about 15 nm in diameter, consisting of subunits of a protein (flagellin) arranged in helical spirals. Unlike eukaryote flagella they are not membrane-bounded, are rigid, and function by a complex rotation of their bases. *Compare* undulipodium.

flame cell (**solenocyte**) A cup-shaped cell that contains a group of cilia and occurs in large numbers within the tissues of various invertebrates (e.g. Platyhelminthes, Rotifera, and some Annelida). The beating of the cilia draws fluid wastes from the surrounding tissues into the flame-cell cavity, which connects with the excretory duct leading to the body surface. *See also* protonephridium.

flatworms *See* Platyhelminthes.

flavin /**flay**-vin/ A derivative of riboflavin occurring in the flavoproteins; i.e. FAD or FMN.

flavin adenine dinucleotide *See* FAD.

flavin mononucleotide /mon-oh-**new**-klee-ŏ-tÿd/ *See* FMN.

flavone /**flay**-vohn/ *See* flavonoid.

flavonoid /**flay**-vŏ-noid/ One of a common group of plant compounds having the

petal

anther

filament

sepal

pedicel

stigma

style

ovary

ovule

receptacle

Typical half-flower

C_6–C_3–C_6 chemical skeleton in which C_6 is a benzene ring. They are an important source of non-photosynthetic pigments in plants. They are classified according to the C_3 portion and include the yellow chalcones and aurones; the pale yellow and ivory flavones and flavonols and their glycosides; the red, blue, and purple anthocyanins and anthocyanidins; and the colorless isoflavones, catechins, and leukoanthocyanidins. They are water soluble and usually located in the cell vacuole. *See* anthocyanin.

flavonol /**flay**-vŏ-nol, -nohl/ A plant pigment that modifies the effects of certain growth substances. *See* flavonoid.

flavoprotein /flay-voh-**proh**-teen, -tee-in/ A conjugated protein in which a flavin (FAD or FMN) is joined to a protein component. They are enzymes in the electron-transport chain.

fleas *See* Siphonaptera.

flies *See* Diptera.

flocculation /flok-yŭ-**lay**-shŏn/ The aggregation of soil particles into crumbs. Compacted structureless clay soils can be flocculated by the addition of neutral salts, particularly of calcium. The addition of

lime to saturated clay soils (*liming*) is a common agricultural practice.

floral formula A series of symbols used to describe the structure of a flower. The various whorls of structures are abbreviated as follows: K (calyx), C (corolla), P (perianth), A (androecium), and G (gynoecium). These letters are each followed by a number indicating the number of parts in the whorl. If the number exceeds 12 then the symbol ∞ is used to denote an indefinite number. Fusion of parts of the whorl is indicated by placing the number of parts in brackets. The position of the gynoecium is shown by a line above or below the letter G denoting an inferior or superior ovary respectively. The formula is preceded by ⊕ to indicate actinomorphic flowers and ·|· for zygomorphic flowers. The floral formula for the buttercup would be written:

⊕K5 C5 A∞ G∞

florigen /**flor**-ă-jĕn, **floh**-ră-/ (**flowering hormone**) A hypothetical plant hormone that has been postulated to account for the transfer of photoperiod stimulus from the leaves to the apex where flowering is induced. Attempts to isolate florigen have so far been unsuccessful. *See* photoperiodism.

flower The characteristic reproductive structure of an angiosperm. It usually consists of an axis or receptacle bearing the sepals, petals, and stamens. The gynoecium is borne either above the receptacle or enclosed within it. Flower structures are extremely variable and show numerous adaptations to promote pollination and seed dispersal. *See also* floral formula.

flowering hormone *See* florigen.

flowering plants *See* Anthophyta.

flukes *See* Trematoda.

fluorine /**floo**-ŏ-reen, -rin/ *See* trace element.

fluorocarbon /floo-ŏ-rŏ-**kar**-bŏn/ A compound derived from a hydrocarbon by replacing hydrogen atoms with fluorine atoms. Fluorocarbons are unreactive and most are stable up to high temperatures. They have a variety of uses – in aerosol propellants, oils and greases, and synthetic polymers such as PTFE. *See also* chlorofluorocarbon.

FMN (**flavin mononucleotide**) A derivative of riboflavin that is a coenzyme in electron-transfer reactions. *See also* flavoprotein.

foetus /fee-tŭs/ *See* fetus.

folic acid /**fol**-ik, **foh**-lik/ (**pteroylglutamic acid**) A water-soluble vitamin belonging to the VITAMIN B COMPLEX. The principal dietary sources of folic acid are leafy vegetables, liver, and kidney. Deficiency of the vitamin exhibits itself in anemia in a similar manner to vitamin B_{12} deficiency, while deficiency during pregnancy increases the risk of birth defects in children.

Folic acid is important in metabolism in various coenzyme forms, all of which are specifically concerned with the transfer and utilization of the single carbon (C_1) group. Before functioning in this manner folic acid must be reduced to either dihydrofolic acid (FH_2) or tetrahydrofolic acid (FH_4). It is important in the growth and reproduction of cells, participating in the synthesis of purines and pyrimidines.

follicle /**fol**-ă-kăl/ 1. (*Botany*) A dry dehiscent fruit formed from one carpel that splits along one edge to release its seed, for example columbine fruit.
2. (*Zoology*) A small cavity or sac within an organ or tissue. Follicles within the ovary, for example, contain developing ova. *See* Graafian follicle; hair follicle.

follicle-stimulating hormone (**FSH**) A gonadotropin, also called follitropin, produced by the anterior pituitary gland. It acts on the ovary to stimulate the growth and maturation of the tissues forming follicles and ova, which, under the action of luteinizing hormone, mature and are released from the ovary. It also stimulates spermatogenesis in males. It has been used in the treatment of female sterility.

fontanelle /fon-tă-**nelle**/ A gap in the cranium where the bone has not yet formed, being closed only by a membrane. For example, the *anterior fontanelle* is found in newborn babies, situated on top of the head between the frontal and two parietal bones. It is completely closed after about 18 months.

food chain The chain of organisms existing in any natural community, through which food energy is transferred. Each link in the food chain obtains energy by eating the one preceding it and is in turn eaten by the organisms in the link following it. At each transfer a large proportion (80–90%) of the potential energy is lost as heat, therefore the number of links in a sequence is limited, usually to four or five. The shorter the food chain, the greater the available energy, so total energy can be increased by cutting out a step in the food chain, for example if people consume cereal grain instead of consuming animals that eat cereal grains.

Food chains are of two basic types: the grazing food chain, which goes from green plants, to grazing herbivores, and finally to carnivores; and the detritus food chain, which goes from dead organic matter, to microorganisms, and then to detritus-feeding organisms. The food chains in a community are interconnected with one another, because most organisms consume more than one type of food, and the interlocking pattern is referred to as a *food web* or *food cycle*. *See* trophic level.

food web *See* food chain.

foramen /fŏ-**ray**-mĕn/ (*pl.* **foramina**) A natural opening in an animal organ or other structure, especially in a bone or cartilage.

foramen magnum /**mag**-nŭm/ The opening in the skull through which the spinal cord passes.

forebrain /**for**-brayn, **fohr**-/ (**prosencephalon**) The most anterior of the three basic anatomical regions of the brain. It consists principally of the cerebrum, with its pair of greatly enlarged lateral outgrowths (cerebral hemispheres), the thalamus, and the hypothalamus. *Compare* hindbrain; midbrain.

foregut /**for**-gut, **fohr**-/ The first part of the alimentary canal of arthropods and vertebrate animals.

In arthropods, it consists of the buccal cavity, esophagus, crop, and gizzard. The foregut is lined with epithelium secreting a protective lining of chitin similar to the exoskeleton. This lining has to be shed at ecdysis. Embryologically it arises from the stomodaeum, which is lined with ectoderm.

In vertebrates, it consists of that part of the alimentary canal anterior to the bile duct. *Compare* hindgut; midgut.

form 1. (*Botany*) The lowest taxonomic group, ranking below the variety level. Subforms may also be recognized.
2. (*Zoology*) A vague term used when the appropriate taxonomic rank is not clear. It may also be applied to seasonal variants and the different forms found in polymorphic series.

formalin /**for**-mă-lin/ A mixture of about 40% formaldehyde (methanal), 8% methyl alcohol (methanol), and 52% water (the methyl alcohol is present to prevent polymerization of the formaldehyde). It is a powerful reducing agent and is used as a disinfectant, germicide, and fungicide and also as a general preserving solution. In contact with the skin formalin may cause irritant dermatitis and ingestion can cause severe abdominal pain.

formic acid /**for**-mik/ (**methanoic acid**) The most simple carboxylic acid, HCOOH. It is present in ant bites and contributes to the nettle sting reaction.

fossil /**fos**-ăl/ The remains of, or impressions left in rocks by, long dead animals and plants. Most fossils consist of hard skeletal material because soft tissues and organs rot away very quickly. Mineral salts from surrounding rocks gradually replace the hard organic material, to give a cast in a process termed *petrification*. Alternatively the organic material dissolves away leaving an impression or mold in the surrounding rocks. Trace fossils (e.g. of dinosaur footprints) provide indirect evidence of prehistoric life forms. *See also* chemical fossils.

founder effect The impact on the genetic make-up of a new population that results from its being founded by a few individuals as an offshoot of an existing population. Because the founder members are an unrepresentative genetic sample of the main population, they are likely to bring only a limited set of alleles, meaning that the offshoot population is genetically different from the main population. If this genetic divergence is maintained, it might ultimately cause the new population to evolve into a new species.

fovea /**foh**-vee-ă/ (*pl.* **foveae**) The point of most acute vision on the retina of the vertebrate eye. The image of an object falls on this region, which is directly opposite the center of the pupil and lens. It consists entirely of densely packed cones for daylight vision. Some birds have two foveae for sharp forward and lateral vision. *See also* retina. *See illustration at* eye.

fragmentation A form of asexual reproduction found in certain metazoan animals such as some aquatic annelids (e.g. *Lumbriculus*) and sea anemones. It describes the breaking up of the body into two or more pieces that subsequently develop into complete organisms. A form of fragmentation (*strobilization*) is seen in the formation of medusae in jellyfish.

frameshift A shift in the READING FRAME that results in the sequence of bases in messenger RNA (mRNA) being misread by the ribosome during protein synthesis. It is caused by a gene mutation involving the insertion or deletion of one or more base pairs in the DNA, with the error being carried over into mRNA during transcription. The presence of an extra or a missing base means that the ribosome binds incorrectly to all subsequent base triplets (codons) during TRANSLATION, resulting in an abnormal protein or premature termination of protein synthesis.

fraternal twins (**dizygotic twins**) Two offspring born to the same mother at the same birth, resulting from the fertilization of two eggs at the same time. They may be of unlike sex and are no more genetically similar than any two siblings. *Compare* identical twins. *See also* freemartin.

freemartin /**free**-mar-tin/ A masculinized but genetically female calf. In some mammals (e.g. cattle) FRATERNAL TWINS share a common blood circulation in the uterus and this may result in a female twin receiving male hormones from her brother. The freemartin shows some male characters and is sterile.

free radical An atom or group of atoms that possesses an unpaired electron and is highly reactive. Free radicals, such as the superoxide anion O_2^-, are formed in living tissues during normal metabolism, or following exposure to infection or toxins, and are generally 'mopped up' in reactions with ANTIOXIDANTS to prevent them from damaging tissue.

freeze fracturing A method of preparation of material for electron microscopy, particularly useful for studying membranes. Material is frozen rapidly (e.g. by immersion in liquid nitrogen) thus preserving it in lifelike form. It is then fractured, usually with a sharp knife. The fracture plane tends to follow lines of weakness, such as between the two lipid layers of membranes, revealing their internal surfaces. Replicas made of the surfaces are shadowed for examination in the electron microscope. In *freeze etching* the fractured surface is etched, i.e. some ice is allowed to sublime away, before shadowing. This exposes further structure, such as the outer surface of the membrane. In this way membranes have been shown to contain particles, e.g. quantasomes. *See* shadowing.

frogs *See* Anura.

frond A term usually applied to large well-divided leaves as found in ferns, palms, and cycads. The leaflike thalli of certain algae and lichens may also be termed fronds. *See also* megaphyll.

fructose /**fruk**-tohs, **frûk**-/ A sugar ($C_6H_{12}O_6$) found in fruit juices, honey, and cane sugar. It is a ketohexose, existing in a pyranose form when free. In combination (e.g. in sucrose) it exists in the furanose form. *See* furanose; pyramose.

fruit The ripened ovary of a flower that is usually formed following fertilization of the ovule. It may consist of the ripened ovary only or include other parts of the flower. Fruits vary according to the method of seed dispersal, succulent fruits normally being distributed by animals while dry fruits may be dispersed by wind or water. Fruits are classified according to how the ovary wall (pericarp) develops, depending on whether it becomes fleshy or hard. Fruits are further classified according to whether or not the fruit wall opens to release the seeds. *See also* composite fruit; pseudocarp.

FSH *See* follicle-stimulating hormone.

fucoxanthin /fyoo-koh-**zan**-theen, -thin/

A xanthophyll pigment of diatoms, brown algae, and golden brown algae. The light absorbed is used with high efficiency in photosynthesis, the energy first being transferred to chlorophyll *a*. It has three absorption peaks covering the blue and green parts of the spectrum.

Fucus /**fyoo**-kŭs/ A genus of parenchymatous marine brown algae commonly found between high- and low-tide marks. The gametophyte thallus is differentiated into a holdfast, which anchors the plant to the substrate, a stalk or stipe, and a large blade upon the ends of which the reproductive organs are borne. Reproduction is oogamous, the large female gamete being nonmotile. The life cycle is haplobiontic, the gametophyte being diploid and the zygote giving rise directly to the new gametophyte generation.

fumaric acid /fyoo-**ma**-rik/ An unsaturated dicarboxylic acid, which occurs in many plants. The fumarate ion participates in several important metabolic pathways, e.g. the Krebs cycle, purine pathways, and the urea cycle. *See also* butenedioic acid.

Funaria /fyoo-**nair**-ee-ă/ *See* Musci.

functional group A group of atoms in a compound that is responsible for the characteristic reactions of the type of compound. Examples are:

alcohol –OH
aldehyde –CHO
amine –NH_2
ketone =CO

carboxylic acid –CO.OH
acyl halide –CO.X (X = halogen)
nitro compound –NO_2
sulfonic acid –SO_2.OH
nitrile –CN
diazonium salt –N_2^+
diazo compound –N=N–

Fungi /**fun**-jÿ/ A kingdom of nonphotosynthetic mainly terrestrial organisms that are now regarded as quite distinct from plants or other living kingdoms. They are characterized by having cell walls made chiefly of chitin, not the cellulose of plant cell walls, and they all develop directly from spores without an embryo stage. Moreover, undulipodia (cilia or flagella) are never found in any stage of their life cycles. Fungi are generally saprophytic or parasitic, and may be unicellular or composed of filaments (termed hyphae) that together comprise the fungal body or mycelium. Hyphae may grow loosely or form a compacted mass of pseudoparenchyma giving well-defined structures, as in toadstools. *See also* ascomycetes; basidiomycete; zygomycetes.

Fungi Imperfecti /im-per-**fek**-tÿ/ *See* Deuteromycota.

funicle /**fyoo**-nă-kăl/ The stalk attaching the ovule to the placenta in angiosperm ovaries. *See illustration at* ovule.

furanose /**fyoo**-ră-nohs/ A sugar that has a five-membered ring (four carbon atoms and one oxygen atom). *See also* sugar.

fusiform initial /**fjoo**-ză-form/ *See* initial.

GABA *See* gamma-amino butyric acid.

galactose /gǎ-**lak**-tohs/ A sugar found in lactose and many polysaccharides. It is an aldohexose, isomeric with glucose. *See also* sugar.

gall An abnormal growth or swelling on a plant caused by infection (e.g. by bacteria or fungi) or attack by insects, mites, nematodes, or other predators. The shape and structure of the gall and its location on the plant vary according to the organism responsible. Some galls are simple and undifferentiated, while others have multiple layers of differentiated gall tissue. The mechanism of gall formation is understood only in certain cases. For instance, the soil bacterium *Agrobacterium tumefaciens* is responsible for crown gall, a tumor on the stem just below the soil surface. The bacterium transfers some of its DNA into cells of its plant host, transforming them into tumor cells. *See Agrobacterium*.

gall bladder /gawl/ A saclike extension of the bile duct that occurs in many vertebrates, situated between the liver lobes. It serves as a temporary store for bile, releasing it in response to food in the duodenum. The release of bile is controlled by the hormone cholecystokinin. *See illustration at* alimentary canal.

gametangium /gam-ĕ-**tan**-jee-ŭm/ (*pl.* gametangia) A structure in which sexual cells (gametes) are produced, the term commonly being used with reference to reproduction in the algae and fungi.

gamete /gam-eet/ A cell capable of fusing with another cell to produce a zygote, from which a new individual organism can de-velop. Gametes may have similar structure and behavior (*isogametes*), as in many simple organisms, but are usually dissimilar in appearance and behavior (*anisogametes*). The typical male gamete is small, motile, and produced in large numbers. The typical female gamete is large because of the food reserves it contains, immotile, and is produced in small numbers. Fusion of gametes results in the nucleus of the zygote having exactly twice the number of chromosomes present in the nucleus of each gamete. *See also* ovum; spermatozoon.

gametogenesis /gam-ĕ-toh-**jen**-ĕ-sis/ The formation of sex cells or gametes, i.e. ova or spermatozoa. *See also* oogenesis; spermatogenesis.

gametophyte /gǎ-**mee**-tŏ-fÿt/ The generation of a plant life cycle that is haploid and produces sex organs. It is the main generation in the life cycle of bryophytes, in which the sporophyte is completely or partially dependent upon it. The prothallus of pteridophytes is the gametophyte. *Compare* sporophyte. *See also* alternation of generations; diploid; haploid.

gamma-amino butyric acid /byoo-**ti**-rik/ (GABA) An inhibitory neurotransmitter derived from the amino acid glutamate. It occurs in the central nervous system of vertebrates, and at neuromuscular junctions of crustaceans and annelids.

gamma globulin *See* globulin.

ganglion /**gang**-glee-ŏn/ (*pl.* ganglia) A collection of nerve cell bodies, usually bound by a sheath or capsule. In vertebrates the ganglia are located chiefly outside the central nervous system; in invertebrates ganglia occur along the major nerve cords and are the centers of nervous integration.

ganoid scales /**gan**-oid/ Hard ridged rhomboidal scales found in the skin of sturgeons and gars. The outer layer consists of *ganoin*, a calcareous substance secreted by the dermis; the inner layer consists of *isopedin*. *Compare* cosmoid scales. *See* scales.

gap junction A type of junction that connects close-packed cells in animal tissues and consists of a cluster of hollow cylinders spanning the intercellular gap. The wall of each cylinder is made of proteins (called connexons), and encloses a water-filled channel that is continuous with the cytoplasm of the neighboring cells. The channels permit the passage of ions and small molecules, and are important in cell–cell communication, for example in coordinating the contraction of smooth muscle cells.

gas chromatography A technique widely used for the separation and analysis of mixtures. Gas chromatography employs a column packed with either a solid stationary phase (*gas-solid chromatography* or *GSC*) or a solid coated with a non-volatile liquid (*gas-liquid chromatography* or *GLC*). The whole column is placed in a thermostatically controlled heating jacket. A volatile sample is introduced into the column using a syringe, and an unreactive carrier gas, such as nitrogen, passed through it. The components of the sample will be carried along in this mobile phase. However, some of the components will cling more readily to the stationary phase than others, either because they become attached to the solid surface or because they dissolve in the liquid. The time taken for different components to pass through the column is characteristic and can be used to identify them. The emergent sample is passed through a detector, which registers the presence of the different components in the carrier gas.

Two types of detector are in common use: the katharometer, which measures changes in thermal conductivity, and the flame-ionization detector, which turns the

volatile components into ions and registers the change in electrical conductivity.

gas–liquid chromatography *See* gas chromatography.

gas–solid chromatography *See* gas chromatography.

gastric juice A mixture of digestive enzymes and hydrochloric acid secreted by gastric glands situated in the thick stomach wall. It contains two main enzymes: pepsin, which breaks proteins down into short polypeptide chains, and, in milk-feeding young mammals, chymosin (rennin), which coagulates caseinogen to form casein. The mechanical and chemical stimulation of the stomach lining by food itself causes secretion of gastric juice and of a hormone (gastrin). This hormone circulates in the blood and causes the gastric glands to secrete hydrochloric acid, thus generating the low (acid) pH of the stomach.

gastric mill *See* gizzard.

gastrin /**gas**-trin/ A polypeptide hormone, secreted by the stomach, that stimulates secretions of gastric acid and pepsin in the stomach. It is released in response to the presence of components of the meal in the stomach.

Gastropoda /gas-**trop**-ŏ-dă/ A large class of mollusks containing the terrestrial slugs (e.g. *Limax*) and snails (e.g. *Helix*), which lack gills and have an air-breathing lung, as well as many marine and freshwater members, e.g. *Patella* (limpet) and *Limnaea* (pond snail).

Gastropods are characterized by a well-developed head with tentacles and eyes, a single shell, and a large flat foot. They always undergo torsion during their development, i.e. the visceral hump twists through 180° so that the mantle cavity, gills, and anus are anterior and the other organs are asymmetrically arranged. The shell and visceral hump are often spirally coiled.

gastrula /**gas**-trû-lă/ *See* gastrulation.

gastrulation /gas-trû-**lay**-shŏn/ The stage of animal embryonic development at which the gut cavity and germ layers first appear. In most animals gastrulation follows cleavage and precedes neurulation (development of the nervous system); the embryo at this stage is called a *gastrula*. In most animals gastrulation is the stage when the embryo's main features are established or determined by interaction of the primary organizer with other tissues of the embryo. In some animals (e.g. nematodes), however, there are no such interactions, and gastrulation is simply a mechanical folding in of gut and the other internal structures. *See* organizer.

gel /jel/ A lyophilic colloid that is normally stable but may be induced to coagulate partially under certain conditions (e.g. lowering the temperature). This produces a pseudo-solid or easily deformable jelly-like mass, called a gel, in which intertwining particles enclose the whole dispersing medium. Gels may be further subdivided into elastic gels (e.g. gelatin) and rigid gels (e.g. silica gel).

gel filtration (**gel-permeation chromatography**) A chromatographic method using a column packed with porous gel particles. It is a standard technique used for separating and identifying macromolecules of various sizes, e.g. proteins or nucleic acids. A solution of the mixture of macromolecules is added to the top of the column and allowed to flow through by gravity. The smaller molecules are hindered in their passage down the column because they are better able to penetrate the hydrated pores within the particles of the gel. Molecules too large to penetrate the pores are excluded, and thus flow more rapidly through the column. By analyzing the liquid that drips from the bottom of the column (the eluate) at set intervals and comparing it with a standard (obtained by running a known macromolecule through the column) information about the sizes and molecular weights of the components

of the mixture is gathered. The most frequently used commercial gel is Sephadex.

gemma /**jem**-ă/ (*pl.* **gemmae**) An organ of vegetative reproduction produced in mosses, liverworts, and certain pteridophytes, e.g. *Lycopodium selago*. They often form in groups in receptacles called *gemma-cups* and eventually become detached from the parent to form new plants.

gemmation /jĕ-**may**-shŏn/ **1.** (*Botany*) A type of asexual reproduction, seen in mosses and liverworts, involving the production of a group of cells (a *gemma*) that develops into a new individual, before or after separation from the parent. **2.** (*Zoology*) *See* budding.

gene /jeen/ In classical genetics, a unit of hereditary material located on a chromosome that, by itself or with other genes, determines a characteristic in an organism. It corresponds to a segment of the genetic material, usually DNA (although the genes of some viruses consist of RNA). Genes may exist in a number of forms, termed *alleles*. For example, a gene controlling the characteristic 'height' in peas may have two alleles, one for 'tall' and another for 'short'. In a normal diploid cell only two alleles can be present together, one on each of a pair of homologous chromosomes: the alleles may both be of the same type, or they may be different. The segregation of alleles at meiosis and their dominance relationships are responsible for the particulate nature of inheritance. Genes can occasionally undergo changes, called MUTATIONS, to new allelic forms.

Although the DNA molecules of the chromosomes account for the great majority of genes, genes are also found in certain DNA-containing cell organelles (e.g. mitochondria, plastids).

In classical terms, a gene can be defined as the smallest hereditary unit capable either of recombination or of mutation or of controlling a specific function. These three definitions do not necessarily describe the same thing and a unit of function, a *cistron*, may be much larger than a unit of recombination or mutation. In molecular terms, a gene is the entire nucleic acid sequence that is required for the cell to manufacture a given product, whether a polypeptide or an RNA. This definition includes not only the coding region (the cistron), which is transcribed into RNA, but other transcription-control regions, which may be distant from the cistron.

There are essentially three types of gene: (i) *structural genes*, which code for polypeptides of enzymes and other proteins; (ii) *RNA genes*, which code for ribosomal RNA and transfer RNA molecules used in polypeptide assembly; and (iii) *regulator genes*, which regulate the expression of the other two types. *See also* operon.

gene cloning (DNA cloning) A technique of genetic engineering whereby a gene or other DNA segment is replicated, giving many identical copies. The gene sequence is isolated by using restriction endonucleases, or by making a complementary DNA from a messenger RNA template using a reverse transcriptase. It is then inserted into the circular chromosome of a cloning vector, i.e. a plasmid or a bacteriophage. The hybrid is used to infect a bacterium, usually *Escherichia coli*, and is replicated within the bacterial cell. A culture of such cells produces many copies of the gene, which can subsequently be isolated and purified.

Since its introduction in the mid-1970s, gene cloning has transformed study of the molecular structure of genes. Moreover, expression of the cloned genes enables the host cell, whether prokaryote or eukaryote, to produce novel proteins, which is the basis of GENETIC ENGINEERING.

genecology /jeen-ĕ-kol-ŏ-jee/ The study of population genetics in relation to the environment.

gene family A set of genes with similar structures that encode proteins with similar but nonidentical amino-acid sequences. Such genes often lie close to each other on the same chromosome, probably having arisen by duplication of a single ancestral gene. Over time the duplicate genes acquire mutations that alter the amino-acid sequence and structure of the ancestral protein, so changing its functional capabilities. This results typically in a cluster of closely linked genes whose proteins have evolved distinct but related functions. The classic example is the human β-globin gene family, whose proteins are components of the blood pigment hemoglobin. This family comprises a cluster of five functional genes and two nonfunctional PSEUDOGENES on chromosome 11. The five different β-globins encoded by this gene family have distinct properties that suit them for different physiological roles.

gene flow The movement of alleles between populations through interbreeding.

gene frequency The proportion of an allele in a population in relation to other alleles of the same gene. *See* allele frequency.

gene knockout (knockout) A technique for selectively inactivating a certain gene or genes within a living cell or organism. It involves replacing the normal alleles with mutant nonfunctional alleles, and can be applied to experimental organisms such as yeasts, nematodes, and mice, to study how normal function is disrupted by loss (i.e. 'knockout') of the gene. The target gene is cloned and then inactivated by making changes in its nucleotide sequence. Using a suitable vector, copies of the inactivated gene are introduced to normal cells, and in a small fraction of cases the gene is integrated into the chromosome in place of the normal allele. Pure breeding lines of knockout mice, homozygous for particular knockouts, are produced by a complex procedure that requires the isolation of successfully targeted cells, embryo manipulation, and selective breeding. These animals have many uses, for example to study human genetic diseases and to investigate the effectiveness of potential new drugs. *See also* gene silencing.

gene library (DNA library) A collection of cloned DNA fragments, usually held in the form of petri-dish cultures of bacterial colonies or plaques of phage-cloning vectors. A *genomic DNA library* represents the entire genome of a particular species, whereas a *cDNA library* contains complementary DNA (cDNA) copies of messenger RNA (mRNA) molecules, for example from a particular tissue or cell type. The genetic material of the organism is first broken up randomly into fragments using restriction enzymes, for example. Then each fragment is cloned using a vector (e.g. plasmid or bacteriophage) inside a suitable host, such as the bacterium *E. coli*. Gene libraries are screened generally using radioactively labeled DNA or RNA probes, having first denatured the cloned DNA into single-stranded form. The probe hybridizes with any complementary DNA sequence, enabling identification of the relevant plaque or bacterial colony, which can be replated and cultured for further analysis.

gene pool The total number and variety of genes existing within a breeding population or species at a given point in time.

gene probe *See* DNA probe.

generation time The average time between the cell division of parent and daughter cells within a population of cells.

generative nuclei The two gametic nuclei found in the pollen tube. One fuses with the egg cell to form a zygote while the other either degenerates or, in certain angiosperms, fuses with the polar nucleus to give the primary endosperm nucleus.

gene sequencing Determination of the order of bases of a DNA molecule making up a gene. Sequencing requires multiple cloned copies of the gene; long DNA sequences are cut into more manageable lengths using restriction enzymes. Since these cleave DNA at specific points, it is possible to reconstitute the overall sequence once the constituent fragments have been analyzed individually by the methods outlined below.

There are two methods of sequencing DNA. One is the *chemical cleavage method*, or *Maxam–Gilbert method*. This involves firstly labeling one end of the DNA of the gene or DNA segment with radioactive ^{32}P. The segment is then subjected to a chemical reaction that cleaves the sequence at positions occupied by one of the four bases, say, adenine. Starting with numerous cloned DNA segments, the result is a set of radioactive fragments extending from the ^{32}P label to each successive position of adenine in the segment. This process is repeated for the other three bases, and the four sets of fragments are then separated according to the number of nucleotides they contain by gel electrophoresis, in adjacent lanes on the gel. The sequence can then be deduced directly from the autoradiograph of the gel.

The second method is the *chain-termination* or *dideoxy method* (also called the *Sanger method*). A single-stranded segment of DNA taken from the gene is used as a template to replicate a new DNA strand using the enzyme DNA polymerase. The enzyme is provided with the four normal nucleoside triphosphates (ATP, GTP, CTP, TTP), plus the dideoxy (dd) derivative of one of them, say ddATP. Incorporation of the dideoxy derivative causes replication of the new strand to cease at that point. Hence, the result is a set of new strands of varying length, terminating at all the different positions where adenine, say, normally occurs in the sequence. The process is repeated in turn for each of the three remaining bases, and the set of fragments from each incubation are separated according to size by gel electrophoresis. The sequence of bases in the newly synthesized strand can be deduced directly from the gel.

Automation of both procedures, for example by using laser scanning of fluorescent dye markers instead of autoradiography, has greatly increased the speed with which DNA can be sequenced, and made possible the analysis of entire genomes, including the human genome.

gene silencing Suppression of gene expression, so that the manufacture of gene product(s) by the affected cell is reduced or even 'switched off' completely. Silencing can operate at various stages of transcription or translation, and can occur naturally or be induced artificially. For example, transcription can be prevented by the binding of repressor proteins to certain regulatory sites (silencers) within the DNA as one of the normal cellular mechanisms of gene regulation. Silencing at stages following transcription, or *post-transcriptional gene silencing* (PTGS), can be induced by introduced 'foreign' genes (transgenes), certain viruses, or by the presence of small RNA molecules, particularly short double-stranded RNAs called small interfering RNAs (siRNAs) – the phenomenon called RNA INTERFERENCE). The latter is now widely used experimentally for investigating gene function.

gene splicing 1. The joining of exons after the intron sequences have been removed, to produce functional messenger RNA. In the nucleus this is performed by a special assemblage of RNA and proteins called a *spliceosome*.
2. In genetic engineering, the joining of DNA fragments by the action of the enzyme DNA ligase.

genetically modified organism (GMO) An organism whose genetic material has been altered by GENETIC ENGINEERING so that it incorporates and expresses one or more foreign genes.

genetic code /jĕ-**net**-ik/ The sequence of bases in either DNA or messenger RNA that conveys genetic instructions to the cell. The basic unit of the code consists of a group of three consecutive bases, the base triplet or codon, which specifies instructions for a particular amino acid in a polypeptide, or acts as a start or stop signal for translation of the message into polypeptide assembly. For example, the DNA triplet CAA (which is transcribed as GUU in mRNA) codes for the amino acid valine. There are 64 different triplet combinations but only 20 amino acids; thus many amino acids can be coded for by two or more triplets. The code is said to be *degenerate*, since it appears that only the first two bases, and in certain cases only one base, are necessary to insure the coding of a specific amino acid. Three triplets, termed 'nonsense triplets', do not code for

any amino acid and have other functions, e.g. marking the beginning and end of a polypeptide chain. *See also* codon.

genetic drift (Sewall Wright effect) The fluctuation of allele frequencies in a small population due entirely to chance. If the number of matings is small then the actual numbers of different types of pairing may depart significantly from the number expected on a purely random basis. Genetic drift is one of the factors that can disturb the Hardy–Weinberg equilibrium. The effect is named for the American statistician and geneticist Sewall Wright (1889–1988).

genetic engineering (recombinant DNA technology) The direct introduction of foreign genes into an organism's genetic material by micromanipulation at the cell level. Genetic engineering techniques bypass crossbreeding barriers between species to enable gene transfer between widely differing organisms. Gene transfer can be achieved by various methods, many of which employ a replicating infective agent, such as a virus or plasmid, as a vector (*see* gene cloning). Other methods include microinjection of DNA into cell nuclei and direct uptake of DNA through the cell membrane (*see* transfection). Recognizing whether or not transfer has occurred may be difficult unless the new gene confers an obvious visual or physiological characteristic. Consequently the desirable gene may be linked to a marker gene, e.g. a gene conferring resistance to an antibiotic in the growth medium. The transferred gene must also be linked to appropriate regulatory DNA sequences to insure that it works in its new environment and is regulated correctly and predictably.

Initial successes in DNA transfer were achieved with bacteria and yeast. Human genes coding for medically useful proteins have been transferred to bacteria. Human insulin, growth hormone, and interferon are now among a wide range of therapeutic substances produced commercially from genetically engineered bacteria. Genetically engineered vaccines have also been produced by transfer of antigen-coding genes to bacteria.

Modified microorganisms are grown in large culture vessels and the gene product harvested from the culture medium. Genetic manipulation of higher animals and plants has been achieved more recently. Transgenic mammals, including mice, sheep, and pigs, have been produced by microinjection of genes into the early embryo, and it is also possible to clone certain mammals from adult body cells (*see* clone; nuclear transfer). Such technology has found application chiefly in the production of useful proteins – e.g. therapeutic proteins – in the milk of farm livestock species. Dicotyledonous plants have been transfected using the natural plasmid vector of the soil bacterium *Agrobacterium tumefaciens* (*see* Agrobacterium). Genes have been introduced to crop plants for various reasons, for instance to reduce damage during harvest or to make them resistant to the herbicides used in controlling weeds. There is also hope that in future many genetic diseases will be treatable by manipulating the faulty genes responsible. However, genetic engineering raises many legal and ethical issues, and the introduction of genetically modified organisms into the environment requires strict controls and monitoring. *See also* recombinant DNA.

genetic fingerprinting A technique for identifying individuals by means of their DNA. The DNA being tested is extracted from cells (from blood, semen, tissue fragments, etc.) and broken into fragments of 600–700 bases each, using restriction enzymes. The human genome contains many loci where short base sequences are repeated in tandem, with great variation between individuals in the number of such repeats. These so-called *variable number tandem repeats* (VNTR) can be identified using special DNA probes, thus providing a virtually unique set of markers for any given individual. This technique is used in veterinary and human medicine to establish the parentage of individuals, and in forensic science to identify individuals from traces of body tissue or fluids. Even minute amounts of DNA can now be amplified, using the POLYMERASE CHAIN REAC-

TION, to provide sufficient material for genetic fingerprinting.

genetic map *See* chromosome map.

genetics /jĕ-**net**-iks/ The term coined by Bateson to describe the study of inheritance and variation and the factors controlling them. Today the subject has three main subdivisions – Mendelian genetics, population genetics, and molecular or biochemical genetics.

genome /**jee**-nohm/ The genetic information carried by set or sets of chromosomes of an organism. Haploid organisms have one set of chromosomes, diploid organisms have two sets, polyploid organisms have many sets sometimes from the same ancestor (autopolyploids) and sometimes from different ancestors (allopolyploids).

genomics /jee-**noh**-miks, -nom-iks/ The branch of genetics involved with the study of genomes, especially by using computer-based systems to collect and analyse the data generated by automated sequencing of genes and proteins. Several subdisciplines have emerged since the advent of genomics in the 1980s. *Structural genomics* is concerned primarily with mapping and sequencing the entire DNA of any particular organism, and with determining three-dimensional structures of nucleic acid and protein molecules. *Functional genomics* attempts to unravel the complex mechanism of gene expression, and how it is controlled and integrated. A central strategy involves analysing the entire set of transcripts (i.e. the TRANSCRIPTOME) of a given cell under particular conditions, and seeing how this changes as conditions change, for example in disease states. *Comparative genomics* compares the genomes of different species to identify sequence similarities. This can provide valuable clues about the possible evolutionary relationships of different organisms and the processes of gene evolution. It also enables predictions about the functions of newly characterized genes based on evidence of the function of similar genes in other organisms. *See also* bioinformatics.

genotype /jee-nŏ-tÿp/ The genetic make-up of an organism. The actual appearance of an individual (the phenotype) depends on the dominance relationships between alleles in the genotype and the interaction between genotype and environment.

genus /jee-nŭs/ (*pl.* **genera**) A collection of similar species. Genera may be subdivided into subgenera, and also, especially in plant taxonomy, into sections, subsections, series, and subseries. Similar genera are grouped into families.

geological time scale A system of measuring the history of the earth by studying the rocks of the earth's crust. Since new rocks are generally deposited on top of existing material, those lower down are oldest. The strata of rock are classified according to their age, and a time scale corresponding to this can be constructed. The main divisions (eras) are the Paleozoic, Mesozoic, and Cenozoic. These are further subdivided into periods and epochs. *See illustration overleaf. See also* Paleozoic; Mesozoic; Cenozoic.

geophyte /jee-ŏ-fÿt/ *See* cryptophyte.

geotaxis /jee-oh-**taks**-is/ A movement of a motile microorganism or cell in response to the force of gravity.

geotropism /jee-**ot**-rŏ-piz-ăm/ (**gravitropism; geotropic movement**) A directional growth movement of part of a plant in response to gravity. Primary roots (tap roots) grow vertically towards gravity (positive geotropism) whereas primary shoots grow vertically away from gravity (negative geotropism), though the direction of shoot growth may also be modified by light. Dicotyledon leaves and some stem structures (e.g. rhizomes and stolons) grow horizontally (*diageotropism*). Secondary (lateral) roots and stem branches may grow at an intermediate angle with respect to gravity (*plagiogeotropism*).

Geotropic responses are believed to involve hormones, notably auxins. It is proposed that if a shoot or coleoptile is lying on its side, these hormones move to the lower surface in response to gravity, stimulate growth, and cause upward growth of the organ. In a horizontally placed root, the same high level of hormones inhibits growth of the lower surface, resulting in downward curvature. However, other physiological gradients may play a part, e.g. pH, electrical potential, or water potential. These gradients may arise owing to the detection of gravity by large starch grains (called *statoliths* or *amyloplasts*). The statoliths sediment towards the lower wall of the cell exerting a pressure that in some way causes the gravitational response. This is termed the *statolith hypothesis. See also* clinostat; tropism.

germ cell Any of the cells in animals that give rise to the gametes.

germination /jer-mă-**nay**-shŏn/ The first outward sign of growth of a reproductive body, such as a spore or pollen grain. The term is most commonly applied to seeds, in which germination involves the emergence of the radicle or coleoptile through the testa. Both external conditions (e.g. water availability, temperature, and light) and internal biochemical status must be appropriate before germination can occur. Seed germination may be either *epigeal*, in which the cotyledons appear above ground, or *hypogeal*, in which the cotyledons remain below ground.

germ layers The three major body layers – ectoderm, mesoderm, and endoderm – that develop in the embryos of most animals during gastrulation. These layers do not include special cells or groups of cells that may be migratory (e.g. neural crest cells of vertebrates) or perform special functions (e.g. germ cells). *See* ectoderm; endoderm; mesoderm.

germ line The lineage of cells from which gametes arise, which is continuous through generations.

germ plasm /plaz-ăm/ **1.** The part of an organism that, according to Weissmann at the beginning of the 20th century, passed its characters on to the next generation. It

ERAS	Periods	Years x 10^6	Angiosperms	Gymnosperms	Pteridophytes	Bryophytes	Bacteria, Algae, Fungi	Protoctists, Poriferans, Simple Metazoa	Ammonites / Trilobites	Myriapods	Insects, Mollusks	Echinoderms	Fish	Amphibians	Reptiles	Birds	Mammals
CENOZOIC	Tertiary	65															
MESOZOIC	Cretaceous	145															
MESOZOIC	Jurassic	215															
MESOZOIC	Triassic	250															
PALEOZOIC	Permian	280															
PALEOZOIC	Carboniferous	355															
PALEOZOIC	Devonian	405															
PALEOZOIC	Silurian	440															
PALEOZOIC	Ordovician	510															
PALEOZOIC	Cambrian	590															

Geological time scale

is now known that most of this information is carried by DNA in the chromosomes.

2. The special cytoplasm of the eggs of most animals that becomes the germ cells when provided with nuclei. It lies at one end of the eggs of insects, under the gray crescent of amphibian eggs, and in the endodermal area of amniotes.

gestation /jes-**tay**-shŏn/ The period of time between fertilization and birth in a viviparous animal. It is normally nine months in humans. The length of the gestation period tends to vary with the type of placenta and the size of the species; those with a highly developed placenta, i.e. fewer layers, and a smaller size have a shorter gestation.

giant chromosome *See* polytene.

giant fiber (**giant axon**) A nerve fiber that has a relatively large diameter, enabling the rapid conduction of a nerve impulse. Giant fibers occur in many invertebrate groups and usually supply the muscles used in a protective response, such as the end-to-end contraction in earthworms.

gibberellic acid /jib-ĕ-**rel**-ik/ (**GA₃**) A common GIBBERELLIN and one of the first to be discovered. Together with GA₁ and GA₂ it was isolated from *Gibberella fujikuroi*, a fungus that infects rice seedlings causing abnormally tall growth.

gibberellin /jib-ĕ-**rel**-in/ A plant hormone involved chiefly in shoot extension. Gibberellins are diterpenoids; their molecules have the gibberellane skeleton. More than 60 of these hormones have been isolated, the first and one of the most common being gibberellic acid, GA₃.

Gibberellins stimulate elongation of shoots of various plants, especially the extension to normal size of the short internodes of genetically dwarf pea or maize plants. Increased gibberellin levels can mimic or mediate the effect of long days. Thus they stimulate internode extension and flowering in long-day plants such as lettuce and spinach. They are also effective in inhibiting tuber development or breaking tuber dormancy, e.g. in the potato, and breaking bud dormancy in woody species. They have similar effects in substituting for chilling in some species with a vernalization requirement; e.g. causing bolting in biennials at the rosette stage.

Synthesis of α-amylase and certain other hydrolytic enzymes in barley aleurone layers is regulated by gibberellin produced by the embryo. This initiates germination by mobilizing endosperm food reserves. Gibberellins may be produced in both shoots and roots and travel in both xylem and phloem.

gill /gil/ **1.** (*Zoology*) An organ in aquatic animals that effects the exchange of respiratory gases between the blood or body fluids of the animal and the water in which it lives. Gills usually consist of many flattened lobes or filaments, providing a large surface area through which diffusion of dissolved gases can occur. Usually a plentiful supply of blood or body fluids is pumped to the gill surface. *External gills* (e.g. in amphibian larvae) trail in the water so that the water around them is renewed as the animal swims along. Most groups have evolved a means of mechanically renewing the water supply; for example by respiratory movements of appendages in crustaceans and aquatic insect larvae. The *internal gills* of fish are situated in gill slits. They are ventilated by forcing water from the pharynx past the gills and out through the gill slits.

2. (*Botany*) (lamella) One of many thin platelike spore-producing structures radiating outwards from the stalk on the undersurface of the cap (*see* pileus) of agaric fungi, such as mushrooms and toadstools (*see* basidiomycetes). The spores (basidiospores) are produced on the outer layer (hymenium). Not every gill extends to the edge of the cap.

gill bar One of a number of skeletal structures in the wall of the pharynx in fish that supports the tissue separating successive gill slits. They bear numerous fine filaments and are richly supplied with blood, into which oxygen diffuses from the water.

Gill bars are also present in lower chordates, such as *Branchiostoma* (*see* Cephalochordata).

gill books Respiratory organs of the king crabs (class Paleostracha), consisting of groups of thin plates attached to the abdominal appendages. They contain circulating blood. Their structure is similar to the LUNG BOOKS of terrestrial arachnids, indicating an evolutionary relationship between these groups.

gill cleft (**visceral cleft**) An opening through which water leaves the body of a cartilaginous fish after passing through the gills. There are normally six pairs of gill clefts situated laterally on the head of the fish. The anterior pair (called spiracles) are usually smaller and situated close behind the eye.

gill pouch (**visceral pouch**) One of several pairs of compartments lying laterally to the respiratory tube in cyclostomes (e.g. lampreys). They contain gill filaments.

gill slits Openings in the pharyngeal (throat) region of the alimentary canal of aquatic vertebrates, leading to the gills. In cartilaginous fish the first gill slit is modified as the spiracle. Traces of gill slits appear in the embryos of all vertebrates but persist only as the Eustachian tube in adult terrestrial vertebrates.

gingiva /jin-**jÿ**-vă/ (*pl.* **gingivae**) The gum that covers the jaw bones and is continuous with the lining of the mouth. It consists of stratified squamous epithelium covering a dermal layer of connective tissue containing nerves, lymph vessels, and many blood capillaries, which give a pink color to the gum. The gingiva is continuous with the alveolar membrane and with the pulp cavity through the pulp canal. *See also* alveolus; teeth.

Ginkgo /**gink**-oh/ A genus comprising only one species, *G. biloba* (maidenhair tree), this being the sole representative of the phylum Ginkgophyta. It is a deciduous dioecious tree with fan-shaped leaves and motile male sperms, native to the Far East. *See also* gymnosperm.

gizzard (**gastric mill**) Part of the alimentary canal of certain animals that are unable to chew their food. Leading from the crop, it has thick muscular walls and a very tough lining. The gizzards of birds and earthworms contain small stones or grit, and the muscular contractions of the walls grind the food between the stones. The gizzards of arthropods do not contain stones, but have stiff teeth and spines of tough chitin to break up the food. In birds the gizzard is the posterior part of the stomach.

gland 1. (*Zoology*) An organ that synthesizes a specific chemical substance and secretes this either through a duct into a tubular organ or onto the surface of the body or directly into the bloodstream.
2. (*Botany*) A specialized cell or group of cells concerned with the secretion of various substances produced as by-products of plant metabolism. The secretions may pass to the exterior or be contained in cavities or canals in the plant body. Ethereal oils, tannins, and resins are usually retained by the plant and give aromatic plants their characteristic scents. The hydathodes of leaves exude a watery solution onto the surface of the leaf in the process termed guttation, and the nectaries of flowers exude sugary substances to attract insects. Glandular hairs develop from the epidermis of many plants, e.g. stinging nettle and geranium. *See also* laticifers.

GLC *See* gas-liquid chromatography.

glenoid cavity /**glee**-noid/ A cuplike depression in the scapula of tetrapods into which the head of the humerus fits in a ball-and-socket joint.

glia /**glÿ**-ă/ *See* neuroglia.

global warming See greenhouse effect.

globulin /**glob**-yŭ-lin/ One of a group of proteins that are insoluble in water but will dissolve in neutral solutions of certain salts. They generally contain glycine and coagulate when heated. Three types of globulin are found in blood: *alpha* (α), *beta* (β), and *gamma* (γ). α and β globulins are made in the liver and are used to transport non-protein material. γ globulins are made in reticuloendothelial tissues, lymphocytes, and plasma cells and most of them have antibody activity (*see* immunoglobulin).

glomerulus /glŏ-**me**-rû-lŭs, -ryû-lŭs/ (*pl.* **glomeruli**) A small knot of capillaries within the cortex of the vertebrate kidney, which is surrounded by a cup-shaped Bowman's capsule. Together a glomerulus and a capsule form a MALPIGHIAN BODY. The glomerulus is supplied with blood by an afferent arteriole, which branches from the renal artery. In birds and mammals, a smaller efferent arteriole drains away the blood and branches into capillaries surrounding the uriniferous tubules before joining the renal vein. *See also* nephron.

glossopharyngeal nerve /gloss-oh-fă-**rin**-jee-ăl/ (**cranial nerve IX**) One of the pair of nerves that arises from the mid-region of the vertebrate hindbrain to supply the posterior part of the mouth cavity, including the tongue. It carries mainly sensory nerve fibers. *See* cranial nerves.

glottis /**glot**-is/ The opening through which air passes from the pharynx to the trachea (windpipe). It is situated in front of the opening to the esophagus. *See also* epiglottis.

glucagon /**gloo**-kă-gon/ A polypeptide hormone produced by the A-cells of the islets of Langerhans of the pancreas in response to somatotropin. Its action opposes that of INSULIN, causing an increase in blood glucose by promoting the breakdown of glycogen to glucose in the liver.

glucans /**gloo**-kănz/ *See* glycan.

glucocorticoid /gloo-koh-**kor**-tă-koid/ A type of steroid hormone produced by the adrenal cortex. Glucocorticoids (e.g. corticosterone and hydrocortisone) accelerate the formation of glucose from protein (gluconeogenesis) and the breakdown of glyco-

gen. They also inhibit inflammation, for example by depressing T-cell activity. *See also* corticosteroid.

gluconeogenesis /gloo-koh-nee-ŏ-**jen**-ĕ-sis/ A metabolic process occurring in the liver by which glucose or other carbohydrates can be manufactured from lactic acid, glycerol, or non-carbohydrate precursors, such as amino acids. During prolonged fasting, it is the route by which amino acids derived from the breakdown of muscle proteins are converted into usable energy. The reactions are essentially those of glycolysis in reverse, and the pathway is stimulated by the hormone glucagon.

glucosamine /gloo-kohs-ă-**meen** -**am**-in/ *See* amino sugar.

glucose /**gloo**-kohs/ (**dextrose; grape sugar**) A monosaccharide ($C_6H_6O_6$) occurring widely in nature as D-glucose. It occurs as glucose units in glycogen (hydrolysis to glucose) and sucrose (hydrolysis to glucose and fructose) as well as in starch and cellulose. Its significance to mammals lies in its participation in energy-storage and energy-release systems. Glucose is made available by hydrolysis of starch in foods initially by salivary amylase and later by pancreatic amylase to give maltose (2-glucose units), then by intestinal maltase to glucose itself. *See also* sugar.

glumes /gloomz/ *See* bract.

glutamic acid /gloo-**tam**-ik/ *See* amino acids.

glutamine /**gloo**-tă-meen, -min/ *See* amino acids.

glutathione /gloo-tă-**th'ÿ**-ohn, -th'ÿ-**ohn**/ A tripeptide of cysteine, glutamic acid, and glycine, widely distributed in living tissues. It takes part in many oxidation–reduction reactions, due to the reactive thiol group (–SH) being easily oxidized to the disulfide (–S–S–), and acts as an antioxidant, as well as a coenzyme to several enzymes.

gluten /**gloo**-tĕn/ A mixture of proteins found in wheat flour. It is composed mainly of two proteins (gliaden and glutelin), the proteins being present in almost equal quantities. Certain people are sensitive to gluten (celiac disease) and must have a gluten-free diet.

glycan /**glÿ**-kan/ 1. A polysaccharide made up of a single type of sugar unit (i.e. >95%). As a class the glycans serve both as structural units (e.g. cellulose in plants and chitin in invertebrates) and energy stores (e.g. starch in plants and glycogen in animals). The most common homoglycans are made up of D-glucose units and called *glucans*.
2. Any polysaccharide.

glyceride /**gliss**-ĕ-rÿd, -rid/ (**acylglycerol**) An ester of glycerol and one or more fatty acids. They may be mono-, di-, or triglycerides according to the number of –OH groups esterified. The fat stores of the body consist mainly of triglycerides. These can form a source of energy when carbohydrate levels are low, being broken down by lipases into fatty acids, which can enter metabolic pathways. *See also* lipid.

glycerol /**gliss**-ĕ-rol, -rohl/ (**glycerin; 1,2,3-propanetriol**) An alcohol with three OH groups. Glycerol is biologically important as the alcohol involved in LIPID formation (these particular lipids being called *glycerides*).

glycine /**glÿ**-seen, glÿ-**seen**/ *See* amino acids.

glycocalyx /**glÿ**-koh-**kay**-liks/ 1. (**cell coat**) A carbohydrate-rich layer that coats the outer surface of the plasma membrane of eukaryotic cells. It consists chiefly of the side chains of membrane glycoproteins and glycolipids, and in some cases also contains mucus. The glycocalyx tempers the movement of materials into and out of the cell, and can affect their nature. For example, in intestinal epithelial cells the glycocalyx contains enzymes that help to complete the digestion of nutrients before they enter the cell. The carbohydrate side chains of the glycocalyx act as antigens and are crucial in determining cellular recognition by antibodies and immune cells.
2. A sticky outer layer found in certain bacteria. It varies in chemical composition and form, but is usually rich in polysaccharides and glycoproteins. In some cases it forms a rigid *capsule*, whereas in others it is more easily deformed and is called a *slime layer*. The glycocalyx can assist the adherence of bacterial cells, and the attachment of pathogenic bacteria to their host cells.

glycogen /**glÿ**-kŏ-jĕn/ (**animal starch**) A polysaccharide that is the main carbohydrate store of animals. It is composed of many glucose units linked in a similar way to starch. Glycogen is readily hydrolyzed in a stepwise manner to glucose itself. It is stored largely in the liver and in muscle but is found widely distributed. After a meal, most of the glucose contained in food is absorbed via the intestine and blood and converted to glycogen in the liver (*glycogenesis*). The concentration of glucose in the blood is then normally regulated by conversion of glycogen back to glucose (*glycogenolysis*). The liver can store about 100 g glycogen.

glycogenesis /**glÿ**-kŏ-**jen**-ĕ-sis/ *See* glycogen.

glycogenolysis /**glÿ**-kŏ-jĕ-**nol**-ă-sis/ *See* glycogen.

glycolipid /**glÿ**-koh-**lip**-id/ Any of a group of compounds that consist of a carbohydrate attached to a lipid. The lipid is usually a mono- or diacylglycerol, sphingosine, or a ceramide, and the carbohydrate is a sugar, most commonly glucose or galactose. Glycolipids are found in the outer surface of plasma membranes and in certain other types of cell membranes.

glycolysis /**glÿ**-**kol**-ă-sis/ (**Embden–Meyerhof pathway; Emden–Meyerhof–Parnas pathway**) The conversion of glucose into pyruvate, with the release of some energy in the form of ATP. Glycolysis occurs in cell cytoplasm. In ANAEROBIC RESPIRATION, breakdown proceeds no further and pyruvate is converted into ethanol or lactic acid

(starch, glycogen)   (glucose)

(glucose 1-phosphate)   — ATP
                                         → ADP

(glucose 6-phosphate)

fructose 6-phosphate
                                         — ATP
                                         → ADP
fructose 1,6-diphosphate

2 x (glyceraldehyde
3-phosphate)
                                         — 2 NAD⁺
                                         → 2 NADH
2 x (1,3-diphosphoglycerate)
                                         — 2 ADP
                                         → 2 ATP
2 x (3-phosphoglycerate)

2 x (2-phosphoglycerate)

2 x (phosphoenolpyruvate)
                                         — 2 ADP
                                         → 2 ATP
2 x (pyruvate)

Production of fructose diphosphate

This stage requires energy and the conversion of ATP to ADP

The 6C fructose diphosphate breaks into two 3C molecules

This stage yields energy and 2 molecules of pyruvate

ADP is converted to ATP. NADH is also formed, and may participate in subsequent reactions

Glycolysis: some of the main stages in converting glucose (or starch or glycogen) to pyruvate, with an overall yield of ATP

for storage or elimination. In AEROBIC RESPIRATION, glycolysis is followed by the Krebs cycle. Glycolysis alone yields only two molecules of ATP per molecule of glucose in anaerobic respiration. In aerobic respiration there is a net yield of six (the conversion of NADH back to NAD yields a further four ATP molecules, and can occur only when oxygen is present). The pathway is named for the German biochemists Gustav Emden (1874–1933), Otto Meyerhof (1884–1951), and Karol Parnas (1884–1949).

glycoprotein /glÿ-koh-**proh**-teen, -tee-in/ A conjugated protein formed by the combination of a protein with carbohydrate side chains. Certain antigens, enzymes, and hormones are glycoproteins.

glycosaminoglycan /glÿ-kŏ-sam-ă-noh-glÿ-kan/ (**GAG**) One of a group of compounds, sometimes called *mucopolysaccharides*, consisting of long unbranched chains of repeating disaccharide sugars, one of the two sugar residues being an amino sugar – either N-acetylglucosamine or N-acetylgalactosamine. These compounds are present in connective tissue; they include heparin and hyaluronic acid. Most glycosaminoglycans are linked to protein to form proteoglycans (sometimes called *mucoproteins*). *See also* glycoprotein.

glycoside /**glȳ**-kŏ-sȳd/ A derivative of a pyranose sugar (e.g. glucose) in which there is a group attached to the carbon atom that is joined to the –CHO group. In a glycoside the C–OH is replaced by C–OR. The linkage –O– is a *glycosidic link*; it is the link joining monosaccharides in polysaccharides.

glyoxylate cycle /glȳ-**koks**-ă-layt/ A modification of the Krebs cycle occurring in some microorganisms, algae, and higher plants in regions where fats are being rapidly metabolized, e.g. in germinating fat-rich seeds. In plants, part of the cycle takes place in cell organelles called glyoxysomes (*see* microbody). The cycle essentially converts the insoluble fatty acids into sucrose, which is soluble and can be easily transported from storage cells to actively metabolizing cells in the seedling.

glyoxysomes /glȳ-**koks**-ă-sohmz/ *See* microbody.

GMO *See* genetically modified organism.

Gnathostomata /nay-thoh-**stom**-ă-tă/ A superclass used in some classifications and containing all the vertebrates that possess jaws, i.e. the Chondrichthyes and Osteichthyes (sometimes grouped as the Pisces), the Amphibia, Reptilia, Aves, and Mammalia. Members typically have paired olfactory organs and nostrils and paired limbs, although these are secondarily lost in some groups. The notochord is not retained throughout life.

gnotobiosis /noh-toh-bȳ-**oh**-sis/ The use of microbiologically monitored environments in experimental work; for example, laboratory animals may be reared in a germ-free environment or in the presence of known microorganisms.

goblet cell A cell that secretes mucus onto a surface or into a cavity. In columnar epithelium some cells produce mucus as a droplet, which enlarges until it distends the upper part of the cell; the lower end, containing the nucleus, remains narrow. This gives the cell the appearance of a wine glass or goblet. The mucus is discharged from the free surface of the cell to lubricate and protect it. Goblet cells are found in the lining of the alimentary canal and in the skin of such animals as earthworms.

Golgi apparatus /**gol**-jee/ (**Golgi body**; **Golgi complex**) An organelle of eukaryotic cells discovered by the Italian anatomist, cytologist, and neurologist Camillo Golgi (1843–1926) in 1898. It is associated with the endoplasmic reticulum but lacks ribosomes. It consists of stacks of flattened membrane-bounded sacs (cisternae) associated with vesicles (*Golgi vesicles*). Secretory cells are rich in Golgi apparatus. The cisternae are either spread randomly (as dictyosomes) as in plant cells or form a single network as in most animal cells. In the cisternae, newly synthesized proteins and other materials (e.g. polysaccharides) are processed and packaged in Golgi vesicles for transport, often to the plasma membrane for secretion, for storage, or for transport to LYSOSOMES.

Processes involving the Golgi apparatus include formation of zymogen granules; synthesis and transport of secretory polysaccharides, e.g. cellulose in cell plate formation or secondary wall formation and mucus in goblet cells; assembly of glycoproteins; and packaging of hormones in nerve cells carrying out neurosecretion. *See* zymogen granule. *See illustration at* cell.

Golgi vesicle *See* Golgi apparatus.

gonad /**gon**-ad, **goh**-nad/ The reproductive organ of animals. It produces the sex cells (gametes) and sometimes hormones. The female gonad, the ovary, produces ova; the male gonad, the testis, produces spermatozoa. Some invertebrates have both male and female gonads. *See also* ovotestis.

gonadotropic hormone /gon-ad-ŏ-**trop**-ik/ *See* gonadotropin.

gonadotropin /gon-ă-dŏ-**trop**-in, gŏ-nad-ŏ-/ (**gonadotropic hormone**) A hormone that acts on the gonads (ovary and testis). The *pituitary gonadotrophins* are follicle-stimulating hormone, luteinizing hormone, and prolactin. They are involved in the initiation of puberty, regulation of the menstrual cycle, and lactation in females, and in the control of spermatogenesis in males. They are used in the treatment of infertility. In women such treatment may lead to multiple pregnancies if not carefully controlled. *See also* chorionic gonadotrophin.

G protein Any of a class of proteins that play a crucial role in relaying signals from certain types of cell-surface receptors (*G protein-coupled receptors*) to activate signal pathways inside the cell. G proteins are involved in signal transduction from receptors for various hormones and neurotransmitters, as well as light-activated receptors in the eye and odor receptors in the nose. Essentially, binding of the ligand (hormone, neurotransmitter, etc.) to the exterior region of the receptor changes its interior (cytosolic) region, which activates an associated G protein. The activated G protein in turn might activate or inhibit a further enzyme that produces a SECOND MESSENGER, or change the permeability of certain ion channels and alter membrane potential. For example, in epinephrine and glucagon receptors, the activated G protein binds to and activates the enzyme adenylyl cyclase, which catalyzes the synthesis of the second messenger cyclic AMP (cAMP).

Graafian follicle /grah-fee-ăn/ A fluid-filled ball of cells within the mammalian ovary, in which an ovum develops. A Graafian follicle matures periodically during the active reproductive years from one of the enormous number of follicles present in the ovary at birth (about 400 000 in humans). It enlarges and migrates through the ovary tissue coming to lie at the surface. Finally it bursts and releases the ovum (OOCYTE) to the Fallopian tube. The follicle then becomes a solid body, the corpus luteum. The growth of Graafian follicles is under the influence of a hormone (follicle-stimulating hormone, FSH) released by the pituitary gland. The follicle is named for the Dutch anatomist Regnier de Graaf (1641–73). *See also* estrous cycle.

graft The transplantation of an organ or tissue in plants and animals.

In plants, grafting is an important horticultural technique in which part (the scion) of one individual is united with another of the same or different species. Usually the shoot or bud of the scion is grafted onto the lower part of the stock. Incompatibility between species is much less likely to occur in plants than in animals.

In animals, a graft is a transplantation of an organ or tissue, either on the same individual or on different individuals (i.e. from a donor to a recipient). Antibody mechanisms of the recipient recognize a graft of dissimilar tissue and tend to cause its rejection. The closer the relationship between donor and recipient, the greater the chance of a successful graft. Grafts may be from one place to another on the same individual (*autograft*) or between different individuals. A graft between individuals of the same species is a *homograft* (or *allograft*); between individuals that are genetically identical (as between identical twins) it is an *isograft* (or *syngraft*); and between different species it is a *heterograft* (or *xenograft*).

graft hybrid *See* chimaera.

grain *See* caryopsis.

gram calorie *See* calorie.

Gram's stain A stain containing crystal violet and safranin used for bacteria and the basis for the division into Gram positive and Gram negative bacteria. The former retain the deep purple color of crystal violet; the latter are counterstained red with safranin. The stain is named for the Danish physician and bacteriologist Hans Christian Joachim Gram (1853–1938).

granulocyte /gran-yŭ-lŏ-sÿt/ A white blood cell (leukocyte) that has granules in the cytoplasm. Granulocytes are sometimes called *polymorphonuclear leukocytes* (*polymorphs*) because the nucleus is lobed. The three types of granulocytes are NEUTROPHILS (70% of all leukocytes), EOSINOPHILS (1.5%), and BASOPHILS (0.5%).

granum /gray-nŭm/ (*pl.* **grana**) A stack of membranes (resembling a pile of coins) in a CHLOROPLAST. With the light microscope these stacks are just visible as grains (grana).

grape sugar *See* dextrose.

graticule /grat-ă-kyool/ *See* micrometer.

gravitropism /grav-it-rŏ-piz-ăm/ *See* geotropism.

gray Symbol: Gy The SI unit of absorbed energy dose per unit mass resulting from the passage of ionizing radiation through living tissue. One gray is an energy absorption of one joule per kilogram of mass. The symbol is named for the British physician L. Harold Gray (1905–65).

gray matter Nerve tissue that consists mainly of nerve cell bodies and their connections, giving it a grayish color. It occurs in the core of the spinal cord and in many parts of the brain, especially the cerebral cortex.

green algae *See* Chlorophyta.

green bacteria *See* Cyanobacteria.

greenhouse effect The rise in temperature of the atmosphere, analogous to that in a greenhouse. Solar (short-wave) radiation passes easily through the atmosphere (or glass in a greenhouse) and is absorbed by the earth's surface. It is re-emitted in the form of infrared (long-wave) radiation, which is absorbed by water vapor and carbon dioxide in the atmosphere with a consequent increase in the atmospheric temperature. Increasing atmospheric pollution by carbon dioxide is leading to a rise in global temperatures (global warming), which is already affecting other aspects of climate and causing profoundly damaging effects on natural ecosystems.

grooming An action of self-care, such as preening in birds and fur care in mammals. Mutual grooming is an important behavior in many primates, forming a means of friendly contact between individuals and helping to form social bonds between the animals in a group or colony. In dominance hierarchies it is a placatory gesture that plays an important part in maintaining the hierarchy.

ground meristem The region of the apical meristem from which the ground tissues of pith, cortex, medullary rays, and mesophyll differentiate. In root meristem it is also called *periblem*.

ground substance *See* extracellular matrix.

ground tissues (conjunctive tissues; fundamental tissues) The tissues that are found in any region of the plant not occupied by the specialized tissue of vascular bundles, cambium, epidermis, etc. The pith and cortex of the root and stem are ground tissue, as are the mesophyll layers of the leaf. Ground tissue generally consists of parenchyma cells, but other cell types, e.g. collenchyma and sclerenchyma, are also often present.

group selection A theory, proposed in 1962 by the British ethologist V.C. Wynne-Edwards, that natural selection can operate at the level of a group or population rather than at the level of individuals, as held by orthodox neo-Darwinians. It was an attempt to explain the apparently altruistic behavior of individual members of certain animal species. The most striking example is provided by colonies of social insects, in which the sterile workers forego their own reproduction for the benefit of the colony as a whole. The theory was soon superseded by that of KIN SELECTION.

growth An irreversible increase in size and/or dry weight. It excludes certain developmental processes which involve no size change, for example cleavage, and uptake of water by seeds (imbibition). Growth involves cell division and cell expansion through synthesis of new materials, and is closely related to subsequent developmental processes. If some measure of growth of an organism, such as height or weight, is plotted throughout its life, a characteristic S-shaped (sigmoid) *growth curve* is obtained for most organisms. In some organisms growth never stops entirely, though it may become extremely slow (*indeterminate growth*). Man and certain other vertebrates stop growing once adult (*determinate growth*). Arthropods show *intermittent growth* wherein growth is restricted by the hard exoskeleton to periods immediately following ecdysis (molting) when the new exoskeleton is still soft enough to allow expansion. *See* allometric growth.

growth hormone (**GH; somatotropin**) A polypeptide hormone produced by the anterior pituitary (adenohypophysis) that has a range of metabolic and developmental effects. It stimulates gluconeogenesis, inhibits uptake of glucose by cells, and promotes the mobilization and use of fatty acids as an energy source. It thus tends to shift metabolism away from glucose toward fats, and indirectly increases blood glucose concentration, thereby opposing the effects of insulin. It is normally released several hours after a meal, when the supply of nutrients from the meal has begun to wane. GH also promotes the growth of tissues, especially bone and cartilage. It stimulates the synthesis of RNA and protein, and causes cell proliferation, both directly and indirectly by triggering the liver to produce INSULIN-LIKE GROWTH FACTORS. The release of GH is controlled by two neurohormones produced by the hypothalamus: growth-hormone releasing hormone (GRH) and SOMATOSTATIN.

growth ring *See* annual ring.

GSC Gas-solid chromatography. *See* gas chromatography.

guanine /gwah-neen, -nin/ A nitrogenous base found in DNA and RNA. Guanine has a purine ring structure. *See illustration at* DNA.

guanosine /gwah-nŏ-seen, -sin/ (**guanine nucleoside**) A nucleoside present in DNA and RNA and consisting of guanine linked to D-ribose via a β-glycosidic bond.

guard cell A specialized kidney-shaped epidermal cell, located to the side of a stoma. Two guard cells together encircle each stoma and control the opening and closing of the stomatal aperture. The control is effected through changes in turgidity. The wall of the guard cell bordering the pore is heavily thickened while the opposite wall is comparatively thin. Thus when the guard cell is turgid the thin wall becomes distended, bulging out away from the pore, and causes the thickened wall, which cannot distend, to be drawn outwards with it. This results in an aperture being formed between adjacent guard cells. When osmotic pressure of the guard cell drops the pore closes.

gum One of a group of substances that swell in water to form gels or sticky solutions. Similar compounds that produce slimy solutions are called *mucilages*. Gums and mucilages are not distinguishable chemically. Most are heterosaccharides, being large, complex, flexible, and often highly-branched molecules. Gums may be formed in plants as a result of injury, for example gum arabic from *Acacia* species. Gums and mucilages sometimes form part of the cell wall matrix in plants.

gut The alimentary canal of an animal.

guttation /gut-**tay**-shŏn/ Loss of water as liquid from the surface of a plant. Water is normally lost as vapor during transpiration but, if the atmosphere is very humid, water may also be forced from the leaves through hydathodes.

gymnosperm /**jim**-nŏ-sperm/ Any vascular plant bearing naked seeds. Gymnosperms include the cycads, ginkgo, conifers, and the gnetophytes (e.g. *Gnetum* and *Welwitschia*), which in some older classifications constitute the class Gymnospermae. These groups show considerable diversity. For instance, cycads and ginkgo have motile sperm, unlike gnetophytes and conifers. Most gymnosperms retain archegonia except for the most advanced genera *Gnetum* and *Welwitschia*. These genera and *Ephedra* also have advanced tracheids with a structure reminiscent of angiosperm vessels, whereas in other gymnosperms the xylem is composed solely of more primitive tracheids. *See* coniferophyta; *Cycas*; ginkgo.

gynodioecious /jÿ-nŏ-dÿ-**ee**-shŭs/ Denoting a species that has female and hermaph-rodite flowers on separate plants. *Compare* androdioecious.

gynoecium /jÿ-**nee**-see-ŭm/ (*pl.* **gynoecia**) The CARPEL or carpels of a flower. The term pistil or compound pistil is also used to mean one or many carpels respectively.

gynomonoecious /jÿ-nŏ-mŏn-**ee**-shŭs/ Denoting a species that has female and hermaphrodite flowers on the same plant. *Compare* andromonoecious.

H

habitat The place where a particular organism lives, described in terms of its climatic, vegetative, topographic, and other relevant factors. Habitats vary both in space and time. For example in a forest the conditions at ground level are very different from those in the leaf canopy. Conditions also vary between seasons. However the conditions found in a specific habitat at a given time are unique to that habitat even though they may resemble conditions found in other similar habitats. The term microhabitat describes a small area, perhaps only a few square millimeters or centimeters in size, e.g. the undersurface of a log or stone.

habituation /hă-bich-û-**ay**-shŏn/ A simple type of learning in which an animal shows a diminishing response to a repeated or continuous stimulus that does not produce reinforcement, i.e. it is neither harmful nor rewarding.

Hadean /hay-dee-ăn/ *See* Precambrian.

haem /hem/ *See* heme.

hair 1. Any of the narrow threadlike outgrowths from the skin of mammals, consisting of dead cells containing large amounts of horny keratin (a protein) and granules of pigment. The presence of hair reduces heat loss from the skin. *See also* hair follicle; sebaceous gland.
2. *See* trichome.
3. *See* paraphysis.

hair follicle A tubular pocket, formed by an ingrowth of epidermal cells into the dermis, enclosing the root of a hair. At the lower end, the cells are modified to produce hair. Muscles attached to the outside of the follicles can erect the hairs to increase thermal insulation. Nerve endings attached to the follicles can detect when objects make contact with the hairs. Sebaceous glands open into the follicle.

hallux /**hal**-ŭks/ (*pl.* **halluces**) The first digit of the hindfoot of tetrapods; it forms the big toe in humans. In the typical pentadactyl limb it contains two phalanges, however there are reductions and modifications to this general plan and in some mammals, e.g. the rabbit, it is absent. It is directed backwards in most birds as an adaptation to perching. *Compare* pollex.

halophyte /**hal**-ŏ-fÿt/ A plant that grows in soils with a high concentration of salt, as found in salt marshes. Examples are species of *Spartina*.

halosere /**hal**-ŏ-seer/ A series of successional stages leading to a climax, originating in a saline area. *See also* sere; succession.

haltere /**hal**-ter/ One of a pair of club-shaped structures protruding from the sides of the thorax of flies (Diptera). They are highly modified hind wings and vibrate rapidly during flight to provide information for the maintenance of balance.

haplobiontic /hap-loh-bÿ-**on**-tik/ Describing life cycles in which only one type of somatic body is formed, which may be either haploid or diploid. Haplobiontic is thus a collective term for haplontic and diplontic. *Compare* diplobiontic.

haploid /**hap**-loid/ (**monoploid**) A cell or organism containing only one representative from each of the pairs of homologous chromosomes found in the normal diploid cell. Haploid chromosomes are thus unpaired and the haploid chromosome number (*n*) is half the diploid number (2*n*). Meiosis, which usually precedes gamete formation, halves the chromosome number to produce haploid gametes. The diploid condition is restored when the nuclei of two gametes fuse to give the zygote. In humans there are 46 chromosomes in 23 pairs and thus the haploid egg and sperm each contain 23 chromosomes. Gametes may develop without fertilization, or meiosis may substantially precede gamete formation. This is especially true in plants, and leads to the formation of haploid organisms, or haploid phases in the life cycles of organisms. *See also* anther culture.

Various multiples of the haploid number, e.g. the tetraploid (4*n*), hexaploid (6*n*), and octaploid (8*n*) conditions, are common in some plant groups, especially in certain cultivated plants.

haplont /**hap**-lont/ A haploid organism that represents the vegetative stage in life cycles in which diploidy is restricted to the zygote. *Haplontic* life cycles are typical of the filamentous green algae. *Compare* diplont.

haplontic /hap-**lon**-tik/ *See* haplont.

haplostele /**hap**-lŏ-steel, -stee-lee/ A type of protostele having a solid strand of stele with the xylem on the inside encircled by phloem, pericycle, and endodermis.

hapteron /**hap**-tĕ-ron/ **1.** *See* elater. **2.** *See* holdfast.

haptonasty /**hap**-toh-nas-tee/ (**haptonastic movements**) A NASTIC MOVEMENT in response to contact.

haptotropism /hap-**tot**-tŏ-piz-ăm/ *See* thigmotropism; tropism.

hard palate *See* palate.

Hardy–Weinberg equation /**har**-dee wÿn-berg/ If a pair of alleles, A and a, have the frequencies p and q in a population,

and p + q = 1, then random crossing among individuals in the population will give genotypes AA, Aa, and aa in the frequencies p^2, pq, q^2 respectively. The Hardy–Weinberg equation $p^2 + 2pq + q^2 = 1$ is obtained from the expansion of $(p+q)^2$, the total of the frequencies making up the gene pool being unity. The equation is named for the British mathematician Godfrey Harold Hardy (1877–1947) and the German physician Wilhelm Weinberg (1862–1937). *See* Hardy–Weinberg equilibrium.

Hardy–Weinberg equilibrium The situation in a large randomly mating population in which the proportion of dominant to recessive genes remains constant from one generation to the next. It is described by the equation $p^2 + 2pq + q^2 = 1$, where p^2 and q^2 are the frequencies of the double dominant and double recessive respectively, and 2pq is the frequency of the heterozygote. The law was formulated in 1908 and disproved the then current theory that dominant genes always tend to increase in a population at the expense of their equivalent recessive alleles. The equilibrium only holds providing that the population is sufficiently large to avoid chance fluctuations of allele frequencies in the gene pool (genetic drift) and providing there is no mutation, selection, or migration. The fact that allele frequencies may be seen to change fairly rapidly in large populations that show minimal mutation and migration, emphasizes the important role natural selection must play. Until the Hardy–Weinberg law was formulated, the extent of natural selection was not fully appreciated.

Hatch–Slack pathway *See* C_4 plant.

haustorium /haw-**stor**-ee-ŭm, -**stoh**-ree-/ (*pl.* **haustoria**) A specialized outgrowth of certain parasitic plants, e.g. fungi, that penetrates into and withdraws food material from the cells of the host plant.

Haversian canals /hă-**ver**-shăn/ A series of fine interconnecting canals that run usually longitudinally through dense bone of

pulmonary artery to right lung
right subclavian artery
carotid arteries
left subclavian artery
pulmonary vein from right lung
pulmonary artery to left lung
anterior vena cava
pulmonary vein from left lung
aorta
right atrium
left atrium
3 semilunar valves
mitral valve
tricuspid valve
thick muscular wall
right ventricle
left ventricle
less muscular wall
posterior vena cava
dorsal aorta
→ oxygenated blood
- - -▶ deoxygenated blood

Mammalian heart in longitudinal section

vertebrates. They contain the blood and nerve supply, each canal serving a series of concentric rings of bone units (lamellae), which surround it. Each series of lamellae with its canal is termed a *Haversian system*. A lamella is comprised of a ring of bone cells (osteoblasts), which lie in spaces (lacunae) and link with other bone cells by fine protoplasmic threads (canaliculi).

H band *See* sarcomere.

HCG (**human chorionic gonadotropin**) *See* chorionic gonadotropin.

heart A muscular organ – essentially a specialized region of a blood vessel – that pumps blood around the body. In mammals, it consists of four chambers (two upper atria and two lower ventricles) with the right and left sides totally separate. Deoxygenated blood is carried to the right atrium via the venae cavae and oxygenated blood from the lungs is carried to the left atrium by the pulmonary vein. Contraction of both atria forces blood into the respective ventricles, which in turn contract forcing blood into arteries: the pulmonary artery carries deoxygenated blood from the right ventricle to the lungs and the aorta transports oxygenated blood from the left ventricle to the body (*see* cardiac cycle). Valves prevent the backflow of blood. The rhythmic contractions of the heart (cardiac) muscle are basically automatic (*see* pacemaker). In a resting human contractions occur about 72 times per minute.

The form of the heart varies greatly throughout the animal kingdom (reaching its greatest complexity in birds and mammals). In annelids there are a number of lateral contractile vessels known as hearts. Crustaceans have a single heart with openings (ostia) possessing valves. In insects the heart is a long dorsal tube divided into 13 chambers, also with ostia. Mollusks and fish have two-chambered hearts. Amphibians have two atria and one ventricle. In reptiles there is one ventricle, with a septum that completely separates the ventricle into two in crocodiles.

heartwood (**duramen**) The hard central region of a tree trunk made up of xylem vessels that are no longer involved in water transport. Such vessels are often blocked by substances (e.g. resins and tannins) that give the wood a darker color. *Compare* sapwood.

helical thickening /**hel**-ă-kăl/ *See* spiral thickening.

heliotropism /hee-lee-**ot**-rŏ-piz-ăm/ *See* phototropism.

Helix /**hee**-liks/ (**snail**) *See* Gastropoda.

helophyte /**hee**-lŏ-fÿt/ A perennial marsh plant that retains its overwintering buds under water.

hematoxylin /hee-mă-**toks**-ă-lin, hem-ă-/ *See* staining.

heme /heem/ (**haeme**) An iron-containing porphyrin that is the prosthetic group in HEMOGLOBIN, myoglobin, and cytochromes.

hemerythrin /hem-ĕ-**rith**-rin/ A red oxygen-carrying blood pigment similar to hemoglobin in containing iron and found in various invertebrates.

hemicellulose /hem-ee-**sel**-yŭ-lohs/ One of a group of substances that make up the amorphous matrix of plant cell walls together with pectic substances (and occasionally, in mature cells, with lignin, gums, and mucilages). They are heteropolysaccharides, i.e. polysaccharides built from more than one type of sugar, mainly the hexoses (mannose and galactose) and the pentoses (xylose and arabinose). Galacturonic and glucuronic acids are also constituents. They vary greatly in composition between species. The hemicelluloses and pectic substances interact with the cellulose microfibrils of the cell wall to give extra strength and at the same time retain some plasticity. In some seeds (e.g. the endosperm of dates) hemicelluloses are a food reserve.

Hemichordata /hem-ee-kor-**day**-tă/ A small phylum of marine invertebrates. Hemichordates have an unsegmented body divided into three regions, the proboscis, collar, and trunk, each containing part of the coelom. The collar bears gill slits. Development is via a tornaria larva similar to that of echinoderms. The phylum includes the free-living wormlike Enteropneusta (acorn worms), e.g. *Balanoglossus*, which live in sand and mud, and the often colonial Pterobranchia, e.g. *Cephalodiscus*, which live in transparent tubes and have only one pair of gill slits and tentacles for feeding. *See* tornaria.

hemicryptophyte /hem-ee-**krip**-tŏ-fÿt/ A perennial plant, usually non-woody, with its over-wintering buds at soil level. The buds are protected by the soil or surface litter. *See also* Raunkiaer's plant classification.

hemin /**hee**-min/ The hydrochloride form of heme. Hemin is the crystalline form in which heme can be isolated and studied in the laboratory. The iron present is the trivalent state (iron(III)). Hemin can be made to crystallize by heating hemoglobin gently with acetic acid and sodium chloride. A variety of crystal forms are known.

Hemiptera /hem-**ip**-tĕ-ră/ A very large order of insects, the bugs, characterized by piercing and sucking mouthparts modified into a beak (*rostrum*). The plant bugs (e.g. *Aphis*) feed on plant sap and are serious pests, both damaging plants and carrying disease. They have uniform fore wings and membranous transparent hind wings, which are folded over the back when resting. Wingless parthenogenetic generations are common. The blood-sucking bugs, e.g. *Cimex* (bed bug), and the carnivorous water bugs, e.g. *Gerris* (pond skater), have fore wings with tough leathery bases, which are folded over the transparent hind wings when resting.

hemizygous /hem-ee-**zȳ**-gŭs/ Describing genetic material that has no homologous counterpart and is thus unpaired in the diploid state. Both single genes and chromosome segments may be hemizygous; for example, the X chromosome in the heterogametic sex, and whole chromosomes in aneuploids.

hemocoel /**hee**-mŏ-seel/ *See* coelom.

hemocyanin /hee-mŏ-**sȳ**-ă-nin, hem-ŏ-/ A blue copper-containing blood pigment found in many mollusks and arthropods. Hemocyanin is the second most abundant blood pigment after hemoglobin and functions similarly in acting as an oxygen-carrier in the blood.

hemoglobin /hee-mŏ-**gloh**-bin, hem-ŏ-, **hee**-mŏ-gloh-bin/ The pigment of the red blood cells (erythrocytes) that is responsible for the transport of oxygen from the lungs to the tissues. It consists of a basic protein, globin, linked with heme. Its molecular weight is 64 500 and it contains four heme groups, and hence four iron atoms, per molecule. The most important property of hemoglobin is its ability to combine reversibly with one molecule of oxygen per iron atom to form *oxyhemoglobin*, which has a bright red color. The iron is present in the divalent state (iron(II)) and this re-

mains unchanged with the binding of oxygen. Oxygen molecules, diffusing across the red cell membrane, are very readily attached to hemoglobin in the lungs and equally readily detached in the tissues. This is the mechanism by which blood transports oxygen through the body.

There are variations in the polypeptide chains, giving rise to different types of hemoglobins in different species. The binding of oxygen depends on the oxygen partial pressure; high pressure favors formation of oxyhemoglobin and low pressure favors release of oxygen. It also depends on pH. The affinity of oxygen decreases as the pH is lowered (more acid, as a result of dissolved carbon dioxide). This dependence is known as the *Bohr effect*. Carbon dioxide can also combine with hemoglobin at its amino groups.

hemolysis /hi-**mol**-ă-sis/ The release of hemoglobin from red corpuscles due to rupture of the cell membrane. It may be caused by such factors as toxins and incompatible blood transfusion.

hemophilia /hee-mŏ-**fil**-ee-ă, hem-ŏ-/ An inherited sex-linked condition caused by an abnormal gene on the X chromosome resulting in a deficiency of clotting Factor VIII. Haemophiliacs bleed profusely after the slightest wound or injury. As the abnormal gene is on the X chromosome, a female is affected only if she is homozygous for the condition, but a male is affected if his single X chromosome carries the gene. In this type of inheritance (X-linked recessive inheritance) the condition is transmitted either by an affected male or by a heterozygous female (the carrier). Treatment is by the administration of Factor VIII in a concentrated form.

hemopoiesis /hee-mŏ-poi-**ee**-sis, hem-ŏ-/ (**hematopoiesis**) The formation of blood cells. In the fetus hemopoiesis occurs in the spleen and liver; in the adult it occurs in the bone marrow (erythrocytes and polymorphonuclear white cells) and lymphoid tissue (lymphocytes and monocytes). All blood cells are ultimately derived from a population of hemipoietic stem cells in red bone marrow. Division of these gives rise to various cell lines, each of which undergoes proliferation and maturation via a series of cell divisions.

hemostasis /hi-**mos**-tă-sis/ The stoppage of bleeding following injury to a blood vessel. Initially, contraction of smooth muscle fibers in the vessel wall partly closes the wound. Platelets in the blood that encounter the damaged region become 'sticky' and bind to each other and to the injured vessel, forming a platelet plug. The sticky platelets release various substances, including thromboxane A_2, which attract other platelets to the site. Substances released by the damaged tissue and by the sticky platelets trigger the events that lead to formation of a blood clot at the injury site. *See* blood clotting.

heparin /**hep**-ă-rin/ A substance that prevents blood clotting by neutralizing prothrombin and stopping the action of thrombin. It is present in tissues of mammals, and is secreted by some blood-sucking animals.

Hepaticae /hi-**pat**-ă-see/ In some classifications, a class of the phylum BRYOPHYTA, containing the liverwort. These are now usually placed in their own phylum, HEPATOPHYTA.

Hepatophyta /hi-**pat**-ŏ-fÿ-tă/ (**liverworts**) A phylum containing prostrate thallose dichotomously branching plants bearing unicellular rhizoids. They are simpler than the mosses with leafy axes only developing in more advanced species. Like the mosses, liverworts have no roots or vascular tissues. Similarly, they show alternation of generations in which the gametophyte is dominant; the sporophyte is reduced to a spore capsule borne on a stalk and is wholly dependent on the gametophyte. Liverworts were formerly classified in the same phylum, BRYOPHYTA, as the mosses. There are seven orders of liverworts, including the Metzgeriales, e.g. *Pellia*, and the Marchantiales, e.g. *Marchantia*. *See also* Anthocerophyta.

hepatic portal system /hi-**pat**-ik/ A venous pathway comprising the hepatic portal vein, which carries blood rich in absorbed food materials, such as glucose and amino acids, from the intestine to the liver. There the materials may be stored, converted, or released to the general circulation via the hepatic vein.

herbaceous perennial /her-**bay**-shŭs/ *See* perennial.

herbivore /**her**-bă-vor, -vohr/ A plant-eating animal, especially one of the herbivorous mammals, such as cows, rabbits, etc. There may be various modifications associated with this diet, e.g. to the teeth and digestive system. *Compare* carnivore; omnivore.

heritability /he-ri-tă-**bil**-ă-tee/ The proportion of phenotypic variation due to genetic factors. It can be estimated from measurements of individuals from different generations, and is used in animal and plant breeding to predict how successful genetic selection will be in improving a particular trait.

hermaphrodite /her-**maf**-rŏ-dÿt/ (**bisexual**) 1. An animal possessing both male and female reproductive organs. The earthworm is a common example.
2. A plant bearing stamens and carpels in the same flower. *Compare* monoecious; dioecious.

herpesvirus /**her**-peez-vÿ-rŭs/ One of a group of DNA-containing viruses, about 100 nm in diameter, that cause such diseases as herpes and chickenpox. These viruses can remain latent in their host cells for long periods until triggered to produce symptoms of disease. *See* latent virus.

herpetology /her-pĕ-**tol**-ŏ-jee/ The branch of vertebrate zoology that is concerned with the study of snakes.

hesperidium /hes-pĕ-**rid**-ee-ŭm/ (*pl.* **hesperidia**) *See* berry.

heterocercal tail /het-ĕ-rŏ-**ser**-kăl/ The type of tail found in the cartilaginous fish, in which the vertebral column extends into the tail, bending upwards as it does so. The caudal fin below the vertebral column is much larger than that above; when the fish is swimming the lower caudal fin contributes lift, preventing the fish from sinking (which it does immediately it stops swimming). *Compare* homocercal tail; diphycercal tail.

heterochromatin /het-ĕ-rŏ-**kroh**-mă-tin/ *See* chromatin.

heteroecious Denoting rust fungi that require two host species to complete their life cycle. An example is *Puccinia graminis*, the stem rust of cereals and grasses, which overwinters on the barberry. *Compare* autoecious.

heterogametic sex /het-ĕ-roh-gă-**met**-ik/ The sex with dissimilar sex chromosomes, one (in mammals the Y chromosome) being shorter than the other (the X chromosome). *See* sex chromosomes; sex determination.

heterogeneous nuclear RNA (hnRNA) An assortment of RNA molecules found associated with proteins to form *heterogeneous ribonucleoprotein* (hnRNP) particles in the nuclei of eukaryotic cells. hnRNA consists of newly transcribed pre-messenger RNA (pre-mRNA) plus other nuclear RNAs. The proteins bind to different regions of the pre-mRNA, and to each other. They are thought to facilitate subsequent processing of the pre-mRNA by the enzymes of the spliceosome to produce the mature mRNA transcript, for example, by preventing base pairing between complementary regions of the RNA. They might also function in transporting the mRNA out of the nucleus.

heterograft /het-ĕ-rŏ-graft, -grahft/ (xenograft) A type of graft from one organism to another of a different species. *See* graft.

heterokaryosis /het-ĕ-roh-ka-ree-**oh**-sis/ The presence of two or more nuclei with differing genotypes within a single cell. Fusion of fungal hyphae of differing genetic complements leads to heterokaryosis. Induced fusion of different animal or plant cells is also effective in causing heterokaryosis. *Compare* homozygous; heterozygous.

heteromorphism /het-ĕ-rŏ-**mor**-fiz-ăm/ The existence of more than one form, used especially with reference to life cycles in which the alternating generations are markedly different morphologically, as in ferns and jellyfish. *Compare* isomorphism.

heterophylly /**het**-ĕ-rŏ-fil-ee/ Denoting plant species that have more than one form of foliage leaf on the same individual. Where young leaves differ from adult leaves, as in ivy, this is termed *developmental heterophylly*. If different leaf types form in response to changes in habitat, as in *Sagittaria*, the phenomenon is termed *environmental heterophylly*. A few species exhibit heterophylly at random.

heterosis /het-ĕ-**roh**-sis/ (hybrid vigor) The condition in which the expression of a characteristic is greater in the heterozygous offspring than in either of the homozygous parents. The effect arises from an accumulation of dominant genes in the F_1. Thus, if height is controlled by two genes, A and B, and tall and short forms are determined by dominant and recessive alleles respectively, then the cross AAbb × aaBB would give an F_1 AaBb, containing both dominant genes for tallness. Usually the more unlike the parents are the more hybrid vigor is released, but the effect diminishes in subsequent generations as more recessive homozygotes reappear.

heterospory /het-ĕ-**ros**-pŏ-ree/ The production of two different sizes of spore: microspores and megaspores. Heterospory is found in *Selaginella* species and shows this genus to be more advanced than the *Lycopodium* species and most of the ferns. The microspores develop into male gametophytes whereas the megaspores produce female gametophytes. Moreover, gametophyte development is completed inside the spore and the gametophytes of both sexes are totally dependent nutritionally on the sporophyte. The evolution of heterospory is seen as a significant stage in the development of the seed habit. *Compare* homospory.

heterostyly /het-ĕ-**ros**-tă-lee/ A dimorphism in which the styles of flowers of the same species are of different lengths, thus dividing the species into groups. An example is the primrose, which is divided into pin-eyed and thrum-eyed types. Pin-eyed plants have long styles and short stamens and thrum-eyed plants conversely have short styles and long stamens. Such differences promote pollination between groups. Heterostyly alone will not prevent self-pollination. Less obvious differences, like stigma epidermal cell size and pollen grain proteins, assist in maintaining incompatibility within a group. *See* dimorphism; incompatibility.

heterothallism /het-ĕ-rŏ-**thal**-iz-ăm/ A condition found in algae and fungi in which sexual reproduction occurs only between genetically different self-incompatible mating types (strains) of the same species. The strains may vary morphologically or in the size of the gametes they produce. When heterothallism is purely physiological, morphologically identical strains exist (often designated plus and minus strains). Physiological heterothallism is the most primitive form of sexuality. Generally heterothallism is less complex in algae than fungi. *Compare* homothallism.

heterotherm /**het**-ĕ-rŏ-th'erm/ An animal that generates its own body heat but does not maintain its body temperature within a narrow range. A wide range of organisms fit this description, including some 'warm-blooded' birds and mammals. *Temporal heterotherms* allow their body temperature to vary appreciably over time. For example, hummingbirds enter a state of torpor every night, when their metabolic rate declines and body temperature falls toward that of their surroundings. Many mammals

show a more seasonal torpor and temperature decline associated with hibernation. Among reptiles, fishes, and flying insects are numerous so-called *regional heterotherms*, organisms that are able to generate sufficient heat through muscular activity to raise their core body temperature, even though their extremities remain close to the ambient temperature. *Compare* ectotherm; endotherm.

heterotrophism /het-ĕ-rŏ-**troff**-iz-ăm/ A type of nutrition in which the principal source of carbon is organic, i.e. the organism cannot make all its own organic requirements from inorganic starting materials. Most heterotrophic organisms are chemotrophic (i.e. show *chemoheterotrophism*); these comprise all animals and fungi and most bacteria. A few heterotrophic organisms are phototrophic (i.e. show *photoheterotrophism*). The nonsulfur purple bacteria, for instance, require organic molecules such as ethanol and acetate. *Compare* autotrophism. *See* chemotrophism; phototrophism.

heterozygous /het-ĕ-rŏ-**zy**-gŭs/ Having two different alleles at a given locus. Usually only one of these, the dominant allele, is expressed in the phenotype. On selfing or crossing heterozygotes some recessives may appear, giving viable offspring. Selfing heterozygotes halves the heterozygosity, and thus outbreeding maintains heterozygosity and produces a more adaptable population. *Compare* homozygous.

hexacanth /**heks**-ă-kanth/ (**onchosphere**) The six-hooked larva of a cestode worm (tapeworm). The hooks are used by the larva for boring its way out from the intestine of the intermediate host and into the blood or lymph vessels. It is then carried in the blood or lymph to the muscles, where it develops into a bladderworm.

Hexapoda /heks-**ap**-ŏ-dă/ *See* Insecta.

hexose /**heks**-ohs/ A sugar that has six carbon atoms in its molecules. *See* sugar.

hexose monophosphate shunt *See* pentose phosphate pathway.

hibernation A state of sleep and greatly reduced metabolic rate that enables certain mammals to survive prolonged periods of low temperature and food scarcity. Stored body fat supplies enough energy for their bodies to work slowly and maintain their temperatures just higher than their surroundings. Some, such as bats, wake and feed on warm days.

Day length or food shortage may stimulate the hibernation mechanism, but in temperate and arctic animals the stimulus is cold. Temperature regulation is maintained but at a much lower level, often just above air temperature. When hibernation ends, body temperature rises spontaneously, starting at the body core. *See also* dormancy; metabolism; sleep; estivation; diapause.

Hill reaction The reaction, first demonstrated by the British plant biochemist Robert (Robin) Hill (1899–1991) in 1937, by which isolated illuminated chloroplasts bring about the reduction of certain substances with accompanying evolution of oxygen. For example, the blue dye dichlorophenol indophenol (DCPIP) may be reduced to a colorless substance. The reaction involves part of the normal light reaction of photosynthesis. Electrons from water involved in noncyclic photophosphorylation are used to reduce the added substance. It provided support for the idea that a light reaction preceded reduction of carbon dioxide in photosynthesis.

hilum /hȳ-lŭm/ (*pl.* hila) 1. A scar on the testa of a seed marking the point at which it was attached to the ovary wall by the funicle. It is a feature that distinguishes seeds from fruits.
2. The center of a starch grain around which the layers of starch are deposited.
3. A small projection at the base of a basidiospore near its attachment to the sterigma.

hindbrain /hȳnd-brayn/ (rhombencephalon) The posterior anatomical region of the brain, consisting of the medulla oblongata, the pons, and the cerebellum. *Compare* forebrain; midbrain.

hindgut /hȳnd-gut/ The last part of the alimentary canal of arthropods and vertebrate animals.

In arthropods it consists of the ileum, colon, and rectum and is lined with epithelium secreting a protective lining of chitin, similar to the exoskeleton. This lining has to be shed at ecdysis. Embryologically it is derived from the proctodaeum, which is lined with ectoderm. In vertebrates it consists of the posterior part of the colon. *Compare* foregut; midgut.

hip girdle *See* pelvic girdle.

hippocampus /hip-ŏ-**kam**-pŭs/ (*pl.* hippocampi) A ridge extending over the floor of each lateral ventricle of the vertebrate brain, linked together by a band of nerve fibers, the hippocampal commissure. It functions in short-term memory, and the expression of instinct and mood.

Hirudinea /hi-ryû-**din**-ee-ă/ The class of the Annelida that contains the leeches, mostly terrestrial or freshwater carnivorous or bloodsucking invertebrates (e.g. *Hirudo medicinalis*, the European medicinal leech). Leeches have a flat body, each segment being subdivided externally into narrow rings (*annuli*), and anterior and posterior suckers for attachment to a host. Leeches can swim strongly. They are hermaphrodite, the eggs developing in cocoons produced by the clitellum.

histamine /**his**-tă-min, -meen/ An amine formed from the amino acid histidine by decarboxylation and produced mainly by the mast cells in connective tissue as a response to injury or allergic reaction. It causes contraction of smooth muscle, stimulates gastric secretion of hydrochloric acid and pepsin, and dilates blood vessels, which lowers blood pressure and produces inflammation, itching, or allergic symptoms (such as sneezing).

histidine /**his**-tă-deen, -din/ *See* amino acids.

histiocyte A wandering ameboid cell, capable of ingesting foreign particles, found in the matrix of connective tissue. *See* macrophage.

histochemistry /his-toh-**kem**-iss-tree/ The location of particular chemical compounds within tissues by the use of specific staining techniques, for example phloroglucinol to stain lignin.

histocompatibility /his-toh-kŏm-pat-ă-**bil**-ă-tee/ The extent to which an organism's immune system will tolerate tissue grafts from another organism. Normally, a graft from an unrelated individual is recognized as foreign by the recipient's white blood cells because the marker molecules (self-antigens) on the surface of the foreign cells differ from the recipient's marker molecules. Hence, the white cells are stimulated to mount an immune response against the foreign tissue. Only certain close relatives share the same self-antigens, and can tolerate grafts of each other's tissues. The most important of these self-antigens are proteins coded by a complex cluster of genes called the MAJOR HISTOCOMPATIBILITY COMPLEX (MHC). Two classes of these proteins are important in histocompatibility and immune responses: class I MHC proteins, which occur on most body cells and 'present' viral antigens to cytotoxic T-CELLS; and class II MHC proteins, which occur only on certain immune system cells, such as macrophages and B-CELLS, and are essential for activating T-helper cells and consequently antibody secretion by B-cells.

histogenesis /his-tŏ-**jen**-ĕ-sis/ The development of the special characters of the tissues of an embryo or of the developing organ of an adult. For example, histogenesis involves the formation of multinucleate fibers and striations of muscle and of collagen and fibroblasts in skin.

histogen theory /**his**-tŏ-jĕn/ A theory, proposed by Hanstein, in which the apical meristem is considered to consist of three main zones, the *dermatogen, periblem,* and *plerome,* which differentiate into the epidermis, cortex, and stele respectively. This concept has now been replaced, for stem apices, by the tunica–corpus theory. In roots, however, the concept is still applied, and in some angiosperm roots a fourth histogen zone is recognized, the calpytrogen, which gives rise to the root cap. *Compare* tunica–corpus theory.

histology /his-**tol**-ŏ-jee/ The study of tissues and cells at microscopic level.

histone /**his**-tohn/ One of a group of relatively small proteins found in chromosomes, where they organize and package the DNA (*see* chromatin). When hydrolyzed, they yield a large proportion of basic amino acids. They dissolve readily in water, dilute acids, and alkalis but do not coagulate readily on heating.

HIV (human immunodeficiency virus) A RETROVIRUS that causes AIDS in humans by infecting and ultimately destroying certain cells, T-helper cells and macrophages, that are vital for immunity against infections. The viral particles are spherical and contain genes in the form of RNA. The virus recognizes a suitable host cell by its surface markers, called CD4 proteins, and fuses with the cell membrane. Infected T-CELLS can bud off new virus particles from their surface. However, they can also fuse with uninfected cells carrying the same surface markers, forming multinucleate cells that are ineffective as immune cells. As increasing numbers of T-helper cells become infected, the effectiveness of the immune system diminishes and the patient becomes more and more prone to infections. There are two types of HIV: HIV-1, which occurs worldwide; and HIV-2, which is found mainly in Africa.

HLA system (human leukocyte-associated antigen system) A group of histocompatibility antigens found on the surface of human body cells and encoded by the MAJOR HISTOCOMPATIBILITY COMPLEX of genes. It is the most important system of self-antigens in humans, and is crucial in determining the compatibility of tissue transplants between different persons, for example.

hnRNA *See* heterogeneous nuclear RNA.

Holarctic /hol-**ark**-tik/ *See* Arctogaea.

holdfast (hapteron) In algae, the cell or organ that attaches the plant to the substrate. It is often disclike and is particularly prominent in the brown algae (e.g. *Fucus*).

holoblastic /hol-ŏ-**blas**-tik/ Describing the type of cleavage seen in certain eggs in which the first cleavages (animal-to-vegetal in orientation) divide the egg cytoplasm and yolk completely, as in sea urchins and frogs. Sometimes, as in most fishes, the cytoplasm but not the yolk is cleaved; this is not holoblastic cleavage. *Compare* meroblastic.

Holocene /**hol**-ŏ-seen/ (**Recent**) The present epoch in the geological time scale, being the second epoch of the Quaternary period, dating from the end of the last glaciation, about 10 000 years ago, to the present day.

holoenzyme /hol-oh-**en**-zÿm/ A catalytically active complex made up of an apoenzyme and a coenzyme. The former is responsible for the specificity of the holoenzyme whilst the latter determines the nature of the reaction.

holophytic /hol-ŏ-**fit**-ik/ The type of nutrition in which complex organic molecules are synthesized from inorganic molecules using light energy. It is another term for *photoautotrophic*.

Holothuroidea /hol-ŏ-thû-**roi**-deez/ The class of the Echinodermata that contains the sea cucumbers (e.g. *Cucumaria*), which have a long cylindrical body showing secondary bilateral symmetry and covered with tough leathery skin in which the skeletal plates are reduced to spicules. There are no arms, but tube feet modified as food-catching tentacles surround the mouth.

Other tube feet are suckered for locomotion or pointed for burrowing.

holozoic /hol-ŏ-**zoh**-ik/ (**heterotrophic**) Designating organisms that feed on other organisms or solid organic matter, i.e. most animals and insectivorous plants. *Compare* holophytic.

homeobox /**hoh**-mee-ŏ-boks, **hom**-ee-/ A segment of DNA found in many so-called homeotic genes concerned with controlling the development of organisms. It consists of 180 base pairs, and the sequence of bases is remarkably similar across a wide range of species, from yeasts to human beings. This suggests it arose early in evolutionary time and has been little changed since. The homeobox encodes a 60-amino acid 'homeodomain' (peptide sequence) that enables the parent protein to bind to DNA. This is consistent with the role of homeotic proteins as transcription factors, binding to control elements of genes to control their expression. *See* differentiation.

homeostasis /hoh-mee-ŏ-**stay**-sis, -hom-ee-/ The maintenance of a constant internal environment by an organism. It enables cells to function more efficiently. Any deviation from this balance results in reflex activity of the nervous and hormone systems, which tend to negate the effect. The degree to which homeostasis is achieved by a particular group, independent of the environment, is one measure of evolutionary advancement.

homeotic gene /hoh-mee-**ot**-ik, hom-ee-/ Any of a class of genes that are crucial in determining the DIFFERENTIATION of tissues in different parts of the body during development. Mutations of such genes result in *homeosis* – the substitution of one body part by a quite different one. Homeotic genes encode transcription factors that regulate the expression of other genes by binding to DNA. This binding capability can be pinpointed to a characteristic base sequence known as a HOMEOBOX. Homeotic genes have been intensively studied in the fruit fly *Drosophila*. They come into play when the basic pattern of body segments has been established in the fly embryo, and direct the development of particular groups of cells in each segment. In *Drosophila* there are two clusters of homeotic genes on chromosome 3: the *antennapedia* complex controls development of the head and front thoracic segments, while the *bithorax* complex governs the fate of cells in more posterior segments. The physical order of the genes in these complexes corresponds to the order in which they are expressed from anterior to posterior in the developing embryo. Differentiation in each segment requires expression of the homeotic gene regulating that particular segment in combination with the gene for the preceding segment. Mammals such as mice and humans have four clusters of homeotic genes, or *Hox genes* (from homeobox), on separate chromosomes. The mammalian Hox genes and *Drosophila* homeotic genes show great similarity in both their base sequences and relative chromosomal order.

hominid /hom-ă-nid/ A member of the primate family Hominidae, which includes modern humans (*Homo sapiens*) as well as extinct forms found in great number and variety as fossils. Diagnostic features of fossil hominids include smaller canines, hyperbolic tooth rows, and smaller lower jaws than in fossil apes. The presence of stone tools associated with a fossil can also help in identification. *See also* Australopithecus; Cro-Magnon man; Homo; Neanderthal man.

Homo /hoh-moh, hom-oh/ A genus of the family Hominidae of which modern man, *Homo sapiens*, is the only surviving species. The earliest *Homo* fossils were found in east Africa, contemporaneous with *Australopithecus*, and are generally identified as *H. habilis* or *H. rudolfensis*; living over 2 million years ago, these hominids had a bipedal gait and were capable of tool production and use. Fossils of *H. erectus*, dating from about 1 million years ago, have been found widely distributed throughout the Old World. *H. erectus* had larger teeth and brain than *H. habilis*. *H.*

heidelbergensis appeared in Africa about 600 000 years ago, evolving into *H. neanderthalensis* (*see* Neanderthal man) and *H. sapiens*. Early modern man (*H. sapiens*) evolved independently in several parts of the world about 250 000 years ago. They had larger brains than their *H. erectus* ancestors. Modern man (*H. sapiens sapiens*), with a more rounded skull and higher forehead, evolved at least 70 000 years ago, possibly in Africa, and replaced Neanderthal man in Europe about 30 000 years ago (*see* Cro-Magnon man).

homocercal tail /hoh-mŏ-**ser**-kăl, hom-ŏ-/ The type of tail found in adult bony fish, in which the vertebral column does not extend to the end of the tail and the caudal fin has two lobes of equal size. *Compare* diphycercal tail; heterocercal tail.

homodont /**hoh**-mŏ-dont, hom-ŏ-/ A type of dentition in which the teeth are all alike. This is found in such animals as frogs, in which all the teeth are similar small conical structures, cemented to the maxillae, premaxillae, and vomers of the upper jaw.

homogametic sex /hoh-mŏ-gă-**met**-ik/ The sex with homologous sex chromosomes, in mammals, designated XX. *See* sex chromosomes; sex determination.

homogamy /hŏ-**mog**-ă-mee/ The condition in flowers in which the anthers and stigmas ripen at the same time, so encouraging self-pollination. Homogamy occurs in the closed cleistogamous flowers that appear late in the season in certain plants. *Compare* dichogamy.

homograft /**hoh**-mŏ-graft, hom-ŏ-, -grahft/ (allograft) A type of graft between individuals of the same species. *See* graft.

homoiothermy /hŏ-**moi**-ŏ-th'er-mee/ The maintenance of the body temperature at a constant level, irrespective of environmental conditions. Birds and mammals are homoiothermic ('warm-blooded'). *See* endotherm. *Compare* poikilothermy.

homologous /hŏ-**mol**-ŏ-gŭs/ 1. (*Biology*) Describing structures or other traits that, though in different species, are believed to have the same origin in a common ancestor. Thus the forelimbs and hindlimbs of all land vertebrates are said to be homologous, being constructed on the same five-digit (pentadactyl) pattern. *See also* analogous.
2. (*Biochemistry*) Describing large molecules (e.g. proteins or nucleic acids) that have similar sequences of chemical groups (e.g. amino acids or bases) at corresponding positions in the molecule because of a common evolutionary origin.
3. (*Genetics*) *See* homologous chromosomes.

homologous chromosomes Chromosomes that pair at meiosis. Each carries the same genes as the other member of the pair but not necessarily the same alleles for a given gene. During the formation of the germ cells only one member of each pair of homologs is passed on to the gametes. At fertilization each parent contributes one homolog of each pair, thus restoring the diploid chromosome number in the zygote. With the exception of the sex chromosomes, for example in mammals the Y chromosome is much smaller than the X chromosome, the members of each homologous pair are similar to one another in size and shape.

homoplastic /hoh-mŏ-**plas**-tik, hom-ŏ-/ Describing similarity due to convergent or parallel evolution. *Compare* patristic.

homoplasy /hŏ-**mop**-lă-see, hoh-/ A trait that appears similar in two or more species but is the result of CONVERGENT EVOLUTION rather than inheritance from a common ancestor. Homoplastic traits are therefore not HOMOLOGOUS, and are potentially misleading when attempting to construct phylogenetic trees. *Compare* apomorphy.

homospory /hoh-**mos**-pĕ-ree/ The production of only one kind of asexual spore, which then develops into a hermaphrodite gametophyte. In the vascular plants the condition is seen in the ferns and *Lycopodium* species. *Compare* heterospory.

homostyly /hoh-**mos**-tă-lee/ The usual condition found in plants in which the styles of all flowers of the same species are about the same length. The term is not widely used, except as a comparison with heterostyly. *See* heterostyly.

homothallism /hoh-mŏ-**thal**-iz-ăm, hom-ŏ-/ A condition found in algae and fungi in which each thallus is self-compatible. Homothallic species may produce distinctly different sizes of gametes and are thus effectively hermaphrodite. *Compare* heterothallism.

homozygous /hoh-moh-**zy**-gŭs, hom-oh-/ Having identical alleles for any specified gene or genes. A homozygote breeds true for the character in question if it is selfed or crossed with a similar homozygote. An organism homozygous at every locus produces offspring identical to itself on selfing or when crossed with a genetically identical organism. Homozygosity is obtained by inbreeding, and homozygous populations may be well adapted to a certain environment, but slow to adapt to changing environments. *Compare* heterozygous. *See* pure line.

hormone 1. (*Zoology*) A chemical messenger liberated by a certain type of gland (endocrine gland) or group of cells and transported in the blood to a specific (target) organ or tissue where it acts to control growth, metabolism, sexual reproduction, and other body processes. Hormones may be steroids, polypeptides, or amines. They are recognized by specific molecules or receptors in the cells of target organs. These receptors are usually proteins located at the membrane (e.g. for insulin, glucagon, and epinephrine) or in the cytoplasm (e.g. for estrogens and progesterone). The hormones exert their effects on enzymes or nucleic acids.
2. (*Botany*) *See* plant hormone.

hornworts /**horn**-werts/ *See* Anthocerotophyta.

horsetails *See* Sphenophyta.

host An organism used as a source of nourishment by another organism, the parasite, which lives in or on the body of the host. In a *definitive host* the parasite reaches sexual maturity. In an *intermediate host* the resting stage or young of the parasite are supported. *See* parasitism.

Human Genome Project An international project launched in 1989 with the aim of mapping and sequencing the entire human genome. A draft sequence was published in February 2001, and the high-quality sequence was completed in April 2003, two years earlier than originally planned. The precise number of human genes remains uncertain, with some studies indicating less than 25 000 protein-coding genes, much fewer than earlier estimates of 100 000, and not a world of difference from the 20 000 or so genes of the nematode worm *Caenorhabditis elegans*. The results will help in the diagnosis and possibly the treatment of a wide range of diseases. These include not only hereditary disorders, such as Huntington's disease, but also many common ailments with a genetic component, such as heart disease and breast cancer. Sequencing the genes responsible for the development of these diseases enables the design of DNA probes that identify susceptible individuals, so allowing preventive measures or routine check-ups. However, such knowledge has profound ethical and commercial implications. For example, insurance companies may insist on a full genome check before agreeing terms for life insurance, even in healthy applicants. *See* chromosome map.

human immunodeficiency virus *See* HIV.

humerus /**hyoo**-mĕ-rŭs/ (*pl.* **humeri**) The long bone of the upper forelimb in tetrapods; the upper arm bone in humans, extending from the shoulder to the elbow. Its rounded upper head articulates with the glenoid cavity of the scapula in a ball-and-socket joint. The lower end is modified to form an articular surface (condyle) for the

radius and ulna, which produces the hinge joint of the elbow. *See illustration at* pentadactyl limb.

humus /**hyoo**-mŭs/ The nonliving finely divided organic matter in soil derived from the decomposition of animal and plant substances by soil bacteria. Humus consists of 60% carbon, 6% nitrogen, and small amounts of phosphorus and sulfur, and is valued by horticulturalists and farmers as it improves the fertility, water holding capacity, and workability of the soil. Different types of humus can be recognized depending upon the types of organisms involved in its decomposition, the vegetation from which it is derived, and the degree of incorporation into the mineral soil. *Mull* humus is found in deciduous and hardwood forests and grasslands in warm humid climates. It is alkaline and bacteria, worms, and larger insects are abundant. Decay is rapid and layers are not distinguishable. *Mor* or raw humus is usually acidic and characteristic of coniferous forest areas. Few microorganisms or animals exist in this type of humus, small arthropods and fungi being the most common organisms.

hyaloplasm /**hy**-ă-lŏ-plaz-ăm/ (**cell matrix; ground substance**) *See* cytoplasm.

hyaluronic acid /hy-ă-lû-**ron**-ik/ A type of organic acid that has the properties of a lubricant. It is found, for example, in the synovial fluid of joints and vitreous humor of the eye.

hybrid An organism derived from crossing genetically dissimilar parents. Thus most individuals in an outbreeding population could be called hybrids, but the term is usually reserved for the product of a cross between individuals that are markedly different. If two different species are crossed the offspring is often sterile; for example, the mule, which results from a cross between a horse and a donkey. The sterility results from the non-pairing of the chromosomes necessary for gamete formation. In plants this is sometimes overcome by the doubling of the chromosome number, giving an allopolyploid.

hybridoma /hȳ-bră-**doh**-mă/ A hybrid cell resulting from the fusion of a cultured cancer cell (e.g. mouse myeloma) with a normal lymphocyte. Such cells are potentially immortal, and can be cultured indefinitely. They can also be primed to produce specific antibodies, e.g. in the production of monoclonal antibodies. *See* monoclonal antibody.

hybrid swarm A very variable series of organisms resulting from the continual crossing, recrossing, and backcrossing of the hybrid generations of two species.

hybrid vigor *See* heterosis.

hydathode /**hȳ**-dă-thohd/ A specialized leaf structure involved in the removal of excess water from plants. It may be a modified stoma with the guard cells permanently open, or a glandular hair. Hydathodes are found at the leaf tips or along the leaf margins. *See* guttation.

hydatid cyst /**hȳ**-dă-tid/ The bladderworm stage of certain cestodes (tapeworms). It consists of a large fluid-filled cyst containing numerous secondary cysts, each with several tapeworm heads inside. The body of the host attempts to protect itself by producing a layer of connective tissue around the bladder, the whole being known as a hydatid cyst. Owing to its large size it may cause severe damage to the intermediate host; for example, hydatid cysts of the dog tapeworm (*Echinococcus*) in the human brain.

Hydra /**hȳ**-dră/ A genus of solitary freshwater cnidarians. *Hydra* has a cylindrical body attached to the substratum. The mouth is surrounded by a ring of tentacles bearing stinging cells (cnidoblasts) for food capture. Although usually sedentary, *Hydra* can move by undulating or somersaulting. *Hydra* reproduces both asexually by budding and sexually in cold weather and possesses great powers of regeneration. *See also* Hydrozoa.

hydranth /**hȳ**-dranth/ A POLYP of a colonial hydrozoan, specialized for feeding. It has a mouth surrounded by tentacles. Food taken in by hydranths is shared with the reproductive polyps of the colony.

hydrocortisone /hȳ-droh-**kor**-tă-zohn/ (**cortisol**) A steroid hormone produced by the adrenal cortex having glucocorticoid activity. *See* glucocorticoid.

hydrogen /**hȳ**-drŏ-jĕn/ An essential element in living tissues. It enters plants, with oxygen, as water and is used in building up complex reduced compounds such as carbohydrates and fats. Water itself is an important medium, making up 70–80% of the weight of organisms, in which chemical reactions of the cell can take place. Hydrogenated compounds, particularly fats, are rich in energy and on breakdown release energy for driving living processes.

hydrogen bond A type of bond occurring between molecules. Hydrogen bonding takes place between oxygen, nitrogen, or fluorine atoms on one molecule, and hydrogen atoms joined to oxygen, nitrogen, or fluorine on the other molecule. The attraction is due to electrostatic forces. Hydrogen bonding is responsible for the properties of water. It is important in many biological systems for holding together the structure of large molecules, such as proteins and DNA.

hydrolase /**hȳ**-drŏ-layss/ An enzyme that catalyzes a hydrolysis reaction. Digestive enzymes are an example. Hydrolases play an important part in rendering insoluble food material into a soluble form, which can then be transported in solution.

hydrolysis /hȳ-**drol**-ă-sis/ In general, a reaction between a compound and water, particularly one involving H^+ or OH^- ions.

hydrophily /hȳ-**droff**-ă-lee/ Pollination in which water carries the pollen from anther to stigma. It occurs in some pondweeds.

hydrophyte /**hȳ**-drŏ-fȳt/ A plant found growing in water or in extremely wet areas.

Hydrophytes show certain adaptations to such habitats, notably development of aerenchyma, reduction of cuticle, root system, and mechanical and vascular tissues, and divided leaves. Examples are *Sagittaria* and water lilies. *Compare* mesophyte; xerophyte.

hydroponics /hÿ-drŏ-**pon**-iks/ The growth of plants in liquid culture solutions rather than soil. The solutions must contain the correct balance of all the essential mineral requirements. The method is used commercially, especially for glasshouse crops, and also in experimental work in determining the effects of mineral deficiencies. Support may be provided by using beds of gravel through which the aerated solution is pumped.

hydrostatic skeleton /hÿ-drŏ-**stat**-ik/ A type of skeleton found in many soft-bodied invertebrates that relies on internal compartments containing a fixed volume of incompressible fluid. The contraction of muscles in the wall of a compartment exerts pressure on the fluid, which transmits the forces generated by the muscles to change the shape of the compartment. This is the principle by which earthworms extend and compress their segmented bodies when burrowing through soil, for example (*see* Annelida), and one adopted by many other wormlike animals for locomotion. Other animal groups also employ hydrostatic principles to move certain of their body parts, for instance the foot of mollusks and the limbs in arachnids.

hydrotropism /hÿ-**drot**-rŏ-piz-ăm/ (**hydrotropic movement**) A TROPISM in which the stimulus is water. It is a special kind of chemotropism. Roots are positively hydrotropic, the stimulus of water being stronger than the stimulus of gravity in determining response.

5-hydroxytryptamine /hÿ-droks-ee-**trip**-tă-meen/ *See* serotonin.

Hydrozoa /hÿ-drŏ-**zoh**-ă/ A class of the Cnidaria in which alternation of generations of polyps and medusae typically occurs in the life cycle. Most are marine, with colonial sedentary polyps (e.g. *Obelia*). The polyps reproduce asexually, forming either new polyps or free-swimming sexually-reproducing medusae. The best-known exception, the freshwater *Hydra*, exists as a solitary polyp with no medusa phase. It reproduces both sexually and asexually.

hymenium /hÿ-**mee**-nee-ŭm/ (*pl.* **hymenia**) A layer of the fruiting body of certain ascomycete and basidiomycete fungi in which the asci or basidia are borne. The hymenium may be directly exposed to the air, as in the gills of the mushroom (a basidiomycete) and the apothecia of certain ascomycetes (e.g. the Pezizales), or may open into a flask-shaped cavity, as occurs in the perithecia of the ascomycete order Sphaeriales.

Hymenoptera /hÿ-mě-**nop**-tě-ră/ A large order of insects containing the bees (e.g. *Apis*), wasps (e.g. *Vespula*), and ants (e.g. *Formica*). Most are carnivorous, with biting mouthparts, although some (e.g. bees) have additional sucking mouthparts. The hind wings are coupled to the larger fore wings by small hooks for a more stable flight, and the thorax is usually joined to the abdomen by a narrow waist. The female's ovipositor may be modified as a saw, drill, or sting. The larvae are either caterpillar-like, feeding on plants, or legless and helpless, being cared for by the adults. Metamorphosis is complete. Although some Hymenoptera are solitary the order also includes social insects, such as honeybees, ants, and termites, living in highly organized colonies.

hyoid arch /**hÿ**-oid/ The second visceral arch, lying between the jaws and spiracle. In tetrapods, its ventral elements form the *hyoid apparatus*, which supports the tongue. In fish, its dorsal element is modified to form the hyomandibular.

hyomandibular /hÿ-oh-man-**dib**-yŭ-ler/ One of a pair of cartilages or bones in fish that attaches the ends of the upper and lower jaws to the rest of the skull. In tetrapods, with the changes in jaw articulation, the hyomandibular is modified to form an ear ossicle. *See* columella auris; stapes.

hyperplasia /hȳ-per-**play**-zhǎ/ Enlargement of a tissue due to an increase in the number of its cells. For example, if part of the liver is removed, the remaining part may undergo hyperplasia in order to regenerate. *Compare* hypertrophy.

hyperpolarization /hȳ-per-poh-lǎ-ri-**zay**-shŏn/ An increase in the polarity of the potential difference across a membrane of a nerve or muscle cell, i.e. an increase in the resting potential. It is caused by the pumping of ions across the membrane so that differential concentrations are created on either side – the inside becoming more negative. As a result, a stronger stimulus is needed to evoke a response.

hypersensitivity /hȳ-per-sen-sǎ-**tiv**-ǎ-tee/
1. (*Immunology*) Any excessive or inappropriate immune response, including the various forms of ALLERGY and AUTOIMMUNITY. *See also* anaphylaxis.
2. (*Botany*) The reaction of plant tissues to invasion by pathogenic organisms, such as fungi, bacteria, or viruses. The hypersensitive reaction can involve strengthening of cell walls; synthesis by tissue cells of proteins that act to 'disarm' the invading organisms, for example by degrading fungal cell walls or inhibiting pathogenic enzymes; and ultimately APOPTOSIS, or programmed cell death, at the site of the invasion, to slow the spread of infection.

hypertonic /hȳ-per-**tonn**-ik/ Designating a solution with an osmotic pressure greater than that of a specified other solution, the latter being HYPOTONIC. When separated by a semipermeable membrane (e.g. a cell membrane) water moves by osmosis into the hypertonic solution from the hypotonic solution. *Compare* isotonic.

hypertrophy /hȳ-per-**tro**-fee/ Enlargement of a tissue or organ due to an increase in the size of its cells or fibers. An example is the enlargement of muscles as a result of exercise. *Compare* hyperplasia.

hypha /**hȳ**-fǎ/ (*pl.* **hyphae**) In fungi, a fine non-photosynthetic tubular filament that spreads to form a loose network termed a *mycelium* or aggregates into fruiting bodies (e.g. toadstools). Hyphae may be branched or unbranched and may or may not have cross walls (*septa*) dividing them into cells. They are parasitic or saprophytic and the tips secrete enzymes to digest and penetrate the food supply. The hyphal walls of most species of fungi differ from those of plants in being composed of a nitrogenous compound called chitin or a form of fungal cellulose. Hyphae also differ in lacking plastids, any pigment being contained in the walls, cytoplasm, or oil globules. Hyphae are also found in the algae, in which they may form compact pseudoparenchymatous tissues as in the thalloid seaweeds.

hypocotyl /hȳ-pŏ-**kot**-ǎl/ The stem below the cotyledons, occupying the region between the cotyledon stalks and the point where lateral roots arise. Rapid elongation of the hypocotyl after germination pushes the cotyledons above ground in plants showing epigeal germination. The arrangement of tissues in the hypocotyl shows a transition between primary root and primary stem structure. *Compare* epicotyl.

hypodermis /hȳ-pŏ-**der**-mis/ One or more layers of cells that may be found immediately below the epidermis of plants. It may be composed of thin-walled colorless cells and functions as water-storing tissue as in certain succulent leaves and the aerial roots of epiphytes. Alternatively in some species, the hypodermal cells possess heavily thickened walls and assist in mechanical protection of internal tissues as in pine leaves.

hypogeal germination /hȳ-pŏ-**jee**-ǎl, hip-ŏ-/ Germination in seed plants during which the cotyledons remain underground, as in broad bean. The cotyledons thus act only as food storage organs and not as photosynthetic organs and the hypocotyl does not elongate. *Compare* epigeal germination.

hypoglossal nerve /hȳ-pŏ-**gloss**-ăl, hip-ŏ-/ (**cranial nerve XII**) One of the pair of nerves that arises from the posterior region of the medulla oblongata in the brain of higher vertebrates and carries motor nerve fibers to the muscles of the tongue. *See* cranial nerves.

hypogyny /hi-**poj**-ă-nee/ The simple arrangement of flower parts in which the receptacle is expanded at the top of the pedicel in such a way that the androecium and the perianth arise from beneath the gynoecium, giving rise to a superior ovary. *Compare* epigyny; perigyny. *See illustration at* receptacle.

hyponasty /**hȳ**-pŏ-nă-stee, **hip**-ŏ-/ (**hyponastic movements**) The curving of a plant organ upwards and towards the axis, caused by greater growth on the lower side. *See* nastic movements.

hypophysis /hȳ-**poff**-ă-sis/ *See* pituitary gland.

hypostasis /hȳ-**pos**-tă-sis, hi-/ The situation in which the expression of one gene (the *hypostatic gene*), is prevented in the presence of another, nonallelic, gene (the *epistatic gene*). *See* epistasis.

hypothalamus /hȳ-pŏ-**thal**-ă-mŭs/ (*pl.* **hypothalami**) Part of the vertebrate forebrain that is concerned primarily with regulating the physiological state of the body. Blood temperature and chemical composition are monitored by the hypothalamus, which also regulates body temperature, drinking, eating, water excretion, and other metabolic functions, largely by its influence on the release of hormones by the pituitary gland. Heart rate, breathing rate, blood pressure, and sleep patterns are also controlled by the hypothalamus via its connections to centers in the cerebral cortex and the medulla oblongata. The hypothalamus connects directly to the pituitary gland via the infundibulum.

hypotonic /hȳ-pŏ-**tonn**-ik/ Designating a solution with an osmotic pressure less than that of a specified other solution, the latter being hypertonic. When separated by a semipermeable membrane (e.g. a cell membrane) water is lost by osmosis from the hypotonic to the hypertonic solution. *Compare* hypertonic; isotonic.

hypoxia /hȳ-**poks**-ee-ă/ A deficiency of oxygen, for example in tissues or in the environment.

IAA (indole acetic acid) A naturally occurring AUXIN.

I band *See* sarcomere.

Ice Age A period in the latter part of the Pleistocene characterized by successive coolings and warmings of the earth. In at least four major glaciations (cold periods), ice caps spread south from the Arctic and north from the Antarctic. Large areas of Britain, Europe, and North America were covered by ice from the North. The cause of the Ice Ages is not known, but it is assumed that others will occur in the future.

ichthyology /ik-th'ee-**ol**-ŏ-jee/ The branch of vertebrate zoology that is concerned with the study of fishes.

Ichthyosauria /ik-th'ee-ŏ-**saw**-ree-ă/ The only order of the extinct subclass of reptiles, the Ichthyopterygia, which lived from the Triassic to the Cretaceous and are known as the fish-lizards. *Ichthyosaurus*, common in the Jurassic, was a large predaceous marine reptile with a long snout, numerous teeth, a streamlined sharklike body, four paddle-shaped limbs, and dorsal and caudal fins. Ichthyosaurs were probably ovoviviparous and never left the water.

ICSH (interstitial-cell-stimulating hormone) *See* luteinizing hormone.

identical twins (monozygotic twins) Two offspring, produced during one birth, resulting from the division of a single fertilized egg. They are of the same sex and otherwise genetically identical, but may differ because of differences in nutrition, injuries, etc., either before or after birth. Identical triplets, quads, and quins are known, but multiple births in human beings usually result from the simultaneous fertilization of several eggs.

idioblast /**id**-ee-ŏ-blast/ Any specialized plant cell that is dispersed among cells of a different kind. Idioblasts may contain a variety of materials, e.g. tannins, oils, crystals, and waste products. Isolated sclereids can be called idioblasts.

idiogram /**id**-ee-ŏ-gram/ *See* karyogram.

IF *See* intermediate filament.

ileum /**il**-ee-ŭm/ (*pl.* **ilea**) The distal part of the small intestine in mammals between the jejunum and the colon. Digestion of food and absorption of the soluble products takes place in the ileum. The lining is thrown into folds covered with millions of microscopic villi that increase the surface area for absorption. Between the bases of the villi are the openings of simple glands that secrete intestinal juice containing digestive enzymes. The ileum has two layers of muscle – longitudinal and circular – whose rhythmic contractions cause peristaltic waves to move the contents along. *See illustration at* alimentary canal.

ilium /**il**-ee-ŭm/ (*pl.* **ilia**) One of a pair of bones that form the dorsal part of the tetrapod pelvic girdle; they are the fan-shaped part of the hip bones in humans. Each ilium is fused at its posterior side to the processes of the sacral vertebrae.

imaginal disk Any of several paired clusters of undifferentiated cells in an insect larva that give rise to the structures of the adult insect (imago) inside the pupa. The cells of the imaginal disks are thus 'set aside' during embryological and larval development. When METAMORPHOSIS is triggered by increased levels of the molting hormone ECDYSTERONE, the disk cells start to differentiate and form wings, legs, and other adult organs, according to their location within the pupa.

imago /i-**may**-goh/ (*pl.* **imagines**) The final or adult stage in the life cycle of an insect. Unlike the earlier stages, the imago can fly (except in flightless species) and reproduce.

imbibition /im-bă-**bish**-ŏn/ The phenomenon in which a substance absorbs a liquid and swells, but does not necessarily dissolve in the liquid. The process is reversible, the substance contracting on drying. Water is imbibed by many biological substances: cellulose, hemicelluloses, pectic substances, lignin (all plant cell wall constituents); starch; certain proteins, especially in seeds; etc. Dry seeds absorb water by imbibition, initially via the testa, and as seed volume increases great imbibitional pressures develop. Imbibition combined with osmosis is responsible for water uptake in growing plant cells and water retention may be aided by swelling of mucilaginous materials, as occurs in most succulent plants.

immune clearance /i-**myoon**/ (**immune elimination**) The rapid removal of antigen introduced into the body of an immune individual, as a result of its complexing with antibody.

immune response A response to the introduction of antigen into the body involving production of specific antibodies or lymphocytes, which combine with the antigen. It is the basic mechanism of active IMMUNITY, and of hypersensitivity, e.g. allergies.

immunity /i-**myoo**-nă-tee/ The ability of plants and animals to withstand harmful infective agents and toxins. It may be due partly to a number of non-specific mechanisms, such as inflammation and phagocytosis or an impervious skin (*non-specific*

immunity). In vertebrates it is largely the result of a specific mechanism, whereby certain substances (ANTIBODIES) or LYMPHO-CYTES present in the body combine with an introduced foreign substance (antigen) – *specifically acquired immunity*. Specifically acquired immunity includes *passive immunity*, where the antibody has been derived from another individual (e.g. from the mother to offspring), and *active immunity*, where the antibody is produced following stimulation with antigen (e.g. by vaccination or by exposure to infection). *Cell-mediated immunity* is achieved by T-cells, which detect and destroy virus-infected body cells (*see* T-cell). *Humoral immunity* involves the circulation of free antibody in the blood. *See also* immunoglobulin; specific acquired resistance.

immunization /im-yû-ni-**zay**-shŏn/ The process of making an animal resistant to infection or harmful agents. *See* immunity.

immunoassay /i-myoo-noh-**ass**-ay, -ă-**say**/ Any of various techniques for measuring biological substances that depend on the substance acting antigenically and binding with a specific antibody. Most also require the addition of labeled antigen or antibody, bearing a radioisotope, fluorescent molecule, or enzyme, for example, to reveal the extent of the antigen-antibody binding, and enable the amount or concentration of the test sample to be established. *See* ELISA; radioimmunoassay.

immunoelectrophoresis /i-myoo-noh-i-lek-troh-fŏ-**ree**-sis/ *See* electrophoresis.

immunoglobulin /i-myoo-noh-**glob**-yŭ-lin/ (**Ig**) Any of a group of proteins that take part in the immune responses of higher animals. They are produced by B-lymphocytes (*see* B-cell), and act mainly as antibodies by binding to foreign antigens, such as bacteria, viruses, or toxins. Each consists of four polypeptide chains, two heavy chains and two light chains, arranged to form a Y-shaped structure. The two arms of the 'Y' each bear an antigen-binding site. There are five classes, distinguished by the structure of their heavy chains.

Immunoglobulin A (IgA) is found mainly in secretions such as saliva and tears, and its main role is to neutralize viruses and bacteria on the external mucous surfaces of the body. *Immunoglobulin D* (IgD) occurs on the surface of lymphocytes and is thought to control the activation and suppression of B-cell activity. *Immunoglobulin E* (IgE) is normally present in very low concentrations in blood and connective tissue underlying epithelia, but its level is raised in allergies. It binds to mast cells, and in the presence of antigen causes histamine release from mast cells, with consequent inflammation and the other common symptoms of allergies.

Immunoglobulin G (IgG) is the main immunoglobulin of blood and tissue fluid. It binds to microorganisms, enhancing their engulfment by phagocytic cells, and neutralizes viruses and bacterial toxins. It also activates COMPLEMENT, leading to lysis of target cells, and can cross the placenta, affording protection to the fetus. *Immunoglobulin M* (IgM) is a star-shaped molecule comprising five of the basic Y-shaped units. It occurs within blood vessels and is most prominent early in the immune response, mopping up microorganisms or other antigens with its formidable array of binding sites. It can also activate complement. *See also* immunity.

immunological tolerance /i-myoo-nŏ-**loj**-i-kăl/ The failure of the antibody response to an antigen, usually one to which the animal has been exposed previously.

implantation /im-plan-**tay**-shŏn/ (**nidation**) The attachment of the developing mammalian egg to the wall of the uterus. The human egg enters the uterus from the Fallopian tube, where it has been fertilized four days earlier. Cells on its outer surface destroy the cells of the uterine wall and invade the mother's tissues, anchoring the growing embryo and making way for the development of the placenta. Implantation occurs at the BLASTOCYST stage of embryonic development. In women, implantation prevents the next menstrual period from occurring, probably by the action of human chorionic gonadotropin (HCG), which maintains the secretion of progesterone by the corpus luteum during pregnancy. *See also* trophoblast.

imprinting A form of learning in the first few hours of a young animal's life in which it fixes its attention on the first object with which it has contact (visual, auditory, tactile, or olfactory) and afterwards follows that object. The period during which the animal is susceptible to imprinting is short but the actual stimuli seem irrelevant. In nature, an animal nearly always imprints on its parent and this early experience becomes critical in later determining its choice of mate.

inbreeding Breeding between closely related individuals. The most extreme form of inbreeding is self-fertilization, which occurs in some plants. In animals mating between siblings or between parents and offspring is generally the closest form of inbreeding. Inbreeding increases homozygosity so that deleterious recessive genes are expressed more often in the phenotype, and decreases heterozygosity and hence the potential genetic variability of the population. There is also a general lowering of vigor in inbred stock (*inbreeding depression*), which is especially pronounced amongst normally outbreeding populations. In human societies there are usually cultural restraints on marriage between close relatives. *Compare* outbreeding.

incipient plasmolysis The condition of a cell that results when it is surrounded by a solution having the same osmotic pressure as the cell contents. *See* plasmolysis.

incisor /in-**sÿ**-zer/ A tooth in the very front of the jaw of a mammal. It has a single crown and root and is usually chisel-shaped with a sharp cutting edge. Incisors are used for biting off a portion of food, and, in rodents, they are used for gnawing. The incisors of rodents continue to grow throughout life, to compensate for being worn away by continuous gnawing, and therefore have open pulp canals. Most mammals have four incisors on the upper and four on the lower jaw. *See also* teeth.

incompatibility   1. The rejection of grafts, transfusions, or transplants between animals or plants of different genetic composition.
2. A mechanism in flowering plants that prevents fertilization and development of an embryo following pollination by the same or a genetically identical individual. It is due to interaction between genes in the pollen grain and those in the stigma, in such a way that the pollen is either unable to grow or grows more slowly on the stigma. It results in self-sterility, thus preventing inbreeding.
3. A genetically determined mechanism in some fungi that prevents sexual fusion between individuals of the same race or strain. *See also* heterothallism.

incomplete dominance   *See* co-dominance.

incubation   1. The period during which the eggs of birds and some reptiles must be maintained at a certain temperature to ensure their normal development after laying. Birds incubate their eggs using their own body heat, often with the aid of a nest, while reptiles may bury their eggs in mud or sand to be warmed by the sun.
2. (incubation period) The period of time between initial infection by a pathogenic organism and the appearance of the first signs of disease.
3. The process of maintaining a culture of microorganisms in conditions that produce optimal growth.

incus /**ink**-ŭs/ (*pl.* **incudes**) (**anvil**)   The anvil-like bone forming the middle ear ossicle in mammals. It is homologous with the quadrate bone of other vertebrates. *See illustration at* ear.

indefinite inflorescence   *See* racemose inflorescence.

indehiscent /in-di-**hiss**-ĕnt/   Describing a fruit or fruiting body that does not open at maturity to release the seeds or spores. The

fruit wall either decays releasing the seeds *in situ*, or may be adapted in various ways for dispersal by birds, mammals, insects, wind, or water. *Compare* dehiscent.

independent assortment   The law, formulated by Gregor Mendel, that genes segregate independently at meiosis so that any one combination of alleles is as likely to appear in the offspring as any other combination. The work of T. H. Morgan later showed that genes are linked together on chromosomes and so tend to be inherited in groups. The law of independent assortment therefore applies only to genes on different chromosomes. *See* linkage; Mendel's laws.

indeterminate growth   *See* growth.

indicator   A substance used to test for acidity or alkalinity of a solution by a color change. Examples are litmus and phenolphthalein. A *universal indicator* shows a range of color changes over a wide range from acid to alkaline, and can be used to estimate the pH.

indicator species   An organism that can be used to measure the environmental conditions that exist in a locality. Lichen species are indicators of levels of pollution, as different species are sensitive to different levels and types of pollutants. *Tubifex* worms indicate low levels of oxygen and stagnant water.

indigenous   Describing an organism that is native to an area, rather than introduced.

indole acetic acid (**IAA**)   A naturally occurring AUXIN.

inducible enzyme /in-**dew**-să-băl/   *See* adaptive enzyme.

induction /in-**duk**-shŏn/   1. (*Embryology*) The process by which cells influence the development of other cells. Induction is crucial in laying down the basic body plan of an early embryo, and in differentiation of tissue cells during subsequent development. Chemical signals (inducers) produced by one cell or group of cells (*see* organizer) influence the development of recipient cells. Some signals act only on neighboring cells, while others diffuse through the extracellular matrix to affect cells further away. Moreover, signaling is often a reciprocal process, and the responsiveness of cells varies according to their location and the stage of development. **2.** (*Biochemistry*) The production of enzymes by prokaryote cells in response to the presence of a certain substance (the *inducer*) in the cell's environment. The inducer acts to 'switch on' the genes encoding the enzymes by binding to and inactivating a REPRESSOR molecule. **3.** (*Obstetrics*) The artificial stimulation of labor to accomplish childbirth.

indusium /in-**dew**-zee-ŭm/ (*pl.* **indusia**) The covering that encloses the developing sporangia in the sorus of a fern.

industrial melanism An increase in dark forms, for example in the moth *Biston betularia*, in industrial soot-polluted environments. Natural selection against normal pale forms by predators results in dark offspring being at a selective advantage in such environments. This results in an increase in the numbers of the better camouflaged dark forms. *See also* polymorphism.

infection The diseased condition arising when pathogenic microorganisms enter the body, establish themselves and multiply.

inferior Below. In botany, the term is generally applied to the position of the ovary of a flower in which the sepals, petals, and stamens arise above it. The ovary appears to have sunk into and fused with the cup-shaped receptacle. This condition occurs only in epigynous flowers. In a floral formula, an inferior ovary is denoted by a line above the gynoecium symbol and number. *Compare* superior. *See also* epigyny.

inflammation /in-flă-**may**-shŏn/ A defensive reaction of animal tissues to injury, infection, or irritation, characterized by redness, swelling, heat, and pain. An *acute inflammatory response* typically follows

bacterial infection, for example, and is triggered by complement proteins. These bind to the microorganisms and also act as chemotactic agents, attracting phagocytic cells (neutrophils and macrophages) to the infection site: complement components increase the permeability of blood capillaries in the vicinity, so allowing an influx of neutrophils from the blood. The same components trigger tissue mast cells to release further chemical mediators of inflammation, including histamine, leukotrienes, and tumor necrosis factor (TNF). Tissue macrophages also release mediators, including interleukins and TNF, which sustain the inflammation. The process is regulated and resolved by complement regulatory proteins, and by cytokines such as prostaglandin E$_2$ and transforming growth factor b. Fibrin forms a clot to seal off the site, and phagocytic cells engulf microorganisms, thus destroying them.

inflorescence /in-flŏ-**ress**-ĕns/ A collection of flowers sharing a common stalk. Each flower usually arises in the axil of a leaf or bract. There are many types of inflorescence, determined mainly by the method of branching. *See* cymose inflorescence; racemose inflorescence.

infundibulum /in-fŭn-**dib**-yŭ-lŭm/ (*pl.* **infundibula**) The funnel-shaped process that connects the pituitary gland with the hypothalamus in the floor of the brain. The term may be applied to other funnel-shaped structures, such as the ciliated funnel of the oviduct.

inhibition The reduction or prevention of activation of an effector by means of inhibitory nerve impulses. Inhibitory synapses release a neurotransmitter that opens potassium (K$^+$) or chloride (Cl$^-$) channels in the postsynaptic membrane, causing either an outflow of K$^+$ ions from the postsynaptic cell or an inflow of Cl$^-$ ions. This increases the polarization of the membrane and makes depolarization, and therefore formation of an action potential, less likely. Inhibitory synapses play an important part in central-nervous-system control of motor activity.

inhibitory postsynaptic potential /in-**hib**-ă-tor-ee, -toh-ree; pohst-si-**nap**-tik/ (**IPSP**) A localized hyperpolarization of the postsynaptic membrane at an inhibitory synapse or a neuromuscular junction that tends to inhibit production of an action potential in the postsynaptic nerve fiber. It is due to an inhibitory neurotransmitter, which is released from the presynaptic membrane on stimulation by an impulse and causes an increase in membrane permeability to certain ions (*see* inhibition). The size of the IPSP depends on the amount of neurotransmitter released. *Compare* excitatory postsynaptic potential.

initial A cell permanently in a meristem, perpetuating itself while adding new cells to the plant body. An initial never becomes differentiated. Two basic groups exist: apical initials at root and shoot apices, and lateral meristem initials whose position depends on the location of the meristem, e.g. vascular cambium between xylem and phloem. Apical initials are rarely single cells except in certain lower plants, and initiate more than one structure, e.g. lateral shoots, flowers, and leaves, as well as new stem tissues. Vascular cambium possesses two distinct types of initials, *ray initials* forming medullary rays, and *fusiform initials* producing different xylem and phloem elements.

initiation *See* transcription; translation.

innate behavior Any behavior that is performed instinctively. *See* instinct.

inner ear The innermost and sensory region of the vertebrate ear, situated in the auditory capsule of the skull. In most tetrapods, it is connected to the middle ear by two membranes: the oval window (fenestra ovalis) and the round window (fenestra rotunda). It is filled with a fluid (perilymph), in which is suspended the membranous labyrinth, which is responsible for hearing and balance. *See* cochlea; labyrinth; sacculus; semicircular canals; utriculus. *See illustration at* ear.

innervation /in-ner-**vay**-shŏn/ **1.** The number, type, and distribution of nerves that supply an organ or part of the body. **2.** The nervous stimulation of an organ or part of the body.

innominate artery /in-**nom**-ă-nit, -nayt/ An artery in mammals and some birds that arises from the arch of the aorta and divides to form the right carotid and right subclavian arteries. In fish, the innominate arteries arising from the ventral aorta divide into the first and second pairs of afferent branchial arteries.

innominate bone The mass of bone forming each half of the pelvic girdle in reptiles, birds, and mammals. It results from fusion of the ilium, ischium, and pubis bones.

inositol /i-**noh**-să-tol, -tohl/ An optically active cyclic sugar alcohol. It is a component of the vitamin B complex and is required for growth in certain animals and microorganisms. The stereoisomer *myo-inositol* is a component of the phospholipid phosphatidylinositol, an important constituent of cell membranes and a precursor of several SECOND MESSENGERS, including inositol 1,4,5 trisphosphate (IP_3) and diacylglycerol (DAG).

inquilinism /ing-**kwil**-ă-niz-ăm/ An association between organisms of two different species in which one (the *inquiline*) obtains shelter by living on or in the body of the other (the host), or by sharing the host's nest. In some cases the inquiline might also take some of the host's food. It is a specialized form of COMMENSALISM, in which the inquiline benefits without any serious consequences for the host. For example, the nests of colonial insects sometimes provide shelter for other insect species.

Insecta /in-**sek**-tă/ (**Hexapoda**) The largest class of arthropods and the largest in the animal kingdom. Most insects can fly. The body is characteristically divided into a head, thorax, and abdomen. The head bears a pair of antennae, compound eyes, and simple eyes (ocelli). The mouthparts are modified according to the diet. The thorax bears three pairs of five-jointed legs and, typically, two pairs of wings. The abdomen is usually limbless. Most insects are terrestrial and respiration is carried out by tracheae with segmentally arranged spiracles. Excretion is by Malpighian tubules. Usually the life history includes metamorphosis but in some metamorphosis is incomplete – the larvae (nymphs) resemble the adult and there is no pupal stage. Many insects are beneficial, being pollinators of flowers and predators of pests; others are harmful, being pests of crops, disease carriers, and destroyers of clothes, furniture, and buildings. *See also* mouthparts; Anoplura; Coleoptera; Dermaptera; Dictyoptera; Diptera; Hemiptera; Hymenoptera; Lepidoptera; Odonata; Siphonaptera.

Insectivora /in-sek-**tiv**-ŏ-ră/ An order of small primitive insectivorous or omnivorous and generally nocturnal mammals, e.g. *Sorex* (shrew), *Erinaceus* (hedgehog), and *Talpa* (mole). Insectivora have a long tapering snout with sensitive vibrissae and numerous small teeth with pointed cusps for crushing insects. Most have five-clawed digits and locomotion is plantigrade.

insectivore /in-**sek**-tă-vor, -vohr/ Any animal that feeds mainly on insects. Most insectivores are small mammals of the order Insectivora.

insertion 1. (*Anatomy*) The end of a muscle where it attaches to the part (e.g. a bone) moved by contraction of the muscle. 2. (*Genetics*) A type of MUTATION in which a base pair is added to the DNA of a gene, resulting in misreading of the genetic code by the ribosome during translation. *See* frameshift.

instar /in-star/ The stage between successive molts of insects (e.g. locusts and aphids) that develop by incomplete metamorphosis. The early instars, in which the insect is small, sexually immature, flightless, and often differently marked, are called *nymphs*. In the later instars, wing buds may appear, but these do not become capable of flight until the final molt, at which the imago (adult) emerges.

instinct A response to an external stimulus that an animal is born with and performs involuntarily. Instinct provides an animal with adaptive responses that have evolved over a long time and appear the very first time the stimulus is perceived. Usually instinctive responses are fixed stereotyped movements that are the same in all individuals of the species every time they are performed. Instinctive responses are most important in animals with short lifespans and little or no parental care and which have little opportunity to modify their behavior as a result of experience. Also, when it is essential for immediate action to be taken in response to a particular stimulus, it is an advantage for these actions to be instinctive. For example, the alarm calls of birds given when a predator approaches are performed instinctively whereas the courtship song is learnt by listening to other birds.

insulin /in-sŭ-lin/ A hormone that controls the metabolism of glucose. Lack of insulin results in diabetes mellitus, but excess insulin leads to coma. Insulin is a polypeptide produced by the B-cells of the islets of Langerhans of the pancreas. Its secretion is stimulated by high blood levels of glucose and amino acids after a meal. Glucose uptake is then stimulated by the action of insulin on various tissues (e.g. muscles, liver, and fat). It also stimulates glycogen and fat synthesis. Insulin is used therapeutically in the treatment of diabetes mellitus. *See* diabetes.

insulin-like growth factor (IGF) Any of a group of polypeptides that are structurally related to insulin and act as growth-promoting agents. They stimulate cell division and protein synthesis, and also promote glycogen formation in muscle and fat deposition. Their production is itself stimulated by growth hormone (somatotropin), and they include the factors designated IGF-I (somatomedin C) and IGF-II (somatomedin A).

integration The process by which sensory input to the brain is coordinated to produce an effector output that is appropriate to the input but not simply a function of it. It allows adaptive behavior. Integration is made possible largely because of the many complex synaptic connections between nerve cells. The various postsynaptic responses interact in different ways, causing the relay of different information.

integument /in-**teg**-yŭ-mĕnt/ **1.** (*Botany*) A layer surrounding the nucellus in the ovules of gymnosperms and angiosperms. Most angiosperm ovules possess two integuments, while gymnosperms usually have only one. Enclosure of the nucellus by the integuments is incomplete, the micropyle remaining to allow access to the embryo sac or, in gymnosperms, to the archegonium. *See illustration at* ovule. **2.** (*Zoology*) A body covering consisting of one or more tissues, serving to insulate and protect the body from its environment. Examples are a cuticle and the skin.

intention movement The first action of a series of behavioral actions that triggers off the subsequent actions. For example, in nestmaking by sticklebacks the intention movement is the initial digging of a pit and this action triggers off the chain of events that follows in completing the nest. Intention movements may be interpreted by other individuals, for example those in threat behavior, which may then modify their behavior accordingly.

intercalary meristem /in-**terk**-ă-lair-ee, in-ter-**kay**-lă-ree/ A region of actively growing primary tissue clearly separate from the apical meristem. Intercalary meristems occur at the internode and leaf sheath bases (joints) of many monocotyledons including grasses. In very young internodes all the cells may be meristematic, but the upper ones soon differentiate leaving only those at the base still actively dividing. In still older internodes all meristematic activity ceases but activity may be regained to re-elevate the stem if it has been flattened. Intercalary (non-localized) growth is also seen in algae. For ex-

ample certain filamentous algae (e.g. *Ulothrix*) may extend by division of cells at any point along the filament.

intercellular /in-ter-**sell**-yŭ-ler/ Describing materials found and processes occurring between cells. *Compare* intracellular.

intercostal muscles /in-ter-**kos**-t'l/ A set of muscles connecting adjacent ribs of reptiles, birds, and mammals. They are responsible for rotating the ribs to increase and decrease the volume of the rib cage, thus effecting breathing movements. The *external intercostal muscles* contract to rotate the ribs anteriorly (inspiration); the *internal intercostal muscles* contract to rotate the ribs posteriorly (expiration).

interfascicular cambium /in-ter-fă-**sik**-yŭ-ler/ A single layer of actively dividing cells between the vascular bundles in stems. It is formed when parenchyma cells resume meristematic activity. The interfascicular and INTRAFASCICULAR CAMBIUM link into a complete cambium cylinder that cuts off secondary xylem tissue to the inside, secondary phloem tissue to the outside, and parenchyma cells both sides (forming medullary rays).

interferon /in-ter-**feer**-on/ Any of a group of proteins produced by animal cells in response to infection by viruses or other agents. They act to inhibit viral replication within the cell; some also have antibacterial and anticancer properties. They are produced commercially for various therapeutic applications, using recombinant DNA technology.

interleukin /in-ter-loo-kin/ (IL) Any of a class of cytokines that act as chemical messengers between various types of white blood cells (leukocytes). All are soluble proteins or glycoproteins. For example, interleukin-1 (IL-1) is secreted by antigen-presenting macrophages, and stimulates T-cells. These in turn secrete IL-2, which promotes proliferation of T-cells and differentiation of B-cells to antibody-secreting plasma cells. Around 20 other interleukins have been described. *See* cytokine. *Compare* lymphokine.

intermediate filament (IF) Any of a group of fibrous proteins that provide support for cells as components of the CYTOSKELETON. They typically have a diameter of about 10 nm, intermediate between the narrower microfilaments and the wider microtubules. Various types are found in different cell types. For example, IFs made of keratin occur in epithelial cells. When such cells die, the keratin IFs form hair, nails, and wool. Other types include vimentin IFs in leukocytes and fibroblasts, and desmin IFs in muscle cells. In general, each IF consists of helically coiled polypeptides, forming rodlike arrays that are joined end to end in long protofibrils. Each IF consists of four protofibrils, arranged side by side. IFs stabilize cell membranes, and help in forming an internal framework for the cell. For example, the inner surface of the nuclear envelope is supported by a network of IFs composed of the protein lamin; other IFs extend through the cytoplasm and run underneath the plasma membrane. They also participate in certain cell junctions (*see* desmosomes).

internal energy (**intrinsic energy**) Symbol: U The quantity of energy possessed by a given substance or system. The internal energy is the sum of the kinetic and potential energies of the atoms or molecules of the system. In practice, changes in internal energy are important in chemical reactions. *See* enthalpy.

internal environment The medium surrounding the body cells of multicellular animals, i.e. the intercellular fluid. In vertebrates its composition is kept relatively constant by the mechanisms of homeostasis.

interneuron /in-ter-**new**-rohn, -ron/ A neuron that connects sensory neurons and motor neurons. Interneurons are generally located in the central nervous system.

internode /**in**-ter-nohd/ **1.** (*Botany*) The region of a plant's stem that is situated between two nodes. *See* node; intercalary meristem.

2. (*Zoology*) The region of a medullated nerve fiber between two nodes of Ranvier. It is covered with a myelin sheath.

interphase /**in**-ter-fayz/ The stage in the cell cycle when the nucleus is not in a state of division. Interphase is divisible into various stages each characterized by a differing physiological activity. *See* cell cycle.

interpositional growth /in-ter-pŏ-**zish**-ŏ-năl/ *See* intrusive growth.

intersex /**in**-ter-seks/ An abnormal organism with physical characteristics that are intermediate between the male and female of the species. Intersexes may or may not have both male and female reproductive organs and are usually sterile. They are usually the result of changes or malfunctions of the sex chromosomes and/or sex hormones during development.

interstitial-cell-stimulating hormone (**ICSH**) *See* luteinizing hormone.

intestine The tube that conveys food from the stomach to the anus. During its passage along the intestine, digestion and absorption take place. In vertebrates, the intestine is divided into small (narrow) and large (wide) intestines. *See* large intestine; small intestine.

intine /**in**-tin, -tÿn/ (**endosporium**) The inner layer of the cell wall surrounding the pollen grains of angiosperms and gymnosperms. By contrast with the outer cuticularized layer (exine), the intine is thin and composed of cellulose. The pollen tube, which emerges during germination of the pollen grain, is an outgrowth of the intine.

intracellular /in-tră-**sell**-yŭ-ler/ Describing the material enclosed and processes occurring within a cell membrane. *Compare* intercellular.

intrafascicular cambium /in-tră-fă-**sik**-yŭ-ler/ (**fascicular cambium; vascular cambium**) A region of meristematic cells between the xylem and phloem of a vascular bundle in a plant stem. *Compare* interfascicular cambium.

intraspecific /in-tră-spĕ-**sif**-ik/ Occurring within a species. For example, *intraspecific selection* is natural selection within a species that favors individuals of a particular type, which tends to produce exaggeration of the features characterizing that type. For instance, deer with large antlers breed more successfully than ones with small antlers.

intrinsic energy *See* internal energy.

intrinsic factor A glycoprotein secreted by the stomach lining that is required for the absorption of vitamin B_{12} (cobalamin). It binds the vitamin in the small intestine and is then itself bound to a specific receptor on the intestinal mucosa, where the vitamin is released and transported across the mucosal cell into the bloodstream. Deficiency of intrinsic factor leads to malabsorption of vitamin B_{12} and hence pernicious anemia.

intrinsic rate of natural increase *See* biotic potential.

introgression /in-trŏ-**gresh**-ŏn/ *See* introgressive hybridization.

introgressive hybridization /in-tŏ-**gress**-iv hŷ-bră-di-**zay**-shŏn/ (**introgression**) The introduction of genetic material from one gene pool to another by hybridization and subsequent back-crossing to one or other of the parents. It forms the basis of most livestock- and crop-breeding programs.

intromittent organ /in-trŏ-**mit**-ĕnt/ The male copulatory organ in animals with internal fertilization. It is used to introduce spermatozoa into the reproductive tract of the female. Examples of intromittent organs are claspers and the penis.

intron /**in**-tron/ A noncoding DNA sequence that occurs between coding sequences (EXONS) of a gene. Introns are found in most eukaryote nuclear genes, in

some cytoplasmic genes, in some genes of prokaryotes (especially archaebacteria and cyanobacteria), and even in certain viral genes. Mature functional messenger RNA (mRNA) does not contain introns, these being removed from the initial transcript (pre-mRNA) during processing. In each case the pre-mRNA is cleaved, the intron is removed as a loop, and the adjoining exons are spliced together. In eukaryotes the process is generally performed by a complex of enzymes called a spliceosome. In other cases, however, the pre-mRNA can act as its own enzyme (*see* ribozyme) to catalyze splicing of its introns.

The function and evolution of introns is a highly contentious subject. One view is that by acting as 'spacers' for exons in mRNA they enable *exon shuffling* – recombination or rearrangement of exons – and hence the rapid evolution of proteins with different combinations of functional groups. Another view is that they represent selfish DNA, which confers no advantage on the host and can move from place to place within the genome (*see* transposon).

introrse /in-**trors**/ Denoting anthers in which dehiscence lines are towards the center of the flower. Pollen is consequently shed towards the carpels, which favors self pollination. The direction of anther dehiscence is usually determined by the position of the connective and is uniform throughout a flowering family. *Compare* extrorse.

intrusive growth (**interpositional growth**) A method of plant growth in which the cells enlarge and force their way between other cells, causing their disruption. *Compare* sliding growth; symplastic growth.

intussusception /in-tŭs-sŭ-**sep**-shŏn/ **1.** (*Botany*) The incorporation of cellulose molecules into the existing cell wall, giving an increase in wall area. *Compare* apposition.
2. (*Medicine*) The telescoping inside one another of two adjacent sections of a tubular organ, such as the intestine.

inulin /**in**-yŭ-lin/ A polysaccharide food reserve of some higher plants, particularly the Compositae, e.g. *Dahlia* root tubers. It is a polymer of fructose.

invagination /in-vaj-ă-**nay**-shŏn/ The formation of a pocket of tissue during embryonic development, especially the intucking of the early gut (archenteron).

inversion /in-**ver**-shŏn/ *See* chromosome mutation.

invertebrate /in-**ver**-tă-brit/ Any animal that does not possess a vertebral column. Invertebrates (the term is a general one, not used in zoological classification) range from the Porifera (sponges) to the simple chordates – the cephalochordates (e.g. *Branchiostoma*) and urochordates (tunicates) – and include such widely differing animals as cnidarians, mollusks, worms, and arthropods (including insects).

invert sugar *See* sucrose.

in vitro /in **vee**-troh/ Literally 'in glass'; describing experiments or techniques performed in laboratory apparatus rather than in the living organism. Cell tissue cultures and *in vitro* fertilization (eventually to produce test-tube babies) are examples. *Compare* in vivo.

in vivo /in **vee**-voh/ Literally 'in life'; describing processes that occur within the living organism. *Compare* in vitro.

involucre /**in**-vŏ-loo-ker/ A protective structure. It is commonly seen in angiosperms with condensed inflorescences (e.g. the capitulum and umbel) as a ring of bracts arising below the inflorescence. An involucre is also seen in bryophytes as an outgrowth of the tissue that protects the archegonia.

involuntary muscle *See* smooth muscle.

involution /in-vŏ-**loo**-shŏn/ **1.** A decrease in the size of an organ, either as a result of normal ageing processes, or following enlargement, as in the shrinking of the womb after pregnancy.
2. The production of abnormal forms of microorganisms in unfavorable conditions.
3. The inturning of the dorsal lip of the blastopore during gastrulation in some vertebrate embryos.

iodine /ÿ-ŏ-dÿn, -deen/ A trace element essential in animal diets mainly as a constituent of the thyroid hormones. Deficiency of iodine causes the thyroid gland to enlarge, giving the condition known as goiter. Iodine is not essential to plant growth although it is accumulated in large amounts by the brown algae. *See also* staining.

ionizing radiation Any high-energy radiation that causes ionization of the medium through which it travels. It includes both short-wavelength electromagnetic radiation (e.g. ultraviolet, X-rays, and gamma rays) and high-energy particles (i.e. alpha and beta particles). Biological tissues are seriously damaged by such radiation, mainly through the effects of highly reactive FREE RADICALS generated by irradiation of the water molecules.

iris /ÿ-ris/ The pigmented circular area in the front of the eye of vertebrates and cephalopods. It has a central hole, the pupil. The iris acts as a diaphragm to vary the size of the pupil and so control the amount of light entering the eye. It is continuous with the ciliary body and the choroid layer and lies between the cornea and the lens. It is bathed in the aqueous humor. Fine particles of brown pigment (melanin) in the iris scatter the light rays and so make it look blue, and additional pigments give a gray, brown, or other color to the iris. The circular muscles round the pupil contract to make it smaller in bright light, or for close viewing. In dim light these muscles relax, and the radial muscles contract, to enlarge the pupil and let in more light. The muscles are involuntary and show a reflex response to the light intensity, through the autonomic nervous system. *See illustration at* eye.

iron /ÿ-ern/ An essential nutrient for animal and plant growth. It is contained in the protein hemoglobin, which gives the color to red blood cells and is responsible for oxygen transport from the lungs. Iron deficiency leads to anemia.

Iron is also found in other porphyrins and in cytochromes, which are important components of the electron-transport chain. It is also required as a cofactor for certain enzymes, e.g. catalase and peroxidase. Certain iron-containing proteins are essential for the fixation of nitrogen by bacteria.

ischium /iss-kee-ŭm/ (*pl.* **ischia**) A bone forming the posterior part of the ventral side of the tetrapod pelvic girdle. In humans, it is L-shaped, passing down from the acetabulum and turning forward to meet the pubis.

islets of Langerhans /ÿl-lĕts/ (**pancreatic islets**) Clusters of endocrine cells within the pancreas (described by Langerhans in 1869) that produce various hormones, notably insulin, which controls blood-sugar level. The islets consist of three types of cells: A-cells, which secrete glucagon; B-cells, which secrete insulin; and D-cells, which secrete somatostatin.

isoantigen /ÿ-soh-**an**-tă-jĕn/ A type of antigen that induces antibody production in genetically different individuals of the same species but not in the individual itself.

isoenzyme /ÿ-soh-**en**-zÿm/ (**isozyme**) An enzyme that occurs in different structural forms within a single species. The isomeric forms all have the same molecular weight but differing structural configurations and properties. Large numbers of different enzymes are known to have isomeric forms; for example, lactate dehydrogenase has five forms. Variations in the isoenzyme constitution of individuals can be distinguished by electrophoresis.

isogamy /ÿ-**sog**-ă-mee/ The sexual fusion of gametes of similar size and form. It occurs in fungi and some protoctists. *Compare* anisogamy.

isograft /ÿ-sŏ-graft, -grahft/ (**syngraft**) A type of GRAFT between individuals that are genetically identical.

isolating mechanism Structural, physiological, behavioral, genetic, geographical, or other factors that restrict the interbreeding of one population with another. The development of isolating mechanisms promotes the formation of new varieties and new species.

isoleucine /ÿ-sŏ-**loo**-seen, -**loo**-sin/ *See* amino acids.

isomers /ÿ-sŏ-merz/ Two or more molecules that have identical molecular formulas but differ in either the nature of their chemical bonding (*structural isomers*) or the arrangement of their atoms or groups in space (*stereoisomers*, or *optical isomers*). Many organic compounds possess a carbon atom – an *asymmetric*, or *chiral atom* – that is bound to four dissimilar atoms or groups. The bonds formed by such an atom can exist in alternative three-dimensional arrangements that are mirror images of each other and cannot be superimposed; i.e. the chiral carbon atom produces two stereoisomers. The stereoisomers of any given compound usually have very different biological properties. For example, all amino acids except glycine have two stereoisomers, denoted D (dextro) and L (levo) isomers. Glucose and other sugars also have D and L isomers. Generally, only L amino acids occur naturally in proteins, while only the D isomers of sugars are used by living organisms.

isomorphism /ÿ-sŏ-**mor**-fiz-ăm/ A condition seen in certain algae (e.g. *Ulva*, the sea lettuce) in which the alternating generations of the life cycle are morphologically identical. *Compare* heteromorphism.

isotonic /ÿ-sŏ-**tonn**-ik/ Designating a solution with an osmotic pressure or concentration equal to that of a specified other solution, usually taken to be within a cell. It therefore neither gains nor loses water by osmosis. *Compare* hypertonic; hypotonic.

isotope /ȳ-sŏ-tohp/ One of two or more atoms of the same element that differ in atomic mass, having different numbers of neutrons. For example ^{16}O and ^{18}O are isotopes of oxygen, both with eight protons, but ^{16}O has eight neutrons and ^{18}O has ten neutrons. A natural sample of most elements consists of a mixture of isotopes.

Many isotopes are radioactive and can be used for labeling purposes. The isotopes of an element differ in their physical properties and can therefore be separated by techniques such as fractional distillation, diffusion, and electrolysis.

isozyme /ȳ-soh-zȳm/ *See* isoenzyme.

J

jejunum /ji-**joo**-nŭm/ (*pl.* **jejuna**) The part of the small intestine between the duodenum and the ileum.

jellyfish *See* Scyphozoa.

joint The point where two or more bones meet. Joints allow different degrees of movement. *Fibrous joints*, as between the bones of the skull, allow no movement; *cartilaginous joints*, as between successive vertebrae, have a pad of cartilage against which the bones can move slightly; *synovial joints*, as between limb bones, allow considerable movement. Synovial joints also show a variety of movements, for example, in one direction (hinge joint), all directions (ball-and-socket joint), where two bones rotate against each other (pivot joint), or slide over one another (sliding or gliding joint).

jugular vein /**jug**-yŭ-ler/ One of a pair of veins in mammals, each with an internal and external portion, that carry deoxygenated blood away from the head and neck. They unite with the subclavian veins and lead into the anterior venae cavae.

jumping gene *See* transposon.

junk DNA *See* selfish DNA.

Jurassic /jû-**rass**-ik/ The middle period of the Mesozoic era, 215–145 million years ago. During the Jurassic dinosaurs were becoming large and abundant and bony fishes (teleosts) were also evolving rapidly. Fossils of the earliest known bird *Archaeopteryx* and of the first mammals are found in the late Jurassic. *See also* geological time scale.

juvenile hormone (neotonin) A hormone secreted by endocrine glands associated with the brain (*see* corpora allata) in insects that prevents metamorphosis into the adult form and maintains the presence of larval characteristics.

The exact mechanism of juvenile hormone action is not clear, but it appears to modify the effect of the molting hormone, ecdysterone. The concentration of juvenile hormone decreases gradually during development, declining below a threshold level during the final instar to allow ecdysterone to promote formation of a pupa.

K

kairomone /**kair**-ŏ-mohn/ A chemical messenger emitted by an individual of a species which causes a response in an individual of another species. This may be detrimental to the producer of the kairomone, for example many parasites are attracted to their hosts by an excreted kairomone. *See also* pheromone.

karyogamy /ka-ree-**og**-ă-mee/ The fusion of two nuclei that exist within a common cytoplasm, as occurs in the formation of the zygote from two gametes. The process also occurs within the multinucleate plasmodium of slime molds belonging to the Myxomycota. *See also* fertilization; plasmogamy.

karyogram /**ka**-ree-ŏ-gram/ (**idiogram**) The formalized layout of the karyotype of a species, often with the chromosomes arranged in a certain numerical sequence.

karyokinesis /ka-ree-oh-ki-**nee**-sis/ *See* mitosis.

karyotype /**ka**-ree-oh-tÿp/ The physical appearance of the chromosome complement of a given species. A species can be characterized by its karyotype since the number, size, and shape of chromosomes vary greatly between species but are fairly constant within species.

keel (**carina**) A large ventral platelike extension of the breast bone (sternum) in birds and bats, to which the wing muscles are attached.

keratin /**ke**-ră-tin/ One of a group of fibrous insoluble sulfur-containing proteins (scleroproteins) found in ectodermal cells of animals, as in hair, horns, and nails.

Leather is almost pure keratin. There are two types: α keratins and β keratins. The former have a coiled structure, whereas the latter have a beta pleated sheet structure. *Cytokeratins* form part of the filamentous cytoskeleton of cells.

keratinization /ke-ră-tă-ni-**zay**-shŏn/ (**cornification**) A process occurring in vertebrate epidermis and epidermal structures in which keratin replaces the cytoplasm of a cell. For example, the cornified outer layer of the epidermis of the skin consists of dead horny cells. Hairs and nails also consist of keratinized cells. *See* keratin.

ketogenesis /kee-toh-**jen**-ĕ-sis/ The formation of ketone bodies.

ketohexose /kee-toh-**heks**-ohs/ A ketose sugar with six carbon atoms. *See* sugar.

ketone /**kee**-tohn/ A type of organic compound with the general formula RCOR, having two alkyl or aryl groups bound to a carbonyl group. They are made by oxidizing secondary alcohols (just as aldehydes are made from primary alcohols). Simple examples are propanone (acetone, CH_3COCH_3) and butanone (methyl ethyl ketone, $CH_3COC_2H_5$).

The chemical reactions of ketones are similar in many respects to those of aldehydes.

ketone body One of a group of organic substances formed in fat metabolism, mainly in the liver. Examples are acetoacetic acid and acetone. If the body has little or no carbohydrate as a respiratory substrate, *ketosis* occurs, in which more ketone bodies are produced than the body can use.

ketopentose /kee-toh-**pen**-tohs/ A ketose sugar with five carbon atoms. *See* sugar.

ketose /**kee**-tohs/ A sugar containing a ketone (=CO) or potential ketone group. *See* sugar.

kidney One of a pair of major excretory organs of vertebrates, which may also function in osmoregulation. They are made up of excretory units (*see* nephron), which are responsible for the filtration and selective reabsorption of materials (water, mineral salts, glucose, etc.) and the production of waste. In mammals, the kidneys are red-brown oval structures, which are attached to the dorsal side of the abdominal cavity. They receive oxygenated blood by the renal artery and are drained of deoxygenated blood by the renal vein. A collecting duct, the ureter, conveys excess water, salts, and nitrogenous compounds (urea and uric acid) as urine from each kidney to the urinary bladder and hence to the exterior. *See also* pronephros; mesonephros; metanephros.

kidney tubule (**uriniferous tubule**) A long narrow tube forming the part of the excretory unit (nephron) of the vertebrate kidney that is responsible for selective reabsorption of useful substances. As filtrate leaves the Bowman's capsule it passes through a series of coiled loops (the proximal convoluted tubule), where glucose, amino acid, and some water are absorbed. It then passes through a long straight loop (loop of Henle) – the major site of water reabsorption – which may reach far down into the medulla before returning to the cortex. The loop is named for the German physician, anatomist, and pathologist Friedrich Gustav Jacob Henle (1809–85). There is a second series of coils (the distal convoluted tubule), concerned with salt and water reabsorption. The remaining liquid enters a collecting duct as urine. Some animals (e.g. amphibians) have very few or no loops of Henle and so do not produce concentrated urine.

killer cell *See* natural killer cell; T-cell.

kilobase Symbol: kb A unit of length of a polynucleotide (i.e. DNA or RNA) equal to 1000 bases of a single strand, or 1000 base pairs (bp) of a double strand. *Compare* bp.

kilocalorie *See* Calorie.

kilogram calorie *See* Calorie.

kinase /**ky**-nayss, **kin**-ayss/ **1.** Any enzyme that transfers a phosphate group, usually from ATP.
2. An enzyme that activates the inactive form of other enzymes. For instance, when trypsinogen, the inactive form of trypsin, comes in contact with enterokinase, active trypsin is released.

kinesis /ki-**nee**-sis/ Locomotory response to a stimulus by an organism or part of an organism, in which the rate of movement or turning is dependent on the intensity of the stimulus but is unaffected by its direction. For example, woodlice move quickly when they are in dry conditions and slowly in damp conditions.

kinetin /**kin**-ĕ-tin/ (**6-furfurylaminopurine**) An artificial cytokinin found in extracts of denatured DNA; the first of the CYTOKININS to be isolated.

kinetochore /ki-**net**-ŏ-kor, -kohr, ki-**neet**-/ *See* centromere.

kinetosome /ki-**net**-ŏ-sohm/ *See* basal body.

kingdom A category used in the classification of living organisms and containing related phyla; for example, all animal phyla can be placed in the kingdom Animalia. In the Five-Kingdom system of classification, all organisms are allocated to one of five kingdoms: Monera (Bacteria); Protoctista (including protozoa and algae); Fungi; Plantae; and Animalia. This widely used system is based on comparing various features of the organisms, including their anatomy, physiology, development, and biochemistry. These five kingdoms can be further organized into two superkingdoms: Prokarya (Prokaryotae), which contains

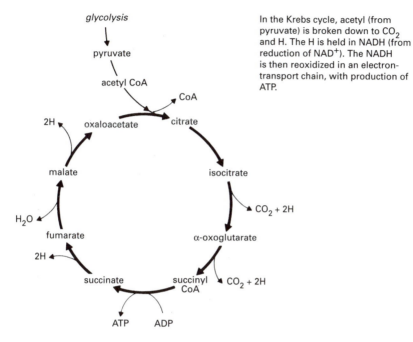

glycolysis

↓

pyruvate

acetyl CoA

CoA

2H ← oxaloacetate — citrate

malate

isocitrate

H_2O ←

fumarate — $CO_2 + 2H$

2H ←

α-oxoglutarate

succinate — succinyl CoA — $CO_2 + 2H$

ATP   ADP

In the Krebs cycle, acetyl (from pyruvate) is broken down to CO_2 and H. The H is held in NADH (from reduction of NAD^+). The NADH is then reoxidized in an electron-transport chain, with production of ATP.

Krebs cycle

the prokaryotes (i.e. Monera); and Eukarya (Eukaryotae), which contains all eukaryote organisms (i.e. the four remaining kingdoms). *Compare* domain; def. 1.

kinin /kÿ-nin/ **1.** Any of a class of peptides that are formed from precursors in blood or other tissues and function as local hormones. They include the angiotensins, which constrict blood vessels and raise blood pressure; bradykinin, which dilates blood vessels and lowers blood pressure; and tachykinins, which have a similar effect and also stimulate salivation and tear production. *See* angiotensin.
2. A former name for cytokinin.

kinomere /kÿ-nŏ-meer/ *See* centromere.

kin selection Natural selection resulting from altruism.

klinostat /klÿ-noh-stat/ *See* clinostat.

kneecap *See* patella.

knockout *See* gene knockout.

Krebs cycle /krebz/ (**citric acid cycle; tricarboxylic acid cycle**) A complex cycle of reactions in which pyruvate, produced by glycolysis, is oxidized to carbon dioxide and water, with the production of large amounts of energy. It is the second stage of AEROBIC RESPIRATION, requires oxygen, and occurs in the matrix of mitochondria.

2-carbon acetate, derived from pyruvate by decarboxylation, reacts with 4-carbon oxaloacetate to form 6-carbon citrate, which is then decarboxylated to reconstitute oxaloacetate. Some GTP (equivalent to ATP) is produced by direct coupling with cycle reactions, but most ATP production is coupled to the ELECTRON-TRANSPORT CHAIN. The operation of this depends on the generation of reduced coenzymes, NADH and $FADH_2$, by the Krebs cycle. Each molecule of glucose oxidized by glycolysis gives two molecules of pyruvate. Under aerobic conditions, these give two 'turns' of the Krebs cycle, which (including the conversion of pyruvate to acetyl CoA) yields: 8 NADH, 2 $FADH_2$, 2 GTP (equivalent to 2 ATPs), and 6 CO_2. (Added to this is the net yield of 2 ATP and 2 NADH from glycolysis.) The reduced coenzymes NADH and $FADH_2$ subsequently enter the electron-transport chain to generate ATP.

The cycle is named for the German-born British biochemist Sir Hans Adolf Krebs (1900–81).

K selection A form of natural selection that favors individuals or species that are adapted to living at or near their maximum population size, or CARRYING CAPACITY, for a particular habitat. Hence K-selected species (*K strategists*) are typically successful competitors for the limited resources, such as space and food, tend to be long-lived, and produce relatively small numbers of offspring with extended development periods. For example, long-lived trees such as redwoods are K strategists. *Compare* r selection.

L

labeling The technique of using isotopes (usually radioactive isotopes) or other recognizable chemical groups to investigate biochemical reactions, or to identify the location of labeled compounds within cells. For instance, a compound can be synthesized with one of the atoms replaced by a radioactive isotope of the element. The radioactivity can then be used to follow the course of reactions involving this compound. Other commonly used labels include enzymes, which can be detected by their catalytic activity, and fluorescent reagents, such as fluorescein. Electron-dense labels (e.g. ferritin) can be attached specifically to certain structures in order to reveal their location under the electron microscope.

labium /**lay**-bee-ŭm/ (*pl.* **labia**) *See* mouthparts.

labrum /**lay**-brŭm/ (*pl.* **labra**) *See* mouthparts.

labyrinth /**lab**-ă-rinth/ (**membranous labyrinth**) The membranous system of cavities and canals that occurs in the vertebrate inner ear and contains a fluid (endolymph). It consists of two cavities (the utriculus and the sacculus) and three semicircular canals, which act as organs of balance, and a spirally coiled canal (the cochlea) containing the organ of hearing. The labyrinth is surrounded by perilymph and protected by a cartilaginous bony casing (the *cartilaginous* or *bony labyrinth*), which forms part of the auditory capsule.

Lacerta /lă-**ser**-tă/ (**lizard**) *See* Squamata.

lacrimal gland /**lak**-ră-măl/ A gland associated with the eye of many vertebrates. It lies beneath the upper eyelid; the fluid (tears) produced from this gland continually washes the front of the eye and drains through the lacrimal duct into the nose. This secretion of small amounts of sterile, slightly antiseptic fluid helps keep the cornea moist.

lactam /**lak**-tam/ A type of organic compound containing the –NH.CO– group as part of a ring in the molecule. Lactams can be regarded as formed from a straight-chain compound that has an amine group (–NH$_2$) at one end of the molecule and a carboxylic acid group (–COOH) at the other; i.e. from an amino acid. The reaction of the amine group with the carboxylic acid group, with elimination of water, leads to the cyclic lactam, which is thus an internal amide.

Lactam

lactation /lak-**tay**-shŏn/ The secretion of milk from the mammary glands of female mammals, which occurs after parturition and is initiated by hormone activity, particularly by an increase in the level of prolactin. Another hormone, oxytocin, stimulates the ejection of milk. During pregnancy, estrogen and progesterone cause an increase in the amount of milk-producing tissue in the breasts, but inhibit prolactin. The levels of both these hormones fall after birth allowing prolactin to act. The sucking action of the young stimulates the continued production of prolactin and oxytocin, so that lactation continues for a prolonged period. *See also* colostrum.

lacteal /**lak**-tee-ăl/ A microscopic blind-ending tube containing lymph, found in each villus of the lining of the small intestine. Digested fat is absorbed into the lacteals and forms an emulsion, which makes the lymph look milky. The lacteals are connected to the lymphatic capillaries and larger lymph vessels of the intestine. Through them the fat from the lacteals reaches the blood system via the thoracic duct, which opens into the jugular vein.

lactic acid /**lak**-tik/ A syrupy liquid occurring in sour milk as a result of fermentation by lactobacillii. It is produced (L-form only) during anaerobic respiration in animals as the end product of glycolysis.

lactic acid bacteria A group of bacteria that ferment carbohydrates in the presence or absence of oxygen, with lactic acid always a major end product. They have a high tolerance of acid conditions. Lactic acid bacteria are involved in the formation of yoghurt, cheese, sauerkraut, and silage. They can occur as spoilage organisms and some are pathogenic, causing infections of the nasopharynx.

lactone /**lak**-tohn/ A type of organic compound containing the group –O.CO– as part of a ring in the molecule. A lactone can be regarded as formed from a compound with an alcohol (–OH) group on one end of the chain and a carboxylic acid

(–COOH) group on the other. The lactone then results from reaction of the –OH group with the –COOH group; i.e. it is an internal ester.

lactose /**lak**-tohs/ (**milk sugar**) A sugar found in milk. It is a disaccharide composed of glucose and galactose units.

lacuna /lă-**kyoo**-nă/ (*pl.* **lacunae**) 1. An empty space in plant tissues resulting from cell breakdown, tissue splitting, or organized formation. In hydrophytes a well-arranged system of these air-filled cavities may exist to provide buoyancy. A large lacuna occupies the center of many stems. The term is infrequently used, being replaced by space, cavity, or canal, sometimes prefixed by air.
2. *See* leaf gap.

Lagomorpha /lag-ŏ-**mor**-fă/ An order of herbivorous mammals that includes the hares (e.g. *Lepus*) and rabbits (e.g. *Oryctolagus*). Lagomorphs resemble rodents, differing principally in the possession of two pairs of incisor teeth in the upper jaw, a small pair of peglike teeth lying behind the larger pair of chisel-like teeth.

Lamarckism /lă-**mar**-kiz-ăm/ Lamarck's theory of evolution (1809) postulating that acquired characteristics can be inherited, so resulting in permanent changes in populations. This theory, which was widely accepted during the 19th century, fell from favor with the rediscovery in 1900 of Mendel's work on inheritance. This and subsequent genetic discoveries showed that characteristics are inherited at fertilization, and that inheritance is thus unaffected by the life of the organism. The theory is named for the French biologist Jean Baptiste Pierre Antoine de Monet, Chevalier de Lamarck (1744–1829).

lamella /lă-**mel**-ă/ (*pl.* **lamellae**) 1. A layer of photosynthetic membranes (thylakoids) in CHLOROPLASTS or in the cytoplasm of photosynthetic bacteria. It may comprise one or more thylakoids, e.g. the three-thylakoid lamella of brown algae.

Lactone

2. A general term applied to any thin plate-like structure, e.g. the layers of calcified matrix in bone, the spore-bearing structures in agaric fungi, or the gill membranes of fish.

lamellipodium /lă-mel-ă-**poh**-dee-ă/ (*pl.* lamellipodia) A broad flaplike protrusion of a cell, seen typically at the leading edge of moving cells such as fibroblasts and keratinocytes.

lamina /lam-ă-nă/ (*pl.* laminae) 1. (of angiosperms) A thin usually flat foliage LEAF blade, commonly attached to the stem by a petiole. In most plants the leaf laminae are the main photosynthetic organs. A lamina is termed *simple* if complete, and *compound* if divided into leaflets. Internal organization shows a wide photosynthetic mesophyll layer permeated by veins and bounded by an epidermis. Leaf variation is mainly due to the structural diversity of the lamina. Age may determine the complexity of lamina shape, but light and humidity control many features including cuticle thickness, number of chloroplasts and stomata, possession of air and water storage tissue, and lamina size. The most bizarre lamina modifications due to habitat occur in insectivorous plants. The great variation of leaf laminae is used in the classification of plants.
2. (of algae) The bladelike part of the thallus of certain algae, notably the brown algae (e.g. kelps and bladderwrack). Intercalary meristems cause lamina growth and regeneration but apical initials are active in some species. The lamina may produce air bladders and reproductive organs. Internally three layers exist – a photosynthetic epidermis and cortex surrounding a medulla of elongated cells separated by mucilage.

laminarin /lam-ă-**nair**-in/ The chief carbohydrate food reserve of the brown algae. It is a polymer of glucose, is soluble, unlike starch, and closely resembles the callose of higher plant sieve tubes.

lampbrush chromosome An extended chromosome structure found in the oocytes of certain animals, notably amphibians, during the prophase of meiosis. In those species that show a great increase in nuclear and cytoplasmic volume during prophase, the lampbrush chromosomes may measure up to 1 mm in length and 0.02 mm in width. Such chromosomes consist of two central strands along which fine loops extend laterally. The loops are associated with an RNA matrix and are sites of active transcription.

lamp shells *See* Brachiopoda.

lancelets /lans-lit, lahns-/ *See* Cephalochordata.

large calorie *See* Calorie.

large intestine The last part of the alimentary canal, consisting of the colon and the rectum. It is called 'large' because it has a much wider diameter than the small intestine. It receives the undigestible remains of the food and prepares it for evacuation through the anus.

larva /lar-vă/ (*pl.* larvae) The young immature stage into which many animals hatch from the fertilized egg. Larvae are independent and self-sustaining but differ appreciably from the adult in structure and mode of life and are usually incapable of sexual reproduction. Development into the adult is by metamorphosis. Most invertebrates have a larval stage (examples include the caterpillar of butterflies and moths and the ciliated planktonic larvae of many marine species); the tadpole of frogs is an example of a vertebrate larva.

larynx /la-rinks/ (*pl.* larynges) A structure situated at the top of the trachea (windpipe) of tetrapods. In mammals, it contains cartilage, which keeps it open to the air. During swallowing it is closed off by the epiglottis. The vocal cords are folds in the lining. *See also* syrinx. *See illustration at* alimentary canal.

latent period The time that elapses between stimulation of an irritable tissue (e.g. a nerve or muscle) and the production of a detectable response. *Compare* reaction time.

latent virus A virus that can remain inactive in its host cell for a considerable period after initial infection. The viral nucleic acid becomes integrated in the host chromosome and multiplies with it. Eventual replication inside the host cell may be triggered by such factors as radiation and chemicals. An example of a latent virus is herpes simplex. *See also* provirus.

lateral-line system *See* acoustico-lateralis system.

latex /lay-teks/ A liquid found in some flowering plants contained in special cells or vessels called laticifers (or laticiferous vessels). It is a complex variable substance that may contain terpenes (e.g. rubber), resins, tannins, waxes, alkaloids, sugar, starch, enzymes, crystals, etc. It is often milky in appearance (e.g. dandelion and lettuce) but may be colorless, orange, or brown. Its function is obscure, but may be involved in wound healing as well as a repository for excretory substances. Commercial rubber comes from the latex of the rubber plants *Ficus elastica* and *Hevea brasiliensis*. Opium comes from alkaloids found in the latex of the opium poppy.

laticifers /lay-tiss-ă-ferz/ Latex-containing structures found in certain plants, e.g. rubber, poppy, and euphorbia. Laticifers may be formed by fusion of cells to give vessels, or by the elongation and branching of a single coenocytic cell. Latex consists of many substances in solution and colloidal suspension, some of which are important in medicine and industry.

Law of Independent Assortment *See* Mendel's laws.

Law of Segregation *See* Mendel's laws.

LD$_{50}$ Median lethal dose, i.e. the dose of toxin at which 50% of exposed animals are killed. It is used as a standard measure of toxicity.

L-DOPA /el-doh-pă/ (L-3,4-dihydroxyphenylalanine) An intermediate in the synthesis of dopamine, norepinephrine, and epinephrine and in the conversion of tyrosine to melanin pigments. L-DOPA is used to treat Parkinson's disease, a primary cause of which is a deficiency of dopamine in the brain cells. Dopamine itself cannot be administered because it cannot pass from the blood to the brain, so L-DOPA is taken orally, passes via the bloodstream to the brain, and is converted by decarboxylation to dopamine.

leaching The removal of soil nutrients by water moving down the soil profile. It makes soils more acid since cations (e.g. potassium and magnesium) are replaced by hydrogen ions. Leaching leads to the formation of podsolized and lateritic soils. Leaching of nutrients, especially nitrates, from agricultural land can lead to serious pollution of water supplies.

leaf A flattened appendage of the stem that arises as a superficial outgrowth from the apical meristem. Leaves are arranged in a definite pattern, have buds in their axils, and show limited growth. Most foliage leaves are photosynthetic, bilaterally symmetrical, and externally differentiated into LAMINA, midrib, and petiole. Many variations occur. Not all foliage leaves are photosynthetic as they may be modified wholly to bud scales and spines or partly to form tendrils. The main characteristics of leaves, variations of which are used for plant descriptions are: form of the lamina; shape of the lamina margin; shape of the leaf or leaflets; mode of attachment to stem; venation; texture. *See also* bract; microphyll; megaphyll; phyllotaxis.

leaf buttress A leaf primordium appearing as a protuberance on the side of the stem apex, produced by periclinal division of the tunica and corpus, and associated with a procambium. The leaf axis arises from the leaf buttress and maintains procambial continuity in the central region. This differentiates into the vascular tissue of the developing leaf.

leaf gap (**lacuna**) A region of parenchyma differentiated in the stem vascular cylinder immediately above a diverging leaf trace. Lateral connections insure no break in the vascular system. A leaf gap is always associated with a leaf trace. They are typical of angiosperms and pteridophytes, but are sometimes difficult to recognize due to the interconnecting vascular system.

leaf trace A vascular bundle or group of vascular bundles connecting the vascular systems of a leaf and a stem. It stretches from the leaf base to the vascular ring in the stem axis.

learning The alteration of behavior by an individual as the result of experience. Learning is the ability of an animal to discover which of its actions gives the best results in a particular set of circumstances, and to modify its behavior to its best advantage. Learnt behavior is diverse and differs between the individuals; it is best developed in animals with a long lifespan and a long period of parental care. *Compare* instinct.

lecithin /**less**-ă-thin/ *See* phosphatidylcholine.

leeches *See* Hirudinea.

legume /**leg**-yoom, li-**gyoom**/ **1.** (**pod**) A dry dehiscent single or many seeded fruit formed from a single carpel, that dehisces mechanically by splitting, often explosively, along both sides. The opposite tensions set up by oblique fibers in the drying pericarp cause dehiscence. The valves may continue twisting and remove any remaining seeds. It is the typical fruit of the Leguminosae family.
2. Any plant of the family Leguminosae.

Leishman's stain /**lÿsh**-mănz/ The stain is named for the British bacteriologist Sir William Boog Leishman (1865–1926). *See* staining.

lemma /**lem**-ă/ *See* bract.

lens In the vertebrate eye, a transparent biconvex disk. It consists of a firm but elastic jelly composed of concentric layers of fibrous tissue enclosed in a skin (capsule) and it is attached to the CILIARY BODY by suspensory ligaments. Circular muscles in the ciliary body adjust the curvature of the lens and vary its focusing power (accommodation). The lens and the cornea together form the image on the retina. In aquatic vertebrates, the lens has a fixed shape and only refracts light. It grows throughout life using glucose from the aqueous humor for its metabolism. *See illustration at* eye.

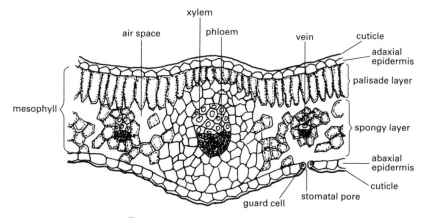

Transverse section through a leaf blade

lenticel /**len**-tă-sel/ A raised pore in the bark of a woody stem allowing gaseous exchange between internal tissues and the atmosphere. Parts of the phellogen (lenticel phellogen) produce cork cells that remain unsuberized and round off leaving conspicuous intercellular spaces. They eventually rupture the epidermis forming typical raised dusty pores. The number and shape of lenticels vary according to species.

Lepidoptera /lep-ă-**dop**-tĕ-ră/ A large order of insects containing the butterflies and moths, characterized by a covering of scales, often brightly colored, over their wings and bodies. Mandibles are usually absent and the maxillae form a tube (proboscis) for sucking nectar or fruit juices. The wings are coupled together in flight. The larvae (caterpillars) are mostly herbivores; some are serious plant pests. Metamorphosis is complete, with a pupal stage (the chrysalis). Butterflies are diurnal, have slim bodies and clubbed antennae, and rest with the wings folded over the back. Moths are mostly nocturnal, never have clubbed antennae, and rest with the wings in various positions.

leptosporangiate /lep-toh-spŏ-**ran**-jee-it, -ayt/ Describing the condition, found in certain pteridophytes (e.g. Filicales), in which the sporangium develops from a single initial cell. *Compare* eusporangiate.

leptotene /**lep**-tŏ-teen/ In meiosis, the stage in early prophase I when the chromosomes, already replicated, start to condense and appear as fine threads, although sister chromatids are not yet distinct. The spindle starts to form around the intact nucleus.

leucine /**loo**-seen/ *See* amino acids.

leukocyte /**loo**-kŏ-sÿt/ (**white blood cell**) A type of blood cell that has a nucleus but no pigment. White cells are larger and less numerous than red cells (about 6000–8000 per cubic millimeter of blood). They are important in defending the body against disease because they devour bacteria and produce antibodies. They are all capable of ameboid movement. There are several types of leukocytes. They can be divided into two groups, granulocytes and agranulocytes, according to the presence or absence of granules in the cytoplasm. The most numerous are the NEUTROPHILS (70%) and LYMPHOCYTES (25%). Leukocytes have a very short lifespan and are continuously produced in the myeloid tissue of the red marrow. *See also* monocyte; myeloid tissue; basophil; eosinophil.

leukoplast /**loo**-kŏ-plast/ A colorless plastid, i.e. one not containing chlorophyll or any other pigment. *See* plastid.

leukosin /loo-**koh**-sin/ One of the structurally stable scleroproteins found in wheat.

leukotrienes /loo-koh-**trÿ**-eenz/ A group of substances that are produced in tissues to serve as local hormones. They are derived from polyunsaturated fatty acids, and all contain at least three double bonds. For example, leukotriene B_4 promotes increased permeability of blood vessels and infiltration of certain types of white blood cells (basophils, neutrophils, and eosinophils), while leukotriene C_4 causes constriction of blood capillaries in various tissues, including the heart and brain.

levorotatory /lee-voh-**roh**-tă-tor-ee. -tohree/ Describing compounds that rotate the plane of polarized light to the left (anticlockwise as viewed facing the oncoming light). *Compare* dextrorotatory. *See* optical activity.

LH *See* luteinizing hormone.

lice *See* Anoplura.

lichens Symbiotic associations between an alga or blue-green bacterium (the *photobiont*) and a fungus (the *mycobiont*). They are slow-growing but can colonize areas too inhospitable for other plants. Usually the fungus is an ascomycete but occasionally it is a basidiomycete. Reproduction in lichens may be asexual by soredia (algal cells enclosed by fungal hyphae) or by sexual fungal spores, which can only survive if some algal cells are also present. Examples of lichens are *Peltigesa* and *Xanthoria*.

life cycle The sequence of changes making up the span of an organism's life from the fertilization of gametes to the same stage in the subsequent generation. The cycle may involve only one form of the organism, as in higher animals and plants. In lower plants, and some animals, two or more different generations exist and there is an alternation of generations, usually between haploid and diploid forms. Various terms exist to describe the different types of life cycle and take into account which generation is dominant and whether the generations differ morphologically. *See also* polymorphism; diplobiontic; diplont; haplobiontic; haplont.

ligament A tough band or capsule of connective tissue that connects two bones together at a joint. It has a high proportion of elastic fibers and white collagen fibers; therefore it combines strength with elasticity to control movement at the joint. The fibers of a ligament penetrate the surface tissue of the bones, making a very strong connection.

ligase /**lig**-ayss, **lÿ**-gayss/ An enzyme that catalyzes the bond formation between two substrates at the expense of the breakdown of ATP or some other nucleotide triphosphate. The degree of bond formation by ligases is proportional to the amount of ATP available in the cell at a particular moment in time.

light-dependent reactions The reactions of photosynthesis that convert light energy into the chemical energy of NADPH and ATP. *See* photosynthesis.

light green *See* staining.

light microscope *See* microscope.

lignin /**lig**-nin/ One of the main structural materials of vascular plants. With cellulose it is one of the main constituents of wood,

where it imparts high tensile and compressive strengths, making it ideal for support and protection. Lignified tissues include sclerenchyma and xylem. Lignin is deposited during secondary thickening of cell walls. The degree of lignification varies from slight in protoxylem to heavy in sclerenchyma and some xylem vessels, but values of 25–30% lignin and 50% cellulose are average. It is a complex variable polymer, derived from sugars via aromatic alcohols. Phenyl propane (C_6–C_3) units are linked in various ways by oxidation reactions during polymerization. Lignin is characteristically stained yellow by aniline sulfate or chloride, and red by phloroglucinol with hydrochloric acid.

ligule /**lig**-yool/ 1. A scalelike outgrowth, varying in shape and size, of certain angiosperm leaves. In grasses the ligule occurs at the junction of the leaf sheath and lamina. Although membranous in most species, it may be only a fringe of hairs.
2. A very small tongue-shaped flap of tissue inserted on the upper surface of the leaves and sporophylls of certain pteridophytes (e.g. *Selaginella* and *Isoetes*).
3. A toothed strap-shaped structure formed by the extension of one side of the corolla tube in certain Compositae florets. The teeth indicate the number of fused petals. In some species (e.g. dandelion), all florets of a capitulum are ligulate but in others (e.g. daisy), only the ray florets have ligules.

Liliopsida (**Monocotyledonae**) A class of the flowering plants (phylum Anthophyta) characterized by having a single cotyledon in the seed. They are usually herbaceous plants and do not show secondary growth. Examples of monocotyledons are the grasses and lilies. Generally the flower parts are borne in threes or multiples thereof and the leaf veins are parallel. The vascular tissue occurs as scattered bundles in the stem giving an atactostele.

limbic system /**lim**-bik/ The parts of the cerebral cortex in the brain that are concerned primarily with experience and expression of emotions, motivation, and mood. The limbic system includes the cingulate gyrus, hippocampus, amygdaloid nucleus, and other structures, which are located on the medial (inner) surface of each cerebral hemisphere. These connect to various other parts of the brain. The limbic system is also important in memory.

liming *See* flocculation.

limiting factor Any factor in the environment that alone governs the behavior of an organism or system by being above or below a certain level. In general, the behavior of a system depends on a number of different factors; under certain conditions, one of these can limit the behavior. For instance, plant growth is limited by low temperature and increases with rising temperature to an optimum, beyond which growth rate decreases.

limiting layer *See* meristoderm.

limnology /lim-**nol**-ŏ-jee/ The scientific study of freshwater, and its flora and fauna. Limnology also includes studying the chemical and physical aspects of inland water. It may be divided into the study of standing water habitats, e.g. lakes and ponds, and running water habitats, e.g. rivers and brooks.

linkage The occurrence of genes together on the same chromosome so that they tend to be inherited together and not independently. Groups of linked genes are termed *linkage groups* and the number of linkage groups of a particular organism is equal to its haploid chromosome number. Linkage groups can be broken up by crossing over at meiosis to give new combinations of genes. Two genes close together on a chromosome are more strongly linked, i.e. there is less chance of a cross over between them, than two genes further apart on the chromosome. Linked genes are symbolized Ab...Y/aB...y, indicating that Ab...Y are on one homolog while aB...y are on the other homolog. *See* cross-over value.

linkage map *See* chromosome map.

linoleic acid /lin-ŏ-**lee**-ik, lă-**noh**-lee-ik/ A common unsaturated fatty acid occurring as glycerides in linseed oil, cottonseed oil and other vegetable oils. It is an essential nutrient in the diet of mammals. *See* essential fatty acids.

linolenic acid /lin-ŏ-**lee**-nik, -**len**-ik/ An unsaturated fatty acid occurring commonly in plants as the glyceryl ester, for example in linseed oil and poppy-seed oil. The biological function of linolenic acid is similar to that of linoleic acid and administration of linolenic acid is also used to cure fat deficiency in animals. It is an essential nutrient in the diet of mammals. *See* essential fatty acids.

lipase /**lỹ**-payss, **lip**-ayss/ Any of various enzymes that catalyze the hydrolysis of fats to fatty acids and glycerol. Lipases are present in the pancreatic juice of vertebrates.

lipid /**lip**-id, **lỹ**-pid/ Any of a diverse group of substances that are characteristically soluble in organic solvents, such as ether and benzene, but insoluble in water. Lipids play various roles in living organisms. TRIGLYCERIDES, which are esters of glycerol and fatty acids, make up the fats and oils that serve as energy reserves in plants and animals. WAXES are esters of long chain monohydric alcohols and fatty acids; they provide a waterproof covering for the external surfaces of many organisms. Phospholipids form another biologically

A triglyceride – an ester of glycerol and carboxylic acids. R indicates an organic group

important class of lipids. These are diglycerides, in which only two of the hydroxyl groups of glycerol are esterified with a fatty acid, the third being linked to a phosphate group. Phospholipids are major components of cell membranes. Membranes often contain glycolipids as well. These are similar to phospholipids but contain the sugar galactose instead of the phosphate group. STEROIDS, and their alcohol derivatives (sterols), are also considered as lipids. The most abundant example is cholesterol, which is found in animal cell membranes and is the precursor for the production of steroid hormones and vitamin D. *See also* lipoprotein.

lipidoplast /**lip**-ă-doh-plast/   *See* elaioplast.

lipoic acid /li-**poh**-ik/ A sulfur-containing fatty acid found in a wide variety of natural materials. It is an essential as a coenzyme for certain dehydrogenase enzymes, notably pyruvate dehydrogenase. This catalyzes the dehydrogenation of pyruvic acid to form acetyl-CoA, as the initial step in the Krebs cycle. It has not yet been demonstrated to be required in the diet of higher animals. Lipoic acid is classified with the water-soluble B vitamins.

lipolysis /li-**pol**-ă-sis/ The splitting of the component fatty acids from a lipid; i.e. part of the process of catabolism of lipid molecules. Lipolysis is effected in the body, largely in the gut, by the lipase enzymes.

lipopolysaccharide /lip-ŏ-pol-ee-**sak**-ă-rÿd, -rid/ A conjugated polysaccharide in which the non-carbohydrate part is a lipid. Lipopolysaccharides are a constituent of the cell walls of certain bacteria.

lipoprotein /lip-ŏ-**proh**-teen, -**proh**-tee-in/ Any conjugated protein formed by the combination of a protein with a lipid. In the blood of humans and other mammals, cholesterol, triglycerides, and phospholipids associate with various plasma proteins to form lipoproteins. These are particles with diameters in the region of 7.5–70 nm, and are placed in several classes. The largest lipoproteins in this range are the *very low-density lipoproteins* (VLDLs), which are formed in the liver and contain up to about 20% cholesterol. *Low-density lipoproteins* (LDLs) are formed in plasma from VLDLs and contain over 50% cholesterol. LDLs transport cholesterol from the liver to peripheral tissues, and are thought to be a factor in the development of fatty arterial deposits that can lead to cardiovascular disease. *High-density lipoproteins* (HDLs), with about 20% cholesterol, are the smallest of the plasma lipoproteins, and apparently function in transporting cholesterol from tissues to the liver.

The largest of all plasma lipoproteins are the *chylomicrons*, which act as vehicles for the absorption of fat from the intestine. Measuring up to 100 nm in diameter, they are formed in the intestinal mucosa and contain mostly triglycerides, with relatively small amounts of cholesterol and protein. They enter the lacteals of the intestinal villi and are conveyed via the lympatic system to the bloodstream.

liposome /**lip**-ŏ-sohm, lÿ-pŏ-/ A microscopic spherical sac consisting of a lipid envelope enclosing fluid. They are made in the laboratory by adding an aqueous solution to a gel of complex lipids, and have a wall consisting of a double layer of lipids similar to that of cell membranes. Liposomes are used experimentally as models of cells or cell organelles, and also in medicine to deliver toxic drugs to target tissues in the body.

lipotropin /lip-ŏ-**troh**-pin, lÿ-pŏ-/ (**lipotropic** or **lipolytic hormone; LPH**) Either of two polypeptide hormones produced in mammals by the pituitary gland, that stimulate lipolysis. Both hormones occur in human blood but their precise physiological role remains uncertain; they may simply be by-products of the formation of corticotropin from its precursor, proopiomelanocortin.

litmus paper Red or blue acid–alkali indicator papers. In acids blue litmus paper turns red and in alkalis red litmus paper turns blue. The paper is prepared by soaking absorbent paper in litmus solution and then drying it. Litmus solution has a pH range from 4.5–8.3. It is obtained from a lichen (*Lecanora tartarea*).

littoral /lit-ŏ-răl/ 1. The zone of the seashore between the high and low tide mark. The term is also applied to organisms living in this zone. *Compare* benthic; sublittoral.
2. The zone between the water's edge and a depth of about six meters in a pond or lake. Rooted hydrophytes, both emergent and submergent, are found in this zone. *Compare* profundal; sublittoral.

liver A large dark red organ, made up of several lobes, lying close to the stomach in vertebrates. It accounts for one fifth of the whole contents of the abdomen, and all the blood in the body passes through the liver every two minutes. Its main function is to regulate the chemical composition of the blood and to act as a depot where the nutrients obtained by digestion or from body stores are converted into the materials that fulfill the body's metabolic requirements. It is supplied with oxygenated blood by the hepatic artery, but receives 80% of its blood in the hepatic portal vein from the intestine. After the digestion of food, this blood is rich in glucose and amino acids. The liver removes the surplus glucose (over 0.1%) and stores it as glycogen until it is needed. It changes the surplus amino acids to urea in a process of deamination. Its other functions are:
1. The production of bile, stored in the gall bladder and then passed into the duodenum along the bile duct.
2. The removal of damaged red corpuscles from the blood.
3. The storage of iron.
4. The manufacture of vitamin A from carotene, and the storage of vitamins A and D.
5. The manufacture of some of the proteins of blood plasma.
6. The manufacture of blood-clotting factors, prothrombin and fibrinogen.
7. The removal of poisons from the blood (detoxication).

8. The conversion of fats into compounds suitable for oxidation during starvation, and the conversion of carbohydrates to fat when there is too much to be stored as glycogen in the liver.
The microscopic structure of the liver insures that every cell is in direct contact with the blood (there are no endothelial linings in the capillaries) so that diffusion of molecules in and out of the cells is very rapid. *See illustration at* alimentary canal.

liverworts /liv-er-werts/ *See* Hepatophyta.

living fossil Modern organisms with anatomical or physiological features that are normally characteristic of extinct ancestral species. They are often associated with highly restricted, remote, and almost unchanging environments, and so evolve very slowly. Examples are the deep-sea fish, the coelacanth, and the *Ginkgo* tree.

lizards *See* Squamata.

loam A medium-textured soil containing a mixture of large and small mineral particles. Loams are easy soils to work and combine the good properties of sandy and clay soils.

loculus /lok-yŭ-lŭs/ (*pl.* **loculi**) 1. An air-filled compartment in an ovary in which the ovules develop. The term loculus can also be applied to any other cavity in an organ in which other structures develop, for example the loculi of anthers in which pollen is formed. When all the carpel edges meet at the middle of a syncarpous ovary there are as many loculi as carpels, and the ovary is designated bi-, tri-, quadri-, or multilocular according to number. Adjoining carpel walls form septa that separate loculi. A unilocular condition occurs in monocarpellary ovaries and syncarpous ovaries lacking septa.
2. (*Zoology*) A small chamber or cavity; for example, one of the shell chambers of a foraminiferan.

locus /loh-kŭs/ (*pl.* **loci**) The position of a gene on a chromosome. Alleles of the same gene occupy the equivalent locus on homologous chromosomes.

lomasome /loh-mă-sohm/ An infolding of the plasma membrane found particularly in fungal hyphae and spores, and also in some algae and higher plants. Lomasomes may be concerned with the formation or breakdown of the cell wall, or with secretion or endocytosis.

loment /lŏ-**ment**/ (lomentum) A jointed dry indehiscent seed pod of certain leguminous plants, such as crown vetch, that breaks transversely into separate seed-bearing segments.

long-day plant (LDP) A plant that flowers in response to a light period longer than a critical minimum. *See* photoperiodism; critical day length.

loop of Henle /**hen**-lee/ *See* kidney tubule.

lophotrichous /lŏ-**fot**-ri-kŭs/ Describing bacteria that possess a tuft of flagella.

luciferin /loo-**sif**-ĕ-rin/ Any substance that acts as substrate for an enzyme (*luciferase*) that catalyzes a light-emitting reaction in living organisms (*see* bioluminescence). Luciferens vary widely in different plant and animal organisms.

lumbar vertebrae /**lum**-ber/ The bones of the lower-back region of the vertebral column, between the thoracic region and the sacral region. In mammals the five lumbar vertebrae have large transverse processes for the attachment of the muscles of the hind limbs. *See* vertebral column.

Lumbricus /**lum**-bră-kŭs/ A genus of earthworms that burrow in soil and whose only appendages are a few chaetae. The burrow walls are stiffened with mucus secreted from the skin and the worms feed by swallowing soil and digesting the organic matter in it. These activities are important in soil aeration and drainage. Earthworms are hermaphrodite but cross-fertilization occurs, two worms binding themselves together with mucus so that the reproductive

segments of each are opposite the clitellum of the other and sperm exchange can take place. *See also* Oligochaeta.

lumen /**loo**-měn/ (*pl.* lumens or lumina) 1. (*Zoology*) The central cavity or canal within a tube, duct, or similar structure. 2. (*Botany*) The central space that remains, surrounded by rigid cell walls, in a cell that has lost its living contents (e.g. in xylem elements).

lung The respiratory organ of air-breathing vertebrates, including aquatic forms (such as turtles and whales). A pair of lungs is situated in the thorax. Air enters and leaves each lung through a bronchus. The lung contains a thin moist membrane of large area, folded so that it occupies relatively little volume. Gases can diffuse readily through this membrane, between the air on one side and the blood in the capillaries that are on the other side. In amphibians and reptiles the lung is a simple elastic sac with folds on its inner walls (deeper in reptiles) to increase the surface area. In birds and mammals the lung has a spongy texture. In mammals the bronchus branches repeatedly into bronchioles, which end in clusters of alveoli, where the main exchange of gases occurs. Lungs contain no muscular tissue. They are inflated either by air being pumped from the mouth cavity (amphibians), by the action of intercostal muscles (reptiles, birds, and mammals), or by the action of the diaphragm (mammals). Lungs are deflated by their own elasticity.

In lungfish the lung is an outgrowth from the pharynx, thought to be homologous with the swim bladder of teleosts. In birds, tubes from the bronchi lead to air sacs between the organs of the body and in the larger bones. This provides highly efficient ventilation of the lungs during flight. The mantle and mantle cavity of terrestrial gastropods is also termed a lung and has the same function. *See* alveolus; bronchiole; bronchus; respiration. *See illustration at* alimentary canal.

lung books The respiratory organs of arachnids, consisting of a cavity containing leaflike folds of the body wall through which blood circulates and between which air circulates. Some spiders have tracheae instead of or as well as lung books. *See also* gill books.

lungfish *See* Dipnoi.

lutein /**loo**-tee-in/ The commonest of the xanthophyll pigments. It is found in green leaves and certain algae, e.g. the Rhodophyceae. *See* photosynthetic pigments.

luteinizing hormone /**loo**-tee-ă-nÿ-zing/ (LH; interstitial-cell-stimulating hormone; ICSH) A glycoprotein hormone secreted by the anterior pituitary lobe under regulation of the hypothalamus. In female mammals it stimulates secretion of estrogen, ovulation, and formation of corpora lutea. In male mammals it stimulates interstitial cells in the testes to secrete androgens. *See also* gonadotropin.

luteotropic hormone /loo-tee-ŏ-**trop**-ik/ (luteotrophic hormone; LTH) *See* prolactin.

lyase /**lÿ**-ayss/ An enzyme that catalyzes the separation of two parts of a molecule with the formation of a double bond in one of them. For example, fumerase catalyzes the interconversion of malic acid and fumaric acid.

Lycophyta /lÿ-kŏ-pŏ-**dof**-ă-tă/ A phylum of spore-bearing vascular plants containing about 1000 living species, most of which constitute two genera: *Lycopodium* and *Selaginella*. There are five orders, three of which – the Lycopodiales, Selaginellales, and Isoetales – contain both living and fossil representatives. The remaining orders – Lepidodendrales and Pleuromeiales – are represented only by fossils. The extinct trees of the genus *Lepidodendron* were once distributed widely and contributed largely to the coal seams of the Carboniferous. Living club mosses are typically small plants that favor damp habitats; some are epiphytes. They have shoots bearing microsporophylls and may be either homosporous or heterosporous. They differ from the whisk ferns (*see* Psilophyta) in having roots, and differ from other vascular plants in having a dichotomous rather than monopodial branching system.

Lycopodium /lÿ-kŏ-**poh**-dee-ŭm/ (**club moss**) A genus of lycopods (*see* Lycophyta). Species of *Lycopodium* are generally found in moist habitats, and have both creeping and erect stems. Numerous small leaves (microphylls) are borne spirally and the fertile leaves (sporophylls) are borne in strobili.

lymph /limf/ The fluid contained within the vessels of the lymphatic system. It is derived from tissue fluid that is drained from intercellular spaces, and is similar to plasma but with a lower protein concentration and contains cells (mainly lymphocytes), bacteria, etc. It is colorless except in the region of the small intestine where absorbed fat gives the lymph a milky appearance. *See* lacteal.

lymphatic system /lim-**fat**-ik/ A series of vessels (lymphatic vessels) and associated lymph nodes that transports lymph from the tissue fluids into the bloodstream and the heart. Tissue fluid not returned to the circulation via blood capillaries is drained as lymph into the blind-ending thin-walled lymphatic capillaries, which also occur between cells. These join to form larger vessels, which eventually unite into major ducts (the right lymphatic duct and the THORACIC DUCT) and empty into the large veins entering the heart. The flow of lymph is achieved by muscular and respiratory movements in mammals and by the pumping of *lymph hearts* in other vertebrates. It is unidirectional: some vessels contain valves to prevent any backflow. The lymphatic system is also the main route by which fats reach the bloodstream from the intestine. *See also* lacteal; spleen.

lymphatic tissue *See* lymphoid tissue.

lymph heart *See* lymphatic system.

lymph node (**lymph gland**) One of a large number of flat oval structures distributed along the lymphatic vessels and clustered in certain regions, such as the neck, armpits, and groin. Lymph nodes are composed of lymphatic tissue and contain special white blood cells (lymphocytes), some of which produce antibodies. They act as defense posts against the spread of infection, their lymphocytes engulfing bacteria and other foreign materials from the lymph; the nodes may become inflamed and enlarged as a result.

lymphocyte /**lim**-fŏ-sÿt/ A type of white blood cell (LEUKOCYTE) with a very large nucleus, rich in DNA, and a small amount of clear cytoplasm. They comprise 25% of all leukocytes and produce antibodies, important in defense against disease (*see* immunity). Lymphocytes are made in myeloid tissue in red bone marrow, lymph nodes, thymus, tonsils, and spleen. During infection, antigens stimulate certain lymphocytes (B-lymphocytes, or B-CELLS) in the lymphoid tissue to multiply rapidly, and the resulting lymphocytes, called PLASMA CELLS, are released into the bloodstream to produce the appropriate antibody. *See also* T-cell.

lymphoid tissue /**lim**-foid/ (**lymphatic tissue**) Tissue in which lymphocytes are produced, found in the lymph nodes, tonsils, spleen, and thymus. It consists of a delicate network of cells through which lymph flows continuously. Lymphocytes have a life span of only a few days and must be constantly replaced. When an antigen enters lymphoid tissue, it is 'recognized' by one particular type of lymphocyte, which then multiplies rapidly; the resulting PLASMA CELLS circulate in the blood, producing the necessary ANTIBODY for that antigen. Lymphoid tissue also contains numerous MACROPHAGES, which ingest foreign particles, especially bacteria, hence the lymph nodes act as filters to remove bacteria from the lymph.

lymphokine /**lim**-fŏ-kÿn/ Any of various CYTOKINES that are released by lymphocytes when activated by encountering their specific antigen. They serve as chemical signals to other cell types involved in the immune response, such as macrophages, neutrophils, and basophils. Examples include *macrophage migration inhibition factor*, which inhibits the movement of macrophages, and *tumor necrosis factor*, which stimulates a range of responses in body cells. Soluble factors that act as signals for other lymphocytes are known as INTERLEUKINS.

lyophilic /lÿ-ŏ-**fil**-ik/ Solvent attracting. When the solvent is water, the word *hydrophilic* is often used. The terms are applied to:
1. Ions or groups on a molecule. In aqueous or other polar solutions ions or polar groups are lyophilic. For example, the –COO⁻ group on a soap is the lyophilic (hydrophilic) part of the molecule.
2. The disperse phase in colloids. In lyophilic colloids the dispersed particles have an affinity for the solvent, and the colloids are generally stable. *Compare* lyophobic.

lyophobic /lÿ-ŏ-**foh**-bik/ Solvent repelling. When the solvent is water, the word *hydrophobic* is used. The terms are applied to:
1. Ions or groups on a molecule. In aqueous or other polar solvents, the lyophobic group will be non-polar. For example, the hydrocarbon group on a soap molecule is the lyophobic (hydrophobic) part.
2. The disperse phase in colloids. In lyophobic colloids the dispersed particles are not solvated and the colloid is easily solvated. Gold and sulfur sols are examples. *Compare* lyophilic.

lysine /**lÿ**-seen, -sin/ *See* amino acids.

lysis /**lÿ**-sis/ (**degeneration**) The death and subsequent breakdown of a cell. Under normal conditions such cells are engulfed by phagocytes and degraded by their lysosomes. Only under rare conditions are such cells degraded from within by lysosomes. During some morphological changes (e.g. regression of the tadpole tail) internal lysosomal action does not initiate degradation but simply participates together with phagocytotic action.

lysogeny /lȳ-**soj**-ĕ-nee/ **1.** The formation of an intercellular space in plants by dissolution of cells. *Compare* schizogeny. **2.** A phage–bacteria relationship in which lysing of the bacteria does not occur. The phage (known as a *temperate phage*) penetrates the host cell and its nucleic acid becomes integrated into the bacterial DNA. In this state the phage is termed a *prophage*; most of the viral genes are repressed and both bacteria and phage reproduce together, producing infected daughter cells. In a process called *induction*, certain environmental factors can cause the phage to leave the host DNA and resume the lytic cycle. *See* phage.

lysosome /lȳ-sŏ-sohm/ An organelle, found in animal cells and certain protists, that contains various hydrolytic enzymes responsible for degrading aged or defective cellular components, or materials taken in by the cell from outside. Lysosomes are bounded by a single membrane, and can fuse with other membranous vesicles. Hence, they can contribute enzymes to food vacuoles, as in *Amoeba*, or to similar vacuoles formed in white blood cells during phagocytosis. The interior of a lysosome is maintained at a pH of about 4.8, providing the acidic conditions under which the lysosomal enzymes function best. If these enzymes escape, they are virtually inactive at the neutral pH of the cytosol, thereby safeguarding the cell's contents from destruction. The lysosomal membrane contains proton pumps, which pump hydrogen ions into the interior of the lysosome, and also chloride channels, which allow the entry of chloride ions. Consequently, the net result is the influx of hydrochloric acid (HCl), which gives the acidic conditions.

Lysosomal enzymes are synthesized in the rough endoplasmic reticulum, from whence they enter the Golgi apparatus. Here they are sorted and labeled for despatch to lysosomes, via transport vesicles. *Primary lysosomes* are roughly spherical and do not contain debris. They fuse with other organelles that contain material destined from digestion, forming a *secondary lysosome*, in which debris is apparent and enzymic digestion takes place. For example, a bacterial cell taken in by phagocytosis is first incorporated in a phagosome, which then fuses with a primary lysosome. Food material internalized by endocytosis enters endosomes, which subsequently undergo fusion with lysosomes. In plant cells, the VACUOLE has an acid pH and contains an array of hydrolytic enzymes, enabling it to perform the role played by lysosomes in animal cells. *See also* autophagy.

lysozyme /lȳ-sŏ-zȳm/ An enzyme present in saliva, tears, egg white, and mucus, discovered by Alexander Fleming in 1922. It destroys bacteria by hydrolysis of the cell walls. *See also* lysis.

macroevolution /mak-roh-ev-ŏ-**loo**-shŏn/ The evolution of living organisms on a broad scale, as in the emergence of entirely new groups such as the seed plants or the giant reptiles. *Compare* microevolution.

macromolecule /mak-roh-**mol**-ĕ-kyool/ A very large molecule, usually a polymer, having a very high molecular weight. Proteins and nucleic acids are examples.

macronucleus /mak-roh-**new**-klee-ŭs/ (*pl.* **macronuclei**) (**meganucleus**) The larger of the two nuclei found in certain protoctists, particularly ciliates (e.g. *Paramecium*). The smaller of the two is the *micronucleus*. The meganucleus contains multiple copies of the DNA needed for normal (nonreproductive) cell metabolism (i.e. is polyploid). It contains nucleoli, is variable in form, and divides amitotically. It degenerates during sexual reproduction and is reconstituted from the micronuclear chromosomes that produce the zygote. The multiple DNA copies are probably a means of controlling a relatively large volume of cytoplasm.

The diploid *micronucleus* contains much less DNA and can undergo normal nuclear division (mitosis or meiosis). It is involved in sexual reproduction when two individuals unite by cytoplasmic bridges (conjugation) and exchange micronuclei.

macronutrient /mak-roh-**new**-tree-ĕnt/ A nutrient required in more than trace amounts by an organism. *See* essential element. *Compare* micronutrient.

macrophage /**mak**-rŏ-fayj/ A large ameboid cell that can engulf, ingest, and destroy bacteria, damaged cells, and worn-out red blood cells. This process is called *phagocytosis* and is an important part of the body's defense against disease. Macrophages are found free ('wandering') in the tissues, in the blood (as monocytes), in connective tissue (as histiocytes), in the lining of the blood sinusoids of the liver (as Kupffer cells), and in lymphoid tissue. Macrophages can also promote inflammation by releasing cytokines, such as interleukin 1 and tumor necrosis factor. They make up the mononuclear phagocyte system. *See also* lymphoid tissue.

macrosclereids /mak-roh-**skleer**-ee-idz/ Elongated rod-shaped sclereids that form a close outer protective layer in the seed testas and fruit walls of some plants.

macula /**mak**-yŭ-lă/ (*pl.* **maculae**) 1. A small area of sensory epithelium in the sacculus and utriculus of the inner ear. It contains sensory hairs embedded in an otolith – a gelatinous mass containing crystals of calcium carbonate. Tilting of the head causes the crystals to bend the hair cells, which are connected to the auditory nerve and register the movement of the head. 2. A region rich in cones surrounding the fovea in the vertebrate eye.

'mad cow disease' *See* bovine spongiform encephalopathy.

madreporite /mad-rĕ-**por**-ÿt, -**poh**-rÿt, mad-rĕ-pŏ-rÿt/ (**sieve plate**) *See* water vascular system.

magnesium /mag-**nee**-zee-ŭm, -**seee**-ŭm/ An element essential for plant and animal growth. It is contained in the chlorophyll molecule and is thus essential for photosynthesis. In animals it is found in bones and teeth. As magnesium carbonate it is found in large quantities in the skeletons of certain marine organisms, and is found in smaller quantities in the muscles and nerves of higher animals. It is an essential cofactor for certain phosphate enzymes, e.g. phosphohydrolase and phosphotransferase. High concentrations of magnesium ions, Mg^{2+}, are needed to maintain ribosome structure.

major histocompatibility complex (MHC) A large cluster of related genes in humans that encode cell-surface proteins (MHC proteins) that play several vital roles in the immune system. They form the markers of the HLA system of self-recognition, which prevents cells of the immune system (lymphocytes) attacking self tissue, and they serve to 'present' foreign antigens to lymphocytes. There are three classes of genes, with protein products having different roles. *Class I MHC proteins* are found on virtually all cells. Within the cell they attach themselves to processed foreign antigen, e.g. from an invading virus, and migrate to the cell surface where they are recognized by cytotoxic T-cells, which are stimulated to destroy the infected cell. *Class II MHC proteins* occur only on certain cells of the immune system, e.g. macrophages and B-cells. They combine with foreign antigen such as debris from bacteria phagocytosed by the macrophage, and are recognized by another set of T-cells called T-helper cells. The latter stimulate the macrophages to destroy the foreign antigens they contain, and also activate antibody secretion by B-cells. *Class III MHC proteins* are components of the complement system.

A large number of alleles exist for the MHC genes, which creates immense genetic variability in the MHC proteins. Hence, only close relatives are likely to have similar class I MHC proteins, i.e. show some degree of histocompatibility. As a consequence, a graft from such a relative is more likely to be tolerated by the immune system. Tissue grafts from unrelated individuals have different class I MHC proteins, and are recognized as foreign and therefore killed by the recipient's cytotoxic T-cells, causing rejection of the graft. *See* B-cell; T-cell.

maleic acid /mǎ-**lee**-ik/ *See* butenedioic acid.

malic acid /**mal**-ik/ A colorless crystalline carboxylic acid, which occurs in acid fruits such as grapes and gooseberries. In biological processes malate ion is an important part of the Krebs cycle.

malleus /**mal**-ee-ŭs/ (*pl.* **mallei**) (**hammer**) The hammer-shaped bone attached to the tympanic membrane (eardrum), which is the first and largest of the ear ossicles in mammals. It is homologous with Meckel's cartilage. *See illustration at* ear.

Malpighian body /mal-**pig**-ee-ăn/ (**Malpighian corpuscle; renal corpuscle**) The part of the excretory unit (NEPHRON) within the cortex of the vertebrate KIDNEY. It comprises a knot of blood capillaries (glomerulus) and a surrounding cup-shaped Bowman's capsule. High pressure created within the glomerulus results in the filtration of water, salts, nitrogenous wastes, etc., across the capillary walls into the capsule, and hence to a kidney tubule for reabsorption or excretion. The Malpighian body is named for the Italian anatomist Marcello Malpighi (1628–94).

Malpighian layer A layer of dividing cells at the base of the epidermis of vertebrates. Its cells contain granules of the pigment melanin, which protects the body against ultraviolet radiation. The cells produced by the Malpighian layer form a layer to the outside, the granular layer. These cells gradually move outwards and become hardened, forming the STRATUM CORNEUM. Pockets of the Malpighian layer dip into the epidermis, forming hair follicles and sebaceous glands. *See illustration at* skin.

Malpighian tubules Slender blind-ending tubes that open into the anterior end of the hindgut of insects, spiders, millipedes, and centipedes. They act as excretory organs, extracting waste products (mainly uric acid) from the surrounding blood and passing them into the hindgut for discharge with the feces.

maltose /môl-tohs/ A sugar found in germinating cereal seeds. It is a disaccharide composed of two glucose units. Maltose is an important intermediate in the enzyme hydrolysis of starch. It is further hydrolyzed to glucose.

Mammalia /mă-**may**-lee-ă/ The class of vertebrates that contains the most successful tetrapods. They are homoiothermic, with an insulating body covering of hair and usually with sweat and sebaceous glands in the skin. The socketed teeth are differentiated into incisors, canines, and grinding premolars and molars. Mammals have a relatively large brain and an external ear (pinna), and three auditory ossicles in the middle ear. Oxygenated and deoxygenated blood are separated in the four-chambered heart and a diaphragm assists in respiratory movements. Typically, the young are born alive and are suckled on milk secreted by the mammary glands. A bony secondary palate allows the retention of food in the mouth while breathing. Mammals evolved from active carnivorous reptiles in the Triassic. There are two subclasses: Prototheria, which comprises a single order (MONOTREMATA) containing all the egg-laying mammals; and Theria, which contains all the mammals that bear live young. Subclass Theria is divided into two infraclasses: METATHERIA, which comprises the marsupials (pouched mammals); and EUTHERIA, which contains the placental mammals.

mammary gland /mam-ă-ree/ The milk-producing gland in female mammals (it is the breast in women). There may be one or more pairs, depending on the species, situated ventrally. Each gland consists of fatty tissue in which are embedded lobules consisting of clusters of milk-producing alveoli. The alveoli lead into tiny ducts that converge with ducts from adjacent alveoli to form *lactiferous tubules*. The tubules lead to the mammary papilla (the *nipple* or *teat*). On each lactiferous tubule, there is a swelling known as an *ampulla* just beneath the nipple in which milk is stored. Milk is usually produced after the birth of the young (*see* lactation). The Monotremata

have mammary glands without nipples scattered over the abdomen and the young acquire the milk by lapping rather than by suckling.

mandible /man-dă-băl/ 1. One of a pair of feeding appendages (mouthparts) of various arthropods, such as crustaceans, insects, millipedes, and centipedes.
2. The extended upper and lower jaws of birds; they form a beak.
3. The lower jaw of vertebrates. In many species it is comprised mainly of a paired membrane bone, the dentary, which in mammals bears the lower set of teeth. *See also* Meckel's cartilage.

mandibular arch /man-**dib**-yŭ-ler/ The first visceral arch; it is modified to form the upper and lower jaws in Chondrichthyes (cartilaginous fish).

Mandibulata /man-dib-yŭ-**lay**-tă/ In some classifications, a phylum of arthropods containing the insects, centipedes, and millipedes. They are characterized by having mouthparts (mandibles) adapted for crushing and grinding food, a feature they share with crustaceans. Unlike crustaceans, they possess one not two pairs of antennae. *See* Arthropoda.

manganese /mang-gă-neez, -nees/ *See* trace element.

mannitol /man-ă-tol, -tohl/ A soluble sugar alcohol (carbohydrate) found widely in plants and forming a characteristic food reserve of the brown algae. It is a hexahydric alcohol, i.e. each of the six carbon atoms has an alcohol (hydroxyl) group attached. In medicine it is used as a diuretic to treat fluid retention.

mannose /man-ohs/ A simple sugar found in many polysaccharides. It is an aldohexose, isomeric with glucose.

mantle 1. (**pallium**) A tissue covering most of the body of mollusks. It secretes the shell(s) and in shell-less mollusks it is tough and protective. The mantle is folded to enclose the *mantle cavity*, which con-

tains the respiratory organs. In squids the mantle cavity has muscular walls; it contracts to force water out of the mantle cavity, which propels the animal rapidly through the water.
2. Part of the body wall of brachiopods and tunicates.

marker gene A gene of known location and function which can therefore be used to establish the relative positions and functions of other genes. During gene transfer, a marker gene may be linked to the transferred gene to determine whether or not the transfer has been successful. *See* chromosome map; genetic engineering.

marl A soil containing a high proportion of calcium carbonate (chalk), as in limestone areas.

marsupials /mar-**soo**-pee-ăl/ *See* Metatheria.

marsupium /mar-**soo**-pee-ŭm/ (*pl.* **marsupia**) A pouch on the abdomen of marsupials and some monotremes, consisting of a fold of skin supported by the epipubic bones of the hip girdle. It covers the mammary glands and serves to protect the young, which migrate there after birth to complete development.

mass flow A hypothesis put forward by Münch (1930) to explain the mechanism of phloem transport. The movement of substances is believed to be the result of changes in osmotic pressure. Thus in an actively photosynthesizing region (source) the osmotic pressure is high and water is taken in. Conversely, in regions (sinks) where photosynthetic products are being used up or converted to storage compounds there is a lowering of osmotic pressure and water is lost. A system is then set up in which there is mass flow from source to sink. Water is carried back in the other direction in the transpiration stream of the plant's xylem.

mass spectroscopy A technique for identifying the nature and abundance of molecules in a sample according to their characteristic mass spectra. The sample is placed in a vacuum, then vaporized and ionized, causing molecules to split into charged fragments. The fragments are then accelerated in a mass spectrometer and subjected to electrical and magnetic fields, which deflect the fragments according to their mass. The resulting pattern impinges on a detector as a series of peaks of variable intensity, corresponding to the mass spectrum. Each molecule has its own characteristic mass spectrum, and such spectra can be used as 'fingerprints' to detect the presence of particular molecules.

mast cell A type of white blood cell (leukocyte) with granular cytoplasm found within connective tissue, e.g. beneath the skin and around blood vessels. Mast cells bind certain immunoglobulin antibodies (IgE) to their cell membrane, and when a specific antigen is encountered this triggers the mast cell to release chemical mediators, such as histamine, tumor necrosis factor, and leukotriene B$_4$, from its granules. These substances increase vascular permeability at the site, causing inflammation and attracting other types of immune cell. Mast cells are also activated by complement proteins, and are responsible for causing the symptoms of various allergies and other hypersensitivity reactions. *See also* basophil.

mastigonema /mas-tă-**gon**-ĕ-mă/ Small projections occurring laterally on certain types of undulipodia (flagella).

Mastigophora /mas-tă-**goff**-ŏ-ră/ (**Flagellata**) In some older classifications, a class of the phylum Protozoa whose members possess one or more undulipodia (flagella) for locomotion. In more recent classifications they are placed in the kingdom Protoctista, in several phyla, especially Zoomastigota.

mating type The equivalent of 'male' and 'female' sexes in various fungi, protoctists, bacteria, and some plants. Mating can only take place between individuals of different mating types. The mating types are generally indistinguishable in their physical appearance, but each has certain genetically determined physiological features that define their mating type. For example, the fission yeast *Saccharomyces cerevisiae* has two mating types designated α and a. Cells of type α will only mate with type a cells. Similarly in the orange bread mold *Neurospora crassa*, mating can only occur between hyphae of opposite mating types, denoted A and *a*. Some species have numerous mating types.

maxilla /maks-**il**-ă/ (*pl.* **maxillae**) **1.** One of a pair or two pairs of feeding appendages (mouthparts) of various arthropods, such as crustaceans, insects, millipedes, and centipedes.
2. One of a pair of large bones of the upper jaw of vertebrates. In mammals they bear the molar and premolar teeth.

maxilliped /maks-**il**-ă-ped/ Any of up to three pairs of appendages arising from the most anterior thoracic segments of a crustacean that function as accessory feeding organs. Located behind the mandibles, they transfer food to the mouth.

mean (**average**) The sum of a set of *n* values or numbers divided by *n*; i.e. the *arithmetic mean.*

meatus /mee-**ay**-tŭs/ A passage or channel in the body, such as the *external auditory meatus* (ear canal) in birds, mammals, and some reptiles, which leads from the external opening of the ear to the eardrum. *See also* outer ear.

mechanoreceptor /mek-ă-noh-ri-**sep**-ter/ A receptor that responds to a mechanical stimulus, e.g. touch, pressure, sound, etc.

Meckel's cartilage /**mek**-ĕl/ A paired cartilage forming the lower jaw in cartilaginous fish, such as sharks, skates, and dogfish. In bony fish (Osteichthyes), reptiles, and birds it is ossified to form the articular bone. In mammals it persists as an ear ossicle, the malleus. The cartilage is named for the German anatomist and embryologist Johann Friedrich Meckel the Younger (1781–1833).

median /**mee**-dee-ăn/ The middle number or value in a set of numbers or values.

median eye An eye in the middle of the head, found in some crustaceans, such as the microscopic pond animal *Cyclops*. It is a simple light receptor (ocellus). Some insects, such as the locust, have a median ocellus as well as a pair of compound eyes. The New Zealand lizard, *Sphenodon*, has a median third eye that is functional. *See also* pineal eye.

mediastinum /mee-dee-as-**tÿ**-nŭm/ (*pl.* **mediastina**) **1.** A membranous septum in the midline of the thorax of mammals that separates the two pleural cavities ventrally.
2. The space between the pleural membranes surrounding the lungs, that contains the heart, thymus, esophagus, and trachea.

medulla /mi-**dul**-ă/ **1.** The central region of an animal organ, when this differs in structure or function from the outer regions. An example is the medulla of the kidney. *Compare* cortex.
2. *See* pith.

medulla oblongata /ob-long-**gay**-tă, -**gah**-tă/ A region of the hindbrain that is concerned with the functioning of the visceral organs, e.g. the stomach, lungs, and heart. It is continuous with the spinal cord and, in addition to the tracts of nerve fibers passing from higher brain regions down the spinal cord, it contains centers of gray matter controlling respiratory rhythm, blood circulation, the reflex movements of the eye muscles, and other involuntary functions. Many of the cranial nerves arise from the medulla.

medullary plate /**med**-ŭ-lair-ee, -lă-ree/ *See* neural plate.

medullary ray (**pith ray**) In young plants, and plants not showing secondary thickening (most monocotyledons), the undifferentiated parenchyma tissue found between the vascular bundles and connecting the pith and cortex. In older dicotyledons and gymnosperms in which secondary tissues are formed, cells of the medullary ray differentiate to form interfascicular cambium. Secondary xylem and phloem are developed to either side of the interfascicular cambium thus taking the place of the medullary ray. However certain cells of the cambium, the ray initials, produce parenchyma cells that form narrow secondary rays between the secondary xylem and phloem at right angles to the axis. The ray parenchyma cells store food (e.g. starch) and may also contain tannins and crystals.

medullated protostele /med-ŭ-lay-tid/ A type of protostele in which there is a central nonvascular pith (medulla) surrounded successively by xylem, phloem, pericycle, and endodermis. *See* stele.

medusa /mĕ-**dew**-să, -ză/ (*pl.* **medusae**) A stage in the life cycle of cnidarians in which the body is shaped like a bell or inverted saucer with a fringe of tentacles around the rim and a mouth beneath. The medusa is the free-swimming dispersal stage of the life cycle; in scyphozoans (jellyfish) it is the only form. Medusae have sex organs (male or female); sperms swim from the male to fertilize eggs in the female. The fertilized egg develops into a planula. *See also* polyp.

megabase Symbol: Mb A unit of length used for measuring polynucleotides (i.e. DNA or RNA) and equal to 106 bases or base pairs. *Compare* kilobase.

meganucleus /meg-ă-**new**-klee-ŭs/ (*pl.* **meganuclei**) *See* macronucleus.

megaphyll /**meg**-ă-fil/ (**macrophyll**) A foliage LEAF with a branched system of veins in the blade. It is typical of ferns, gymnosperms, and angiosperms. The large often pinnately divided megaphyll of ferns, often termed a *frond*, contrasts with the generally much smaller leaves (microphylls) of other pteridophytes. The leaf trace to a megaphyll leaves a leaf gap in the stele. *Compare* microphyll.

megasporangium /meg-ă-spŏ-**ran**-jee-ŭm/ (*pl.* **megasporangia**) A sporangium

that produces megaspores. In *Selaginella* the megasporangium is borne in the axis of a sporophyll located in a strobilus. Usually all the spore mother cells degenerate except one, which forms a tetrad of cells. One or more of these may develop into megaspores, which, when shed, develop into the female gametophyte. The nucellus of seed plants may be considered equivalent to the megasporangium of the pteridophytes. *Compare* microsporangium.

megaspore /**meg**-ă-spor, -spohr/ The larger of the two types of spores in heterosporous pteridophytes and in seed plants that produces the female gametophyte. Megaspores are released from the sporophyte in pteridophytes to insure fertilization, but internal fertilization techniques developed by gymnosperms and angiosperms make this unnecessary in these plants. The megaspores of pteridophytes and gymnosperms produce a female gametophyte called a prothallus that forms two or more archegonia each containing a haploid female gamete. In angiosperms the megaspore becomes the embryo sac, lacking an obvious prothallus but containing eight nuclei, one of which is organized as the female gamete. *Compare* microspore.

megasporophyll /meg-ă-spor-ŏ-fil, -spohr-/ A leaf or modified leaf that bears the megasporangium. Simple megasporophylls are the fertile ligulate leaves of heterosporous lycopods (e.g. *Selaginella*). They are usually grouped in a strobilus that may also contain microsporophylls or vegetative leaves. The carpel of angiosperms and the ovuliferous scale of gymnosperms are modified megasporophylls. *Compare* microsporophyll.

meiosis /mÿ-**oh**-sis/ (*pl.* **meioses**) The process of cell division leading to the production of daughter nuclei with half the genetic complement of the parent cell. Cells formed by meiosis give rise to gametes and fertilization restores the correct chromosome complement.

Meiosis consists of two divisions during which the chromosomes replicate only once. Like mitosis the stages prophase,

Prophase

Leptotene

Chromosomes appear as single uncoiled threads

Zygotene

Homologous chromosomes attract each other, coming together to form bivalents

Pachytene

Chromosomes shorten by coiling and individual chromatids become distinguishable, giving tetrads

Diplotene — Diakinesis

Homologous chromosomes repel each other at the centromeres, remaining attached only at chiasmata

Metaphase I

Nuclear membrane breaks down, spindle forms, and bivalents align themselves along the spindle equator

Anaphase I

Homologous chromosomes continue to repel each other, the homologues of each pair moving to opposite ends of the spindle

Telophase I

The haploid number of chromosomes gathers at either end of the spindle

Metaphase II — Anaphase II

Two spindles form at right angles to the first and chromatids separate

Telophase II

A nuclear membrane forms around each group of daugher chromosomes to form four haploid nuclei and the cytoplasm divides, forming four gametes

Stages in meiosis

metaphase, and anaphase can be recognized. However during prophase homologous chromosomes attract each other and become paired forming bivalents. At the end of prophase genetic material may be exchanged between the chromatids of homologous chromosomes. Meiosis also differs from mitosis in that after anaphase, instead of nuclear membranes forming, there is a second division, which may be divided into metaphase II and anaphase II. The second division ends with the formation of four haploid nuclei, which develop into gametes.

melanin /mel-ă-nin/ One of a group of pigments found in animals and plants, derived from the amino acid tyrosine. The colors range from black through brown to yellow, orange, or red. In animals melanin occurs in *melanophores* (pigment cells) in the skin, usually below the epidermis. It gives color to the skin, hair, and eyes of animals and causes color in various seedlings and roots of plants. The absence of the enzyme tyrosinase in animals leads to a condition known as *albinism*, in which no pigment develops in the eyes, skin, or hair.

melanism /mel-ă-niz-ăm/ The possession of a dark appearance due to the presence of a dark brown pigment, melanin. *See also* industrial melanism.

melanocyte-stimulating hormone /mel-ă-nŏ-sÿt/ (MSH) A peptide hormone produced by the anterior pituitary gland in vertebrates. It has a marked action on pigmentation in the skin of amphibians and reptiles, but its physiological role in mammals is unclear. *Compare* melatonin.

melatonin /mel-ă-toh-nin/ A hormone, produced by the PINEAL GLAND in the brain, that communicates changes in day length to the rest of the body. Hence, it is responsible for triggering seasonal alterations in biological rhythms, and for controlling the onset of reproduction in seasonally breeding animals such as sheep, horses, and deer. Melatonin is synthesized from the neutrotransmitter serotonin, and secreted into the bloodstream during the dark phase of the day–night cycle. Synthesis is inhibited during daylight hours, hence the duration of melatonin secretion is proportional to the length of the dark period. It mediates its effects via melatonin receptors in various parts of the brain, particularly the hypothalamus. Melatonin controls reproduction by inhibiting secretion of certain gonadotropic hormones, thus repressing activity of the gonads. Injections of mela-tonin can be given to sheep and deer to artificially manipulate their reproductive cycles. Sleep patterns are also influenced by melatonin, and it is sometimes used to alleviate sleep disorders and the symptoms of jet lag. In embryonic fish and larval amphibians it induces marked lightening of the skin.

membrane A structure consisting mainly of lipid and protein (lipoprotein) surrounding all living cells as the *plasma membrane*, or *plasmalemma*, and also found surrounding organelles within cells. Membranes function as selectively permeable barriers, controlling passage of substances between the cell and its organelles, and the environment, either actively or passively. Membranes are typically 7.5–10 nm in thickness with two regular layers of lipid molecules (a *bilayer*) containing various types of protein molecules. Some proteins traverse the membrane completely, others are associated with the inner or outer face; some float freely over the surface while others remain stationary. Various types of transport proteins and protein-lined channels convey ions and small molecules across the membrane – either actively with the expenditure of energy (usually in the form of ATP), or passively. Larger molecules or particles enter or leave cells by endocytosis or exocytosis, respectively. The outer face of plasma membranes also contains receptor proteins, which bind signal molecules from the cell's environment and activate signal transduction proteins on the inner face or in the cell interior.

The lipids are mostly phospholipids. These are polar molecules, i.e. one end (the phosphate end) is *hydrophilic* (water-loving) and faces outwards, while the other end (two fatty acid tails) is *hydrophobic* (water-hating) and faces inwards. Short chains of sugars may be associated with the proteins or lipids forming glycoproteins and glycolipids. The particular types of carbohydrates, lipids, and proteins determine the characteristics of the membrane, affecting, for example, cell–cell recognition (as in embryonic development and immune mechanisms), permeability, and hormone recognition. Membranes may contain efficient arrangements of molecules involved in certain metabolic processes, e.g. electron transport and phosphorylation (ATP production) in mitochondria and chloroplasts. *See* osmosis; freeze fracturing.

membrane bone (**dermal bone**) A bone that has been formed by ossification of connective tissue instead of cartilage. Such bones are usually flat and thin and include those in the skull. The connective tissue becomes inundated with fine fibers around which calcium phosphate is deposited. Osteoclasts then invade and erode this calcified matrix, followed by osteoblasts, which form bony trabeculae. *See also* bone; osteoblast; osteoclast.

membranous labyrinth /**mem**-bră-nŭs/ *See* labyrinth.

memory cell *See* B-cell.

menaquinone /men-ă-kwă-**nohn**/ *See* vitamin K.

Mendelism /**men**-dĕ-liz-ăm/ The theory of inheritance according to which characteristics are determined by particulate 'factors', or genes, that are transmitted by the germ cells. It is the basis of classical genetics, and is founded on the work of the Austrian Roman Catholic priest Gregor Johann Mendel (1822–84) in the 1860s. *See* Mendel's laws.

Mendel's laws Two laws formulated by Mendel to explain the pattern of inheritance he observed in plant crosses. The first law, the Law of Segregation, states that any character exists as two factors, both of which are found in the somatic cells but only one of which is passed on to any one gamete. The second law, the Law of Independent Assortment, states that the distribution of such factors to the gametes is random; if a number of pairs of factors is considered each pair segregates independently.

Today Mendel's 'characters' are termed genes and their different forms (factors) are called alleles. It is known that a diploid cell contains two alleles of a given gene, each of which is located on one of a pair of homologous chromosomes. Only one homolog of each pair is passed on to a gamete. Thus the Law of Segregation still holds true. Mendel envisaged his factors as discrete particles but it is now known that they are grouped together on chromosomes. The Law of Independent Assortment therefore only applies to pairs of alleles found on different chromosomes.

meninges /mĕ-**nin**-jeez/ (*sing.* **meninx**) The protective membranes that surround the brain and spinal cord in vertebrates. In humans and other mammals there are three: the stiff outer dura mater, the arachnoid membrane, and the soft inner pia mater. The latter two are separated by the subarachnoid space, which is filled with cerebrospinal fluid.

menstrual cycle /**men**-strû-ăl/ A modified form of the ESTROUS CYCLE found in humans, Old World monkeys, and anthropoid apes. The actual estrus ('heat') is not obvious, the female being continuously sexually receptive. There is a massive breakdown and expulsion of the glandular lining of the uterus, together with much bleeding, in the monthly menstrual flow (*menstruation*). This menstrual flow, representing the estrous regression phase, is distinct from the normal slight vaginal bleeding that accompanies estrus in most female mammals.

menstruation /men-strû-**ay**-shŏn/ *See* menstrual cycle.

mericarp /**m'e**-ră-karp/ A one-seeded dehiscent or indehiscent portion of a schizocarpic fruit. It is found for example in the Geraniaceae, whose fruit, the regma, splits into five dehiscent mericarps. The cremocarp of the Umbelliferae splits into two indehiscent mericarps.

meristele /**m'e**-ră-steel, -stee-lee/ A segment of a DICTYOSTELE formed by overlapping leaf gaps. A meristele contains all the tissues of a stele, usually arranged in a concentric pattern. The xylem is surrounded by a layer of phloem, and bounded by a pericycle and endodermis. It is seen in many club mosses and ferns.

meristem /m'e-ră-stem/ A distinct region of actively dividing cells primarily concerned with growth. Numerous meristems occur in plants. In active meristems separation occurs between the cell that remains meristematic (initial) and the cell ultimately being differentiated. Two basic meristematic groups are the primary apical meristems at root and shoot apices, and the secondary lateral meristems, which include vascular and cork cambia. *See also* intercalary meristem.

meristoderm /m'e-ră-stŏ-derm/ (limiting layer) The outermost cellular layer of the thallus of certain brown algae. It consists of small densely packed rectangular cells containing brown pigmented plastids and covered by a mucilaginous layer to prevent desiccation. The meristoderm maintains its meristematic activity and assists the outer cortical layers in adding to the thickness of the thallus.

meroblastic /m'e-rŏ-**blas**-tik/ Describing the type of incomplete cleavage that occurs in very yolky eggs (e.g. of birds and sharks), in which the egg cytoplasm but not the yolk divides. *Compare* holoblastic.

mesarch /**mess**-ark/ Denoting a stele or part of a stele in which the protoxylem is surrounded by metaxylem. *Compare* centrarch; endarch; exarch.

mesencephalon /mess-en-**sef**-ă-lon/ *See* midbrain.

mesenchyme /**mess**-enk-ȳm/ A loose network of cells, usually underlying epithelial layers, in animal embryos. It is sometimes divided into *primary mesenchyme*, the first cells to invade the blastocoel, and *secondary mesenchyme*, later contributions of diffuse cells from other sources. The mesenchyme of glands, gut regions, and skin usually determines the differentiation of the epithelium over it.

mesentery /**mess**-ĕn-tĕ-ree/ 1. A thin transparent membrane consisting of a double layer of peritoneum that surrounds all parts of the alimentary canal and attaches it to the dorsal wall of the abdomen. The blood vessels, lymphatics, and nerves that supply the alimentary canal lie between two layers of mesentery. It consists of areolar connective tissue, covered by an external layer of squamous epithelium.
2. One of the vertical partitions in the coelenteron of sea anemones.

mesocarp /**mess**-ŏ-karp, **mee**-sŏ-/ *See* pericarp.

mesocotyl /mess-ŏ-**kot**-ăl, -mee-sŏ-/ A structure found in certain germinating monocotyledon seeds (e.g. maize) that is regarded as: either the region between the base of the coleoptile and the attachment of the scutellum, in which case it becomes the elongated cotyledonary node; or the first internode of the epicotyl. The distinction is based on whether or not the scutellum and coleoptile are regarded as parts of one cotyledon.

mesoderm /**mess**-ŏ-derm, **mee**-sŏ-/ The germ layer from which muscles, connective tissues, and blood system usually develop. At gastrulation the mesoderm comes to lie between ectoderm on the outside and endoderm lining the gut. In most animals the coelom divides the mesoderm into an outer *somatopleure* under the skin and an inner *splanchnopleure* around the gut; other regions include the somites. *See* germ layers; somite.

mesoglea /mess-ŏ-**glee**-ă, mee-sŏ-/ The layer of jelly-like material that separates and is secreted by the ectoderm and endoderm in sponges and cnidarians. It varies from a thin membrane (e.g. in *Hydra*) to a thick gelatinous mass (e.g. in jellyfish).

mesonephros /mess-ŏ-**nef**-ros, mee-sŏ-/ The second type of vertebrate KIDNEY: it develops after the pronephros, to which it is posterior, forming the functional kidney of adult fish and amphibians. It is comprised of segmentally arranged ducts that end in cup-shaped Bowman's capsules. Sometimes the ducts have open-ended side branches. They are drained by the mesonephric or Wolffian duct, which replaces the pronephric duct. In reptiles, birds, and mammals it is functionally replaced by the metanephric kidney, forming instead the epididymis of the male testis.

mesophilic /mess-ŏ-**fil**-ik, mee-sŏ-/ Designating microorganisms with an optimum temperature for growth between 25–45°C. *Compare* psychrophilic; thermophilic.

mesophyll /**mess**-ŏ-fil, **mee**-sŏ-/ Specialized tissue located between the epidermal layers of the leaf. Veins, supported by sclerenchyma and collenchyma, are embedded in the mesophyll. *Palisade mesophyll* consists of cylindrical cells, at right angles to the upper epidermis, with many chloroplasts and small intercellular spaces. It is the main photosynthesizing layer in the plant. *Spongy mesophyll*, adjacent to the lower epidermis, comprises interconnecting irregularly shaped cells with few chloroplasts and large intercellular spaces that communicate with the atmosphere through stomata allowing gas exchange between the cells and the atmosphere. The distribution of mesophyll tissue varies in different leaves depending on the environment in which the plant lives. *See illustration at leaf.*

mesophyte /**mess**-ŏ-fÿt, **mee**-sŏ-/ A plant that is adapted to grow under adequate conditions of water supply. In drought conditions wilting is soon apparent as the plants have no special mechanisms to conserve water. Most angiosperms are mesophytes. *Compare* hydrophyte; xerophyte.

mesosome /**mess**-ŏ-sohm, **mee**-sŏ-/ An extensive invagination of the plasma membrane of certain bacteria, associated with DNA synthesis and protein secretion. Its precise function is unknown.

mesosporium /mess-ŏ-**spor**-ee-ŭm, **spoh**-ree-/ *See* exo-intine.

mesothelium /mess-ŏ-**thee**-lee-ŭm, mee-sŏ-/ (*pl.* **mesothelia**) The tissue, consisting of one or more layers of cells, that lines a coelomic cavity.

Mesozoic /mess-ŏ-**zoh**-ik, mee-sŏ-/ The middle era in the most recent (Phanerozoic) eon of the geological time scale, dating from about 230–66 million years ago. Known as the 'Age of Reptiles', it is divided into three main periods: the Triassic, Jurassic, and Cretaceous. *See also* geological time scale.

messenger RNA (**mRNA**) The form of RNA that transfers the information necessary for protein synthesis from the DNA in the nucleus to the ribosomes in the cytoplasm. One strand of the double helix of DNA acts as a template along which complementary RNA nucleotides become aligned. These form a polynucleotide identical to the other DNA strand, except that the thymine bases are replaced by uracil. This polynucleotide, which forms the major component of HETEROGENEOUS NUCLEAR RNA, contains both coding and noncoding sequences (*see* exon; intron); the introns are then removed to produce mRNA. The whole process is termed *transcription*. The new mRNA molecule thus has a copy of the genetic code, which directs the formation of proteins in the ribosomes. *Compare* transfer RNA.

metabolism /mĕ-**tab**-ŏ-liz-ăm/ The chemical reactions that take place in cells. The molecules taking part in these reactions are termed *metabolites*. Some metabolites are synthesized within the organism, while others have to be taken in as food. It is metabolic reactions, particularly those that produce energy, that keep cells alive. Metabolic reactions characteristically occur in small steps, comprising a *metabolic pathway*. Metabolic reactions involve the breaking down of molecules to provide energy (catabolism) and the building up of more complex molecules and structures from simpler molecules (anabolism).

metabolite /mĕ-**tab**-ŏ-lÿt/ A substance that takes part in a metabolic reaction, either as reactant or product. Metabolites are thus intermediates in metabolic pathways. Some are synthesized within the organism itself, whereas others have to be taken in as food. *See also* metabolism.

metacarpal bones /met-ă-**kar**-păl/ Rod-shaped bones in the lower forelimb or forefoot of tetrapods; they form the palm of the hand in humans. They articulate with the carpal bones proximally and phalanges distally. In the typical pentadactyl limb there are five, although there are modifications to this basic plan: in fast-running mammals (e.g. the horse) they are greatly elongated and raised off the ground. *Compare* metatarsal bones. *See illustration at* pentadactyl limb.

metacarpus /met-ă-**kar**-pŭs/ (*pl.* **metacarpi**) The collection of metacarpal bones forming part of the forefoot or lower forelimb in tetrapods; the palm of the hand in humans.

metachronal rhythm /met-ă-**kroh**-năl/ A pattern of movement shown by cilia, parapodia of certain polychaetes, etc., in which each beats one after the other in regular succession and gives the appearance of wave motion. A wave passing forwards may propel the surrounding medium backwards or the organism forwards.

metamere /**met**-ă-meer/ *See* metameric segmentation.

metameric segmentation /met-ă-**m'e**-rik/ (**metamerism**; **segmentation**) The repetition of body parts of an animal along the longitudinal axis of the body to produce a series of similar units (called *segments* or *metameres*). Metameric segmentation is most clearly seen (externally and internally) in annelids; for example, in the earthworm, in which most segments contain blood vessels, ganglia, nephridia, and muscle blocks. It is also seen in arthropods, but has been obscured by cephalization at the anterior end. In chordates, external segmentation is lost and internal segmentation is best seen in the embryo, although it

is confined mainly to the muscular, skeletal, and nervous systems.

metamerism /mě-**tam**-ě-riz-ăm/ *See* metameric segmentation.

metamorphosis /met-ă-**mor**-fŏ-sis/ A phase in the life history of many animals during which there is a rapid transformation from the larval to the adult form. Metamorphosis is widespread among invertebrates, especially marine organisms and arthropods, and is typical of the amphibians. It is normally under hormone control and usually involves widespread lysosome-mediated destruction of larval tissues.

metanephridium /met-ă-ně-**frid**-ee-ŭm/ (*pl.* **metanephridia**) An excretory organ in many Annelida, consisting of a tubule that opens into the coelom by a ciliated funnel and conducts waste fluids to the exterior. *See also* nephridium.

metanephros /met-ă-**nef**-ros/ (*pl.* **metanephroi**) The third type of vertebrate kidney. It develops from the mesonephros, to which it is posterior, forming the functional kidney of reptiles, birds, and mammals. The metanephros consists of a concentrated group of ducts drained by a different duct, the ureter, which leads to the cloaca or bladder. *Compare* pronephros; mesonephros. *See also* kidney; nephron.

metaphase /**met**-ă-fayz/ The stage in mitosis and meiosis when the chromosomes become aligned along the equator of the nuclear spindle.

metaphloem /met-ă-**floh**-em/ The primary phloem formed from the procambium after the protophloem. It is found behind the zone of elongation below the meristem, and is more durable than the protophloem. In plants showing no secondary thickening the metaphloem is responsible for the transport of most of the organic materials in the plant, but this function is taken over by the secondary phloem in regions where secondary tissues have differentiated.

metaplasia /met-ă-**play**-zhă/ The change from one tissue type to another as seen in response to certain diseases or abnormal conditions. For example the epithelium of the respiratory tract may show metaplasia in response to irritants, e.g. smoke.

metatarsal bones /met-ă-**tar**-săl/ Rod-shaped bones in the lower hindlimb or hindfoot of tetrapods; they form the arch of the foot in humans. They articulate with the tarsal bones proximally and phalanges distally. In the typical pentadactyl limb there are five, although there are modifications to this basic plan; in fast-running mammals (e.g. the horse) they are greatly elongated and raised off the ground. *Compare* metacarpal bones.

metatarsus /met-ă-**tar**-sŭs/ (*pl.* **metatarsi**) The collection of metatarsal bones forming part of the hindfoot or lower hindlimb in tetrapods; the arch of the foot in humans.

Metatheria /met-ă-**theer**-ee-ă/ The mammalian infraclass that contains the marsupials (pouched mammals). Marsupials are more primitive than the placental mammals. The brain is relatively small and there are often more than three incisor teeth on each side of the jaw. The young, born after a brief gestation period and in a very immature state, typically continue to develop in a pouch (*marsupium*) on the abdomen of the mother, where they are suckled. Epipubic bones in the pelvis assist in supporting the pouch. Marsupials (e.g. kangaroos, koala bears) are confined to Australasia, where they fill the niches occupied elsewhere by the placental mammals, and to North and South America (e.g. opossums). The Australasian marsupials exhibit syndactyly, i.e. the second and third toes of the hind foot are encased in a sheath of skin at their base, forming a comb for grooming. *Compare* Eutheria; Monotremata.

metaxylem /met-ă-**zy**-lŭm/ The primary XYLEM elements that are differentiated from the procambium after the PRO-TOXYLEM. They are found some distance behind the apical meristem beyond the zone of elongation. The secondary cell walls show reticulate and scalariform thickening and are thus inextensible.

Metazoa /met-ă-**zoh**-ă/ In some older classifications, a subkingdom of multicellular animals whose bodies are composed of specialized cells grouped together to form tissues and that possess a coordinating nervous system. This subkingdom included all animals except the Protozoa and Parazoa (sponges). In more recent classifications the term Eumetazoa embraces all members of the kingdom Animalia except the Parazoa.

metestrus /met-**est**-rŭs/ *See* estrous cycle.

methanogen /mě-**th'an**-ŏ-jěn, -/ Any of certain ARCHAEBACTERIA that generate methane, such as *Methanobacterium* and *Methanococcus*. They are obligate anaerobic organisms that live only where oxygen is absent, such as swamps, marshes, sewage sludge, and in the rumen of cattle, sheep, and other ruminants. Most obtain their energy by reducing carbon dioxide to methane using hydrogen as an electron donor: $CO_2 + 4H_2 \rightarrow CH_4 + 2H_2O$. However, various organic compounds, including formate, methanol, and acetate, can also be converted to methane by certain methanogens.

methionine /meth-ÿ-ŏ-neen, -nin/ *See* amino acids.

methylene blue *See* staining.

MHC *See* major histocompatibility complex.

microarray /mÿ-kroh-ă-**ray**/ A glass slide or bead whose surface bears a microscopic gridlike pattern of reagent biomolecules or other material, designed to allow simultaneous assay of numerous target substances or biological activities. Microarrays can carry various types of reagents, including small DNA molecules (i.e. DNA microarrays), proteins, carbohydrates, and living cells. They are used especially in high-throughput automated analysis of complex biological systems, for example in sampling the gene transcripts or proteins expressed by tissue cells over time or under different conditions. In preparing a microarray, the reagent molecules are applied to the glass substrate in a precise manner, so that each particular reagent molecule can be identified according to the coordinates of its 'address' on the grid. Then, if a target substance interacts with a reagent on the microarray, it attaches or activates a label (e.g. a fluorescent dye) at that address. The microarray is then scanned (e.g. by a fluorescence microscope scanner) and the amount of label at each address, and hence the amount of target substance, can be determined.

microbiology /mÿ-kroh-bÿ-**ol**-ŏ-jee/ The study of microscopic organisms (e.g. bacteria and viruses), including their interactions with other organisms and with the environment. Microbial biochemistry and genetics are important branches, due to the increasing use of microorganisms in biotechnology and genetic engineering.

microbody /mÿ-kroh-bod-ee/ A common organelle of plant and animal cells, bounded by a single membrane, spherical, and usually about 0.2–1.5 μm in diameter. Microbodies originate from the endoplasmic reticulum and contain enzymes that oxidize various substrates using molecular oxygen. Unlike aerobic respiration in mitochondria, this oxidation is not linked directly to ATP production. *Peroxisomes* occur in most animal cells and are numerous in photosynthetic cells of plant leaves. Oxidation of organic compounds results in the formation of hydrogen peroxide (H_2O_2), which is highly reactive and toxic. Hence, the peroxisome contains high levels of the enzyme catalase, which breaks down hydrogen peroxide into water and oxygen. Peroxisomes are generally the main site of fatty acid oxidation in eukaryote cells, a process that yields acetyl groups, which serve as building blocks for biosynthetic reactions. In animals, toxic molecules absorbed from the bloodstream can also be degraded using peroxisomal enzymes, especially within cells of the liver and kidney. In plants, peroxisomes are concerned with glycolate metabolism in PHOTORESPIRATION and contain high levels of glycolate oxidase and other associated enzymes. Glycolate comes from chloroplasts and products such as glycine are passed to mitochondria. Hence these three organelles often appear close together.

Another type of microbody, called a *glyoxysome*, is common in seeds, especially in lipid-rich food reserves, for example in the endosperm of castor-oil seeds. Glyoxysomes contain enzymes of the GLYOXYLATE CYCLE, transaminases, and enzymes associated with β-oxidation of fatty acids, and play a major role in conversion of lipids to sucrose during germination and seedling growth.

microdissection /mÿ-kroh-di-**sek**-shŏn/ The technique of dissecting under a microscope using fine mechanically manipulated instruments. Such techniques are often used in microsurgery or when dealing with living organisms at the cellular level.

microevolution /mÿ-kroh-ev-ŏ-**loo**-shŏn/ The evolution of organisms on a relatively fine scale, for example the emergence of new species or races. *Compare* macroevolution.

microfilament /mÿ-kroh-**fil**-ă-mĕnt/ A minute filament, about 6 nm wide, found as a component of the CYTOSKELETON in eukaryotic cells and having roles in cell motion and shape. Microfilaments are made of two helically twisted strands of globular subunits of the protein actin, almost identical to the actin of muscle. They can undergo rapid extension or shortening by subunit assembly or disassembly, or form complex three-dimensional networks. The thin filaments of muscle are modified actin microfilaments, which interact with thick myosin filaments. Microfilaments often occur in sheets or bundles just below the plasma membrane and at the interface of moving and stationary cytoplasm. They are involved in cell movements, e.g. ameboid movement, and movement of subcellular components, e.g. pinocytotic vesicles.

micrograph /**mȳ**-kroh-graf, -grahf/ A photograph taken with the aid of a microscope. *Photomicrographs* and *electron micrographs* are produced using optical microscopes and electron microscopes respectively.

micrometer[1] /**mȳ**-kroh-mee-ter/ In microscopy, a device for measuring the size of an object under the microscope. An eyepiece micrometer (*graticule*) of glass or transparent film, with a scale etched or printed on it, is placed in the eyepiece so that both the object to be measured and the scale are in focus. The scale of the graticule changes at different magnifications and it must therefore be calibrated against a stage micrometer, which is contained in a glass slide and placed on the microscope stage.

micrometer[2] /mȳ-**krom**-ĕ-ter/ Symbol: μm A unit of length equal to 10^{-6} meter (one millionth of a meter). It is often used in measurements of cell diameter, sizes of bacteria, etc. Formerly, it was called the *micron*.

micron /**mȳ**-kron/ *See* micrometer.

micronucleus /mȳ-kroh-**new**-klee-ŭs/ (*pl.* micronuclei) *See* macronucleus.

micronutrient /mȳ-kroh-**new**-tree-ĕnt/ A nutrient required in trace amounts by an organism. For example, a plant can obtain sufficient of the essential trace element manganese from a solution containing 0.5 parts per million of manganese. Micronutrients include trace elements and vitamins. *See* deficiency disease.

microphagous feeding /mȳ-**kroff**-ă-gŭs/ *See* suspension feeding.

microphyll /**mȳ**-krŏ-fil/ A foliage leaf that, even if large, has a single unbranched vein running from base to apex (some fossil club mosses are an exception). It is the typical leaf of club mosses, horsetails, and whisk ferns. The stele remains entire when a leaf trace branches off to a leaf. *Compare* megaphyll.

micropyle /**mȳ**-krŏ-pȳl/ 1. A pore leading to the nucellus formed by incomplete integument growth around the apex of the ovule. Pollen tubes usually pass through it prior to fertilization. In most seeds the micropyle forms a small hole in the testa through which water is absorbed, but in some seeds it is closed. *See illustration at* ovule.
2. *See* chorion.

microsatellite DNA /mȳ-kroh-**sat**-ĕ-lÿt/ *See* satellite DNA.

microscope An instrument designed to magnify objects and thus increase the resolution with which one can view them. *Resolution* is the ability to distinguish between two separate adjacent objects. Radiation (light or electrons) is focused through the specimen by a *condenser lens*. The resulting image is magnified by further lenses. Since radiation must pass through the specimen, it is usual to cut larger specimens into thin slices of material (sections) with a microtome. Biological material has little contrast and is therefore often stained. If very thin sections are required the material is preserved and embedded in a supporting medium.

The *light microscope* uses light as a source of radiation. With a *compound microscope* the image is magnified by two lenses, an *objective lens* near the specimen, and an *eyepiece*, where the image is viewed, at the opposite end of a tube. Its maximum magnification is limited by the wavelength of light. Much greater resolution became possible with the introduction of the *electron microscope*, which uses electrons as a source of radiation, because electrons have much shorter wavelengths than light. However, only dead material can be observed because the specimen must be in a vacuum and electrons eventually heat and destroy the material.

Electron microscopes are of two main types, the *transmission electron microscope* and the more recent *scanning electron microscope*. The former produces an image by passing electrons through the specimen. With the scanning microscope electrons scan the surfaces of specimens rather as a screen is scanned in a TV tube, allowing surfaces of objects to be seen with greater depth of field and giving a 3D appearance to the image. Scanning microscopes cannot operate at such high magnifications as transmission electron microscopes.

microsomes /mÿ-krŏ-sohmz/ Fragments of endoplasmic reticulum and Golgi apparatus in the form of vesicles formed during homogenization of cells and isolated by high speed centrifugation. Microsomes from rough endoplasmic reticulum are coated with ribosomes and can carry out protein synthesis in the test tube.

microsporangium /mÿ-kroh-spŏ-**ran**-jee-ŭm/ (*pl.* **microsporangia**) In heterosporous plants, the sporangium that produces the microspores, located on the microsporophyll. The microsporangium wall splits to disperse the mature microspores. Microsporangia are found in some pteridophytes (e.g. *Selaginella* and many ferns) and are represented by the pollen sacs in gymnosperms and angiosperms.

microspore /mÿ-krŏ-spor, -spohr/ The smaller of the two types of spores produced in large numbers by spermatophytes and heterosporous pteridophytes. In pteridophytes, microspores develop into the male gametophyte generation, but in gymnosperms and angiosperms, the microspores (pollen grains) develop into a very reduced gametophyte represented by the pollen tube and the vegetative nucleus (pollen tube nucleus) and the generative nuclei. The latter are the male gametic nuclei. *Compare* megaspore. *See also* pollen.

microsporophyll /mÿ-krŏ-**spor**-ŏ-fil, -**spoh**-rŏ-/ A leaf or modified leaf on which the microsporangium is borne. Simple microsporophylls include the scales found in the male cones of gymnosperms and the fertile photosynthetic leaves of lycopods, usually grouped in a strobilus. The stamen of angiosperms is a highly modified microsporophyll. *See* megasporophyll.

microtome /mÿ-krŏ-tohm/ An instrument for cutting thin sections (slices a few micrometers thick) of biological material for microscopic examination. The specimen is usually embedded in wax for support and cut by a steel knife. Alternatively it is frozen and a *freezing microtome*, which keeps the specimen frozen while cutting, is used. For electron microscopy, extremely thin (20–100 nm) sections can be cut by an *ultramicrotome*. Here the specimen is embedded in resin or plastic for support and mounted in an arm that advances slowly, moving up and down, towards a glass or diamond knife. As sections are cut they float off on to the surface of water contained in a trough behind the knife.

microtubule /mÿ-kroh-**too**-byool/ A thin cylindrical unbranched tube of variable length found in eukaryotic cells, either singly or in groups. Its walls are made of the protein tubulin. Microtubules are one of the three main components of the CYTOSKELETON, along with microfilaments and intermediate filaments, and help cells to maintain their shape and provide the motive force for some forms of cell movement. They are part of the structure of centrioles, basal bodies, and undulipodia (cilia and flagella); and form the spindle during cell division, bringing about chromosome movement. Microtubules help to orientate materials and structures in the cell, for example cellulose fibrils during the formation of plant cell walls. They also serve as tracks along which cargoes (e.g. vesicles) are translocated within cells, for example along the axons of neurons.

A microtubule is typically 24 nm in diameter, with a wall consisting of 13 longitudinally arranged protofilaments. The protofilaments comprise tubulin subunits, each of which has two similar parts (monomers): α-tubulin, which binds GTP (guanosine triphosphate) irreversibly; and β-tubulin, which binds GTP reversibly, and can split it to GDP (guanosine diphosphate), thereby yielding energy. Microtubule formation is initiated by one or more MICROTUBULE-ORGANIZING CENTERS within the cell (the centrosome in animal cells). The microtubules are rapidly assembled from a pool of tubulin subunits, and can be quickly disassembled, according to the requirements of the cell.

Various proteins bind to microtubules and give them particular properties. Motor proteins include kinesin and DYNEIN; both use ATP to generate the force required to transport vesicles, proteins, or organelles along microtubules. There are several classes of microtubule-associated proteins (MAPS), which regulate the stability of microtubules, and can also bind them to membranes, to each other, or to intermediate filaments.

microvilli /mȳ-kroh-**vil**-ȳ/ (*sing.* **microvillus**) Elongated slender projections (typically 0.5–10 μm long) of the plasma membrane, found especially in secretory and absorptive cells. The closely packed arrangement of microvilli on the free surface of epithelial cells constitutes a *brush border*. Microvilli provide an increased surface area for the exchange of molecules. Numerous microvilli occur on the epithelial cells of the intestine and also in the kidney tubules. In addition to their presence in secretory and absorptive cells, they are commonly observed in many other cells although they may not be permanent structures. *See* epithelium; membrane.

midbrain /**mid**-brayn/ (mesencephalon) One of the three basic anatomical divisions of the brain, connecting the FOREBRAIN and the HINDBRAIN. It is traversed by the cerebral aqueduct. The midbrain roof (tectum) is a dominant center of the brain in fishes and amphibians and may have a pair of optic lobes, especially prominent in birds. The midbrain is less well developed in mammals.

middle ear (**tympanic cavity**) An air-filled cavity in the skull, between the outer and inner ear in most tetrapods. It is connected to the back of the throat by the Eustachian tube. In mammals, it contains three small bones, the ear ossicles (malleus, incus, and stapes), which link the tympanum (ear-

drum) with the oval window (fenestra ovalis) and transmit vibrations to the inner ear. In other tetrapods, it contains only one ear ossicle, the columella auris. *See illustration at* ear.

middle lamella A thin cementing layer holding together neighboring plant cell walls. It consists mainly of PECTIC SUBSTANCES (e.g. calcium pectate). The middle lamella is laid down at the CELL PLATE during cell division. *See illustration at* cell.

midgut /**mid**-gut/ The central part of the alimentary canal of arthropod and vertebrate animals, responsible for digestion and absorption.

In arthropods, it consists of the mesenteron and mesenteric caeca, ending in front of the Malphigian tubules. It is lined with endoderm consisting of tall columnar cells, but has no cuticle.

In vertebrates it is the region between the bile duct and the middle of the colon. *Compare* foregut; hindgut.

migration An instinctive regular two-way movement of part or all of an animal population to and from a given area, usually along well-defined routes. It is closely linked to the cycle of the seasons and is triggered off by seasonal factors such as increasing and decreasing daylengths in spring and autumn. Many birds, hoofed mammals, bats, whales, fish, and insects migrate, often covering immense distances. For example, the Arctic tern breeds on the northernmost coasts of Eurasia and America and winters around the Antarctic pack-ice 11 000 miles to the south. Migratory mammals such as the wildebeest live in habitats with fluctuating climatic conditions and migrate in order to find an adequate food supply.

milk 1. A whitish opaque nutritious fluid that is produced by the mammary glands of female mammals for feeding their young. It contains carbohydrates (mainly lactose), fats, proteins (mainly casein and whey), certain mineral salts, vitamins, and water. It is rich in calcium and phosphorus, vitamins A, D, and riboflavin, but a poor

source of iron, and vitamins C and K. The composition of milk differs between species. For example, cow's milk contains more protein, calcium, phosphorus, and riboflavin than human milk, but less nicotinic acid and lactose. As raw milk contains bacteria and is quickly perishable, it may be treated for human consumption by pasteurization or by ultra-heat treatment (UHT) for a longer storage period. To preserve it practically indefinitely, it may be dried, condensed, or evaporated. *See also* colostrum; rennin.
2. In plants, any of various milklike fluids, such as coconut milk or the latex of certain flowering plants.

milk sugar *See* lactose.

milk teeth *See* deciduous teeth.

millipedes *See* Diplopoda.

mimicry The resemblance of one animal to another by which the mimic gains advantage from its resemblance to the model. For example, in *Batesian mimicry* certain edible insects mimic the warning coloration of noxious insects and so are avoided by their predators. Natural selection produces more accurate mimicry as only those individuals closely resembling the model will be mistaken for it and left alone. In *Müllerian mimicry* a group of poisonous animals resemble each other, for example, bees, wasps, and hornets, increasing the likelihood that potential predators will learn to avoid them. *See also* cryptic coloration.

mineralocorticoid /min-ĕ-ral-ŏ-**kor**-tă-koid/ A type of steroid hormone produced by the adrenal cortex. Mineralocorticoids (e.g. aldosterone and deoxycorticosterone) control salt and water balance by their action on the kidney. *See also* corticosteroid.

Miocene /**mÿ**-ŏ-seen/ An epoch of the Tertiary, 25–7 million years ago. The climate became drier and the grasses evolved and spread rapidly, this perhaps explaining the replacement of early mammals by more modern forms. About half to three-quar-ters of existing mammalian families are represented in rocks of the Miocene.

missense mutation /miss-**sens**/ A MUTATION in which the base sequence of the DNA is altered such that the codon for another amino acid is inserted instead of the normal codon when the DNA is transcribed into messenger RNA. During translation, the variant amino acid is substituted for the normal amino acid in the protein, which might affect the function of the protein.

Mississippian /miss-ă-**sip**-ee-ăn/ The US name for the early Carboniferous subperiod.

mitochondrial DNA /mÿ-tŏ-**kon**-dree-ăl/ (**mt DNA**) A small circular DNA molecule found in mitochondria that carries the genes for certain proteins required for the respiratory functions of mitochondria. Each mitochondrion contains several or numerous copies of mtDNA. For example, human mitochondria have 2 to 10 copies of an mtDNA, each with 37 genes. Mitochondrial DNA is inherited exclusively from the mother in organisms, such as humans, where the egg contributes virtually all the cytoplasm to the zygote following fertilization.

mitochondrion /mÿ-tŏ-**kon**-dree-ŏn/ (*pl.* **mitochondria**) An organelle of all plant and animal cells chiefly associated with aerobic respiration. It is surrounded by two membranes separated by an intermembrane space; the inner membrane forms finger-like processes called *cristae*, which project into the gel-like *matrix*. Mitochondria are typically sausage-shaped, but may assume a variety of forms, including irregular branching shapes. The diameter is always about 0.5–1.0 μm. They contain the enzymes and cofactors of AEROBIC RESPIRATION and therefore are most numerous in active cells (up to several thousand per cell). They may be randomly distributed or functionally associated with other organelles, for example with the contractile fibrils of muscle cells.

Prophase

Chromosomes appear
and shorten and thicken.
Nuclear membrane
disintegrates and
nucleolus disappears

Metaphase

Individual chromosomes,
each comprising two
chromatids, become aligned
along the equator of the
nuclear spindle

Anaphase

Chromosomes split at
the centromere and the
daughter chromatids
move to opposite poles
of the spindle

Telophase

A nuclear membrane
forms around each
group of daughter
chromatids, or new
chromosomes

Stages in mitosis

The reactions of the Krebs cycle take place in the matrix and those of electron transport coupled to oxidative phosphorylation (i.e. the respiratory chain) on the inner membrane. Within the membrane the components of the respiratory chain are highly organized. The matrix is also involved in amino acid metabolism via Krebs cycle acids and transaminase enzymes, and in fatty acid oxidation. It is possible that mitochondria, like chloroplasts, may be the descendents of once independent bacteria-like organisms that early in evolution invaded eukaryotic cells, leading to an extreme form of symbiosis. *See* endosymbiont theory; mitochondrial DNA. *See illustration at* cell.

mitosis /mȳ-**toh**-sis/ (**karyokinesis**) The ordered process by which the cell nucleus and cytoplasm divide in two during the division of body (i.e. non-germline) cells. The chromosomes replicate prior to mitosis and are then separated during mitosis in such a way that each daughter cell inherits a genetic complement identical to that of the parent cell. Although mitosis is a continuous process it is divided into four phases; prophase, metaphase, anaphase, and telophase. *Compare* meiosis; amitosis; endomitosis.

mitral valve /mȳ-trăl/ (**bicuspid valve**) A valve consisting of two membranous flaps or cusps situated between the atrium and ventricle of the left side of the heart in mammals and birds. When the ventricle contracts, blood is forced into the aorta and prevented from returning to the atrium by closure of the valve. *See also* tendinous cords. *See illustration at* heart.

molar /**moh**-ler/ A large cheek tooth, two or more of which are found at the back of jaws of a mammal. The crown has several pointed cusps, or, in herbivorous animals, ridges. These teeth are used for crushing, chewing, or grinding the food. They are not present in the first (milk) dentition and in humans the third molar on each side of the upper and lower jaws does not appear until later in life (these teeth are therefore sometimes referred to as *wisdom teeth*). *See also* teeth.

molarity /moh-**la**-ră-tee/ A measure of the concentration of solutions based upon the number of molecules or ions present, rather than on the mass of solute, in any particular volume of solution. The molarity (M) is the number of moles of solute in one cubic decimeter (litre). Thus a 0.5M solution of hydrochloric acid contains 0.5 × (1 + 35.5)g HCl per dm^3 of solution.

mole Symbol: mol The SI base unit of amount of substance, defined as the amount of substance that contains as many elementary entities as there are atoms in 0.012 kilogram of ^{12}C. The elementary entities may be atoms, molecules, ions, electrons, photons, etc., and they must be specified. The amount of substance is proportional to the number of entities, the constant of proportionality being the Avogadro number. One mole contains 6.022 045 × 10^{23} entities. One mole of an element with relative atomic mass *A* has a mass of *A* grams (this mass was formerly called one *gram-atom* of the element).

molecular clock (**evolutionary clock**) The concept that substitutions in the nucleotide or amino acid sequence of large biomolecules (e.g. DNA or proteins) due to mutation accumulate at a fairly constant rate over evolutionary time. This is because the majority of such substitutions have little impact on the function of the molecules concerned; i.e. their effect is neutral in evolutionary terms (*see* neutral evolution). Hence, if corresponding nucleic acids or proteins are compared in two different species, the degree of difference between the sequences is proportional to the time since those species diverged from a common ancestor. The rate of neutral substitutions, representing the 'ticking' of the molecular clock, can be calibrated in years by referring to the fossil record. However, the rate varies in different types of molecules, and it might also speed up or slow down over time, which may cause errors in interpreting molecular comparisons.

molecular systematics A branch of biology that compares functionally equivalent macromolecules from different organisms as a basis for classification. Sequences of amino acids in proteins (e.g. enzymes) or of nucleotides in nucleic acids (e.g. ribosomal RNA) are determined using automated techniques, and compared statistically using sophisticated computer programs. Essentially, how closely two organisms are related in evolutionary terms is reflected in the degree of similarity of their macromolecules.

molecular weight *See* relative molecular mass.

Mollusca /mŏ-**lus**-kă/ A phylum of bilaterally symmetrical unsegmented invertebrates, including the aquatic bivalves, mussels, octopuses, squids, etc., and the terrestrial slugs and snails. The body is divided into a head, a ventral muscular locomotory organ (foot), and a dorsal visceral hump that houses most of the body organs and is covered by a tissue layer (*mantle*), which typically secretes a calcareous shell into which the head and foot can retract. The mantle extends into folds forming a cavity containing the gills (ctenidia). The rasping radula is used for feeding. The coelom is restricted. Development usually occurs via a trochophore larva. *See also* Cephalopoda; Gastropoda; Pelecypoda.

molting 1. *See* ecdysis.
2. The seasonal loss of hair or feathers by mammals or birds respectively.

molybdenum /mŏ-**lib**-dĕ-nŭm/ *See* trace element.

Monera /mŏ-**neer**-ă/ An alternative name for the kingdom Prokaryotae. It was originally used by the German biologist Ernst

Haeckel to refer to bacteria and blue-green algae (now called cyanobacteria) as a group within the kingdom Protista, in his three-kingdom classification scheme. These prokaryotic organisms were later transferred to their own kingdom, in recognition of the fundamental differences between prokaryotes and all other organisms. *See also* eukaryote; prokaryote.

monochasial cyme /mon-ŏ-**kay**-zhee-ăl, -zee-ăl/ (**monochasium**) A type of CYMOSE INFLORESCENCE in which each flower branch has only one lateral branch. Variations in the inflorescence arrangement occur, according to the direction of the lateral branches, for example in forget-me-not, all the branches arise on the same side of the parent stem while in buttercup, the branches arise on alternate sides of the parent stem.

monoclonal antibody /mon-ŏ-**kloh**-năl/ A specific antibody produced by a cell clone (i.e. one of many identical cells derived from a single parent). The parent cell is obtained by the artificial fusion of a normal antibody-producing mouse spleen cell with a cell from cancerous lymphoid tissue of a mouse. This hybrid cell, or *hybridoma*, multiplies rapidly *in vitro* and yields large amounts of antibody, which comprises only a single species of immunoglobulin molecule. Monoclonal antibodies can identify a specific antigen within a mixture and are now used widely as reagents in immunoassays.

Monocotyledonae /mon-ŏ-ko-tă-lee-**doh**-nee-ee/ *See* Liliopsida.

monocyte /**mon**-ŏ-sÿt/ The largest type of white blood cell (leukocyte). It has nongranular cytoplasm and a large kidney-shaped nucleus. Monocytes are actively phagocytic, devouring foreign particles (such as bacteria). They make up 4–5% of all leukocytes. *See also* leukocyte.

monoecious /mŏ-**nee**-shŭs/ Denoting plants in which the male and female reproductive organs are on the same individual. Monoecious flowering plants bear separate unisexual male and female flowers as seen in maize plants. *Compare* dioecious.

monohybrid /mon-ŏ-**hÿ**-brid/ A hybrid heterozygous at one locus and obtained from crossing homozygous parents with different alleles at a given locus; for example, Mendel's cross between tall (TT) and dwarf (tt) garden peas to give a tall monohybrid (Tt). When a monohybrid is selfed, dominant and recessive phenotypes appear in the offspring in the ratio of 3:1 (the *monohybrid ratio*). *Compare* dihybrid.

mononuclear phagocyte system (**reticuloendothelial system**) The system of MACROPHAGE cells, which are scattered throughout the body and are capable of engulfing foreign particles. The cells are found in connective tissue and in the lining of small blood vessels, and are concentrated especially in the lung, liver, spleen, and lymph nodes. The mononuclear phagocyte system is important in defending the body against disease and in destroying worn-out erythrocytes.

monophyletic /mon-ŏ-fÿ-**let**-ik/ Describing a group that contains all the species that are descended from the most recent common ancestor that is exclusive to the group. Such groups are defined solely by SYNAPOMORPHIES, and represent the ideal goal of the cladistics approach to constructing phylogenetic trees. *Compare* paraphyletic.

monophyodont /mon-ŏ-**fÿ**-ŏ-dont/ Describing a type of dentition in which an animal has only one set of teeth during its lifetime, which are not replaced if they fall out. *Compare* diphyodont; polyphyodont.

monoploid /**mon**-ŏ-ploid/ *See* haploid.

monopodial /mon-ŏ-**poh**-dee-ăl/ Describing the system of branching in plants in which the main axis of the stem (the monopodium) continues to grow indefinitely by the terminal bud. Lateral buds and branches are always subsidiary to the terminal bud, and this may give a very regular branching pattern as seen in conifers.

Monopodial growth is also called indefinite or racemose branching and is typical of the formation of a racemose inflorescence. *Compare* sympodial.

monosaccharide /mon-ŏ-**sak**-ă-rÿd, -rid/ A SUGAR that cannot be hydrolyzed to simpler carbohydrates of smaller carbon content. Glucose and fructose are examples.

monosomy /mŏ-**nos**-ŏ-mee/   *See* aneuploidy.

Monotremata /mon-ŏ-**trem**-ă-tă, -tree-mă-/ (**Prototheria**) The order that contains the most primitive mammals – the only mammals that lay eggs. After hatching the young are transferred to a pouch on the abdomen and are nourished by milk secreted by primitive mammary glands whose ducts do not form nipples. Other primitive features include poor temperature control, and possession of a cloaca and a primitive pectoral girdle, but the brain, hair, heart, and diaphragm are typically mammalian. Monotremes, which include *Ornithorhynchus* (duck-billed platypus), are found in Australia. *Ornithorhynchus* is aquatic, with webbed feet, a flat, paddlelike tail, and a bill for crushing invertebrates. *Tachyglossus* and *Zaglossus* (echidnas, or spiny anteaters) are terrestrial insectivores. *Compare* Eutheria; Metatheria.

monotrichous /mŏ-**not**-ră-kŭs/ Describing bacteria that possess one flagellum, e.g. *Vibrio*.

monozygotic twins /mon-ŏ-zÿ-**got**-ik/ *See* identical twins.

morphogen /**mor**-fŏ-jĕn/ *See* differentiation.

morphogenesis /mor-fŏ-**jen**-ĕ-sis/ The development of form and structure.

morphology /mor-**fol**-ŏ-jee/ The study of the form of organisms. The term may be used synonymously with 'anatomy' although generally the study of external form is termed 'morphology' while the study of internal structures is termed 'anatomy'.

morula /**mor**-yŭ-lă, **mor**-û-/ (*pl.* **morulae**) A loose aggregation of blastomeres resulting from cleavage of the egg of mammals. It develops into the blastocyst.

mosaic A hybrid organism whose cells differ genetically, although they have all arisen from a single zygote. The genetic difference is usually created by irradiating the zygote during early stages of cleavage.

mosquitoes *See* Diptera.

mosses *See* Musci.

moths *See* Lepidoptera.

motor neuron A nerve cell (neuron) that transmits impulses from the brain or spinal cord to a muscle or other effector.

mouthparts Jointed appendages on the heads of arthropods, modified in various ways for dealing with food. They consist of the *labrum* (upper lip), which is a single plate; a pair of *mandibles* (upper jaws), which have serrated edges; and a pair of *maxillae* (lower jaws), which also have serrated regions. In insects the second pair of maxillae is specialized to form the *labium* (lower lip), which is a single plate. The labium and maxillae have sensory feelers (*palps*) concerned with the tasting of food, and in crustaceans the second maxillae may also be used for producing respiratory water currents. The basic system of mouthparts found in the more primitive groups (e.g. cockroaches and locusts) is highly modified in other groups to suit the mouthparts for one particular kind of food. For example, the tubular mouthparts of butterflies and moths are adapted for sucking nectar; the piercing and sucking mouthparts of mosquitoes are adapted to feed on blood or plant juices. *See also* chelicera; pedipalps.

mRNA *See* messenger RNA.

mtDNA *See* mitochondrial DNA.

mucilage /myoo-să-lij/ *See* gum.

mucin /myoo-sin/ The main constituent of mucus. It is a glycoprotein.

mucopolysaccharide /myoo-koh-pol-ee-sak-ă-rÿd, -rid/ *See* glycosaminoglycan.

mucoprotein /myoo-koh-**proh**-teen, **proh**-tee-in/ *See* proteoglycan.

mucosa /myoo-**koh**-să/ (*pl.* mucosae) *See* mucous membrane.

mucous membrane /myoo-kŭs/ (mucosa) The tissue, in vertebrates, that lines many tracts (e.g. the intestinal and respiratory tracts) that open to the exterior. It consists of surface epithelium containing goblet cells, which secrete mucus, and is underlaid by connective tissue.

mucus /myoo-kŭs/ A slimy substance produced by goblet cells in mucous membranes of animals. It is viscous and insoluble, consisting mainly of glycoproteins. Its function is to protect and lubricate the surface on which it is secreted.

Müllerian duct /myoo-**leer**-ee-ăn/ The oviduct of female jawed vertebrates. It develops in both sexes from embryonic mesoderm, in association with the Wolffian duct, but becomes vestigial in the male. In most vertebrates it is paired (single in birds) and extends from a ciliated funnel, which opens into the coelom near the ovary, to the cloaca (when present). In mammals it is usually differentiated into Fallopian tube, uterus, and vagina. Ova entering the funnel are conveyed along the duct towards the exterior by muscular and ciliary movements. If spermatozoa are present at this time, fertilization may occur.

Müllerian mimicry *See* mimicry.

multicellular /mul-ti-**sell**-yŭ-ler/ Consisting of many cells.

multifactorial inheritance /mul-ti-fak-**tor**-ee-ăl, -**toh**-ree-/ *See* quantitative inheritance.

multiple allelism /ă-**lee**-liz-ăm/ The existence of a series of alleles (three or more) for one gene. In humans, for example, there are three alleles (A, B, and O) governing blood type. Only two alleles of the series can be present in a diploid cell. Dominance relationships within an allelic series are often complicated.

murein *See* peptidoglycan.

Musci /muss-ÿ/ In some classifications, a class of the phylum BRYOPHYTA containing the mosses.

muscle /muss-ăl/ Tissue consisting of elongated cells (*muscle fibers*) containing fibrils that are highly contractile. *See* cardiac muscle; skeletal muscle; smooth muscle.

muscle spindle A stretch receptor in skeletal muscle that gives information about the degree of contraction of the muscle and initiates such reflexes as the knee jerk. It is a PROPRIOCEPTOR. The end of the sensory nerve is wrapped spirally around a short noncontractile section of a muscle fiber. When this section is stretched by relaxation of the whole muscle (or, in the knee jerk, by striking the tendon) the nerve ending is stimulated and impulses pass to the spinal cord, resulting in a reflex contraction of adjacent muscle fibers, and the spindle is no longer stimulated (negative feedback). This is important in maintaining posture. For voluntary contraction of the muscle, the sections of fiber on either side of the spindle contract, maintaining the stretch on the spindle.

mutagen /myoo-tă-jĕn/ Any physical or chemical agent that induces MUTATION or increases the rate of spontaneous mutation. Chemical mutagens include ethyl methanesulfonate, which causes changes in the base pairs of DNA molecules, and acridines, which cause base pair deletions or additions. Physical mutagens include ultraviolet light, x-rays, and gamma rays.

mutation (gene mutation) A change in one or more of the bases in DNA, which results in the formation of an abnormal protein. Mutations are inherited only if they occur in the cells that give rise to the gametes; somatic mutations may give rise to chimaeras and cancers. Mutations result in new allelic forms of a gene and hence new variations upon which natural selection can act. Most mutations are deleterious but are often retained in the population because they also tend to be recessive and can thus be carried in the genotype without affecting the viability of the organism. The natural rate of mutation is low, but the mutation frequency can be increased by mutagens. *See also* chromosome mutation; frameshift; missense mutation; mutagen; point mutation; polyploid.

mutualism /**myoo**-chû-ă-liz-ăm/ The close relationship between two or more species of animals or plants in which all benefit from the association. There are two types of mutualism: *obligatory mutualism*, in which one cannot survive without the other, for example the algal/fungal partnership found in lichens; and *facultative* or *nonobligatory mutualism*, in which both species can survive independently, for example marine crabs and their associated invertebrate fauna of sponges, cnidarians, etc., that attach to the crab shell and act as camouflage. Facultative mutualism is sometimes termed *protocooperation*. *Compare* symbiosis.

mycelium /mÿ-**see**-lee-ŭm/ (*pl.* **mycelia**) A filamentous mass comprising the body of a fungus, each filament being called a HYPHA. The mycelium often forms a loose mesh as in *Mucor*, but the hyphae may become organized into definite structures, e.g. the fruiting body of a mushroom. The mycelium produces the reproductive organs of a fungus. The whole thallus of unicellular fungi may be thus employed, but only part of the thallus produces gametangia or sporangia in most species, the rest of the thallus being vegetative.

mycetozoa /mÿ-see-tŏ-**zoh**-ă/ A taxon name used for slime molds in certain classifications. *See* slime mold.

mycoplasmas /mÿ-koh-**plaz**-măz/ (PPLO; **pleuropneumonia-like organisms**) A group of extremely small bacteria that naturally lack a rigid cell wall. They often measure less than 200 nm in diameter and their cells are delicate and plastic. Mycoplasmas can cause pneumonias in humans, bovine tuberculosis, and other mammalian diseases, but they may exist harmlessly in mucous membranes. Mycoplasmas are resistant to penicillin and related antibiotics that work by inhibiting wall growth.

mycoprotein /mÿ-koh-**proh**-teen, -**proh**-tee-in/ Any protein produced by a fungus or bacterium.

mycorrhiza /mÿ-kŏ-**rÿ**-ză/ (*pl.* **mycorrhizae**) The association between the hyphae of a fungus and the roots of a higher plant. Two main types of mycorrhiza exist, *ectotrophic* in which the fungus forms a mantle around the smaller roots, as in trees, and *endotrophic* in which the fungus grows around and within the cortex cells of the roots, as in orchids and heathers.

In ectotrophic mycorrhizae the fungus, which is usually a member of the Agaricales, benefits by obtaining carbohydrates and possibly B-group vitamins from the roots. The trees benefit in that mycorrhizal roots absorb nutrients more efficiently than uninfected roots, and it is common forestry practice to insure the appropriate fungus is applied when planting seedling trees. In endotrophic mycorrhizae the fungus is generally a species of *Rhizoctonia* and again both partners benefit nutritionally from the relationship.

myelinated nerve fiber /mÿ-ĕ-lă-nay-tid/ A nerve fiber that is surrounded by a fatty (myelin) sheath. Most nerves of vertebrates consist of thousands of medullated fibers, which appear white because of the fatty sheaths. *See* myelin sheath.

myelin sheath /mÿ-ĕ-lin/ An insulating covering that surrounds the axon of a neuron. It is composed of the cell membranes of SCHWANN CELLS wound tightly in a spiral around the axon. The membranes consist of a fatty material (myelin). In between each Schwann cell is a short region of bare axon (*node of Ranvier*). Myelinated (medullated) axons occur in most vertebrate neurons but are less common in invertebrates.

myelocyte /mÿ-ĕ-lŏ-sÿt/ A cell in the myeloid tissue of red bone marrow. Myelocytes are formed by cell division of precursor cells (myeloblasts) and they change into granulocytes, which are released into the bloodstream. *See also* granulocyte.

myeloid tissue /mÿ-ĕ-loid/ Tissue that manufactures white blood cells. It occurs in the red bone marrow, surrounding the blood vessels. It contains myeloblast cells, which divide continuously to give myelocytes (which develop into granulocytes), and lymphoblasts and monoblasts, which give rise to agranular leukocytes. Since white blood cells have a very short lifespan (only a few days in some cases), this tissue is very active.

myocardium /mÿ-ŏ-**kar**-dee-ŭm/ The muscular layer of the heart wall, contractions of which are responsible for the pumping action of the heart's chambers. It is thickest around the ventricles, and is lined internally and externally by membranous layers (respectively the *endocardium* and *epicardium*). *See* cardiac muscle.

myofibril /mÿ-ŏ-**fÿ**-brăl/ A very fine fiber (1–2 μm in diameter) many of which are embedded in the sarcoplasm of a muscle fiber. In SKELETAL MUSCLE these fibrils are striated, being divided along their length into a great number of SARCOMERES, which constitute the contractile apparatus of the muscle fiber.

myoglobin /mÿ-ŏ-**gloh**-bin/ A conjugated protein found in muscles (sometimes referred to as 'muscle hemoglobin'). It is similar to hemoglobin in being a heme protein capable of binding oxygen but is structurally simpler, having only one polypeptide chain combined with the heme group. Each molecule of myoglobin can attach one molecule of oxygen.

myoneme /mÿ-ŏ-neem/ A contractile fibril found in certain protoctists (e.g. *Vorticella*). *Compare* myofibril; myosin.

myosin /mÿ-ŏ-sin/ A protein that is able to move along actin filaments and is involved in muscle contraction and other types of movement within cells. Myosin can hydrolyze ATP to cause conformational changes in the myosin molecule that enable it to 'walk' along an actin filament. Hence myosin is a MOTOR PROTEIN, able to convert chemical energy (in the form of ATP) into mechanical energy. There are several types of myosin, the most common being myosin I, involved in cytoskeletal interactions with actin, and myosin II, which occurs mainly in muscle. All myosins comprise a globular 'head' domain, a flexible 'neck' domain, and an extended 'tail' domain. The head domain contains actin-binding sites and ATP-binding sites, and it is this region that walks along the actin filament, with each 'step' coupled to the hydrolysis of one ATP molecule. *See* actin; sarcomere.

myotome /mÿ-ŏ-tohm/ The part of each SOMITE of a vertebrate embryo that differentiates as a muscle block. The muscle myotomes remain segmental in fishes, but in terrestrial vertebrates they lose much of their original pattern and buds from myotomes (with their ventral root nerves) form muscles in limbs, etc.

Myriapoda /mi-ree-**ap**-ŏ-dă/ A group of terrestrial arthropods containing the classes CHILOPODA (centipedes) and DIPLOPODA (millipedes), characterized by a distinct head, bearing antennae, mandibles, and maxillae, and numerous body segments bearing walking legs.

myxobacteria /miks-ŏ-bak-**teer**-ee-ă/ (**slime bacteria**) A group of bacteria in the class Schizomycetes in which individual cells are typically rod-shaped and covered in slime, but which may congregate to form gliding swarms or various upright reproductive structures under certain conditions. For example, the reproductive body of *Stigmatella aurantiaca* consists of a stalk bearing several cysts. When these are wetted, they open to release masses of individual gliding bacteria that move together over the substrate as discrete colonies. In some species of myxobacteria, the reproductive structures are brightly colored and just visible to the naked eye. Myxobacteria are saprophytic soil organisms, common in soil, animal dung, and decaying plant matter.

Myxomycetes /miks-ŏ-**mȳ**-seets/ *See* slime molds.

Myxomycota /miks-ŏ-mȳ-**koh**-tă/ See slime molds.

myxovirus /miks-ŏ-**vȳ**-rŭs/ One of a group of RNA-containing viruses that cause such diseases as influenza, mumps, measles, and rabies.

NAD (nicotinamide adenine dinucleotide) A derivative of nicotinic acid that acts as a coenzyme in electron-transfer reactions (e.g. the electron-transport chain). Its role is to carry hydrogen atoms. The oxidized form, denoted NAD^+, accepts a pair of electrons and one proton to become the reduced form, denoted as NADH.

NADP Nicotinamide adenine dinucleotide phosphate; a coenzyme similar in its action to NAD. As with NAD, the oxidized form of NADP ($NADP^+$) accepts two electrons and a proton to become the reduced form, NADPH. During PHOTOSYNTHESIS, $NADP^+$ generally serves as the electron acceptor that receives electrons from the electron-transport chain in the thylakoid membrane of the chloroplast. The resultant NADPH provides electrons for the fixation of carbon dioxide.

nano- Symbol: n A prefix denoting one thousand-millionth, or 10^{-9}. For example, 1 nanometer (nm) = 10^{-9} meter. *See also* SI units.

nares /**nar**-eez/ *See* nostrils.

nasal cavity /**nay**-zǎl/ A paired cavity in the heads of vertebrates. It contains thin bones covered with a mucous membrane and it is lined with epithelium, some regions of which contain nerve endings sensitive to smells. In mammals the large area of mucous membrane warms and moistens air as it passes through on its way to the lungs. Hairs near the nostrils and mucus from the membrane filter dust and bacteria from the air.

nastic movements /**nas**-tik/ (nasties) Movements of plant parts in which direction of movement is independent of the direction of the stimulus. With *photonasty* the stimulus is light. For instance, at constant temperature crocus and tulip flowers open in the light and close in the dark because of slight growth movements. Opening is caused by *epinasty* – i.e. greater growth of the upper surface of a plant organ – and closed by *hyponasty* – i.e. greater growth of the lower surface of a plant organ. Similarly, at constant light intensity crocus and tulip flowers show *thermonasty*; i.e. they open in warm air and close in cool air. Such day-night rhythms are examples of *nyctinasty* and may also occur in leaves, as in *Oxalis*.

Nongrowth nastic movements also occur and are more rapid. The 'sensitive plant' (*Mimosa pudica*) rapidly closes its leaflets upwards and the petioles droop in response to touch, shock (*seismonasty*), or injury. This plant, and many other legumes, shows nongrowth nyctinastic movements. Movement is the result of osmotic changes in special swollen groups of cells (*pulvini*) at the bases of the moving structures. When the stimulus is contact, the movement is *haptonastic* as in the closure movements of insectivorous plants. The closure of the two halves of the Venus fly-trap leaf is a nongrowth haptonastic movement caused by a loss of turgor in the cells along the midrib following stimulation of the sensitive hairs on the leaf. *See also* taxis; tropism.

nasties /**nas**-teez, **nah**-steez/ *See* nastic movements.

natural killer cell (NK cell) A large lymphocyte with prominent lysosomes that attaches to the surface of virus-infected cells and secretes substances (perforins) that puncture the target cell's membrane. This allows the entry of other substances that trigger the death of the cell. Unlike cytotoxic T-cells, NK cells are not primed to attack specific target cells, but are thought to recognize certain chemical groups that occur on the surface of infected cells. *See also* T-cell.

natural selection The process, which Darwin called the 'struggle for survival', by which organisms less adapted to their environment tend to perish, and better-adapted organisms tend to survive. According to DARWINISM, natural selection acting on a varied population results in EVOLUTION.

nature and nurture The interaction between inherited and environmental factors (*nature* and *nurture* respectively) in determining the observed characteristics of an organism. It is often applied in a discussion of behavioral characteristics, such as intelligence, in which the relative importance of inherited and environmental factors, including such factors as social background, are a matter of great controversy. The term *heritability* is sometimes used as an alternative, meaning the proportion of the total variation in a species caused by genetic influences alone.

nauplius /naw-plee-ūs/ (*pl.* **nauplii**) An early free-swimming larval stage of some crustaceans. It possesses only three pairs of appendages (antennules, antennae, and mandibles) and later molts to form the cypris larva. The nauplius is important in the life cycle of the barnacle as it is the stage during which this sessile animal is dispersed.

Neanderthal man /nee-an-der-thawl, -tahl/ A species of early man, *Homo neanderthalensis*, that lived in Europe from about 150 000 years ago and was replaced by modern man, *H. sapiens*, about 30 000 years ago. They were dominant in western Europe during the first stages of the last glaciation, were cave-dwelling, and made regular use of fire and tools. *See also* Homo.

Nearctic /nee-ark-tik, -ar-tik/ One of the six zoogeographical regions, including North America from the Central Mexican Plateau in the south to the Aleutian Islands and Greenland in the north. The fauna include mountain goat, prong-horn antelope, caribou, and muskrat.

necrosis /ne-kroh-sis/ The death of cells in a tissue or organ. It may be caused by injury, disease, interruption to the blood supply, or exposure to noxious chemicals or toxins.

nectar A sugar-containing fluid secreted by the nectaries in plants.

nectary /nek-tă-ree/ A patch of glandular epidermal cells on the receptacle or other parts of certain flowers, producing a sugary liquid (nectar) that attracts insects. The epidermal, and sometimes underlying, cells of the nectary may or may not be organized into clearly defined structures. Nectaries are an adaptation to encourage cross-pollination by insects.

negative staining A method of preparation of material for electron microscopy used for studying three-dimensional and surface features, notably of viruses, macromolecules (e.g. enzyme complexes), and the cristae of mitochondria. A stain is used that covers the background and penetrates surface features of the specimen, but leaves the specimen itself unstained.

nekton /nek-ton/ Animals of the pelagic zone of a sea or lake that are free-swimming and independent of tides, currents, and waves, such as fish, whales, squid, crabs, and shrimps. Nekton are limited in distribution by temperature and nutrient supply, and decrease with increasing depth. *See also* benthic; plankton.

nematocyst /nem-ă-tŏ-sist/ *See* cnidoblast.

Nematoda /nem-ă-toh-dă/ A large phylum of marine, freshwater, and terrestrial invertebrates, the roundworms. Most are free-living, e.g. *Anguillula* (vinegar eel), but many are parasites, e.g. *Heterodera* (eelworm of potatoes) and *Ascaris* (found in pigs' and human intestines). Some cause serious diseases in humans, e.g. *Wuchereria* (causing elephantiasis). Nematodes are bilaterally symmetrical with an unsegmented smooth cylindrical body pointed at both ends and covered with a tough cuticle.

The body cavity is not a true coelom and there are no blood or respiratory systems. The muscular and excretory systems and embryonic development are unusual. Nematodes are not closely related to any other phylum.

neocortex /nee-oh-**kor**-teks/ (*pl.* neocortices) *See* neopallium.

neo-Darwinism /nee-oh-**dar**-wă-niz-ăm/ Darwin's theory of evolution through natural selection, modified and expanded by modern genetic studies arising from the work of Mendel and his successors. Such studies have answered many questions which Darwin's theory raised, but could not adequately explain because of lack of knowledge at the time it was formulated. Notably, modern genetics has revealed the source of variation on which natural selection operates, namely mutations of genes and chromosomes, and provided mathematical models of how alleles fluctuate in natural populations, thereby quantifying the process of evolution.

Neolithic /nee-ŏ-**lith**-ik/ The recent Stone Age, dating from about 10 000 years ago until the beginning of the Bronze Age. It is characterized by more advanced and often polished stone tools and the development of agriculture.

neopallium /nee-oh-**pal**-ee-ŭm/ (neocortex) A type of nerve tissue that constitutes most of the CEREBRAL CORTEX of mammalian brains. It consists of several layers of nerve cell bodies with a highly complex network of connecting fibers and it is the most advanced type of cerebral cortical tissue in vertebrates.

neoteny /nee-**ot**-ĕ-nee/ The retention of larval or other juvenile features beyond the normal stage in the development of an animal. It may be either temporary, because of climatic or other factors, or permanent, in which case the animal breeds in the larval stage (*see* paedogenesis). Neoteny is thought to be important in the evolution of some groups, including humans, who can be seen to have certain resemblances to young stages of apes.

neotonin /nee-ŏ-**toh**-nin/ *See* juvenile hormone.

Neotropical /nee-ŏ-**trop**-ă-kăl/ One of the six main zoogeographical regions of the earth. It includes South and Central America, the West Indies, and the Mexican lowlands. The characteristic fauna includes sloths, armadillos, anteaters, cavies, vampire bats, llama, alpaca, peccary, rhea, toucan, curassows, and certain species of hummingbirds.

nephridium /ni-**frid**-ee-ŭm/ (*pl.* nephridia) An excretory organ present in many invertebrates (e.g. Platyhelminthes, Rotifera, Mollusca, and Annelida) and *Branchiostoma*. It consists of a single or branched tubule, which forms from an ingrowth of ectoderm and may end blindly in flame cells or open into the coelom by a ciliated funnel. Typically, one pair of nephridia occur per body segment and excretory products diffuse into them for conduction to the exterior. *See also* metanephridium; protonephridium.

nephron /**nef**-ron/ The excretory unit of the vertebrate kidney, comprising a Malpighian body (glomerulus and Bowman's capsule) and a KIDNEY TUBULE. Water, salts, nitrogenous wastes, etc., are filtered across the walls of the glomerulus and collected by the Bowman's capsule. As the filtrate passes through the tubule, useful substances are selectively reabsorbed into surrounding capillaries, which join the renal vein. In mammals, the remaining waste (urine) is conducted via collecting ducts to the renal pelvis and hence to the ureter and on to the bladder.

Nereis /**neer**-ee-is/ A genus of polychaete

worms, the ragworms, found in the intertidal zone, where at low tide they form U-shaped burrows in the mud. *Nereis* can swim and crawl by using the segmentally arranged limblike parapodia. *Nereis* is carnivorous, seizing prey with its toothed pharynx. *See also* Polychaeta.

neritic /ni-**rit**-ik/ The marine environment from low water level to a depth of about 200 m, a zone that in many areas corresponds to the extent of the continental shelf. It makes up less than 1% of the marine environment. Nutrients are relatively abundant in this zone and it is penetrated by sunlight. *Compare* oceanic.

nerve A bundle of nerve fibers surrounded by a protective covering of connective tissue. *Mixed nerves*, such as the spinal nerves, contain both sensory and motor fibers. *See also* neuron.

nerve cell *See* neuron.

nerve cord An enclosed cylindrical tract of nerve fibers that forms a central route for the conduction of nerve impulses within the body. Vertebrates and other chordates have a single hollow nerve cord (the SPINAL CORD) situated dorsally. Invertebrates generally have two or more nerve cords, each lacking a central cavity and with ganglia situated at intervals along its length.

nerve fiber The axon of a NEURON.

nerve impulse The signal transmitted along neurons. All nerve impulses are identical in form and strength and consist of changes in permeability of the axon membrane followed by flows of ions into and out of the cell, thereby producing potential changes that can be detected as the ACTION POTENTIAL passing along the axon. The energy required to pass the impulse is derived from the neuron itself, not from the stimulus causing it.

In its passive state the axon has a resting potential of −70 mV inside the membrane, caused by sodium ions being pumped out of the cell. As the impulse passes, the membrane becomes transiently permeable to sodium ions, which flow into the cell. This causes a change in potential to about +30 mV – the action potential. In myelinated neurons this mechanism operates only at the nodes of Ranvier; the myelin sheath insulates the axon so that the action potential is conducted, more rapidly, from one node to the next (saltatory conduction). There is a refractory period following the passage of an impulse, during which another impulse cannot be transmitted and sodium is pumped back out of the neuron. In living organisms the impulse is triggered by local depolarization at a synapse or a receptor cell, but in isolated axons almost any disturbance of the membrane will set off an impulse. Since the impulse is an all-or-nothing event, the strength of the stimulus is signaled by the frequency and number of identical impulses. *See also* neuron.

nerve net A netlike layer of interconnecting nerve cells that is found in the body wall of certain groups of invertebrate animals; the most primitive type of nervous system. It occurs in cnidarians, echinoderms, and hemichordates.

nervous system A ramifying system of cells, found in all animals except sponges, that forms a communication system between receptors and effectors and allows varying degrees of coordination of information from different receptors and stored memory, producing integrated responses to stimuli. The system consists of neurons, supportive glial cells, and various fibrous tissues surrounding the softer matter. Impulses are transmitted through the neurons, which communicate with each other at specialized junctions, the synapses, which are essentially one-way and are the basis of all integration within the system. The impulse is electrochemical, consisting of a propagated change in the potential on either side of the neuron membrane, the action potential, which travels at between 1 and 120 m/s, depending on the animal and the type of neuron.

At its simplest, as found in the Cnidaria, the nervous system is merely a diffuse net with little concentration of function, but higher animals possess groups of neurons (ganglia), within which integration can take place. The major ganglion develops in the head, as the brain, and becomes increasingly important as a control center in more advanced types. The brain communicates with the body through the spinal cord, which is composed mostly of long axons transmitting impulses to and from the brain but also contains the circuits for body reflexes, and the peripheral nervous system, which contains sensory or motor neurons running from receptors or to effectors. *See also* autonomic nervous system; central nervous system; nerve impulse.

neural arch An arch of bone or cartilage that arises dorsally from the centrum of a vertebra and encloses a canal – the spinal or neural canal – through which runs the spinal cord. It may bear a number of projections, such as a neural spine, for attachment of muscles.

neural crest A crest of specialized cells at the edge of the neural plate in vertebrate embryos that comes to lie above the neural tube when it closes over. The cells of the neural crest migrate separately to form pigment cells, Schwann cells, and gill arches and aggregate to form dorsal root and sympathetic ganglia; the neural crest also contributes most of the head mesenchyme.

neural plate (**medullary plate**) The area of the outer layer of vertebrate embryos that gives rise to the neural tube. The notochordal tissue, which comes to lie beneath this area, produces substances that restrict the fates of the overlying cells to neural elements rather than epidermal.

neural tube The first formed element of the spinal cord and brain of vertebrate embryos. It is usually formed by the neural plate rolling up and sinking beneath the surface. The neural tube usually opens to the outside anteriorly by the *neuropore*; posteriorly it may communicate with the archenteron via the *neurenteric canal* (until the tail is formed). Failure of the neural

tube to close completely gives rise to such congenital abnormalities as spina bifida.

neuroblast /newr-ŏ-blast/ Any of the cells of animal embryos that become or produce nerve cells, frequently by unequal mitotic divisions. Although many of the cells of the neural tube of vertebrates will produce nerve cells, the term neuroblast is usually restricted to those cells whose immediate progeny will differentiate as nerve cells.

neurocranium /newr-ŏ-**kray**-nee-ŭm/ The part of the skull that surrounds and protects the brain and sense organs. In most adult vertebrates it develops from the ossified elements of the CHONDROCRANIUM with overlying membrane bones.

neuroendocrine systems /newr-oh-**en**-dŏ-krin, -krÿn/ The systems involving both nervous and endocrine factors that control functions of the body. Many examples are known, particularly involving the pituitary gland, from which hormones are secreted under direct nervous stimulation. *See* neurosecretion.

neuroglia /newr-**og**-lee-ă/ (**glia**) A specialized tissue, found in the central nervous system of vertebrates, that supports and protects the nerve cells. It consists of various types of cells, including *astrocytes*, which have many fine processes, smaller *oligodendrocytes*, and *ependymal cells*, which may be ciliated and line the brain cavities and spinal canal.

neurohormone /newr-oh-**hor**-mohn/ A hormone that is produced by specialized nervous tissue. Examples are norepinephrine, serotonin, vasopressin, and oxytocin. *See* neurosecretion.

neurohypophysis /newr-oh-hÿ-**pof**-ă-sis, -hi-/ The posterior lobe of the pituitary gland in higher vertebrates. It is derived from a fold in the floor of the brain and stores and releases into the blood the hormones oxytocin and vasopressin. These are manufactured in the hypothalamus by neurosecretory cells that have their endings in the neurohypophysis.

neuromast /**newr**-ŏ-mast/ One of numerous groups of sensory cells that occur in pits or canals, scattered or arranged in rows over the head and along the body (lateral line) of fish and aquatic amphibians. The sensory cells bear hairlike processes, which detect vibrations of a frequency too low to be perceived by the ear. *See also* acoustico-lateralis system.

neuromuscular junction /mewr-ŏ-**muss**-kyŭ-ler/ The specialized region in which a nerve ending makes close contact with a muscle. Impulses arriving at the nerve ending cause it to release a chemical transmitter, which diffuses across the intervening gap and stimulates the muscle to contract. *See also* end plate.

neuron /**newr**-ohn, -on/ (**nerve cell**) A cell that is specialized for the transmission of nervous impulses. It consists of a *cell body*, which contains the nucleus and Nissl granules and has numerous branching extensions (*dendrites*), and a single long fine *axon* (nerve fiber), which has few branches and may be surrounded by a myelin sheath. Dendrites carry nervous impulses towards the cell body and the axon carries them away from the cell body. The end of the axon connects with another neuron at a SYNAPSE or with an effector (e.g. a muscle or gland).

Sensory neurons carry impulses from sense organs to the central nervous system and usually have rounded cell bodies; motor neurons carry impulses from the central nervous system to muscles and usually have star-shaped cell bodies. Interneurons relay impulses between sensory and motor neurons. In more advanced animals the cell bodies are located within the brain, spinal cord, or in ganglia and the fibers collectively form nerves. In primitive animals, such as coelenterates, the neurons form a nerve net. *See also* nerve impulse.

neurosecretion /newr-oh-si-**kree**-shŏn/ The secretion of NEUROHORMONES by specialized nerve cells (*neurosecretory cells*). The axon terminals of these cells, instead of releasing neurotransmitters across synapses, release the neurohormones close to blood vessels, enabling the hormones to enter the circulation. For example, the activity of the anterior pituitary gland (adenohypophysis) is controlled by various releasing hormones produced by neurosecretory cells in the hypothalamus and conveyed via small blood vessels of the hypophyseal portal system to the adenohypophysis.

neurotransmitter /newr-oh-trans-**mit**-er, -tranz-/ A chemical that is released from the endings of nerve cells (neurons) to cause either excitation or inhibition of an adjacent neuron (*see* synapse) or an effector cell, such as a muscle cell or gland. It is stored in minute vesicles near the synapse and released when a nerve impulse arrives. In mammals over 30 compounds can act as neurotransmitters. The main ones are acetylcholine, derived from CHOLINE and found throughout the nervous system; and norepinephrine, occurring in the sympathetic nervous system. Norepinephrine is one of the amine group of neurotransmitters, which also includes serotonin, histamine, dopamine, and epinephrine. Other groups are the amino acid neurotransmitters, such as gamma-aminobutyric acid (GABA), glutamic acid (glutamate), and glycine; and the neuropeptides, such as the enkephalins, endorphins, and vasoactive intestinal peptide (VIP). Particular neurotransmitters can be excitatory or inhibitory, depending on the site in the nervous system. For example, acetylcholine is excitatory at NEUROMUSCULAR JUNCTIONS with skeletal muscle, but has an inhibitory effect on cardiac muscle.

neurula /**newr**-ŭ-lă/ (*pl.* **neurulae**) The stage of vertebrate embryos, following gastrulation, when the neural tube is formed (the process is called *neurulation*).

neuter Describing individuals that lack sex organs. For example, some flowers in the family Compositae lack stamens and pistils.

neutral evolution Evolutionary change resulting from random genetic drift of nonadaptive alleles. The Japanese geneticist

Motoo Kimura (1924–94) proposed in 1968 that many mutations at the molecular level cause amino acid substitutions in proteins that have little or no effect on protein function. They are not subject to natural selection because their inheritance does not affect fitness (reproductive success); i.e., they are selectively neutral. Such neutral mutations form the basis of the MOLECULAR CLOCK concept. It is now widely accepted that much genetic variability is due to neutral evolution, with adaptive evolution driving traits that impinge on fitness.

neutrophil /**new**-trŏ-fil/ A white blood cell (LEUKOCYTE) containing granules that do not stain with either acid or basic dyes. Neutrophils have a many-lobed nucleus and are therefore called *polymorphonuclear leukocytes* or *polymorphs*. Comprising about 70% of all leukocytes in humans and certain other mammals, they engulf and digest foreign particles, such as bacteria, using enzymes from their granules. This is the body's first line of defense against disease. They can pass out of capillaries by an ameboid process (*diapedesis*) and wander in the tissues, gathering in large numbers at the site of an infection, where they may die, forming pus. *Compare* macrophage.

niacin /**nÿ**-ă-sin/ *See* nicotinic acid.

niche /nich, neesh/ *See* ecological niche.

nicotinamide adenine dinucleotide /nik-ŏ-**tin**-ă-/ *See* NAD.

nicotinic acid /nik-ŏ-**tin**-ik/ (**niacin**) One of the water-soluble B-group of vitamins. Its deficiency in man causes pellagra. Nicotinic acid functions as a constituent of two coenzymes, NAD and NADP, which operate as hydrogen and electron transfer agents and play a vital role in metabolism. *See also* vitamin B complex.

nictitating membrane /**nik**-tă-tay-ting/ The third eyelid present in amphibians, reptiles, birds, and some mammals (e.g. rabbits and cats). In addition to the mov-

able upper and lower eyelids, this third eyelid in the inner corner of the eye can be flicked across the eye to wash it. It is a fold of the conjunctiva.

nidation /nÿ-**day**-shŏn/ *See* implantation.

nidicolous /nÿ-**dik**-ŏ-lŭs/ (**altricial**) Describing the condition among certain birds (e.g. starling and pigeon) of being born naked, usually blind, and too weak to support their own weight. A nidicolous hatchling is capable only of gaping for food, which is provided by its parents until it reaches almost adult size. It develops very quickly and matures early. *Compare* nidifugous.

nidifugous /nÿ-**dif**-yŭ-gŭs/ (**precocial**) Describing the condition among birds of being born alert, covered with down, with open eyes and well-developed legs. A nidifugous hatchling is usually able to leave the nest and follow its parents and feed itself soon after birth. Ground-nesting running or swimming birds generally have nidifugous young and the Australian megapodes can fly and lead an independent life on hatching. A nidifugous bird has a long infancy and grows slowly. *Compare* nidicolous.

ninhydrin /nin-**hÿ**-drin/ A reagent used to test for the presence of proteins and amino acids. An aqueous solution turns blue in the presence of alpha amino acids in solution. When dissolved in an organic solvent it is used as a developer to color amino acids on chromatograms. If a chromatogram treated with ninhydrin is heated strongly the amino acids appear as purple spots that can be identified by measuring the R_f value. Ninhydrin is carcinogenic. *See also* R_f value.

Nissl granules /**niss**-ăl/ (*collective term* **Nissl substance**) Densely staining material found in the cell bodies of neurons. They consist of endoplasmic reticulum covered by ribosomes, plus many free ribosomes. The granules are stained by the same basic dyes that stain nuclei. The granules are named for the German neuropathologist Franz Nissl (1860–1919).

nitric oxide /nȳ-trik **oks**-ȳd/ (**nitrogen monoxide**) A colorless gas, NO, that has various physiological roles in a wide range of animals, including humans. It is produced continually by endothelial cells lining blood vessels and causes relaxation of neighboring smooth muscle cells in the vessel wall, thereby widening blood vessels and reducing blood pressure. It is synthesized from the amino acid arginine by the enzyme nitric oxide synthase, and works by triggering formation of the compound cyclic guanosine monophosphate (cGMP), which acts as a second messenger to promote muscle relaxation and inhibit contraction. Nitric oxide also regulates platelet aggregation and inhibits blood clotting. In many cells, especially certain immune-system cells (e.g. macrophages and neutrophils) it acts as a microbicidal agent, producing cytotoxic reactive ions that help to destroy invading microbes.

nitrification /ny-tră-fă-**kay**-shŏn/ The conversion of ammonia to nitrite, and nitrite to nitrate, carried out by certain *nitrifying bacteria* in the soil. The chemosynthetic bacteria *Nitrosomonas* and *Nitrobacter* carry out the first and second stages respectively of this conversion. The process is important in the nitrogen cycle since nitrate is the only form in which nitrogen can be used directly by plants. *Compare* denitrification. *See also* nitrogen cycle.

nitrifying bacteria /nȳ-tră-fȳ-ing/ *See* nitrification.

nitrogen /nȳ-trŏ-jĕn/ An essential element found in all amino acids and therefore in all proteins, and in various other important organic compounds, e.g. nucleic acids. Gaseous nitrogen forms about 80% of the atmosphere but is unavailable in this form except to a few nitrogen-fixing bacteria. Nitrogen is therefore usually incorporated into plants as the nitrate ion, NO_3^-, absorbed in solution from the soil by roots. In animals, the nitrogen compounds urea and

uric acid form the main excretory products. *See also* nitrogen cycle.

nitrogenase /nȳ-**troj**-ĕ-nayss/ (**dinitrogenase**) The enzyme responsible for catalysing the reduction of atmospheric nitrogen (N_2) to ammonia (NH_3) in nitrogen-fixing organisms. It is a large protein complex containing iron and molybdenum atoms, and it requires considerable amounts of energy, in the form of ATP, to accomplish nitrogen reduction. Moreover, it is rapidly inactivated by the presence of oxygen. Therefore, the nitrogenase must either be isolated in a special compartment, as in free-living cyanobacteria, or the free oxygen levels must be kept very low by oxygen-binding proteins, as in the root nodules of legumes. *See* nitrogen fixation.

nitrogen cycle The circulation of nitrogen between organisms and the environment. Atmospheric gaseous nitrogen can only be used directly by certain nitrogen-fixing bacteria (e.g. *Clostridium*, *Nostoc*). They convert nitrogen to ammonia, nitrites, and nitrates, which are released into the soil by excretion and decay. Some are free-living, while others form symbiotic associations with plants (*see* nitrogen fixation). Another method by which atmospheric nitrogen is fixed is by lightning. When plants and animals die, the organic nitrogen they contain is converted back into nitrate in the process termed NITRIFICATION. Apart from uptake by plants, nitrate may also be lost from the soil by DENITRIFICATION and by leaching. The increasing use of nitrogen fertilizers in agriculture and the emission of nitrous oxides in car exhaust fumes are also important factors in the nitrogen cycle.

nitrogen fixation The formation of nitrogenous compounds from atmospheric nitrogen. In nature this may be achieved by electric discharge in the atmosphere or by the activities of certain microorganisms. For example, the symbiotic bacteria, *Rhizobium* species, are associated with leguminous plants, forming the characteristic nodules on their roots. The bacteria contain the NITROGENASE enzyme that cat-

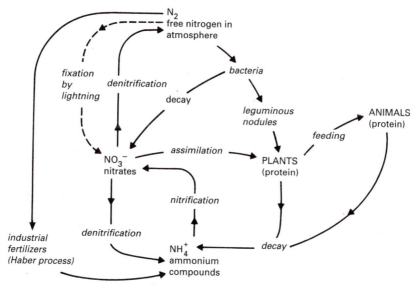

Some of the main stages of the nitrogen cycle

alyzes the fixation of molecular nitrogen to ammonium ions, which the plant can assimilate. In return the legume supplies the bacteria with carbohydrate. Free-living bacteria that can fix nitrogen include members of the genera *Azotobacter* and *Clostridium*. Some sulfur bacteria (e.g. *Chlorobium*), some Cyanobacteria (e.g. *Anabaena*), and some yeast fungi have also been shown to fix nitrogen. In industry the most important method for fixing nitrogen is the Haber process, which is used to make ammonia from nitrogen and hydrogen.

nitrogen monoxide *See* nitric oxide.

NMR *See* nuclear magnetic resonance.

node 1. (*Botany*) The point of leaf insertion on a stem. At the apex of the stem the nodes are very close together but become separated in older regions of the stem by intercalary growth, which forms the internodes. In certain monocotyledons such as those forming bulbs, the nodes are very closely spaced on a condensed stem. Internal stem anatomy is always more complex

at a node due to the presence of leaf gaps and leaf traces.
2. (*Zoology*) A swelling or thickening in an anatomical structure. Examples are the lymph nodes and the sinoatrial node.

node of Ranvier /**ran**-veer, **rahn**-vee-ay/ A region of bare axon that occurs at intervals of up to 2 mm along the length of myelinated nerve axons. The node is named for the French pathologist Louis Antoine Ranvier (1835–1922). *See also* myelin sheath; saltatory conduction.

noncoding strand *See* template strand.

nondisjunction /non-dis-**junk**-shŏn/ The failure of homologous chromosomes to move to separate poles during anaphase I of meiosis, both homologs going to a single pole. This results in two of the four gametes formed at telophase missing a chromosome (i.e. being $n - 1$). If these fuse with normal haploid (n) gametes then the resulting zygote is *monosomic* (i.e. $2n - 1$). The other two gametes formed at telophase have an extra chromosome (i.e. are $n + 1$) and give a *trisomic* zygote (i.e. $2n + 1$) on fusion with a normal gamete. If two gametes deficient for the same chromosome fuse then *nullisomy* (2n − 2) will result, which is almost always lethal, and if two gametes with the same extra chromosomes fuse, *tetrasomy* (2n + 2) results. All these abnormal chromosome conditions are collectively referred to as *aneuploidy*. In humans the condition of Down's syndrome is due to trisomy of chromosome 21.

nonsense mutation /non-sens/ A type of MUTATION in which the base sequence of the DNA is altered so that in the messenger RNA molecule an amino-acid codon is changed into a stop codon. This results in premature termination of translation by the ribosome, and usually a completely nonfunctional protein product.

nontemplate strand /non-tem-playt/ (coding strand; plus strand; sense strand) The strand of a DNA double helix whose base sequence is the same as that of the corresponding RNA molecule produced during TRANSCRIPTION of a gene (except that RNA has uracil instead of thymine). Hence, it mirrors the genetic code carried in the messenger RNA molecule. However, its complementary DNA strand, the TEMPLATE STRAND, is the one actually transcribed by RNA polymerase.

noradrenaline /nor-ă-dren-ă-lin ,-leen/ *See* norepinephrine.

norepinephrine /nor-ep-ă-nef-rin, -reen/ (noradrenaline) A catecholamine, secreted as a hormone by the adrenal medulla, that regulates heart muscle, smooth muscle, and glands. It causes narrowing of arterioles and hence raises blood pressure. It is also secreted by nerve endings of the sympathetic nervous system in which it acts as a neurotransmitter. In the brain, levels of norepinephrine are related to mental function; persistently lowered levels lead to mental depression.

nostrils (nares) Paired openings between the nasal cavity and the exterior in vertebrates. Fish have only *external nostrils* (or *nares*), opening to the exterior, but higher

vetebrates also have *internal nostrils*, opening into the buccal cavity.

notochord /noh-tŏ-kord/ The flexible dorsal supporting rod characteristic of the Chordata; equivalent to the vertebral column in vertebrates. *See* Chordata.

nucellus /new-sell-ŭs/ (*pl.* nucelli) The parenchymatous tissue core of an ovule, enclosing the megaspore or egg cell. The pollen tube gains entry to the nucellus through a gap in the surrounding integuments called the micropyle. In some angiosperm species the nucellus persists as the perisperm providing nourishment for the developing embryo. The nucellus may be considered the megasporangium of angiosperms and gymnosperms. *See* embryo sac. *See illustration at* ovule.

nuclear envelope *See* nucleus.

nuclear magnetic resonance (NMR) A property of atomic nuclei utilized in the analysis of molecular structure. Small changes in the resonance of certain atomic nuclei (e.g. ^{1}H, ^{13}C) can be induced by irradiating them with radio waves in the presence of a strong magnetic field. This property is utilized in a form of spectroscopy called *NMR spectroscopy* to provide information about the composition and structure of complex molecules. It is also exploited in medicine as the basis of *NMR imaging*, which is used to detect tumors, etc. in the body.

nuclear transfer A method used in cloning vertebrates that involves transferring the nucleus from a donor cell into a recipient egg cell. The egg-cell nucleus is removed beforehand, so that subsequent development of the egg cell into an embryo is directed by the donor nucleus. The feasibility of nuclear transfer was established in 1968 by experiments on frog cells, but the first live clones were two sheep born in 1996. The technique, used by a team led by British scientist Ian Wilmut, was to fuse a cell taken from a sheep embryo with an egg cell whose chromosomes had been removed. The resulting fused cell developed into an embryo, which was then implanted into the womb of a surrogate mother to complete its development as a normal pregnancy. Wilmut's team then took tissue cells from the udder of an adult sheep, cultured them in a low-nutrient medium to induce a state of quiescence, and again fused them with recipient egg cells. The outcome was the birth of Dolly, the first mammal to be cloned from an adult body cell. This proved, in spite of a high failure rate, that it is possible to reprogram fully differentiated tissue cells so that they can direct the formation of a new individual.

nuclease /new-klee-ayss/ An enzyme that catalyses the hydrolysis of a nucleic acid into its component nucleotides. *See* endonuclease; exonuclease.

nucleic acid hybridization /new-**klee**-ik/ (**DNA hybridization**) The pairing of a single-stranded DNA or RNA molecule with another such strand, forming a DNA–DNA or RNA–DNA hybrid. In order to achieve hybridization the base sequences of the strands must be complementary. This phenomenon is exploited in many techniques, notably in DNA PROBES, which are designed to bind to particular complementary base sequences among a mass of DNA fragments. Hybridization can also be used in comparative biology to assess the degree of similarity of base sequences between, say, corresponding genes from different organisms. The DNA forms double helices in homologous regions where it is heated and then cooled.

nucleic acids Organic acids whose molecules consist of chains of alternating sugar and phosphate units, with nitrogenous bases attached to the sugar units. They occur in the cells of all organisms. In DNA the sugar is deoxyribose; in RNA it is ribose. *See* DNA; RNA.

nucleoid /new-klee-oid/ The region of a bacterium containing DNA and not enclosed by membranes. It may be associated with the mesosome during cell divisions. *Compare* nucleus.

nucleolar organizer /new-**klee**-ŏ-ler/ *See* nucleolus.

nucleolus /new-**klee**-ŏ-lŭs/ (*pl.* **nucleoli**) A more or less spherical structure found in nuclei of eukaryote cells, and easily visible with a light microscope. One to several per nucleus may occur. Made up of RNA and protein, it is the site of manufacture of ribosomal RNA (rRNA) and is thus most conspicuous in cells making large quantities of protein. It forms around particular loci of one or more chromosomes called *nucleolar organizers*. These loci contain numerous tandem repeats of the genes coding for ribosomal RNA. Nucleoli disappear during cell division.

nucleoplasm /new-klee-oh-plaz-ăm/ The material contained within the NUCLEUS of a cell other than the chromosomes and nucleoli.

nucleoprotein /new-klee-oh-**proh**-teen, -tee-in/ A compound consisting of a protein associated with a nucleic acid. Examples of nucleoproteins are the chromosomes, made up of DNA, some RNA, and histones (proteins), and the ribosomes (ribonucleoproteins), consisting of ribosomal RNA and proteins.

nucleoside /**new**-klee-ŏ-sÿd/ A molecule consisting of a purine or pyrimidine base linked to a sugar, either ribose or deoxyribose. Adenosine, cytidine, guanosine, thymidine, and uridine are common nucleosides.

nucleosome A bead-like particle that is the basic structural unit of chromosomes in their condensed state (*see* chromatin). Each comprises a disk-shaped core of interlocking histone proteins, around which the DNA double helix coils twice, equal to a length of about 146 bp. Consecutive nucleosomes are joined by uncoiled linker DNA of variable length, ranging from some 15 to 55 bp. This 'string of beads' is itself coiled in a spiral some 30 nm in diameter – the solenoid model of condensed chromatin.

nucleotide /new-klee-ŏ-tÿd/ The compound formed by condensation of a nitrogenous base (a purine, pyrimidine, or pyridine) with a sugar (ribose or deoxyribose) and phosphoric acid. The coenzymes NAD and FAD are *dinucleotides* (consisting of two linked nucleotides) while the nucleic acids are *polynucleotides* (consisting of chains of many linked nucleotides).

nucleus /new-klee-ŭs/ (*pl.* **nuclei**) An organelle of eukaryote cells containing the genetic information (DNA) and hence controlling the cell's activities. It is found in virtually all living cells (exceptions include mature sieve tube elements and mature mammalian red blood cells). It is the largest organelle, typically spherical and bounded by a double membrane, the *nuclear envelope* or *nuclear membrane*, which is perforated by many pores (*nuclear pores*) that allow exchange of materials with the cytoplasm. The outer nuclear membrane is an extension of the endoplasmic reticulum. In the nondividing (interphase) nucleus the genetic material is irregularly dispersed as CHROMATIN; during nuclear division (mitosis or meiosis) this condenses into densely staining CHROMOSOMES, and the nuclear envelope disappears as do the nucleoli that are normally present. *Compare* nucleoid. *See* nucleolus; nucleoplasm. *See also* macronucleus.

nullisomy /nŭ-lis-ŏ-mee/ *See* aneuploidy.

numerical taxonomy (**taxometrics**) The assessment of similarities between organisms by mathematical procedures, often involving the use of computers. It involves statistical analysis of some measurable characteristic and uses phenetic rather than phyletic evidence.

nut A dry indehiscent fruit resembling an achene but derived from more than one carpel. It has a hard woody pericarp and characteristically a single seed. Cupules, supporting the nuts, may be distinctive of the species. Examples are the fruits of the beech, hazel, oak, and sweet chestnut. The term 'nut' is frequently misused commercially, e.g. Brazil nuts are seeds and walnuts are drupes. *See* achene.

nutation /new-tay-shŏn/ (**circumnutation**) The spontaneous spiral growth of the shoot tips of certain plants, particularly climbers. The direction of rotation is often constant in a species; for example, *Convolvulus* always rotates in an anticlockwise direction. For climbers nutation increases the likelihood of a solid support being found. It is also marked in many leaf tendrils and to a lesser extent in roots, flower stalks, and the sporangiophores of some fungi. It is an example of autonomic growth.

nyctinasty /nik-tă-mas-tee/ (**nyctinastic movements**) The opening and closing of plant organs, especially flowers and leaves, in response to daily changes in temperature and light. *See* nastic movements.

nymph *See* instar.

Obelia /ŏ-**bee**-lee-ă/ A genus of marine cnidarians found attached to seaweeds in small branching colonies. Each colony consists of a stem bearing feeding polyps and protected by a skeletal covering of chitin. The polyps reproduce asexually, producing either more polyps or free-swimming umbrella-shaped medusae. The medusae produce polyps by sexual reproduction, thus dispersing the species. *Compare* Hydra. *See also* Hydrozoa.

obligate /**ob**-lă-gayt/ Restricted to one particular mode of life or set of conditions. For example, an *obligate anaerobe* is an organism that can live only in the absence of oxygen. *Compare* facultative.

obtect /ob-**tekt**/ Designating the pupae of Lepidoptera in which the wings and legs are fused to the body. *Compare* coarctate; exarate.

occipital condyle /ok-**sip**-ă-tăl/ A rounded prominence on the posterior portion of the tetrapod skull. There are two in mammals; they articulate with the first (atlas) vertebra to allow nodding of the head. There are also two in amphibians, but there is only one occipital condyle in reptiles and birds.

occiput /**ok**-să-pŭt/ **1.** The posterior portion of the vertebrate skull where it joins the vertebral column.
2. A skeletal plate at the back of an insect head.

oceanic /oh-shee-**an**-ik/ The marine environment beyond the continental shelf, which is usually deeper than 200 m. It makes up about 99% of the total marine environment. *Compare* neritic.

ocellus /oh-**sell**-ŭs/ (*pl.* **ocelli**) A simple eye of some invertebrates (e.g. arthropods). It consists of a group of light-sensitive cells (photoreceptors). Ocelli are concerned with perceiving the direction and intensity of light and do not form an image.

oculomotor nerve /ok-yŭ-loh-**moh**-ter/ (**cranial nerve III**) One of a pair of nerves that arise from the ventral midbrain in vertebrates to supply the muscles of the eyeballs. It contains chiefly motor nerve fibers. *See* cranial nerves.

Odonata The order of insects that contains the dragonflies, brightly colored carnivorous insects with a long thin abdomen and two pairs of equal-sized elongated wings. The head bears large compound eyes, reduced antennae, and biting mouthparts and the legs are set well forward for catching the smaller insects that they prey on. The aquatic nymphs are also carnivorous, with an enlarged labrum (the *mask*) for catching prey. Dragonflies have been known from Permian times. Some fossil members had a wingspan of 70 cm.

odontoblast A cell that secretes the DEN-TINE of a tooth. Odontoblasts lie around the outside of the pulp cavity and their long fine processes extend into fine canals (canaliculi) in the dentine. *See also* teeth.

odontoid peg *See* axis.

offset A short runner growing horizontally above ground away from the parent, seen in daisies and houseleeks. It stores no food but turns up at the end and produces a new plant from the apical bud. Like stolons and suckers, offsets are a means of vegetative propagation in angiosperms.

oil immersion A microscopic technique using special high-powered objective lenses. A drop of immersion oil (e.g. cedarwood oil) is placed on the coverslip of a microscope slide and the objective lens carefully lowered into it. The oil has the same refractive index as the lens glass and increases the resolving power obtainable by letting a wider angle of rays enter the objective lens.

olecranon process /oh-**lek**-ră-non, oh-lĕ-**kray**-/ A projection on the upper end of the ulna of tetrapods; it forms the point of the elbow in humans. It is an attachment site for the triceps muscle – the main extensor muscle of the forelimb.

oleic acid /**oh**-lee-ik/ An unsaturated fatty acid occurring as the glyceride in oils and fats. Oleic acid occurs naturally in larger quantities than any other fatty acid. In many organisms oleic acid can be synthesized directly from stearic acid and further enzymatic paths exist for conversion to linoleic acid and linolenic acid. This pathway does not occur in humans and the higher animals so plant sources are an essential dietary element.

oleoplast /**oh**-lee-oh-plast/ *See* elaioplast.

olfactory nerve /ol-**fak**-tŏ-ree, -tree/ (**cranial nerve I**) The nerve that connects the receptor cells of the olfactory organs in the nose with the olfactory bulb at the front of the brain in vertebrates. The cell bodies of its sensory fibers are located not in a ganglion but in the olfactory membrane. *See* cranial nerves.

olfactory organ The organ involved in the detection of smells, which consists of a group of sensory receptors that respond to air- or water-borne chemicals. Vertebrates possess a pair of olfactory organs in the mucous membrane lining the upper part of the nose, which opens to the exterior via the external nares (nostrils). Chemicals from the environment are dissolved in the mucus secreted by the nasal epithelium and information is transmitted to the brain by the receptors via the olfactory nerve. Ol-

factory organs are found on the antennae in insects and in various positions in other invertebrates.

Oligocene /**ol**-ă-goh-seen/ The epoch of the Tertiary period 38–25 million years ago, represented in Britain by deposits around Hampstead in London. It is characterized by the gradual disappearance of earlier mammal groups, including primitive insectivores and primates, and their replacement by more modern forms. *See also* geological time scale.

Oligochaeta /ol-ă-goh-**kee**-tă/ A class of annelid worms containing the terrestrial earthworms (e.g. *Lumbricus*) and many freshwater species. Earthworms are adapted for burrowing and have no parapodia or head appendages and only a few chaetae. They are hermaphrodite but cross-fertilization is usual. The gonads are restricted to a few fixed segments. The eggs develop in cocoons produced by the clitellum, a glandular structure near the reproductive segments. Development is direct, with no larva.

oligotrophic /ol-ă-goh-**troff**-ik/ Describing lakes that are deficient in nutrients and consequently low in productivity. *Compare* eutrophic.

omasum /oh-**may**-sŭm/ (*pl.* **omasa**) The third region of the specialized stomach of ruminants (e.g. the cow). Here surplus water is absorbed. This part of the stomach is very muscular and has a lining of tough cornified stratified squamous epithelium.

ommatidium /om-ă-**tid**-ee-ŭm/ (*pl.* **ommatidia**) One of the units that make up the COMPOUND EYE of arthropod animals. Each ommatidium has a cornea and lens, which focus light onto a sensitive rod (*rhabdome*). A chemical change is brought about in the rhabdome and this stimulates the adjacent retinal cells to send nerve impulses along fibers to the brain. Each ommatidium is surrounded by pigment cells, which isolate it from the others. Thousands of ommatidia together make up the large bulging compound eye, and they give a mosaic image.

omnivore /om-nă-vor, -vohr/ An animal that eats both animal and plant material. Humans are omnivorous. *Compare* carnivore; herbivore.

onchosphere /onk-ŏ-sfeer/ *See* hexacanth.

oncogene /onk-ŏ-jeen/ A gene that is capable of transforming a normal cell into a cancerous cell. Oncogenes are typically mutated forms of normal cellular genes (*proto-oncogenes*). The products of proto-oncogenes are generally involved in some aspect of controlling cell growth and differentiation, such as regulating the cell cycle, or cell signaling and gene expression. The oncogenic mutation can cause the inappropriate expression or over-expression of a normal protein, or give rise to a defective protein (*oncoprotein*) that behaves in a faulty manner. For example, the *ras* oncogene produces a faulty signal transduction protein that cannot 'switch off', and so continuously activates components downstream in its signal pathway. Oncogenes arise from 'gain-of-function' mutations, and are well placed to disrupt the normal activities of cells and cause the changes that lead to cancer.

Oncogenes were originally discovered in RETROVIRUSES. In 1910 the US pathologist Peyton Rous (1879–1970) showed that a cell-free filtrate of a particular tumor (called a sarcoma) found in chickens could induce similar tumors in healthy chickens. The agent causing the tumors was later identified as a virus – now called the *Rous sarcoma virus*. One of the three genes carried by this virus is the oncogene *src* (short for sarcoma). When the virus infects a host cell, the oncogene is inserted into the host's genome. Certain other viruses carry oncogenes that cause cancers in their animal hosts, including the Abelson murine leukemia virus (which affects mice) and the avian erythroblastosis virus (which affects chickens).

oncogenic /onk-ŏ-**jen**-ik/ Causing the production of a tumor, especially a malignant tumor (i.e. cancer). The term is usually applied to viruses known to be implicated in causing cancer; such viruses include some retroviruses, papovaviruses, adenoviruses, and herpesviruses.

one gene–one enzyme hypothesis The theory that each gene controls the synthesis of one enzyme, which was advanced following studies of nutritional mutants of fungi. Thus by regulating the production of enzymes, genes control the biosynthetic reactions catalyzed by enzymes and ultimately the character of the organism. Genes also code for proteins, or polypeptides that form proteins, other than enzymes, so the idea is more accurately expressed as the *one gene–one polypeptide hypothesis*.

ontogeny /on-**toj**-ĕ-nee/ The course of development of an organism from fertilized egg to adult. Occasionally ontogeny is used to describe the development of an individual structure.

Onychophora /on-ă-**koff**-ŏ-ră/ A small invertebrate phylum comprising the velvet worms. They have some annelid and some arthropod features and are often referred to as a 'missing link' between the two groups. Onychophorans (e.g. *Peripatus*) live mainly in tropical forests and have an elongated body covered with a soft cuticle and bearing short unjointed segmentally arranged clawed legs. Cilia, which are absent in arthropods, are present in the reproductive system and the excretory organs are segmentally arranged ciliated coelomoducts. However, in common with arthropods, the body cavity is a hemocoel and respiration is carried out by tracheae.

oocyte /oh-ŏ-sÿt/ A reproductive cell in the ovary of an animal that gives rise to an ovum. The primary oocyte develops from an oogonium, which has undergone a period of multiplication and growth. It divides by meiosis and the first meiotic (or reduction) division produces a secondary oocyte, containing half the number of chromosomes, and a small polar body. The secondary oocyte undergoes the second meiotic division to form an ovum and a second polar body. In many species the second meiotic division is not completed until after fertilization. *See also* oogenesis.

oogamy /oh-**og**-ă-mee/ Sexual reproduction involving the fusion of two dissimilar gametes. The male gamete is usually motile and smaller than the female gamete, which is usually non-motile, contains a food store, and is retained by the parent. The term is generally restricted to descriptions of plants, particularly those that produce female gametes in oogonia. It is an extreme form of anisogamy.

oogenesis /oh-ŏ-**jen**-ĕ-sis/ The formation of ova within the ovary of female animals. Precursor cells in the germinal epithelium multiply by mitosis to form oogonia, even before the animal is born.

In women, there are about 150–500 thousand oogonia present in the ovaries at birth, each surrounded by a small cluster of cells forming a primary follicle. From the onset of sexual maturity until the age of about 45 an oogonium develops into an ovum about once every 28 days in alternate ovaries. (The follicle at this time is called the *Graafian follicle*.) The oogonium grows in size and becomes a primary oocyte, which then undergoes meiosis. The first meiotic division or reduction division results in the formation of a secondary oocyte and a small polar body. The second meiotic division of the secondary oocyte produces an ovum and a second polar body. However, the female gamete may be released at the secondary oocyte stage and the second meiotic division may not be completed until after fertilization.

oogonium /oh-ŏ-**goh**-nee-ŭm/ (*pl.* **oogonia**) **1.** A cell in the ovary of an animal that undergoes a period of multiplication and growth to give rise to an oocyte. *See* oogenesis.
2. The female reproductive organ of certain protoctists and fungi, often distinctly different in shape and size from the male reproductive organ (the antheridium). This unicellular gametangium contains one or more large non-motile haploid eggs called oospheres. These may be liberated prior to fertilization, e.g. *Fucus*, or remain within the oogonium, e.g. *Pythium*.

oomycete /oh-ŏ-**mȳ**-seet/ Any member of a phylum (Oomycota) of fungus-like protoctists that includes the water molds, downy mildews, and white rusts. They are mainly parasites or saprophytes, and feed by extending hyphae into the tissues of their host, releasing digestive enzymes and absorbing nutrients through the hyphal walls. Some are important crop pests, such as *Phytophthora infestans* (potato blight), although most live in fresh water or soil. They can reproduce asexually by means of a sporangium, which develops at the tip of a hypha and from which emerge motile zoospores with two undulipodia (flagella). Each zoospore may germinate to form a new thallus. Sexual reproduction involves male and female structures (antheridia and oogonia), which also develop at the hyphal tips. When these make contact, male nuclei migrate to the oogonium from the antheridium via fertilization tubes. Fertilization of the female egg cells leads to the formation of resistant thick-walled zygotes called oospores. These germinate to produce either a new thallus directly, or zoospores that subsequently give rise to a thallus. The oomyctes were formerly classified as fungi, but are now considered protoctists because they possess undulipodia.

oosphere /oh-ŏ-**sfeer**/ A female gamete or egg cell. In angiosperms it is enclosed in the embryo sac and protected by the integuments of the ovule. In certain oomycetes (e.g. *Pythium*) the oosphere is thin-walled but, after being fertilized, develops a thick resistant wall and becomes an oospore.

oospore /oh-ŏ-spor, -spohr/ A diploid ZYGOTE of certain protoctists, especially oomycetes, produced by fertilization of the female gamete, the oosphere, by the male gamete, the antherozoid. The zygote may form a thick coat and go through a resting period before germination. The term 'oospore' distinguishes a zygote produced oogamously from a zygote produced isogamously or anisogamously and called a zygospore.

OP Osmotic pressure. *See* osmosis.

open reading frame (ORF) **1.** A stretch of DNA running between a start codon and a stop codon that potentially, at least, represents a functional region of a gene. Such stretches are typically hundreds or thousands of bases long. Computerized searches for ORFs in DNA sequence data enable researchers to identify candidate genes in genomic DNA that has been cloned and sequenced.
2. The corresponding sequence in a messenger RNA molecule that is translated into a polypeptide.

operculum /oh-**per**-kyŭ-lŭm/ (*pl.* **opercula**) **1.** The circular lid of the moss capsule, which is characteristic of the class Musci. It covers the peristome and may be forcibly blown off the capsule by pressure developing in the lower portion of the capsule.
2. The bony plate covering the gill chambers of a bony fish.
3. The horny disk that closes the aperture of the shell of some gastropod mollusks.
4. The fold of skin that grows over to enclose the gills during the development of a frog tadpole.

operon /**op**-ĕ-ron/ A genetic unit found in prokaryotes and comprising a group of closely linked genes acting together and coding for the various enzymes of a particular biochemical pathway. At one end is an *operator*, which may under certain conditions be repressed by another gene outside the operon, the *regulator gene*. The regulator gene produces a protein that binds with the operator, renders it inoperative, and so prevents enzyme production. The presence of a suitable substrate prevents this binding, and so enzyme production can commence. Another site in the operon, the PROMOTER, initiates the formation of the messenger RNA that carries the code for the synthesis of the enzymes determined by all the *structural genes* of the operon. *See* repressor molecule.

Ophiuroidea /off-ee-yû-**roi**-dee-ă/ The largest class of the Echinodermata, containing the brittle stars (e.g. *Ophiothrix*). The body is covered by articulating skeletal plates and consists of a small central disk with long fragile arms, which are used in locomotion. Feeding is effected by tube feet, which convey particles of food to the ventral mouth.

opsin /**op**-sin/ The protein component of the retinal pigment rhodopsin, which is localized in the rod cells of the retina. Opsin is released from rhodopsin when light strikes the retina.

opsonin /**op**-sŏ-nin/ Any of various blood proteins that bind to microorganisms or other foreign material and enhance their susceptibility to engulfment by phagocytic cells. The main types of opsonins are immunoglobulin G ANTIBODIES and certain complement proteins.

opsonization /op-sŏ-ni-**zay**-shŏn/ The process by which a host coats the surface of an invading cell with opsonins to render it susceptible to phagocytosis.

optical activity /**op**-tă-kăl/ The property of certain chemical compounds of rotating the plane of plane-polarized light on transmission. Compounds that show optical activity have asymmetric molecules, which can exist in two forms (enantiomers, or optical isomers), identical except that one enantiomer is a mirror image of the other. This property of 'left-' and 'right-handedness' is called *chirality*. One pure enantiomer will rotate plane-polarized light in solution; the other enantiomer rotates it by the same amount in the opposite direction. For example, dextrose and levose are two enantiomeric forms of glucose. In general, enantiomers are designated by the prefixes (+)- or (−)- (formerly *d*- or *l*-) to denote, respectively, dextrorotatory or levorotatory.

optical rotation Rotation of the plane of polarization of plane-polarized light by an optically active substance.

257

D–lactic acid   L–lactic acid

Optical isomers of lactic acid

D–form   L–form   meso–form

Isomers of tartaric acid

Optical isomers

optic chiasma The site in the floor of the vertebrate forebrain at which fibers entering via the optic nerves cross to the opposite side of the brain, forming an X-shaped pattern. In humans and other mammals with good stereoscopic vision, only half the fibers from each side cross over. In lower vertebrates, nearly all the fibers may cross over each other.

optic nerve (cranial nerve II) The tract of sensory nerve fibers that runs from the retina of each eye. It enters the floor of the forebrain via the OPTIC CHIASMA. It is not strictly a true nerve but a connection between two parts of the brain. *See* cranial nerves.

orbit (eye socket) One of two cavities or depressions in the vertebrate skull that contain the eyeballs and their associated muscles, blood vessels, nerves, etc.

order A collection of similar families. Plant orders customarily end in -*ales* (e.g. Liliales) but animal orders do not have any particular ending. Orders may be divided into suborders. Similar orders constitute a class.

Ordovician /or-dŏ-vee-shăn/ The second oldest period of the Paleozoic era, some 510–440 million years ago. It is characterized by an abundance of marine invertebrates (e.g. brachiopods and echinoderms) but an almost total absence of vertebrates apart from some jawless fish. Many of the invertebrates were primitive forms of life that have no living representatives. *See also* geological time scale.

ORF *See* open reading frame.

organ A part of an organism that is made up of a number of different tissues specialized to carry out a particular function. Examples include the lung, stomach, wing, and leaf.

organ culture The maintenance or growth of living organs, usually embryonic, *in vitro*. In animals the organ must be small to enable diffusion of nutrients and excretory products to and from the inner cells and is thus usually embryonic. Mature plant organs, notably roots, may be cultured indefinitely in suitable media. *See also* tissue culture.

organelle /or-gǎ-**nel**/ A discrete subcellular structure with a particular function. The largest organelle is the nucleus; other examples are chloroplasts, mitochondria, Golgi apparatus, vacuoles, and ribosomes. Organelles allow division of labor within the cell. Prokaryotic cells have very few organelles compared with eukaryotic cells.

organizer A part of an embryo whose presence causes neighboring tissue to develop in a particular way. Examples are the eye-cup of vertebrates, which causes lens, and later cornea, to be produced; the gut of snail embryos, which organizes shell gland and mantle; the dorsal lip of the blastopore of frogs, which becomes notochord and organizes all the axial structures of the embryo; and the dermal papilla of a hair, feather, or tooth, which organizes local epidermis and dermis to form the follicle and appendages. The term *primary organizer* is restricted to the first or most important initiator at gastrulation; for example, the dorsal lip of the amphibian blastopore or Hensen's node in mammals and birds. Determinants of major systems (e.g. notochord) are called *secondary organizers*; local centers of developmental activity (e.g. dermal papillae) are *tertiary organizers*.

organ of Corti *See* cochlea.

Oriental One of the six zoogeographical regions of the earth. It includes the southern Asian countries of India, Southeast Asia, and the western Malay archipelago. The characteristic fauna include the Indian elephant, rhinoceros, macaque, gibbon, orang-utan, jungle fowl, and peacock. The boundary between this region and the Australasian region has been the subject of contention in the past. *See* Wallace's line.

origin of life Geological evidence strongly suggests that life originated on earth about 4000 million years ago. The basic components of organic matter – water, methane, ammonia, and related compounds – were abundant in the atmosphere. Energy from the sun (cosmic rays) and lightning storms caused these to recombine into increasingly complex organic molecules. Particular combinations of such complex substances eventually showed the characteristics of living organisms.

ornithine /or-nǎ-th'een/ *See* amino acids.

ornithine cycle (**urea cycle**) The sequence of enzyme-controlled reactions by which urea is formed as a breakdown product of amino acids. It occurs in cells of the liver. The amino acid ornithine is combined with ammonia (from amino acids) and carbon dioxide, forming another amino acid, arginine, which is then split into urea (which is excreted) and ornithine.

ornithology /or-nǎ-**thol**-ǒ-jee/ The branch of vertebrate zoology that is concerned with the study of birds.

orthocampylotropous /or-thoh-kam-pǎ-**lot**-rǒ-pǔs/ *See* campylotropous.

orthogenesis /or-thǒ-**jen**-ě-sis/ An early theory of evolution in which evolutionary change was envisaged as occurring in a definite direction and along a predetermined route, irrespective of natural selection. Orthogenesis therefore conflicts with conventional evolutionary theory such as neo-Darwinism.

orthotropism /or-**thot**-rǒ-piz-ǎm/ *See* tropism.

orthotropous /or-**thot**-rǒ-pǔs/ (**atropous**) Describing the position of the ovule in the ovary, in which the ovule has developed vertically so that the micropylar end is directly over the funicle (stalk) as in *Polygonum*. Compare anatropous; campylotropous. *See illustration at* ovule.

Oryctolagus /or-ik-**tol**-ă-gŭs/ (rabbit) *See* Lagomorpha.

osazones /**oh**-să-zohn/ Distinctly shaped crystals produced by heating monosaccharides with phenylhydrazine hydrochloride and sodium acetate. The osazones are examined microscopically and used for identifying individual monosaccharides; fructose, mannose, and glucose however give identical osazones, but may be distinguished by other tests.

osculum /**os**-kyŭ-lŭm/ (*pl.* oscula) 1. The wide opening at the distal end of a sponge, by means of which water leaves the sponge after it has passed through the chambers and canals inside.
2. Any of the suckers on the head of a tapeworm that help attach it to the gut wall of its host. *See* scolex.

osmiophilic globules /oz-mee-ŏ-**fil**-ik/ (osmiophilic droplets) *See* plastoglobuli.

osmium tetroxide /tet-**roks**-ÿd/ A stain used in electron microscopy because it contains the heavy metal osmium. It also acts as a fixative, i.e. preserves material in a lifelike condition, often being used in conjunction with the fixative glutaraldehyde. It stains lipids, and therefore membranes, particularly intensely.

osmometer /oz-**mom**-ĕ-ter/ An instrument that is used to measure osmotic pressure.

osmoregulation /oz-moh-reg-yŭ-**lay**-shŏn/ The process by which animals regulate their internal osmotic pressure by controlling the amount of water and the concentration of salts in their bodies, thus counteracting the tendency of water to pass in or out by osmosis. In freshwater animals, water tends to enter the body and various methods have been developed to remove the excess, such as the contractile vacuole of protoctists, nephridia and

Malpighian tubules in invertebrates, and kidneys with well-developed glomeruli in freshwater fish. Marine vertebrates prevent excess water loss and excrete excess salts by having kidneys with few glomeruli and short tubules. Terrestrial vertebrates avoid desiccation by having kidneys with long convoluted tubules, which increase the reabsorption of water and salts.

osmosis /oz-**moh**-sis/ The movement of solvent from a dilute solution to a more concentrated solution through a membrane. For example, if a concentrated sugar solution (in water) is separated from a dilute sugar solution by a membrane, water molecules can pass through from the dilute solution to the concentrated one. A membrane of this type (which allows the passage of some kinds of molecule and not others) is called a *semipermeable membrane*. Membranes involved in living systems are not perfectly semipermeable, and are often called *differentially permeable membranes*

Osmosis between two solutions will continue until they have the same concentration. If a certain solution is separated from pure water by a membrane, osmosis also occurs. The pressure necessary to stop this osmosis is called the *osmotic pressure (OP)* of the solution. The more concentrated a solution, the higher its osmotic pressure. Osmosis is a very important feature of both plant and animal biology. Cell walls act as differentially permeable membranes and osmosis can occur into or out of the cell. It is necessary for an animal to have a mechanism of osmoregulation to stop the cells bursting or shrinking. In the case of plants, the cell walls are slightly 'elastic' – the concentration in the cell can be higher than that of the surroundings, and osmosis is prevented by the pressure exerted by the cell walls.

Osmosis is a phenomenon involving diffusion through the membrane; water diffuses from regions of high water concentration to low water concentration. Physiologists now describe the tendency for water to move in and out of cells in terms of WATER POTENTIAL.

osmotic potential /oz-**mot**-ik/ *See* water potential.

osmotic pressure *See* osmosis.

ossification /os-ă-fă-**kay**-shŏn/ (**osteogenesis**) The transformation of embryonic or adult connective tissue (*intramembranous ossification*) or cartilage (*endochondral ossification*) into bone. Bone is produced by the action of special cells (*see* osteoblast), which deposit a network of collagen fibers impregnated with calcium salts; they eventually become enclosed in the bone matrix as bone cells (*see* osteocyte).

Osteichthyes /os-tee-**ik**-th'ee-eez/ The class of vertebrates that contains the bony fishes, characterized by a skeleton of bone and only one external gill opening, which is covered by an operculum. The spiracle is greatly reduced or absent and the body is covered by overlapping scales. Bony fishes are first found as fossils in the Devonian and today are the dominant fishes, invading all types of waters. Primitive forms possessed functional lungs and these are still present in members of the order Dipnoi (lungfish). The Teleostei contains most of the modern bony fishes, in which the lung has become modified to form the swim bladder. *Compare* Chondrichthyes. *See also* coelacanth.

osteoblast /**os**-tee-ŏ-blast/ Any of the cells that form layers of BONE in the early stages of OSSIFICATION. They are at first on the outside of the embryonic cartilage or membrane, but after it has been eroded by OSTEOCLASTS they accompany the ingrowing blood vessels and form temporary trabeculae of bone. Later the osteoblasts lay down the permanent structure of bone, and those that become trapped between the lamellae are called OSTEOCYTES.

osteoclast /**os**-tee-ŏ-klast/ Any of the cells that attack and erode the calcified cartilage or membrane formed in the early stages of ossification of BONE. Blood vessels, preceded by osteoclasts, invade the tissue, and then OSTEOBLASTS lay down the permanent structure of bone.

osteocyte /**os**-tee-ŏ-sÿt/ Any of the cells that secrete the hard matrix of BONE. They are found in small spaces (lacunae) between the concentric lamellae of bone that form the Haversian systems. Each osteocyte has many fine cytoplasmic processes that pass, in fine canaliculae, through the matrix and connect with each other and with blood vessels to maintain supplies of food and oxygen to the living cells. *See also* ossification; osteoblast.

ostium /**os**-tee-ŭm/ (*pl.* **ostia**) A mouthlike opening; for example, any of the lateral openings in the heart of an arthropod or any of the openings through which water enters the body of a sponge.

otic capsule /**oh**-tik, **ot**-ik/ *See* auditory capsule.

otolith /**oh**-tŏ-lith/ One of several granules of calcium carbonate that are contained in a gelatinous mass and attached to hairlike processes of sensory cells within the utriculus and sacculus of the vertebrate inner ear. They respond to changes in the position of the head and so stimulate the sensory cells. *Compare* statocyst. *See also* macula.

outbreeding Breeding between individuals that are not closely related. In plants the term is often used to mean cross-fertilization, and various methods (e.g. stamens maturing before pistils) exist to promote it. In animals behavioral mechanisms often promote outbreeding. The most extreme form – crossing between species – usually results in sterile offspring and there are various mechanisms to discourage it. Outbreeding increases heterozygosity, giving more adaptable and more vigorous populations. *Compare* inbreeding.

outer ear The region of the vertebrate ear that is external to the eardrum (tympanum). It is present in birds, mammals, and some reptiles and consists of a passage, the external auditory meatus, and the pinna (in mammals only).

oval window (**fenestra ovalis**) A membrane that separates the middle and inner ear in tetrapods. It conveys vibrations from the stapes – an ear ossicle of the middle ear – to the liquid (perilymph) of the inner ear. *See illustration at* ear.

ovarian follicle /oh-**vair**-ee-ăn/ *See* Graafian follicle.

ovary 1. (*Botany*) The swollen base of the CARPEL in the gynoecium of plants, containing at least one ovule. The gynoecium of angiosperms may consist of more than one carpel that fuses in certain species forming a complex ovary. After fertilization, the ovary wall becomes the pericarp of the fruit enclosing seeds in its central hollow. *See illustration at* flower.
2. (*Zoology*) The female reproductive organ in animals, which produces egg cells (ova). There is usually a pair of ovaries in vertebrates (in birds, only the left is functional); they also produce sex hormones. In humans, the ovaries are cream-colored oval structures, about 4 cm long, which are attached to the posterior wall of the abdominal cavity, below the kidneys. Each consists of connective tissue, surrounding blood vessels, nerve fibers, etc., and numerous follicles containing immature ova. There are about 150–500 thousand follicles present at birth but only 300–400 ever undergo maturity to become ova. *See also* Graafian follicle; estrogen; oogenesis; ovulation; progesterone.

oviduct /oh-vă-dukt/ A duct in animals that conveys ova from the ovaries to the exterior. It may or may not connect directly with the ovary; in vertebrates, ova are shed into the coelom before entering the open ciliated funnel at its anterior end. *See also* Fallopian tube; Müllerian duct.

oviparity /oh-vă-**pa**-ră-tee/ The production of undeveloped eggs, which are laid or spawned by the female. Fertilization may occur before their release, as in birds and some reptiles, or after, as in most invertebrates, fish, and amphibians. Large numbers of eggs are usually produced because of their poor chances of survival, due to lack of maternal protection. Each egg contains a large yolk store to nourish the developing embryo. *Compare* ovoviviparity; viviparity.

ovipositor /oh-vă-**poz**-ă-ter/ An egg-laying structure at the hind end of the abdomen of female insects, formed from modified paired appendages. It is frequently long and needle-like to enable the piercing of animal and plant tissues to lay eggs. In wasps, bees, and ants it is modified into a sting.

ovotestis /oh-vŏ-**test**-is/ The reproductive organ of certain hermaphrodite animals (e.g. the snail) that functions as both an ovary and a testis.

ovoviviparity /oh-voh-viv-ă-**pa**-ră-tee/ The condition in invertebrates, fish, and reptiles in which eggs are produced and retained within the body of the female during embryonic development. The embryo derives nourishment from the yolk store and so only depends on the mother for physical protection. Ovoviviparity is presumed to be an evolutionary stage leading to VIVIPARITY. *Compare* oviparity.

ovulation /ov-yŭ-**lay**-shŏn/ The release of an egg (ovum) from a Graafian follicle at the surface of a vertebrate ovary. In humans it first occurs at the onset of sexual maturity and a single ovum is released about every 28 days from alternate ovaries until menopause, at the age of about 45. Ovulation actually occurs before the ovum is fully mature; i.e. at the oocyte stage. The process is stimulated by luteinizing hormone (LH) produced by the pituitary gland in the presence of estrogen. *See also* estrous cycle.

ovule /ov-yool/ Part of the female reproductive organs in seed plants. It consists of the nucellus, which contains the embryo sac, surrounded by the integuments. After fertilization the ovule develops into the seed. In angiosperms the ovule is contained within an ovary and may be orientated in different ways being upright, inverted, or sometimes horizontal. In gymnosperms ovules are larger but are not contained within an ovary. Gymnosperm seeds are

micropyle
integuments
embryo sac
nucellus
chalaza
funicle
vascular strand

Orthotropous

Campylotropous

Anatropous

Ovule orientation

true egg. A single ovum is released from the ovary at regular intervals; in humans, about once every 28 days. If fertilized by a spermatozoon it develops into a new individual of the same species. Sometimes fertilization occurs before the ovum is fully developed, i.e. at the oocyte stage. *See also* ovulation.

oxalic acid /oks-**al**-ik/ (**ethanedioic acid**) A dicarboxylic acid, which occurs in rhubarb leaves, wood sorrel and the garden oxalis (hence the name).

oxaloacetic acid /oks-ă-loh-ă-**see**-tik, -**set**-ik/ (**OAA**) A water-soluble carboxylic acid, structurally related to fumaric acid and maleic acid. Oxaloacetic acid forms part of the Krebs cycle, it is produced from L-malate in an NAD-requiring reaction and itself is a step towards the formation of citric acid in a reaction involving pyruvate ion and coenzyme A.

oxidase /oks-ă-dayss/ Any enzyme that catalyzes an OXIDATION–REDUCTION reaction in which electrons are transferred to molecular oxygen (O_2).

oxidation–reduction (**redox**) In a narrow sense, describing any reaction in which one substance loses electrons (i.e. is oxidized) and another substance gains electrons (i.e. is reduced). Because this involves electron transfer, it can only apply to ions. A more general sense, which can also apply to reactions between covalently bonded compounds, is based on the concept of *oxidation number*. The oxidation number is the change in the number of electrons over which an atom has control in a compound compared to the atom in its pure elemental state. Increased control of electrons is indicated by a negative sign, decreased control by a positive sign. For example, if an atom loses control of two electrons, its oxidation number is +2, whereas if it gains control of two electrons the value is −2. Hence, oxidation involves an increase in oxidation number and reduction involves a decrease in oxidation number. In the following reaction: $H_2O + O_2 = 2H_2O$. The oxidation number of the hydrogen in the water is +1

thus naked while angiosperm seeds are contained within a fruit, which develops from the ovary wall.

ovuliferous scale /ov-yŭ-**lif**-ĕ-rŭs/ The megasporophyll found in the axis of the bract scale in the female strobili of the Coniferophyta. It is a large woody structure and bears ovules and later seeds, on its upper surface. The ovuliferous scales are arranged spirally around the central axis and form the bulk of the strobilus.

ovum /oh-vŭm/ (*pl.* **ova**) (**egg cell**) The immotile female reproductive cell (gamete) produced in the ovary of an animal. It consists of a central haploid nucleus surrounded by cytoplasm, containing a variable amount of yolk and a vitelline membrane. Size varies between species; in humans, it is about 0.15 mm in diameter. In chickens it is about 30 mm in diameter and further enlarged by a layer of albumen, more membranes, and a shell to become a and that of the oxygen is −2, so the hydrogen has been oxidized and the oxygen has been reduced. *See also* redox potential.

oxidation–reduction potential *See* redox potential.

oxidative phosphorylation /oks-ă-day-tiv fos-fŏ-ră-**lay**-shŏn/ The production of ATP from phosphate and ADP in aerobic respiration. Oxidative phosphorylation occurs in mitochondria, the energy being provided by steps in the ELECTRON-TRANSPORT CHAIN.

oxidoreductase /oks-ă-doh-ri-**duk**-tayss/ Any of various enzymes that catalyze OXIDATION–REDUCTION reactions. They include DEHYDROGENASES, OXIDASES, and reductases.

oxygen /**oks**-ă-jĕn/ An element essential to living organisms both as a constituent of carbohydrates, fats, proteins, and their derivatives, and in aerobic respiration. It enters plants both as carbon dioxide and water, the oxygen from water being released in gaseous form as a by-product of photosynthesis. Plants are the main if not the only source of gaseous oxygen and as such are essential in maintaining oxygen levels in the air for aerobic organisms.

oxygen debt A physiological state that occurs when a normally aerobic animal is forced to respire anaerobically during a temporary shortage of oxygen (anoxia), e.g. due to violent muscular exertion. Pyruvate, a product of the first stage of cellular respiration, is converted anaerobically to lactic acid, which is toxic and requires oxygen for its breakdown, thereby building up an oxygen debt. The debt is repaid when oxygen is made available and allows oxidation of the lactic acid in the liver. *See also* glycolysis.

oxygen quotient (Q_{O2}) The rate of oxygen consumption of an organism or tissue. It is usually expressed in microliters of oxygen per milligram of dry weight per hour. Small organisms tend to have higher oxygen quotients than larger ones.

oxyhemoglobin /oks-i-hee-mŏ-**gloh**-bin, -hem-ŏ-/ *See* hemoglobin.

oxyntic cell *See* parietal cell.

oxytocin /oks-i-**toh**-sin, -**tos**-in/ A peptide hormone, produced by the hypothalamus and posterior pituitary gland, that acts on smooth muscle. At the end of pregnancy the uterus becomes very sensitive to oxytocin, which promotes labor and also the release of milk from the mammary gland. It is used to induce labor artificially. *Compare* vasopressin.

ozone layer A layer consisting of ozone (O_3) molecules scattered through the stratosphere roughly 15–50 km above the earth's surface. It absorbs about 99% of the harmful ultraviolet radiation entering the earth's atmosphere, and hence provides a shield for living organisms without which life on earth would be impossible. In the early 1980s it was discovered that the ozone layer was becoming thinner, with seasonal 'holes' appearing over Antarctica and elsewhere. This depletion is believed to be caused by increasing atmospheric concentrations of chlorine compounds, especially chlorofluorocarbons (CFCs) – very stable chemicals used as refrigerants and aerosol propellants – which disrupt the delicate balance between ozone production and breakdown in the atmosphere. Such findings led to the introduction in 1987 of restrictions on the use of CFCs and other ozone-depleting substances, with a ban imposed in developed countries by 1996, and in developing countries by 2010. It is now projected that levels of atmospheric chlorine will fall to normal values by about 2050.

P

pacemaker (sinoatrial node; SAN) A small area of specialized cardiac muscle fibers in the wall of the right atrium of the heart in mammals. Its spontaneous rhythmical electrical activity initiates and maintains the contractions of the heart (the heart beat). However, the rate of heart beat is under nervous control. A pacemaker also occurs in the right atrium of birds and reptiles and in the sinus venosus of fish and amphibians.

An *artificial pacemaker* is an electronic device used in certain cases of human heart disease to assume the function of a natural pacemaker.

pachytene /pak-ă-teen/ In meiosis, the stage in mid-prophase I that is characterized by the contraction of paired homologous chromosomes. At this point each chromosome consists of a pair of chromatids and the two associated chromosomes are termed a tetrad. *See* meiosis.

Pacinian corpuscle /pă-sin-ee-ăn/ A type of mechanoreceptor found in the dermis of the skin and in muscles and joints. It consists of a nerve ending encapsulated in numerous concentric layers of connective tissue, resembling the layers of an onion. Pacinian corpuscles respond to changes in pressure, to stretch, and to vibrations, but rapidly adapt to sustained pressure, such as that exerted by clothing on the skin. They are named for the Italian anatomist Filippo Pacini (1812–83).

paedogenesis /pee-doh-jen-ě-sis/ Reproduction in the larval or other juvenile stage by animals exhibiting permanent neoteny. For example, the axolotl – the larva of a salamander – is able to produce offspring similar to itself.

PAGE *See* polyacrylamide gel electrophoresis.

pairing *See* synapsis.

palate /pal-it/ The roof of the mouth; a partition that separates the nasal passage from the buccal cavity in mammals. The anterior portion is supported by bone and is called the *hard palate*. Behind this is the *soft palate* ending in the uvula. The bones forming the hard palate are projections from the premaxillae and maxillae, and posteriorly, the palatines complete the shelf.

In Amphibia, the skin covering the palate is adapted as a respiratory surface.

palea /pal-ee-ă/ (*pl.* **paleae**) *See* bract.

Palearctic /pay-lee-ark-tik, pal-ee-/ One of the six zoogeographical regions, including Europe, the USSR, northern Arabia, and the Mediterranean coastal strip of Africa. The fauna include the hedgehog, wild boar, and fallow and roe deer.

paleobotany /pay-lee-oh-bot-ă-nee, pal-ee-/ *See* paleontology.

Paleocene /pay-lee-ŏ-seen, pal-ee-/ The oldest epoch of the Tertiary, 65–55 million years ago, represented in Britain by marine sand deposits at Thanet. It is characterized by the absence of dinosaurs and the presence of various primitive mammals, now extinct. The first insectivores, rodents, and primates are also found. *See also* geological time scale.

paleoecology /pay-lee-oh-ee-kol-ŏ-jee, pal-ee-/ The investigation of prehistoric ecology as revealed by studying fossils and their artefacts, pollen (palynology) samples and the mineral deposits in which such structures occur.

Paleolithic /pay-lee-ŏ-**lith**-ik, pal-ee-/ The older Stone Age, when stone tools can first be recognized as such, extending from about 2 million years ago to about 10 000 years ago.

paleontology /pay-lee-on-**tol**-ŏ-jee, pal-ee-/ The study of extinct organisms, including their fossil remains, and impressions left by them. Sometimes the subject is divided into *paleobotany*, the study of fossil plants, and *paleozoology*, the study of fossil animals.

Paleozoic /pay-lee-ŏ-**zoh**-ik, pal-ee-/ The first and oldest era in which life became abundant, about 590–230 million years ago. It is divided into six main periods: the Cambrian, Ordovician, Silurian, Devonian, Carboniferous, and Permian. Beginning with aquatic invertebrates and algae, the era ended with the invasion of land by tree ferns and reptiles. *See also* geological time scale.

paleozoology /pay-lee-oh-zoh-**ol**-ŏ-jee, pal-ee-/ *See* paleontology.

palisade mesophyll /pal-ă-**sayd**/ *See* mesophyll.

pallium /**pal**-ee-ŭm/ (*pl.* **pallia**) **1.** *See* mantle.
2. *See* cerebral cortex.

palmella /pal-**mel**-ă/ A stage formed under certain conditions in various unicellular algae in which, after division, the daughter cells remain within the envelope of the parent cell and are thus rendered immobile. The cells may continue dividing giving a multicellular mass, which is contained in a gelatinous matrix. Palmelloid forms may develop undulipodia (flagella) and revert to normal mobile cells at any time. Members of the algal genus *Palmella* typically exist in the palmella condition.

palmitic acid /pal-**mit**-ik/ A saturated fatty acid occurring widely in fats and oils of animal and vegetable origin. *See also* carboxylic acid; oleic acid.

palp *See* mouthparts.

palynology /pal-ă-**nol**-ŏ-jee/ *See* pollen analysis.

pancreas /**pank**-ree-ăs/ A gland lying between the spleen and the duodenum, and having a duct that enters the duodenum. It secretes pancreatic juice which contains various enzymes, including: trypsin for breaking down proteins to amino acids, amylase for converting starch to maltose, and lipase for changing emulsified oils to glycerin and fatty acids. The gland also contains endocrine tissues and produces insulin and glucagon. *See* islets of Langerhans. *See illustration at* alimentary canal.

pangenesis /pan-**jen**-ĕ-sis/ The theory, no longer accepted, introduced by Darwin to explain the inheritance of variation. He postulated that the body fluids carry particles from all over the body to the reproductive cells where they affect the hereditary material and thus the characters inherited by the next generation. Pangenesis was used to explain the erroneous theory of the inheritance of acquired characteristics. *See also* Lamarckism; neo-Darwinism; Weismannism.

panicle /**pan**-ă-kăl/ A compound raceme formed by branching of the peduncle, each branch bearing a raceme, e.g. oat. The term is often applied to any sort of branched racemose inflorescence, for example the horse chestnut, in which each branch is actually a cyme.

pantothenic acid /pan-tŏ-**th'en**-ik/ (vitamin B$_5$) One of the water-soluble B-group of vitamins. Sources of the vitamin include egg yolk, kidney, liver, and yeast. As a constituent of COENZYME A, pantothenic acid is essential for several fundamental reactions in metabolism. A deficiency results in symptoms affecting a wide range of tissues; the overall effects include fatigue, poor motor coordination, and muscle cramps.

paper chromatography A chromatographic method using absorbent paper by which minute amounts of material can be analyzed. A paper strip with a drop of test material at the bottom is dipped into the carrier liquid (solvent) and removed when the solvent front has almost reached the top of the strip. Two-dimensional chromatograms can be produced using square paper and two different solvents. The paper is removed from the first carrier liquid, turned at right angles and dipped into the second. This gives a two dimensional 'map' of the constituents of the test drop. The identity of the constituents may be found by measuring the R_f values.

papilla /pă-**pil**-ă/ (*pl.* **papillae**) 1. (*Botany*) A short sometimes cone-shaped hair often found on petals and giving them their velvety appearance. It is an extension of the outer wall of the epidermal cell.
2. (*Zoology*) A small nipple-like projection from a surface of an animal tissue or organ.

papovavirus /pă-poh-vă-**vÿ**-rŭs/ One of a group of small DNA-containing viruses, about 50 nm in diameter, that cause tumors in animals. The group includes SV40 (simian virus 40), a much-studied virus originally isolated from the African green monkey and belonging to a subgroup called the *polyomaviruses*.

pappus /**pap**-ŭs/ (*pl.* **pappi**) The ring of hairs, scales, or teeth that makes up the calyx in flowers of the Compositae. It persists on the fruit and serves to aid the wind dispersal of seeds. An example is the cypsela of the dandelion in which the pappus remains attached by a long thin stalk and acts as a parachute.

parabiosis /pa-ră-bÿ-**oh**-sis/ The experimental or natural union of two similar animals (each a *parabiont*) so that their blood circulations are continuous. Experimental parabiosis is often performed on insects to study what effects chemicals in one individual have when passed to the other. Natural parabiosis occurs in Siamese twins.

parallel evolution (**parallelism**) The development of similar features in closely related organisms as a result of strong selection in the same direction. There are few examples of this phenomenon. *Compare* convergent evolution.

Paramecium /pa-ră-**mee**-shee-ŭm, -see-ŭm/ A genus of ciliated protoctists belonging to the phylum Ciliophora and common universally in fresh water containing decaying vegetable matter. *Paramecium* is slipper-shaped and covered with cilia, the beating of which produce rapid locomotion. It reproduces asexually by binary fission and sexually by conjugation. There are two contractile vacuoles for osmoregulation and food is taken in through the oral groove and cytopharynx and digested in food vacuoles. There are two nuclei, the meganucleus controlling the vegetative functions and the smaller micronucleus controlling sexual reproduction. *See also* Ciliophora.

paraphyletic /pa-ră-fÿ-**let**-ik/ Describing an incomplete evolutionary group that does not contain all the descendants of a particular ancestor. The best known example is the reptiles, a grouping that excludes the birds and mammals, which also share the same common ancestor. *Compare* monophyletic.

paraphysis /pă-**raf**-ă-sis/ A sterile unbranched multicellular hair found in large numbers between the reproductive organs of certain algae. Club-shaped hyphae in the hymenial layer of certain basidiomycete fungi are also termed paraphyses, and similar structures are also found in the ascocarps of ascomycete fungi.

parapodium /pa-ră-**poh**-dee-ŭm/ (*pl.* **parapodia**) A lobed lateral appendage, pairs of which occur on all segments of a polychaete worm (e.g. *Nereis*) except the first two and the last one. Parapodia act as paddles for swimming and provide a large respiratory surface. The surface of the parapodium is well supplied with blood capillaries to increase the rate of exchange of respiratory gases.

parasexual cycle /pa-ră-**sek**-shû-ăl/ A process involving fusion of nuclei and genetic recombination in the absence of sexual reproduction that occurs in certain fungi, such as wheat rust (*Puccinia graminis*) and *Aspergillus nidulans*. It enables genetic variation and the evolution of new strains without the need for sexual reproduction. The cycle commences with the fusion of genetically different haploid nuclei within a heterokaryon. This results in a diploid nucleus, which subsequently divides by mitosis (not meiosis) alongside the heterokaryotic cells of the parent mycelium. The diploid nuclei undergo mitosis, during which crossing over leads to recombination of parental genes. Also, the diploid cells can give rise to diploid asexual spores (conidiospores) and a wholly diploid mycelium. The haploid state is restored by progressive loss of chromosomes from the diploid nuclei, and the fungus reverts to a heterokaryon, albeit one that contains nuclei with different combinations of the original parental genes.

parasite /**pa**-ră-sÿt/ *See* parasitism.

parasitism /**pa**-ră-sÿ-tiz-ăm/ An association between two organisms in which one, the *parasite*, benefits at the expense of the other, the *host*. The tolerance of the host varies from being almost unaffected to serious illness and often death. An *obligate* parasite can only live in association with a host, whereas a *facultative* parasite can exist in other ways, for example as a saprophyte.

parasitoid /**pa**-ră-sÿ-toid, -si-toid/ A parasite that lives in or on a host for an extended period but eventually causes its death. For example, some insects of the wasp family lay their eggs inside the larvae of other insects. The parasitoid larvae then feed on the tissues of their host as they develop, gradually killing the larval host in the process. This represents a type of feeding that is intermediate between parasitism and predation. Some parasitoid species are introduced deliberately for BIOLOGICAL CONTROL of insect pests.

parasympathetic nervous system /pa-ră-sim-pă-**th'et**-ik/ (**craniosacral nervous system**) One of the two divisions of the AUTONOMIC NERVOUS SYSTEM, which supplies motor nerves to the smooth muscles of the internal organs and to cardiac muscles. Parasympathetic fibers emerge from the central nervous system via cranial nerves, especially the vagus nerve, and a few spinal nerves in the sacral region. Their endings release acetylcholine, which slows heart rate, lowers blood pressure, and promotes digestion, thereby antagonizing the effects of the SYMPATHETIC NERVOUS SYSTEM.

parathyroid glands /pa-ră-**th'ÿ**-roid/ Four small oval-shaped structures embedded in the thyroid gland. The glands are composed of columns of cells with vascular channels between the columns. They produce *parathyroid hormone* and calcitonin, which control the blood calcium level. Secretion of the hormones is controlled by the blood calcium level, through a feedback mechanism. *See* calcitonin.

paratonic movements /pa-ră-**tonn**-ik/ Movements of plants in response to external stimuli. They may be divided into mechanical movements (e.g. the hygroscopic movements of dead cells) and movements caused by the stimulation of sensitive cells (e.g. tropisms and nastic movements). *Compare* autonomic movements.

Parazoa /pa-ră-**zoh**-ă/ A subkingdom of the kingdom Animalia containing the phylum Porifera (sponges). In some recent classifications it also contains the phylum Placozoa, comprising the tiny marine organism *Trichoplax*. *Compare* protozoa; Metazoa.

parenchyma /pă-**renk**-ă-mă, pa-rĕn-**kÿ**-mă/ **1.** (*Botany*) Tissue made up of living thin-walled cells that are not differentiated for any specific function, but in which important metabolic processes are carried out. The leaf mesophyll and the stem medulla and cortex consist of parenchyma. The vascular tissue is also interspersed with parenchyma; for instance the medullary rays of secondary vascular tissue. Aerenchyma, chlorenchyma, collenchyma, and sclerenchyma are all modified forms of parenchyma; storage tissue is also mainly parenchymatous.

2. (*Zoology*) Spongy tissue made of loosely packed cells. In flatworms (Platyhelminthes), such as planarians, parenchyma occurs between the outer skin (ectoderm) and the lining of the gut (endoderm) as a mass of vacuolated and linked mesoderm cells, with fluid in the spaces between the strands of cells. The organs are packed in this tissue and oxygen and food diffuse through it.

parietal cell /pă-rȳ-ĕ-tăl/ (**oxyntic cell**) Any of numerous cells in the stomach wall that secrete hydrochloric acid into the stomach lumen. Parietal cells also secrete INTRINSIC FACTOR, which is required for the absorption of vitamin B_{12} in the small intestine.

parietal eye /pă-rȳ-ĕ-tăl/ A third eye that develops from an anterior outgrowth of the forebrain of early vertebrates.

parthenocarpy /**par**-th'ĕ-noh-kar-pee/ The development of fruit in unfertilized flowers, resulting in seedless fruits. It may occur naturally, as in the banana. It may also be induced artificially by the application of auxin, as in commercial tomato growing.

parthenogenesis /par-th'ĕ-noh-**jen**-ĕ-sis/ Development of unfertilized eggs to form new individuals. It occurs regularly in certain plants (e.g. dandelion) and animals (e.g. aphids). Animals produced by parthenogenesis are always female and, if diploid, look exactly like the parent. *Artificial parthenogenesis* can be induced by pinpricks or treatment with, for instance, cold or acid, especially in eggs shed in water.

Parthenogenesis produces haploid or diploid individuals depending on the genetic state of the ovum when development of the embryo begins. Genetic recombination cannot occur in parthenogenesis and so sexual reproduction occurs occasionally. Aphids show regular alternation of parthenogenesis and heterogamy, whereas

queen bees control parthenogenesis by allowing sperm to fertilize some eggs. *See also* apomixis.

partial dominance *See* co-dominance.

parturition /par-tyû-**rish**-ŏn, -chû-/ The process of giving birth to the fetus at the termination of pregnancy in viviparous animals. Parturition is triggered by hormonal changes initiated by the fetus. In humans, the fetal membranes and endometrium secrete prostaglandins, which stimulate rhythmic contractions of the uterine wall. Stretching of the cervix by the fetus's head triggers the mother's pituitary to secrete oxytocin, which reinforces the uterine muscle contractions of labor, leading to delivery of the fetus.

passage cells Endodermal cells that have Casparian strips but remain otherwise unthickened after deposits of lignin and cellulose have been laid down elsewhere in the endodermis. Passage cells are found opposite the protoxylem and allow transport of water and solutes from the cortex to the stele. They are particularly common in the endodermis of older monocotyledon roots.

passerines /**pass**-ĕ-rin, -rÿn/ (**perching birds**) Birds in which the feet are adapted for grasping branches, having three toes pointing forwards and one pointing backwards. They constitute the largest order (Passeriformes) of birds (class AVES), with over 5000 species, and include the larks, thrushes, warblers, finches, sparrows, and starlings.

pasteurization /pas-chŭ-ri-**zay**-shŏn, -tŭ-/ The partial sterilization of foodstuffs by heating to a temperature below boiling. This kills harmful microorganisms but retains the flavor. It is named after the pioneer of the method, Louis Pasteur, who used it to prevent spoilage of wine and beer. Milk is pasteurized by heating at 65°C for 30 minutes.

patella /pă-**tel**-ă/ (*pl.* **patellae**) The small bone in front of the knee joint between the femur and tibia of the hind limb of most mammals, some birds, and some reptiles; it forms the kneecap in humans. It is a sesamoid bone, having developed in the tendon of the quadriceps femoris muscle.

pathogen /**path**-ŏ-jĕn/ Any organism that is capable of causing disease or a toxic response in another organism. Many bacteria, viruses, fungi, and other microorganisms are pathogenic.

patristic /pă-**tris**-tik/ Describing similarity due to common ancestry. *Compare* homoplastic.

PCR *See* polymerase chain reaction.

peat Partially decomposed plant material that accumulates in waterlogged anaerobic conditions in temperate humid climates, often forming a layer several meters deep. Peat varies from a light spongy material (sphagnum moss) to a dense brown humidified material in the lower layers. If mineral salts are present in the waterlogged vegetation, neutral or alkaline *fen* peat is formed (the salts neutralize the acid produced by decomposition). If there are no mineral salts in the water (as in rain), acid *bog* peat is formed. Peat is used as a fuel and is the first step in coal formation.

peck order *See* dominance hierarchy.

pectic substances /**pek**-tik/ Polysaccharides that, together with hemicelluloses, form the matrix of plant cell walls. They serve to cement the cellulose fibers together. Fruits are a rich source.

They are principally made from the group of sugar acids known as uronic acids. *Pectic acids*, the basis of the other pectic substances, are soluble unbranched chains of α-1,4 linked galacturonic acid units (derived from the sugar galactose). The acid is precipitated as insoluble calcium or magnesium pectate in the middle lamella of plant cells. *Pectinic acids* are slightly modified pectic acids. Under suitable conditions pectinic acids and *pectins* form gels with sugar and acid. Pectins are used commercially as gelling agents, e.g. in jams. Insoluble pectic substances are termed *protopectin* and this is the most important group in normal cell walls. Protopectin is hydrolyzed to soluble pectin by pectinase in ripening fruits, changing the fruit consistency.

pectin /**pek**-tin/ *See* pectic substances.

pectoral fins /**pek**-tŏ-răl/ *See* fins.

pectoral girdle (**shoulder girdle**) A bony or cartilaginous skeletal structure in the anterior region of the vertebrate body, to which the forelimbs or fins are attached. In mammals, it usually consists of two ventral clavicles and two dorsal scapulae. The clavicles are often joined mid-ventrally to the sternum and the scapulae are attached to the vertebral column by muscles, which give mobility to the shoulders. Each scapula bears an articular surface, the glenoid cavity, for the forelimbs. In birds and reptiles the clavicles are functionally replaced by the more prominent coracoids. Lower vertebrates may have both clavicles and coracoids.

The pectoral girdle develops from paired cartilaginous plates in the embryo, which persist in adult cartilaginous fish (Chondrichthyes) as a hoop of cartilage.

pedicel /**ped**-ă-sĕl/ The stalk that attaches individual flowers to the inflorescence axis, the tip of which becomes the receptacle. Some flowers are sessile, having no pedicel.

pedipalps /**ped**-ă-palps/ A pair of appendages on the fourth segment of the head of arachnids. They function in the place of antennae (absent in this group) in spiders and are used for grasping prey in scorpions. The segment nearest to the body bears a hard plate used for chewing food.

peduncle /pi-**dunk**-ăl/ The main axis of an inflorescence on which individual flowers arise in the axils of reduced leaves or bracts. Its shape and size varies according to the type of inflorescence. The stalk of a solitary flower (e.g. tulip) is not considered to be a peduncle.

pelagic /pĕ-**laj**-ik/ Inhabiting the open upper waters rather than the bed of a sea or ocean. Pelagic animals and plants may be divided into the plankton and nekton. Life is found throughout the pelagic zone although the numbers of species and individuals decrease with increasing depth. *Compare* benthic. *See also* photic zone.

Pelecypoda /pel-ĕ-**sip**-ŏ-dă/ (**Bivalvia**) A class of marine and freshwater mollusks, the bivalves, characterized by a laterally compressed body and a shell consisting of two dorsally-hinged valves. Some bivalves are anchored to the substratum by tough filaments (the *byssus*), e.g. *Mytilus* (mussel). Others burrow into sand, e.g. *Cardium* (cockle); rocks, e.g. *Pholas* (piddock); or wood, e.g. *Teredo* (shipworm). Some, e.g. *Pecten* (scallop), swim by clapping the shell valves together. Bivalves have a poorly developed head and large paired gills used for respiration and, in many, for filter feeding.

Pellia /**pel**-ee-ă/ *See* Hepaticae.

pellicle /**pel**-ă-kăl/ A thin flexible transparent outer protective covering of many unicellular organisms, especially undulipodiated protozoan protoctists, e.g. *Euglena* and *Paramecium*. It is made of protein, and maintains the shape of the body. It may thus be regarded as an exoskeleton. In *Paramecium*, the pellicle is perforated by fine pores through which the cilia emerge. Pellicles are also found in many parasitic organisms, e.g. *Monocystis* and *Trypanosoma*.

pelvic fins *See* fins.

pelvic girdle (**hip girdle**) A rigid bony or cartilaginous skeletal structure in the posterior region of the vertebrate body, to which the hindlimbs or fins are attached. In tetrapods, it usually consists of two sides, each consisting of, ventrally, an anterior pubic bone and a posterior ischium and, dorsally, an ilium. They meet at the acetabulum – an articular surface for the hind limb. Each ilium is fused to the sacral vertebrae to permit the transmission of thrust

from the hindlimbs to the body to produce movement, and also to allow the weight of the body to be supported by the girdle. The pelvic girdle develops by ossification in three centers of the cartilaginous stage in the embryo to produce the pubis, ischium, and ilium. The pelvic girdle of fish consists of simple bars of cartilage or bone. *See also* innominate bone; pubis.

pelvis 1. The pelvic girdle.
2. The basin-like cavity produced by the bones of the pelvic girdle, especially in humans and other mammals.
3. (*renal pelvis*) The central cavity of the mammalian kidney into which the urine drains. It forms the expanded upper end of the ureter.

penetrance /**pen**-ĕ-trăns/ The appearance in the phenotype of characteristics that are genetically determined. Certain characters may fail to 'penetrate' even though the individual carries dominant or homozygous recessive genes in their genotype.

penis The male copulatory organ used by many animals with internal fertilization (mammals, some reptiles and invertebrates, etc.) to introduce sperm into the female reproductive tract. In mammals, both urine and semen (the fluid containing spermatozoa) pass, via the urethra, through the penis. The penis consists of a layer of skin and connective tissue surrounding three cylindrical bodies of spongy tissue with numerous blood spaces – two corpora cavernosa dorsally and a single corpus spongiosum ventrally, which surrounds the urethra. Prior to copulation the spaces become filled with blood, making the penis more rigid (i.e. erect).

Pennsylvanian The US name for the late Carboniferous subperiod in the geological time scale.

pentadactyl limb /pen-tă-**dak**-tăl/ A limb having five digits; its bone components have a basic arrangement characteristic of all tetrapod vertebrates. It evolved from modifications to the paired paddle-like fins of the ancestral crossopterygian fish in as-

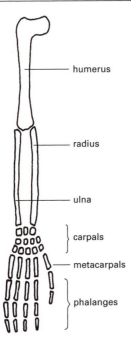

humerus

radius

ulna

carpals

metacarpals

phalanges

The pentadactyl limb

sociation with the transition from water to land. Various alterations have been made to this basic limb type by reduction or fusion of elements as an adaptation for different functions and modes of progression, such as swimming, digging, flying, and running. *See also* digitigrade; plantigrade; unguligrade.

pentose /**pen**-tohs/ A sugar that has five carbon atoms in its molecules. *See* sugar.

pentose phosphate pathway (**hexose monophosphate shunt**) A pathway of glucose breakdown in which pentoses are produced, in addition to reducing power (NADPH) for many synthetic reactions. It is an alternative to glycolysis.

pepo (*pl.* **pepos**) A succulent fruit resembling a berry. Examples of a pepo are cucumber, squash, and pumpkin. The hard outer rind originates from the receptacle of the flower, while the flesh enclosing the seeds is derived from the ovary wall. Typi-

cally the outer coat cannot readily be separated from the ovary wall, unlike a similar fruit, the banana. The pepo is derived from an inferior ovary of three fused carpels.

pepsin /**pep**-sin/ An enzyme that catalyzes the partial hydrolysis of proteins to polypeptides. It is secreted into the stomach by the gastric glands in an inactive form, pepsinogen, and is activated by hydrogen ions. At pH values of 4.6 and less pepsin activates pepsinogen, i.e. it is autocatalytic. Pepsin initiates the digestion of proteins, splitting them into smaller fragments. The extent of this action is proportional to the length of time the protein is in contact with the enzyme.

peptidase /**pep**-tă-dayss/ An enzyme that is responsible for catalyzing the hydrolysis of certain peptide bonds. The peptidases help break down peptides into amino acids.

peptide /**pep**-tÿd/ A type of compound formed of a number of amino acid molecules linked together. Peptides can be regarded as formed by a reaction in which the carbonyl group of one amino acid reacts with the amino group of another amino acid with the elimination of water. This link between amino acids is called a *peptide bond*. According to the number of amino acids linked together they are called di-, tri-, oligo-, or polypeptides. In general, peptides have an amino group at one end of the chain and a carbonyl group at the other. They can be produced by the partial hydrolysis of proteins.

peptidoglycan /**pep**-tÿd-o-**glÿ**-kan/ (**murein**) A substance that forms one or more strong rigid layers in the cell wall of eubacteria. It consists of polysaccharide (glycan) chains linked by short peptides. The glycan typically is formed by alternating sugar units of two main types: N-acetylglucosamine and N-acetylmuramic acid. At intervals the chains are connected by a tetrapeptide consisting of four amino acids joined by peptide bonds. In Gram-negative eubacteria, these amino acids are L-alanine, D-alanine, D-glutamic acid, and either lysine or diaminopimelic acid. Some Gram-positive eubacteria have slightly different amino acids. A similar substance, *pseudopeptidoglycan*, occurs in the cell wall of some archaebacteria; this has N-acetyltalosaminuronic acid instead of N-acetylmuramic acid.

perennation /pe-rĕ-**nay**-shŏn/ A vegetative means of surviving unfavorable seasons, seen in biennial and perennial plants. The metabolic activities are reduced to a minimum, usually by die-back of aerial parts, and food for the next growing season is stored in swollen underground organs. Seeds may also be regarded as perennating organs.

perennial /pĕ-**ren**-ee-ăl/ A plant that may live for several years. Perennials may reproduce in their first growing season or may have to attain a certain age before seed production commences. *Herbaceous perennials* die back each year and survive until the next growing season as tubers (e.g. dahlia), bulbs (e.g. daffodil), rhizomes (e.g. iris), etc. *Woody perennials*, such as trees and shrubs, persist above ground throughout the year but may show adaptations (e.g. leaf fall) to survive unfavorable seasons. *See also* deciduous; evergreen.

perforation plates The remains of the cross walls between the vessel elements in the xylem vessels. The cross wall may have disintegrated completely so that the vessel is effectively one long cylinder (e.g. oak) or parts of the cross wall may remain as bars across the cavity of the vessel (e.g. alder). Most commonly only the centers of the cross walls disappear leaving a distinct rim (e.g. lime).

perianth /**pe**-ree-anth/ The part of the flower that encircles the stamens and carpels. It usually consists of two whorls of leaflike structures which, in dicotyledons, are differentiated into the sepals and petals, i.e. the calyx and corolla. In some flowers, especially those that are wind pollinated, the perianth is reduced (e.g. grasses) or absent (e.g. willow).

periblem /**pe**-ră-blem/ *See* ground meristem; histogen theory.

pericardial cavity /pe-ră-**kar**-dee-ăl/ **1.** In vertebrates, a coelomic space bounded by a membrane (the *pericardium*) and containing the heart. **2.** In some invertebrates (e.g. arthropods), the membrane-bounded cavity enclosing the heart.

pericarp /**pe**-ră-karp/ The ovary wall that becomes the wall of the fruit as the fruit develops. Depending on how the pericarp tissues differentiate, the resulting fruit may be dry or succulent. The outer layer is the *exocarp* (*epicarp*), often a tough skin. The middle layer is the *mesocarp*, whose texture varies with different fruits, e.g. it is juicy in drupes like the plum, and hard in almonds. The mesocarp may be protective or aid in dispersal. The *endocarp* is the innermost layer and forms the stony covering of the seed in a drupe, but in other fruits (e.g. the berry), it is indistinguishable from the mesocarp.

periclinal /pe-ră-**klÿ**-năl/ Describing a line of cell division parallel to the surface of the organ. *Compare* anticlinal.

pericycle /**pe**-ră-sÿ-kăl/ The parenchymatous layer of cells that lies within the endodermis, forming the outermost part of the stele. It remains meristematic in most roots and gives rise to the lateral roots. In most dicotyledonous roots showing secondary growth, the pericycle is also involved in the origination of the vascular cambium and the phellogen. It is not as clear in form or function in the stem.

periderm /**pe**-ră-derm/ The secondary tissue that arises from the activity of the cork cambium.

periderm cambium *See* cork cambium.

perigyny /pĕ-**rij**-ă-nee/ The arrangement in flowers in which the perianth and androecium are inserted around the gynoecium rather than above or below it. The receptacle is extended to a flat or saucer-shaped organ and carries the gynoecium in the middle and the other floral parts around the edge. The OVARY is technically superior even though in extreme forms of perigyny in which the receptacle is cup-shaped (e.g. the wild rose) the floral parts may be inserted at a level above the gynoecium. In perigynous flowers the ovary is not fused with the receptacle. *Compare* epigyny; hypogyny.

perilymph /**pe**-ră-limf/ A fluid that surrounds the structures of the inner ear of vertebrates. *See* labryinth.

perineum /pe-ră-**nee**-ŭm/ (*pl.* **perinea**) The region of the human body between the urethral opening at the front and the anus behind.

periodontal membrane /pe-ree-ŏ-**don**-tăl/ The vascular membrane surrounding the root of a tooth. It is continuous with the periosteum of the jaw bone and the tissue of the gum, filling the narrow space between the tooth and the socket. Interlacing fibers of this membrane pass from the bone to the cement covering the root of the tooth. *See also* alveolus; teeth.

periosteum /pe-ree-**os**-tee-ŭm/ The connective tissue membrane that surrounds a BONE. It is tough and fibrous, with many interlacing bundles of white collagen fibers. It contains OSTEOBLASTS, important in the formation of bone.

peripheral nervous system /pĕ-**rif**-ĕ-răl/ The system of nerves and their ganglia that run from the CENTRAL NERVOUS SYSTEM to the organs and peripheral regions of the body. It constitutes all parts of the nervous system not included in the central nervous system. In vertebrates it comprises the cranial and spinal nerves with their many branches. These convey impulses from sense organs for processing by the central nervous system and transmit the consequent motor impulses to muscles, glands, etc.

periphloic /pe-ră-**floh**-ik/ *See* amphiphloic.

Periplaneta /pe-ră-plan-**ee**-tă/ (cockroach) *See* Dictyoptera.

perisperm /**pe**-ră-sperm/ The nutritive tissue in the seeds of many Caryophyllaceae. It is derived from the nucellus or integuments rather than the embryo sac. Such seeds therefore differ from other angiosperm seeds that have either cotyledons or endosperm tissue acting as the food store.

Perissodactyla /pĕ-riss-ŏ-**dak**-tă-lă/ The order of mammals that contains the odd-toed ungulates, including the horses (with one toe) and the rhinoceroses (with three toes). The middle digit bears the weight of the body. These herbivorous mammals typically have feet encased in a protective horny hoof, lips adapted for plucking, strong cropping incisor teeth, and molars and premolars adapted for chewing. The stomach is simple and bacterial digestion of cellulose occurs in the cecum. *Compare* Artiodactyla.

peristalsis /pe-ră-**stal**-sis/ Waves of muscular contraction that pass along tubular organs of the body, primarily the alimentary canal. It is caused by the sequential contraction of circular muscles in the wall and serves to force the food contents along.

As well as peristaltic waves, localized contractions may occur, which mix the contents together. The rate and force of peristalsis is regulated by autonomic nerves, but the wave itself is an intrinsic property of the muscle tissue.

peristome /**pe**-ră-stohm/ **1.** (*Botany*) A ring of teeth around the opening of the capsule in mosses that is involved in spore dispersal. The teeth twist and bend when subjected to humidity changes, effectively scattering the spores.
2. (*Zoology*) The funnel-like region around the mouth of ciliate protozoan protoctists, such as *Paramecium*, in which food is collected before it is ingested.
3. (*Zoology*) The edge of the opening in a gastropod shell.

perithecium /pe-ră-**th'ee**-shee-ŭm, -see-ŭm/ (*pl.* **perithecia**) *See* ascocarp.

peritoneum /pe-ră-tŏ-**nee**-ŭm/ The lining of the abdomen, continuous with the MESENTERY.

peritrichous /pĕ-**rit**-ră-kŭs/ Describing bacteria that possess flagella all over the cell surface. An example is *Proteus*.

perixylic /pe-ră-**zў**-lik, -**zil**-ik/ *See* amphixylic.

permanent teeth The second set of teeth of most mammals, replacing the deciduous teeth (milk teeth). *See also* diphyodont.

permanent wilting point The point at which soil has dried to the extent that plants can no longer remove the remaining water held on the soil particles, and begin to wilt.

Permian /**per**-mee-ăn/ The most recent period of the Paleozoic era, some 280–250 million years ago. Life became dominated by a few types of reptiles, while amphibians were greatly reduced in number and size. Modern insect groups appeared, and gymnosperm plants largely replaced pteridophytes. *See also* geological time scale.

peroxisome /pĕ-**roks**-ă-sohm/ *See* microbody.

petal One of the usually brightly colored parts of the flower, which together make up the corolla. The petals are pigmented and often scented to attract insects. They are reduced or absent in wind-pollinated flowers. They are thought to be modified leaves, with a much simplified internal structure and vascular system. *See illustration at* flower.

petiole /**pee**-tee-ohl/ The stalk that attaches the leaf blade to the stem. It is similar to the stem except that it is asymmetrical in cross section with the vascular and strengthening tissues arranged in a V shape rather than a circle.

petri dish /**pet**-ree/ A shallow circular glass or plastic container, fitted with a lid, that is used for tissue culture or for growing such microorganisms as bacteria, molds, etc., on nutrient agar or some other medium. It is named for the German bacteriologist J. R. Petri (1852–1921).

pH A measure of the acidity or alkalinity of a solution on a scale 0–14. A neutral solution has a pH of 7. Acid solutions have a pH below 7; alkaline solutions have a pH above 7. The pH is given by $\log_{10} (1/[H^+])$, where $[H^+]$ is the hydrogen ion concentration in moles per liter.

Phaeophyta /fee-**off**-ă-tă/ (**brown algae**) A phylum of protoctists comprising mainly marine algae, notably the macroscopic thallose seaweeds that inhabit the intertidal zones. They contain the pigments chlorophyll *a* and *c*, β carotene, and the xanthophylls, which give the algae their characteristic brown color. Food is stored as mannitol or laminarin and the cell walls contain cellulose or hemicellulose. The Phaeophyta contains nine orders, including the Fucales, or wracks (e.g. *Fucus*), and the Laminariales, or kelps (e.g. *Laminaria*).

phaeophytin /fee-**off**-ă-tin/ A yellow-gray colored pigment of chlorophyll appearing in organic solvent extracts of chlorophyll and often seen during paper chromatography of such extracts.

phage /fayj/ (**bacteriophage**) A virus that infects bacteria. Phages usually have complex capsids composed of a polyhedral head, containing the nucleic acid (DNA or RNA), and a helical tail, through which nucleic acid is injected into the host. After reproduction of the viral nucleic acid the host cell usually undergoes lysis. In genetic engineering, nonviral DNA can be inserted into a phage, which is then used as a cloning vector. *See also* lysogeny; temperate phage; virulent phage.

phagocyte /**fag**-ŏ-sÿt/ A cell that is capable of engulfing particles from its surroundings by a process termed phagocytosis. Examples are the neutrophils and macrophages in vertebrates, which play an important role in protecting the organism against infection. Many other cells are capable of phagocytosis, e.g. intestinal epithelial cells and certain protoctists (*see* protozoa).

phagocytosis /fag-ŏ-sÿ-**toh**-sis/ The process in which a cell's plasma membrane surrounds and engulfs solid particles so that they can pass inside the cell. In a cell (phagocyte) performing phagocytosis, the plasma membrane extends around the particles due to the controlled assembly and disassembly of actin microfilaments of the cytoskeleton, and accompanying changes in viscosity of the cytoplasm (*see* ameboid movement). *Compare* endocytosis; pinocytosis.

phalanges /fă-**lan**-jeez/ (*sing.* **phalanx**) Series of small rod-shaped bones that form the skeleton of the fingers and toes (digits) of tetrapod limbs. In the typical pentadactyl limbs of five digits, there are two phalanges in the first digit and three in each of the others. In some species they may be greatly elongated or reduced. They form hinge joints with each other and with the metacarpals or metatarsals. *See illustration at* pentadactyl limb.

phanerogam /**fan**-ě-rŏ-gam/ In early classifications, any plant that reproduces by seed. The phanerogams are thus equivalent to the spermatophytes of more recent taxonomic systems. *Compare* cryptogam.

phanerophyte /**fan**-ě-rŏ-fÿt/ A perennial plant with persistent shoots and buds well above soil level. *See also* Raunkiaer's plant classification.

Phanerozoic /fan-ě-rŏ-**zoh**-ik/ *See* Precambrian.

pharynx /**fa**-rinks/ (*pl.* **pharynges**) The part of the alimentary canal between the buccal cavity and the esophagus. In mammals, it has openings from the mouth and nasal passage at its anterior end, and to the esophagus and trachea at the posterior end. The Eustachian tubes from the middle ears also open into the pharynx. Masticated food is pushed back by the tongue into the pharynx, which contracts to force it into the esophagus. This contraction also causes the epiglottis to close over the top of the trachea.

In *Branchiostoma*, fish, and amphibian tadpoles, the pharynx is perforated by gill slits, and water, passing out through these, supplies oxygen to the blood in the gill filaments.

In worms, the pharynx is often muscular and aids the ingestion of food.

phellem /**fel**-ĕm/ *See* cork.

phelloderm /**fel**-ŏ-derm/ The inner layer of the periderm.

phellogen /**fel**-ŏ-jĕn/ *See* cork cambium.

phelloid /**fel**-oid/ An unsuberized cork cell.

phenetic /fĕ-**net**-ik/ Describing or relating to the observable similarities and differences between organisms. Phenetic classification systems are based on such characteristics, rather than evolutionary relationships between groups. *Compare* phyletic.

phenocopy /**fee**-noh-kop-ee/ A change in the appearance of an organism caused by the environment, but which is similar in effect to a change caused by gene mutation. Such changes, which are not inherited, are generally caused by environmental factors (e.g. malnutrition, radiation) affecting the organism at an early stage of development.

phenolphthalein /fee-nol-**thal**-een, -nohl-, -fthal-/ An acid-alkali indicator that is colorless in acids and red in alkalis. It has a pH range from 8.4–10.0 and is a frequently used indicator for the detection of pH change in acid-alkali titrations. Phenolphthalein, together with borax, is used to test for saccharide derivatives, e.g. glycerol. Such derivatives turn the solution from red to colorless but on boiling the red color returns.

phenotype /**fee**-nŏ-tӯp/ The observable characteristics of an organism, which are determined by the interaction of the genotype with the environment. Many genes present in the genotype do not show their effects in the phenotype because they are masked by dominant alleles. Genotypically identical organisms may have very different phenotypes in different environments, an effect particularly noticeable in plants grown in various habitats.

phenylalanine /fen-ăl-**al**-ă-neen, fee-năl-/ *See* amino acids.

phenylhydrazine /fen-ăl-**hӯ**-dră-zeen, fee-năl-/ A colorless liquid that reacts with aldehydes and ketones to give phenylhydrazones. These are white solids with definite melting points that can be used to identify the respective aldehydes and ketones. With monosaccharides, phenylhydrazine forms OSAZONES, yellow crystalline compounds, that are distinctive for most monosaccharides.

phenylketonuria /fen-ăl-kee-tŏ-**newr**-ee-ă, fee-năl-/ A genetic disorder resulting in the inability to metabolize phenylalanine to tyrosine and causing severe mental retardation. Phenylketonuria is caused by a defective recessive gene and therefore both parents must be carriers for the child to be affected; it may be diagnosed by the presence of phenylpyruvic acid (a precursor of phenylalanine) in the urine. If detected soon after birth, a diet low in phenylalanine will enable the infant to develop normally.

pheromone /**fe**-rŏ-mohn/ A substance that is excreted by an animal and causes a response in other animals of the same species (e.g. sexual attraction, development). *Compare* kairomone.

phloem /**floh**-em/ Plant vascular tissue in which food is transported from areas where it is made to where it is needed or stored. It consists of sieve tubes, which are columns of living cells with perforated end walls, that allow passage of substances from one cell to the next. The proto- and metaphloem are primary tissues and derived from the procambium while the secondary phloem is formed from the vascular cambium. As well as the sieve element cells there are also companion cells, fibers, and parenchymatous packing tissue in the phloem. *See also* mass flow.

phloroglucinol /flôr-oh-**gloo**-să-nol, -nohl/ *See* staining.

phoresy /**for**-ĕ-see/ A method of dispersal in which an animal clings to a much larger animal to be carried some distance before releasing itself at a new site. For example, some insect PARASITOIDS attach themselves to adult insects of their host species so that they are assured of being present when the host lays its eggs, and thus able to lay their eggs inside the eggs of the host.

phosphagen /**fos**-fă-jen/ Creatine phosphate. *See* creatine.

phosphatase /**foss**-fă-tayss/ An enzyme that catalyzes the removal of a phosphate group from a compound. For example, glucose 6-phosphatase catalyzes the conversion of glucose 6-phosphate to glucose and phosphate, which is the final step in GLUCONEOGENESIS.

phosphatide /**fos**-fă-tӯd, -tid/ A glycerophospholipid. *See* lipid.

phosphatidylcholine /fos-fă-tӯ-dăl-**koh**-leen, -lin/ (**lecithin**) One of a group of phospholipids that contain glycerol, fatty acid, phosphoric acid, and choline and are found widely in higher plants and animals, particularly as a component of cell membranes.

phospholipid /fos-foh-**lip**-id, -**lӯ**-pid/ *See* lipid.

phosphoprotein /fos-foh-**proh**-teen, -tee-in/ A conjugated protein formed by the combination of protein with phosphate groups. Casein is an example.

phosphorescence /fos-fŏ-**ress**-ĕns/ **1.** The absorption of energy by atoms followed by

emission of electromagnetic radiation. Phosphorescence is a type of luminescence, and is distinguished from fluorescence by the fact that the emitted radiation continues for some time after the source of excitation has been removed. In phosphorescence the excited atoms have relatively long lifetimes before they make transitions to lower energy states. However, there is no defined time distinguishing phosphorescence from fluorescence.
2. In general usage the term is applied to the emission of 'cold light' – light produced without a high temperature. The name comes from the fact that white phosphorus glows slightly in the dark as a result of a chemical reaction with oxygen. The light comes from excited atoms produced directly in the reaction – not from the heat produced. It is thus an example of *chemiluminescence*. There are also a number of biochemical examples termed *bioluminescence*; for example, phosphorescence is sometimes seen in the sea from marine organisms, or on rotting wood from certain fungi (known as 'fox fire').

phosphorus /**fos**-fŏ-rŭs/ One of the essential elements in living organisms. In vertebrates, calcium phosphate is the main constituent of the skeleton. Phospholipids are important in cell membrane structure, and phosphates are necessary for the formation of the sugar–phosphate backbone of nucleic acids. Phosphates are also necessary for the formation of high energy bonds in compounds such as ATP. Phosphate compounds are important in providing energy for muscle contraction in vertebrates (creatine phosphate) and invertebrates (arginine phosphate). Phosphorus has many other important roles in living tissues, being a component of certain coenzymes. The phosphate ion, PO_4^{3-}, is an important buffer in cell solutions.

phosphorylation The introduction of a phosphate group (PO_4^{3-}) into a molecule. The phosphorylation of biomolecules, notably proteins and nucleotides, is crucial to many aspects of cellular life, including energy metabolism, uptake and transport of substances, the cell cycle, and cell signaling. High-energy phosphoanhydride bonds are formed by the phosphorylation of certain nucleotides, notably in the formation of ATP from ADP, and provide a means of trapping chemical energy in a usable form. Phosphate groups are transferred from these phosphorylated nucleotides to a host of different proteins by enzymes called PROTEIN KINASES. This protein kinase-catalyzed phosphorylation enables the cell to 'switch on' enzymes and other proteins, for example, by greatly increasing their ability to bind substrates. Conversely, removal of phosphate groups (dephosphorylation), which is catalyzed by phosphorylase enzymes, tends to 'switch off' proteins. *See also* oxidative phosphorylation.

photic zone /**foh**-tik/ The surface layer of an ocean or lake that is penetrated by sunlight and in which the phytoplankton flourish. Red and yellow wavelengths of light penetrate to about 50 m while blue and violet light may reach 200 m. The diatoms, which are the main components of phytoplankton, may be found down to 80 m. Beyond 200 m the water is perpetually dark.

photoautotrophism /foh-toh-aw-tŏ-**troff**-iz-ăm/ *See* autotrophism; phototrophism.

photoheterotrophism /foh-toh-het-ĕ-roh-**troff**-iz-ăm/ *See* heterotrophism; phototrophism.

photolysis /fŏ-**tol**-ă-sis/ Chemical breakdown caused by light. In photosynthesis the process is important in providing hydrogen donors by the splitting of water, as follows:

$$4H_2O \rightarrow 4[H] + 4[OH]$$
$$4[OH] \rightarrow 2H_2O + O_2$$
$$4[H] + CO_2 \rightarrow CH_2O + H_2O$$

photomicrograph /foh-toh-**mỹ**-kroh-graf, -grahf/ *See* micrograph.

photonasty /**foh**-toh-nas-tee/ (**photonastic movements**) A nastic movement in response to change in light intensity. *See* nastic movements.

photoperiodism /foh-toh-**peer**-ee-ŏ-diz-ăm/ The response of an organism to changes in day length (*photoperiod*). In plants, leaf fall and flowering are common responses to seasonal changes in day length, as are migration, reproduction, molting, and winter-coat development in animals. Many animals, particularly birds, breed in response to an increasing spring photoperiod, a long-day response. Some animals (e.g. sheep, goats, and deer) breed in autumn in response to short days so that offspring are born the following spring. The hormone MELATONIN, produced by the pineal gland, is a key regulator of such changes.

Plants are classified as short-day plants (SDPs) (e.g. cocklebur and chrysanthemum) or long-day plants (LDPs) (e.g. cucumber and barley) according to whether they flower in response to short or long days (or, more accurately, long or short nights, respectively). Day-neutral plants (e.g. pea and tomato) have no photoperiodic requirement. The length of the dark period is the critical factor since flowering of SDPs is inhibited by even a brief flash of red light in the dark period, and an artificial cycle of long days and long nights inhibits flowering in LDPs. Thus, it is the interaction between light and dark periods that in some way affects flowering through the mediation of the photosensitive pigment PHYTOCHROME. The P_{FR} form of phytochrome inhibits flowering in SDPs (P_{FR} slowly disappears during long nights) and promotes flowering in LDPs (P_{FR} remains at high levels in short nights). The light stimulus is perceived by the leaves and in some unknown way transmitted to the floral apices.

Changes in daylength control the activities of many invertebrate animals, especially in temperate regions. Precise timing of reproduction and dormancy (diapause) is crucial if these are to synchronize with availability of food and the onset of winter. Insects, for example, often interpret daylength as being either less than or greater than some critical length. Aphids can measure the duration of the dark interval, and only when this is roughly 8 hours long (i.e. representing summer conditions) do they produce wingless females parthenogenetically (characteristic summer forms). Often, though, the photoperiodic responses of insects are modulated by an endogenous BIOLOGICAL CLOCK.

In most high and mid-latitudes, where day length is variable, light is the major synchronizer of the activities described, but temperature, rainfall, and lunar and tidal cycles may reinforce or substitute for the light stimulus in some cases. *See* circadian rhythm; critical day length; thermoperiodism; vernalization.

photophore /foh-toh-for, -fohr/ A light-producing organ found in certain animals, such as fireflies and deep-sea fish. It can contain either the animal's own light-producing cells, or a colony of symbiotic luminescent bacteria. *See* bioluminescence.

photophosphorylation /foh-toh-fos-fŏ-ră-**lay**-shŏn/ (photosynthetic phosphorylation) The conversion of ADP to ATP using light energy. *See* photosynthesis.

photoreceptor /foh-toh-ri-**sep**-ter/ Any light-sensitive organ or organelle. The EYES of vertebrates and the ocelli and compound eyes of insects are photoreceptors, as are the organelles of such protoctists as *Euglena*.

photorespiration /foh-toh-res-pă-**ray**-shŏn/ A light-dependent metabolic process of most green plants that resembles respiration only in that it uses oxygen and produces carbon dioxide. It effectively 'competes' with photosynthesis, and wastes fixed carbon and energy. Photorespiration occurs because of the dual activity of the key enzyme in carbon fixation, RIBULOSE 1,5-BISPHOSPHATE CARBOXYLASE. Besides its carboxylase activity, incorporating carbon dioxide into ribulose bisphosphate (RuBP) as part of the Calvin cycle, this enzyme also has oxygenase activity, splitting RuBP into one molecule of phosphoglycerate and one of phosphoglycolate (a two-carbon compound). Enzymes located in peroxisomes and mitochondria, as well as the chloroplast, constitute a *glycolate cycle*, by which the cell recovers some of

KEY

PSII	photosystem II	PC	plastocyanin
PSII*	excited form of PSII	PSI	photosystem I
PQ	plastoquinone	PSI*	excited form of PSI
Cyt b$_6$–f	cytochrome b$_6$–f complex	Fd	ferredoxin

Light reactions of photosynthesis

the carbon from the excess glycolate by converting it to glycerate. However, some carbon is lost as CO_2. It is estimated that in C_3 plants (*see* C_3 pathway) 40% of the potential yield of photosynthesis is lost through photorespiration. It is therefore economically important and ways of inhibiting the process are sometimes employed. For example, artificially raising the CO_2:O_2 ratio in the air is effective, and CO_2 enrichment of greenhouses is often used for high-value crops such as tomatoes. Yields are increased 30–100%. C_4 plants (*see* C_4 pathway) are more efficient at photosynthesis, and photorespiration is not detectable.

photosynthesis /foh-tŏ-**sin**-th'ĕ-sis/ The conversion of light energy to chemical energy. Various types of living organisms, notably green plants and algae, perform photosynthesis, using chlorophyll pigments to absorb photons of light. This energy is stored in the form of ATP and reduced coenzymes, usually NADPH. Most photosynthetic organisms can manufacture all their organic constituents (carbohydrates, proteins, lipids, etc.) with carbon dioxide as the sole source of carbon. In this process, the carbon dioxide is chemically reduced. ATP supplies the energy, and NADPH the electrons for the reduction. The formation of NADPH depends on reduction of NADP+ by an electron donor. In plants, algae, and certain bacteria (cyanobacteria) this electron donor is water, which is split by light energy to yield electrons, protons, and molecular oxygen. In contrast, photosynthetic purple and green bacteria use various reduced compounds in their environment as electron donors; examples include hydrogen sulfide, elemental sulfur, hydrogen gas, certain organic compounds, and even ferrous iron (Fe^{2+}). These bacteria do not evolve oxygen. Directly or indirectly, photosynthesis is the source of carbon and energy for all except chemoautotrophic organisms.

In green plants, photosynthesis takes place in CHLOROPLASTS, mainly in leaves. The mechanism is complex and involves several stages, sometimes divided into two sets called *light-dependent reactions* and *light-independent reactions* (or 'dark' reactions), although the latter two terms are misnomers. The overall reaction in green plants can be summarized by the equation:

$$CO_2 + 4H_2O \rightarrow [CH_2O] + 3H_2O + O_2$$

Light energy is absorbed by chlorophyll (and other pigments), setting off a chain of chemical reactions in which water is split, electrons are transferred along an ELECTRON-TRANSPORT CHAIN (ultimately to reduce $NADP^+$), oxygen is evolved, and ADP is converted to ATP (in a process known as *photophosphorylation*). The NADPH is used to reduce carbon dioxide to carbohydrates in the subsequent carbon fixation reactions.

The enzymes and other components responsible for the absorption of light are located in the thylakoid membrane of the CHLOROPLAST, where they constitute an electron-transport chain, analogous to that involved in aerobic respiration in mitochondria. The key components of each chain are complexes of chlorophylls, proteins, and carotenoids called photosystems. A chain comprises two such photosystems, designated PSII and PSI, linked by a CYTOCHROME complex. Photons of light are trapped by each photosystem, in clusters of chlorophyll molecules called an *antenna*, and channeled to another cluster of pigment molecules and proteins, the *reaction center*. The process starts at PSII, where electrons enter the electron-transport chain, and water is split into molecular oxygen, electrons, and protons (H^+) – a process called *photolysis*. Electrons are

transferred through the chain to PSI, via a group of cytochrome proteins, called the b_6-f complex, and certain electron-carrier molecules (plastoquinones and plastocyanins). At PSI, absorbed photons generate electrons that reduce ferredoxin, which in turn reduces $NADP^+$ to NADPH. According to the chemiosmotic theory, as electrons flow along the chain they cause protons to be pumped into the thylakoid lumen, creating a chemiosmotic gradient. Protons also enter the lumen due to the photolysis of water by PSII. As these protons diffuse back through the thylakoid membrane into the stroma of the chloroplast they drive ATP production by the F_0F_1 enzyme complex (see ATP SYNTHASE), also located in the thylakoid membrane.

There are two patterns of electron flow. *Non-cyclic electron flow* involves all components of the electron-transport chain, there is photolysis of water yielding oxygen, and both ATP and NADPH are generated. *Cyclic electron flow* involves only PSI, ferredoxin, the b_6-f complex, and plastocyanin. It produces the extra ATP needed for the light-independent reactions, but not NADPH.

Carbon fixation uses ATP and NADPH from the light-dependent reactions to reduce carbon dioxide to carbohydrate. These reactions also require light, and in eukaryotes they take place in the chloroplast stroma. Carbon dioxide is first fixed by combination with the 5-carbon sugar ribulose bisphosphate (RuBP) to form two molecules of phosphoglyceric acid (PGA), a reaction catalyzed by the enzyme RIBULOSE 1,5-BISPHOSPHATE CARBOXYLASE (rubisco). PGA is then reduced to phosphoglyceraldehyde (triose phosphate) using the NADPH and some of the ATP. Some of the triose phosphate and the rest of the ATP is used to regenerate the carbon dioxide acceptor RuBP in a complex cycle involving 3-, 4-, 5-, 6-, and 7-carbon sugar phosphates, usually called the *Calvin cycle*. The rest of the triose phosphate is transported out of the chloroplast into the cytosol, where it can be used to make sucrose, and also amino acids and lipids. *See* C_3 pathway; C_4 pathway; photosynthetic pigments.

photosynthetic bacteria /foh-tŏ-sin-th'et-ik/ Bacteria that can perform PHOTO-SYNTHESIS through possession of various light-absorbing pigments, notably bacteriochlorophylls (which are slightly different from the chlorophylls of plants) and carotenoids. *Anoxygenic* photosynthetic bacteria, including the green sulfur bacteria, purple sulfur bacteria, and purple nonsulfur bacteria, do not use water as an electron donor and thus do not produce oxygen as a product of photosynthesis. In contrast, the CYANOBACTERIA do use water as an electron donor, and are *oxygenic*.

photosynthetic pigments Pigments that absorb the light energy required in photosynthesis. They are located in the chloroplasts of plants and algae, whereas in most photosynthetic bacteria they are located in thylakoid membranes, typically distributed around the cell periphery. All photosynthetic organisms contain chlorophylls and carotenoids; some also contain phycobilins. Chlorophyll *a* is the *primary pigment* since energy absorbed by this is used directly to drive the light reactions of photosynthesis. The other pigments (chlorophylls *b*, *c*, and *d*, and the carotenoids and phycobilins) are *accessory pigments* that pass the energy they absorb on to chlorophyll *a*. They broaden the spectrum of light used in photosynthesis. *See* absorption spectrum.

photosystem /foh-toh-**sis**-tĕm/ *See* photosynthesis.

phototaxis /foh-toh-**taks**-is/ (**phototactic movement**) A TAXIS in response to light. Many motile algae are positively phototactic, e.g. *Volvox*, while cockroaches are examples of negatively phototactic organisms.

phototrophism /foh-toh-**troff**-iz-ăm/ A type of nutrition in which the source of energy for synthesis of organic requirements is light. Most phototrophic organisms are autotrophic (i.e. show *photoautotrophism*); these comprise the green plants, Cyanobacteria, and some photosynthetic bacteria (the purple and green

sulfur-bacteria). A few are heterotrophic (i.e. show *photoheterotrophism*); these are a group of photosynthetic bacteria (e.g. the purple nonsulfur bacteria) and a few algae. *Compare* chemotrophism. *See* autotrophism; heterotrophism. *See also* photosynthesis.

phototropism /foh-**tot**-rŏ-piz-ăm/ (**heliotropism**; **phototropic movement**) A directional growth movement of part of a plant in response to light. The phenomenon is clearly shown by the growth of shoots and coleoptiles towards light (positive phototropism). According to one model, the stimulus is perceived in the region just behind the shoot tip. If light falls on only one side of the apex then auxins produced in the apex tend to diffuse towards the shaded side. Thus more auxin diffuses down the stem from the shaded side of the tip. This results in greater elongation of cells on the shaded side thus causing the stem to bend towards the light source. However, in some cases curvature of the shoot tip can occur without apparent differential transport of auxins; moreover, another group of plant hormones, abscisins, may alo be involved. Most roots are light-insensitive but some (e.g. the adventitious roots of climbers such as ivy) are negatively phototropic. *See also* tropism.

phragmoplast /**frag**-mŏ-plast/ A barrel-shaped body appearing in dividing plant cells during late anaphase and telophase between the two separating groups of chromosomes. It consists of microtubules associated with the spindle, and transports vesicles that coalesce to form the early CELL PLATE.

phycobilins /fÿ-**kob**-ă-linz/ A group of accessory PHOTOSYNTHETIC PIGMENTS found in Cyanobacteria and red algae. Chemically they are linear tetrapyrroles in contrast to chlorophyll, which is a cyclic tetrapyrrole. They absorb light in the middle of the spectrum not absorbed by chlorophyll, an important function in algae living under water where blue and red light are absorbed in the surface layers. They comprise the blue *phycocyanins*, which ab-

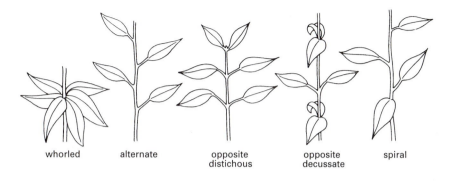

whorled   alternate   opposite distichous   opposite decussate   spiral

plans showing angles of divergence in various leaf arrangements

¹/₂ phyllotaxis
(alternate)

¹/₄ phyllotaxis
(opposite decussate)

²/₅ phyllotaxis
(spiral)

Phyllotaxis

sorb extra orange and red light, and the red *phycoerythrins*, which absorb green light, enabling red algae to grow at depth in the sea. *See also* absorption spectrum.

phycocyanin /fȳ-koh-**sȳ**-ă-nin/ A photosynthetic pigment. *See* phycobilins.

phycoerythrin /fȳ-koh-i-**rith**-rin/ A photosynthetic pigment. *See* phycobilins.

phyletic /fȳ-**let**-ik/ (**phylogenetic**) Relating to or reflecting the evolutionary history of an organism. Some developmental structures or processes, such as gill pouches in mammal embryos, are considered to be phyletic. Phyletic classifications are based on the assumed evolutionary relationships between organisms rather than on any of their observable characteristics. *Compare* phenetic.

phylloclade /**fil**-ŏ-klayd/ A type of CLADODE in which the flattened or globose stem has taken over the photosynthetic function of the leaves, which are reduced to spines or scales. It is often an adaptation to prevent water loss and is seen in certain xerophytic plants, e.g. prickly pear. *Compare* phyllode.

phyllode /**fil**-ohd/ An expanded flattened petiole that acts as the photosynthetic organ if the lamina is missing or very reduced. Phyllodes are seen in the Australian acacia. *Compare* cladode; phylloclade.

phylloquinone /fil-oh-kwă-**nohn**/ *See* vitamin K.

phyllotaxis /fil-ŏ-**taks**-is/ (**phyllotaxy**) The arrangement of leaves on a stem. There may be one, two, or several leaves at each node. When there are three or more leaves forming a circle around the node the arrangement is said to be whorled. When leaves arise singly the arrangement may be spiral or alternate and when they arise in pairs the arrangement is termed opposite. Opposite phyllotaxis may be either distichous or decussate depending on whether the leaf pairs are in the same plane up the stem or arise alternately at right angles to each other.

phylogenetic /fỹ-loh-jĕ-**net**-ik/　*See* phyletic.

phylogeny /fỹ-**loj**-ĕ-nee/ The evolutionary history of a group of organisms.

phylum /**fỹ**-lŭm/ (*pl.* **phyla**) One of the major groups into which a kingdom of organisms is classified. Phyla may be divided into subphyla. In some plant classifications (especially older ones) the term 'division' is used instead of phylum.

physiological saline A solution of sodium chloride and various other salts in which animal tissues are bathed *in vitro* to keep them alive during experiments. It must be isotonic with, and of the same pH as, body fluids. One of the most commonly used is *Ringer's solution* (named for the British physician and physiologist Sydney Ringer (1835–1910)), which contains (in addition to sodium chloride) calcium, magnesium, and potassium chlorides. Other similar solutions may also contain a food supply such as glucose. *See also* tissue culture.

physiological specialization The existence of physiologically distinct but morphologically identical races within a species. Such *physiological races* are important in host-pathogen studies, particularly in planning programs to breed for crop resistance. The cereal rust fungus *Puccinia graminis*, for instance, has over 200 physiological races, which have developed in response to new cereal cultivars as they come on the market.

physiology The way in which organisms or parts of organisms function. *Compare* morphology.

phytoalexin /fỹ-toh-ă-**leks**-in/ A nonspecific antibiotic produced by a plant, usually in response to infection by a fungus or to injury.

phytochrome /**fỹ**-tŏ-krohm/ A proteinaceous pigment found in low concentrations in most plant organs, particularly meristems and dark-grown seedlings. It exists in two interconvertible forms. P_R (or P_{660}) has an absorption peak at 660 nm (red light) and P_{FR} (or P_{730}) at 730 nm (far-red light). Natural white light favors formation of P_{FR}, the physiologically active form. Light intensities required for conversion are very low and it occurs within seconds.

Phytochrome plays a vital role as a photoreceptor in a wide range of light-induced physiological processes: e.g. photoperiodic responses; photomorphogenesis, including leaf expansion, leaf unrolling in grasses and cereals, and greening; and germination of light-sensitive seeds such as lettuce. Phytochrome is thought to act by controlling gene expression in some way. *See* photoperiodism.

phytogeography /fỹ-toh-jee-**og**-ră-fee/ (**plant geography**) The study of the geographical distribution of plant species. Phytogeographical and zoogeographical areas do not necessarily coincide, since barriers and factors affecting growth and distribution are sometimes different for plants and animals. *See also* zoogeography.

phytohormone /fỹ-toh-**hor**-mohn/ *See* plant hormone.

phytoplankton /fỹ-toh-**plank**-tŏn/ *See* plankton.

pia mater /**pỹ**-ă **may**-ter/ The soft delicate innermost membrane that surrounds and protects the brain and spinal cord in vertebrates. *See* meninges.

pico- /**pỹ**-koh/ Symbol: p A prefix denoting one million-millionth, or 10^{-12}. For example, 1 picogram (pg) = 10^{-12} gram. *See* SI units.

picornavirus /pi-kor-nă-**vÿ**-rŭs/ One of a group of small RNA-containing viruses including those responsible for influenza, the common cold, poliomyelitis, and foot-and-mouth disease.

pileus /**pÿ**-lee-ŭs, pil-ee-/ (*pl.* **pilei**) The cap of the mature mushroom (sporophore) in certain basidiomycete fungi (e.g. *Agaricus*).

pilus /**pÿ**-lŭs/ (*pl.* **pili**) A fine straight hair-like protein structure, one or several of which emerge from the walls of certain bacteria (e.g. *E. coli*). They are associated with bacterial CONJUGATION, pulling the conjugating cells together, and are involved in attachment of certain pathogenic bacteria to human tissues. Some viruses specifically adhere to pili. Shorter and more numerous extensions, called *fimbriae* (singular: fimbria), confer the property of 'stickiness' whereby bacteria tend to adhere to one another.

piliferous layer /pÿ-**lif**-ĕ-rŭs/ The region of the root epidermis that gives rise to the root hairs. It is located just behind the zone of elongation and is the main absorptive area of the root.

pineal eye /**pÿ**-nee-ăl/ A structure derived from the PINEAL GLAND and thought to have existed in fossil vertebrates. The only known living example is the tuatara lizard of New Zealand. The eye is situated on the top of the head and seems to be light-sensitive. It has a lens and retina and nerve fibers connected to the brain.

pineal gland A small endocrine gland in the vertebrate brain that detects changes in daylength and secretes the hormone MELATONIN. Sometimes called the 'third eye', it is situated in the midline attached to the roof of the third ventricle, and is shaped like a pine cone, hence its name. During daylight it receives signals from the retina, which inhibit melatonin secretion. These signals are relayed via the hypothalamus and superior cervical ganglia of the sympathetic nervous system. *See also* pineal eye.

pinna /**pin**-ă/ (*pl.* **pinnae**) The external part of the outer ear in mammals, often generally referred to as the 'ear'. It consists of a flap of skin supported by cartilage and surrounding the external opening of the ear, into which it helps deflect sound waves. It is movable in some mammals, e.g. dogs and rabbits.

pinocytosis /pÿ-noh-sÿ-**toh**-sis, pi-/ *See* endocytosis.

Pinus /**pÿ**-nŭs/ *See* Coniferophyta.

Pisces /**pÿ**-seez/ A term sometimes used in classification to include the two classes of fish – OSTEICHTHYES and CHONDRICHTHYES. Fish are poikilothermic aquatic vertebrates with a streamlined body, a powerful muscular finned tail for propulsion, and paired pectoral and pelvic fins for stability and steering. There is usually a body covering of scales. The jaws and pharyngeal gill slits are enlarged to deal with the increased need for oxygen and nutrients brought about by rapid locomotion.

pistil /**piss**-t'l/ In angiosperms, the seed-containing structure. In an apocarpous gynoecium it corresponds to the carpel, while in a syncarpous gynoecium it is made up of two or more carpels.

pit (*Botany*) A gap in the secondary cell wall that enables communication between thickened cells, e.g. tracheids. According to whether or not the secondary wall forms a lip over the pit, pits are described as *bordered* or *simple*, respectively. Usually pits occur in pairs so that the only barrier separating adjacent cells is the middle lamella and the respective primary cell walls. If a pit occurs singly it is termed a *blind pit*.

pith (**medulla**) The central region of the stem and, occasionally, root that is normally composed of parenchymatous tissue. It occurs to the inside of the stele and sometimes contains additional vascular tissue as medullary bundles. *See also* medullated protostele.

middle lamella

primary wall

secondary lignified wall

surface view

longitudinal section

Simple pit

border

torus

Bordered pit

Pit

pitted thickening The most extensive form of thickening found in XYLEM vessels and tracheids, where all the inner wall is thickened apart from small areas called PITS. It is commonly found in secondary xylem and in the last formed metaxylem.

pituitary gland /pi-**too**-ă-tair-ee/ An endocrine gland in the vertebrate brain situated beneath the thalamencephalon behind the optic chiasma. It is regarded as the 'master' endocrine gland because many of its hormones control the secretions of other endocrine glands. The more important pituitary hormones include:

1. Growth hormone (somatotropin), which affects protein metabolism. Excess production leads to gigantism and deficiency results in dwarfism.
2. Vasopressin (antidiuretic hormone), which stimulates reabsorption of water from the kidneys.
3. Adrenocorticotropic hormone (ACTH), which stimulates the secretions of the adrenal gland.
4. Gonadotropic hormones (e.g. follicle-stimulating hormone, luteinizing hormone), which stimulate gonad development.
5. Oxytocin, which stimulates the uterine walls to contract during birth.
6. Prolactin, which stimulates milk production by the mammary glands.
7. Thyrotropin (thyroid-stimulating hormone), which stimulates the secretion of the thyroid glands.

The pituitary gland develops in the embryo from an upgrowth of the stomodeum (the hypophysis, giving rise to the adeno-hypophysis) and a downgrowth from the hypothalamus (the infundibulum, giving rise to the neurohypophysis). In the adult the infundibulum remains connected to the hypothalamus. The neurohypophysis consists of the pars nervosa, the pituitary stalk, and the median eminence (a projection from the stalk into the third ventricle); the pars nervosa secretes oxytocin and vasopressin. The adenohypophysis consists of the pars distalis, pars intermedia, and pars tuberalis; the adenohypophysis secretes the other hormones listed above. Anatomically the pituitary consists of the anterior lobe (pars distalis and pars tuberalis) and posterior lobe (pars intermedia and pars nervosa).

placenta /plă-**sen**-tă/ **1.** (*Zoology*) A disk-shaped organ that develops within the womb (uterus) of a pregnant mammal and establishes a close association between embryonic and maternal tissues for the exchange of materials. It is composed of both embryonic and maternal tissues; embryonic membranes develop numerous finger-like projections (villi) that grow into the highly vascular uterus wall. Into these villi extend embryonic capillaries from the umbilical arteries and vein. This brings the embryonic and maternal circulation into close contact and the fetus is able to obtain oxygen, nutrients, etc., and have waste metabolic products, such as carbon dioxide and nitrogenous compounds, removed. The fetal and maternal blood are never in

direct contact. The placenta is discharged soon after the birth of the young. *See also* afterbirth; umbilical cord.

2. (*Botany*) The region of tissue occurring on the inner surface of the ovary wall of the carpels of flowering plants where the ovules develop. The arrangement of ovules within the ovary (*see* placentation) depends on whether there are one or many carpels and whether the carpels are free or fused.

3. (*Botany*) A central swelling on the abaxial surface of the pinnule of ferns on which clusters of sporangia develop. *See also* sorus.

Placentalia /plas-en-**tay**-lee-ă/   *See* Eutheria.

placentation /plas-en-**tay**-shŏn/   **1.** (*Botany*) The position of the ovule-bearing placentae in angiosperm seeds. Placentation varies according to whether there are one or many carpels. In monocarpellary ovaries, as in pea, placentation is along the ventral suture and is termed marginal. Marginal placentation also occurs in apocarpous polycarpellary ovaries but various types of placentation are seen in syncarpous ovaries. Where the carpels are fused to give a unilocular ovary, the ovules may be found along the two placentae at each line of fusion giving parietal placentation, as in violet. If instead the ovules are borne on a central column, placentation is termed central; a modification of this being free-central placentation in which the column does not extend to the top of the ovary, as in primrose. If each carpel is itself joined before fusing with the other carpels thus giving a multilocular ovary, then the marginal placentae of the carpels fuse in the center of the ovary giving axile placentation, as in tulip. If the walls between the carpels of a multilocular ovary break down to give a unilocular ovary then placentation, although initially axile, appears to be central.

2. (*Zoology*) The degree of union between fetal and maternal tissue in the placenta of mammals. In the early stages of development there are six tissue layers separating fetal and maternal blood. However, de-

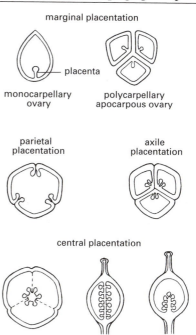

Types of placentation

pending on the species a number of these are later eroded. For example, in humans only the three fetal tissues persist, while in certain rodents (e.g. the rat) only one fetal tissue remains and the fetal capillaries are directly bathed by maternal blood.

placoid scale /**plak**-oid/   *See* denticle.

plagioclimax /play-jee-ŏ-**klÿ**-maks/   A plant community with a stable composition that is in equilibrium under existing environmental conditions, but that has not achieved the natural climax due to the action of biotic factors. For example the continuous pressure of grazing prevents grassland from developing into woodland. *Compare* climax.

plagiogeotropism /play-jee-oh-jee-**ot**-rŏ-piz-ăm/   A plagiotropic response to gravity. *See* tropism.

plagiotropism /play-jee-**ot**-rŏ-piz-ăm/ *See* tropism.

Planaria /plă-**nair**-ee-ă/ A genus of small free-living freshwater platyhelminths (flatworms) of the class Turbellaria. The body is covered with locomotory cilia and the mouth is in the center of the ventral surface at the end of a tubular pharynx, which can be thrust out to seize food and leads into a branched gut. *Planaria* possesses remarkable powers of regeneration – almost any piece of the body can grow into a complete individual – and body organs can be absorbed when food is absent.

plankton A varied collection of aquatic organisms that drift freely, not being attached to any substrate and not possessing any organs for locomotion. The most important components of the plant plankton (*phytoplankton*) are the diatoms upon which the planktonic animals (*zooplankton*) (e.g. crustaceans) feed. The larvae of many species (e.g. cod) make up a large part of the plankton, especially in early summer. The plankton form the basis of the food chain in the sea. *See also* benthic; nekton.

plant An organism that can make its own food by taking in simple inorganic substances and building these into complex molecules by a process termed photosynthesis. This process uses light energy, absorbed by a green pigment called chlorophyll, which is found in all plants but no animals. One major characteristic that distinguishes plants from other plant-like organisms, such as ALGAE or FUNGI, is the possession of an embryo that is retained and nourished by maternal tissue. Fungi and algae lack embryos and develop from spores. Plants are also characterized by having cellulose cell walls, not found in animals, and by the inability to move around freely except for some mobile microscopic plants. Plants also differ from animals by generally responding to stimuli very slowly, the response often taking a matter of days and only occurring if the stimulus is prolonged. The life cycle of plants generally displays ALTERNATION OF GENERATIONS, with both a haploid gametophyte generation and a diploid sporophyte generation. Not all plants make their food by photosynthesis; some are parasites on other plants, and a few are carnivorous, being specialized to capture and digest insects or other small animals.

plant geography *See* phytogeography.

plant hormone (**phytohormone**) One of a group of essential organic substances produced in plants. They are effective in very low concentrations and control growth and development by their interactions. Examples are auxins, gibberellins, cytokinins, abscisic acid, and ethylene.

plantigrade /**plan**-tă-grayd/ Describing the mode of progression in some mammals (e.g. bears, rabbits, and humans) in which the entire sole of the foot, i.e. digits and metatarsals (or metacarpals), is in contact with the ground. *Compare* digitigrade; unguligrade.

planula /**plan**-yŭ-lă/ (*pl.* **planulae**) A small ciliated larva of a cnidarian. After swimming to a suitable site, it settles and develops into a polyp.

plaque /plahk/ A film covering the teeth. It is made up of mucus from saliva with dissolved sugar and other food, and it provides a breeding ground for bacteria. If left on the teeth for any length of time, the bacteria produce acids from the food and these eat into the tooth enamel. This allows bacteria to reach the inner dentine and cause tooth decay.

plasma /**plaz**-mă/ *See* blood plasma.

plasma cell A mature B-lymphocyte (white blood cell) that is programed to secrete just one particular type of ANTIBODY. A single plasma cell can secrete up to 2000 antibody molecules per second. *See also* B-cell; lymphocyte.

plasmagel /**plaz**-mă-jel/ (**ectoplasm**) The gel-like region of cytoplasm located in a thin layer just beneath the plasma membrane of cells that move in an ameboid fashion, such as amebas and macrophages. It consists of a three-dimensional network of actin microfilaments cross-linked by molecules of another protein, filamin, and its semisolid state gives shape to the cell and transmits tension to the substrate. It is thought that the movement of ameboid cells by extension and retraction of cytoplasmic projections (pseudopods) involves reversible changes between plasmagel and the fluid plasmasol in the cell's interior. The gel–sol transition is brought about by dismantling and assembly of the microfilament network through the action of various other proteins, possibly regulated by calcium ion concentration. *See* plasmasol.

plasmagene /**plaz**-mă-jeen/ A gene contained in a self-replicating cytoplasmic particle. Inheritance of the characters controlled by such genes is not Mendelian because appreciable amounts of cytoplasm are passed on only with the female gametes. Mitochondria and plastids contain plasmagenes. *See* cytoplasmic inheritance.

plasmalemma /plaz-mă-**lem**-ă/ *See* plasma membrane.

plasma membrane (cell membrane; plasmalemma) The membrane that surrounds all living cells. *See* membrane.

plasmasol /**plaz**-mă-sol, -sohl/ (endoplasm) The sol-like form of cytoplasm, located inside the plasmagel. It is free-flowing and contains the cell organelles. Ameboid movement involves sol-gel conversions, i.e. the conversion of plasmasol to PLASMAGEL and *vice versa*.

plasmid /**plaz**-mid/ An extrachromosomal genetic element found within bacterial cells that replicates independently of the chromosomal DNA. Plasmids typically consist of circular double-stranded DNA molecules of molecular weight 10^6–10^8. They carry a variety of genes, including those for antibiotic resistance, toxin production, and enzyme formation, and may be advantageous to the cell. Plasmids are widely used as cloning VECTORS in genetic engineering.

plasmin /**plaz**-min/ (fibrinolysin) A proteolytic enzyme that breaks down fibrin in blood clots, restoring the fluidity of the blood. It exists in the blood as an inactive precursor, *plasminogen*, which can be converted to the active form by a variety of factors, including urokinase, trypsin, and leukocyte protease. Plasmin can also lyse other proteins, such as Factor VIII and immunoglobulin.

plasmodesma /plaz-moh-**dez**-mă/ (*pl.* plasmodesmata) A type of plant cell junction that connects the cytosol of adjacent cells, traversing the cell walls. It consists of a channel, typically about 60 nm in diameter, that is lined with plasma membrane and filled with cytosol. Running through the center is an extension of endoplasmic reticulum, called a *desmotubule*. Plasmodesmata permit the passage of various substances, ranging in size from ions, sugars, and amino acids up to proteins, nucleic acids, and even viruses. Soluble molecules pass in the cytosol, whereas membrane-bound molecules can travel via the desmotubule. The permeability is regulated according to developmental, physiological, or environmental factors. The continuity of cytoplasm between cells created by plasmodesmata is called the symplast. *See illustration at* cell.

plasmodium /plaz-**moh**-dee-ŭm/ (*pl.* plasmodia) A multinucleate mass of cytoplasm surrounded by a cell membrane. Such structures are formed during the life cycles of certain SLIME MOLDS

Plasmodium /plaz-**moh**-dee-ŭm/ A genus of parasitic protoctists, some species of which are the cause of malaria in humans. *Plasmodium* is spread by mosquitoes of the genus *Anopheles*. Measures to control the mosquitoes have reduced the incidence of the disease. *Plasmodium* has a complicated life cycle involving asexual reproduction in humans and sexual reproduction in the mosquito. The parasite enters the human bloodstream via the salivary gland of the mosquito (when it bites) and divides asexually, attacking the liver and red blood cells and producing weakness and fever.

The asexual parasite is transmitted to another biting mosquito, in which male and female gametes develop and sexual reproduction occurs. *See also* Apicomplexa.

plasmogamy /plaz-**mog**-ă-mee/ Fusion of protoplasm, usually referring to the fusion of cytoplasm but not nuclei. Plasmogamy in the absence of karyogamy (fusion of nuclei) occurs between fungal mycelia of different strains to form a heterokaryon.

plasmolysis /plaz-**mol**-ă-sis/ Loss of water from a walled cell (e.g. of a plant or bacterium) to the point at which the protoplast shrinks away from the cell wall. The point at which this is about to happen is called *incipient plasmolysis*. Here the cell wall is not being stretched; i.e. the cell has lost its turgidity or become *flaccid* (wall pressure is zero). Wilting of herbaceous plants occurs here. As plasmolysis proceeds parts of the protoplast may remain attached to the cell wall, giving an appearance characteristic of the species. Plasmolysis occurs when a cell is surrounded by a more concentrated solution. More concentrated solutions have lower WATER POTENTIAL and this normally occurs only under experimental conditions.

plastid /**plas**-tid/ An organelle enclosed by two membranes (the envelope) that is found in plants and certain protoctists (e.g. algae), and develops from a proplastid. Various types exist, but all contain DNA and ribosomes. *See* chloroplast; chromoplast; proplastid.

plastocyanin /plas-toh-**sỹ**-ă-nin/ An electron carrier in photosynthesis.

plastogene /**plas**-tŏ-jeen/ A gene present in a self-replicating plastid. *See* plasmagene.

plastoglobuli /plas-tŏ-**glob**-yŭ-lỹ/ (*sing.* **plastoglobulus**) (**osmiophilic globules**) Spherical lipid-rich droplets found in varying numbers inside CHLOROPLASTS. They stain intensely with osmium tetroxide and so appear black and circular with the electron microscope.

plastoquinone /plas-toh-kwă-**nohn**/ A lipid-soluble compound used as an electron carrier in photosynthesis.

plastron /**plas**-tron/ *See* carapace.

platelet /**playt**-lit/ (**thrombocyte**) A tiny particle found in blood plasma. Platelets are 2–3 μm in diameter and there are about 250 000 per cubic millimeter of blood. They are made in red bone marrow, the fragmentation of large cells (called megakaryocytes). When they come into contact with a rough surface, such as a damaged tissue, platelets aggregate to form a plug. They also start the chain of reactions leading to the formation of a blood clot. Platelets release serotonin, which causes constriction of blood vessels, so reducing capillary bleeding; and platelet-derived growth factor, which stimulates tissue cells to grow and repair the wound. *See also* blood clotting.

plate tectonics *See* continental drift.

Platyhelminthes /plat-ă-hel-**min**-theez/ A phylum of primitive wormlike invertebrates, the flatworms, including the classes Turbellaria (aquatic free-living planarians), and the parasitic Trematoda (flukes) and Cestoda (tapeworms). Flatworms are triploblastic bilaterally symmetrical unsegmented animals lacking a coelom and blood system. The flat body provides a large surface area for gaseous exchange. The gut, when present, is often branched and has only one opening (the mouth) and a sucking pharynx. Protonephridia carry out excretion and reproduction is by a complex hermaphrodite system.

plectostele /**plek**-tŏ-steel, -stee-lee/ A form of protostele in which the xylem and phloem exist as alternating bands across the center of the stem, within the pericycle. It is found in *Lycopodium* stems. *See* stele.

pleiomorphism The occurrence of different morphological stages during the life of an organism. Examples are the larval, pupal, and adult forms of an insect, and the different spore forms of the rust fungi. *Compare* polymorphism.

pleiotropism The situation in which one gene is involved in the production of several characters. For example the gene responsible for long petioles in tobacco plants also gives longer calyces, anthers, and capsules.

Pleistocene /plȳs-tŏ-seen/ The first epoch of the Quaternary period, from about two million years ago until the last glaciation ended about 10 000 years ago. The four Ice Ages drove many organisms towards the equator while others (e.g. mammoth) became extinct. Many present-day mammals of South America and Africa resemble pre-Ice Age mammals of Europe. Modern man (*Homo sapiens*) evolved during this period.

plerome /pleer-ohm/ *See* histogen theory; procambial strand.

plesiomorphy /plee-see-ŏ-mor-fee/ A homologous trait that has been inherited from a distant ancestor, and represents a persistent relictual or primitive feature. Such traits are often lost in certain lines of descent, and indicate little about the true evolutionary relationships of the modern descendants of a distant ancestor. For example, the ancestor of cacti was a woody tree with broad flat leaves. Some modern cacti still have ancestral-type leaves, but this plesiomorphic trait provides no useful information about the relationships of modern cacti. *Compare* homoplasy; synapomorphy.

pleura /ploor-ă/ (*pl.* **pleurae**) A double membrane that surrounds the lungs and lines the walls of the thorax in mammals. The narrow space between the two membranes – the pleural cavity – is filled with air and helps cushion the lungs against damage. Cells in the membrane secrete pleural fluid, which lubricates the pleura where they touch, so reducing friction during breathing movements.

pleuron /ploor-on/ The plate, stiffened with chitin, that forms the protective lateral covering on either side of a body segment of an insect. *See also* sternum; tergum.

pleuropneumonia-like organisms /ploor-oh-new-**moh**-nee-ă/ *See* mycoplasmas.

plexus /pleks-ŭs/ (*pl.* **plexuses**) An intricate system of interconnections between nerves, blood vessels, or lymph vessels; for example, the *brachial plexus* of interconnecting spinal-nerve branches supplying the forelimbs of vertebrates.

Pliocene /plȳ-ŏ-seen/ The epoch of the Tertiary period, about 7–2 million years ago, which followed the Miocene. In the Pliocene the hominids, such as *Australopithecus* and *Homo*, became clearly distinguishable from the apes. *See also* geological time scale.

plumule /ploom-yool/ 1. (*Botany*) The shoot apex and first rudimentary leaves in the mature embryo and the seedling. In seedlings showing epigeal germination, the plumule is taken above ground between the cotyledons. When germination is hypogeal, only the plumule emerges from the soil. *Compare* radicle.
2. (**down feather**) (*Zoology*) A small feather in which the barbs are not held firmly together by their barbules. These soft fluffy feathers lie below the contour feathers and provide heat insulation.

plus strand *See* nontemplate strand.

pluteus /ploo-tee-ŭs/ (*pl.* **plutei**) A form of DIPLEURULA larva characteristic of brittle stars and sea urchins, in which the ciliated band is continuous, with a small pre-oral lobe and well-developed post-anal lobes supported by calcareous ribs.

pneumatophore /new-**mat**-ă-for, -fohr/ A specialized negatively geotrophic root produced by certain aquatic vascular plants (e.g. mangrove). The aerial part is covered with pores through which gases can diffuse to and from the highly developed system of intercellular airspaces.

pod *See* legume.

podsol /**pod**-sol, -zol/ The type of soil found under heathland and coniferous forests in temperate climates. It is strongly acid and often deficient in nutrients as a result of leaching.

poikilothermy /poi-kă-loh-**ther**-mee/ The condition of having a body temperature that varies approximately with that of the environment. Most animals other than birds and mammals are poikilothermic ('cold-blooded'). *Compare* ectotherm; homoiothermy. *See also* heterotherm.

point mutation A MUTATION that involves a single base pair in the DNA. There are two main types: *base substitutions*, in which a pair of nucleotides is replaced by another with different bases; and *base additions* or *base deletions*, in which an extra nucleotide pair is inserted, or a pair of nucleotides is deleted. *See* frameshift; missense mutation; nonsense mutation.

polar body (**polocyte**) A minute cell produced during formation of an ovum when the oocyte undergoes two meiotic divisions.

polarized light Light in which the electric and magnetic fields are restricted to single planes. Light is a transverse wave motion; it is composed of electric and magnetic fields vibrating at right angles to the direction of propagation. In 'normal' light the fields vibrate in all directions perpendicular to the propagation direction. Polarized light is produced, for example, by reflection or passage through Polaroid.

polar nuclei The two nuclei found midway along the embryo sac. They may fuse to form the diploid *definitive nucleus*. The endosperm is formed from the fusion of one or both polar nuclei with one of the male gametes from the pollen tube.

pollen /**pol**-ĕn/ The microspores of seed plants, produced in large numbers in the pollen sacs. Pollen grains are adapted according to the method of pollination, those carried by insects often being sticky or barbed while wind pollinated plants generally produce smooth light pollen. Each grain contains male gametes that represent the highly reduced male gametophyte generation. *See also* pollen analysis.

pollen analysis (**palynology**) A means of obtaining information on the composition and extent of past floras by examining the remains of pollen grains in peat and sedimentary deposits. The outer wall (exine) of the pollen grain is very resistant to decay, and reliable quantitative information on the vegetative cover many thousands of years ago can be made. The size and shape of pollen and the patterns on the exine can be used to distinguish genera, and sometimes even species, so qualitative estimates may also be made.

pollen chamber A cavity at the micropylar end of the nucellus in which the pollen collects. It is found in the cycads, gingko, and gnetophytes.

pollen culture *See* anther culture.

pollen mother cell (**PMC**) In angiosperms and gymnosperms, a spore mother cell that gives rise to four haploid pollen grains by meiosis.

pollen sac A chamber in which the pollen is formed in the flowering plants and conifers. In angiosperms there are typically four pollen sacs (constituting the anther), in two pairs at the top of the filament. Conifer species have a variable number of pollen sacs borne on the microsporophylls within the male strobilus.

pollen tube A filamentous outgrowth of the pollen grain that in most seed plants transports the male gamete to the ovule. Germination of the pollen grain to give the pollen tube usually takes place only when the pollen is compatible with the female. The tube is an extension of the intine layer of the pollen grain wall and grows out through a pore in the exine. In angiosperms it grows through the style and nucellus relatively quickly so fertilization occurs soon after pollination. In gymnosperms the tube only grows a short distance before halting at the nucellus where it remains until the female gametophyte is mature, resuming growth the following year. In *Cycas* and *Ginkgo* the pollen tube only has a haustorial function and the male gametes swim to the ovule through a film of moisture by means of undulipodia (flagella).

pollex /**pol**-eks/ (*pl.* **pollices**) A first digit of the forelimbs of tetrapods; it forms the thumb in humans. In the typical pentadactyl limb it contains two phalanges; however, there are modifications and reductions to this general plan and in some mammals it is absent. In humans and other primates it has two phalanges and is in opposition to the fingers to allow grasping. *Compare* hallux.

pollination /pol-ă-**nay**-shŏn/ The transfer of pollen from the anther to the stigma. If the pollen is compatible (i.e. of the right type) then the pollen grains germinate, producing a pollen tube that grows down the style carrying the pollen nuclei to the ovule. Plants may be self-pollinating (e.g. barley), thus ensuring that seed will be set, even in the absence of other members of the same species. However, self-pollination also leads to homozygosity and less adaptable plants; thus, in many plant species mechanisms exist to prevent it and promote cross pollination, either by insects or by wind. *See also* incompatibility; protandry; protogyny.

pollination drop A mechanism that assists pollination in gymnosperms. Pollen grains collect in a drop of liquid secreted at the opening of the micropyle. The pollen is then drawn into the nucellus when the drop is reabsorbed.

pollution Any damaging or unpleasant change in the environment that results from the physical, chemical, or biological side-effects of human industrial or social activities. Pollution can affect the atmosphere, rivers, seas, and the soil.

Air pollution is caused by the domestic and industrial burning of carbonaceous fuels, by industrial processes, and by vehicle exhausts. For example, industrial emissions of sulfur dioxide cause ACID RAIN, and the release into the atmosphere of chlorofluorocarbons, formerly used in refrigeration, aerosols, etc., is linked to the depletion of ozone in the stratosphere (*see* ozone layer). Carbon dioxide, produced by burning fuel and by motor vehicle exhausts, is slowly building up in the atmosphere, and causing an overall increase in the temperature of the atmosphere (*see* greenhouse effect). Vehicle exhausts also contain carbon monoxide and other hazardous substances, such as fine particulate dusts. Lead was formerly a major vehicle pollutant, but the widespread introduction of lead-free petrol has eliminated this problem in most countries. Photochemical smog, caused by the action of sunlight on hydrocarbons and nitrogen oxides from vehicle exhausts, is a problem in many major cities.

Water pollutants include those that are biodegradable, such as sewage effluent, which cause no permanent harm if adequately treated and dispersed, as well as those which are nonbiodegradable, such as certain chlorinated hydrocarbon pesticides (e.g. DDT) and heavy metals, such as lead, copper, and zinc in some industrial effluents. The latter accumulate in the environment and can become very concentrated in food chains. The pesticides DDT, aldrin, and dieldrin are now banned. Water supplies can become polluted by leaching of nitrates from agricultural land, or of a wide range of potentially toxic substances from domestic and industrial waste tips. The discharge of waste heat can cause thermal pollution of the environment, but this is reduced by the use of cooling towers. In the sea, oil spillage from tankers and the inadequate discharge of sewage effluent are the main problems.

Other forms of pollution are noise from airplanes, traffic, and industry, the disposal of radioactive waste, and excessive light from street lamps and other forms of artificial lighting.

polocyte /**pol**-ŏ-sÿt/ *See* polar body.

polyacrylamide gel electrophoresis /pol-ee-ă-**kril**-ă-mÿd/ (PAGE) A type of electrophoresis used for separating the components of mixtures of proteins or nucleic acids according to their molecular size. The mixture is first treated with a detergent (e.g. SDS), then applied to the semi-solid polyacrylamide gel held between two glass plates. An electric current is applied, causing molecules to move through the gel, with smaller ones traveling faster than larger ones. The separated molecules form bands in the gel, and these are stained with dye to visualize them.

polyamine /pol-ee-ă-**meen**, -**am**-in/ An aliphatic compound that has two or more amino and/or imino groups. Polyamines are often found associated with DNA and RNA in bacteria and viruses. This may stabilize the nucleic acid molecule in a way analogous to the action of histones on DNA in eukaryote cells. Examples of polyamines include spermine, spermidine, cadaverine, and putrescine.

Polychaeta /pol-ee-**kee**-tă/ A class of marine annelid worms, the bristle worms. Many are carnivorous and active crawlers. Some, e.g. *Nereis* (ragworm), burrow in sand or mud while others build tubes of sand or mucus, which they rarely leave. Each body segment bears a pair of limblike locomotory *parapodia*, in which numerous stiff hairlike chaetae are embedded. The well-defined head bears sense organs. The sexes are usually separate and development is via a ciliated larva. *See* Nereis.

polyembryony /pol-ee-**em**-bree-ŏ-nee/ The occurrence of many embryos in one seed. Only one of these is a true sexual embryo, the others arising vegetatively. It is a common phenomenon in citrus fruits.

polygene /**pol**-ee-jeen/ A gene with an individually small effect on the phenotype that interacts with other polygenes controlling the same character to produce the continuous quantitative variation typical of such traits as height, weight, and skin color. *See* quantitative inheritance.

polymerase /**pol**-i-mer-ayss/ An enzyme that regulates the synthesis of a polymer. Examples include *RNA polymerases* and *DNA polymerases*. There is only one type of RNA polymerase in prokaryotes, but in eukaryotes there are three different types: type I makes ribosomal RNA, type II makes messenger RNA precursors, and type III makes transfer RNA and 5S ribosomal RNA.

DNA polymerases are involved either in the synthesis of double-stranded DNA from single-stranded DNA or in the repair of DNA by scanning the DNA molecule and removing damaged nucleotides. *See also* reverse transcriptase.

polymerase chain reaction (PCR) A technique for amplifying small samples of DNA rapidly and conveniently. Developed in 1983, it is now used widely in research and forensic science, e.g. to produce a suitable quantity of DNA for genetic fingerprinting from the minute amounts present in traces of blood or other tissue. To amplify a particular segment of DNA it is necessary first to know the sequence of bases flanking it at either end. This enables the construction of short single DNA strands (primers) that are complementary to and will bind with these flanking regions. Then the sample is incubated with the primers, nucleotides, and enzymes, especially DNA polymerase, in a water bath. By varying the temperature precisely and rapidly, amplification proceeds in cycles of DNA denaturation, annealing of primers, and replication of new DNA strands, each lasting about 20 seconds. After 30 cycles, some 10^9 copies of the original DNA are produced.

polymorph /**pol**-ee-morf/ (**polymorphonuclear leukocyte**) A white blood cell (leukocyte) with a lobed nucleus and granules in the cytoplasm. The term can be used for any GRANULOCYTE. *See* neutrophil.

polymorphism /pol-ee-**mor**-fiz-ăm/ A distinct form of variation in which significant proportions of different types of individuals exist within a species. If the differences persist over many generations then there is a *balanced polymorphism*, which is maintained by contending advantages and disadvantages. If one form is increasing at the expense of the other, so the latter is eventually reduced to the status of a rare mutant, then there has been a *transient polymorphism*. Polymorphism usually results from the occurrence of different allelic forms of a gene and balanced polymorphism arises when the heterozygote is at an advantage compared to the homozygotes. The caste system in social insects results, in some cases, from differences in nutrition rather than genotype and is thus an environmental rather than genetic polymorphism. *Compare* pleiomorphism.

polyp /**pol**-ip/ A stage in the life cycle of cnidarians in which the body is tubular, with a mouth surrounded by tentacles at one end; the other end is attached to a fixed surface. In some species the polyps form branching colonies of several or many individuals. In a few of these species the colony floats freely on the water surface, but generally it remains fixed at one place. In the anthozoans (sea anemones and corals) and some hydrozoans (e.g. *Hydra*) the polyp is the only existing form. It can reproduce asexually (by budding or splitting) and sexually. In other hydrozoans (e.g. *Obelia*) specialized reproductive polyps develop buds, which break away as free-swimming medusae. In the scyphozoans (jellyfish) the polyp is absent or much reduced. *See also* hydranth; medusa.

polypeptide /pol-ee-**pep**-tÿd/ A compound that contains many amino acids linked together by peptide bonds. *See* peptide.

polyphyletic /pol-ee-fÿ-**let**-ik/ Describing a taxon some of whose members are thought to have distinct evolutionary histories. *Compare* monophyletic.

polyphyodont /pol-ee-**fÿ**-ŏ-dont/ A type of dentition in which the teeth are replaced throughout the animal's lifetime if damaged or broken. It is found in frogs and lizards. *Compare* diphyodont; monophyodont.

polyploid /**pol**-ee-ploid/ The condition in which a cell or organism contains three or more times the haploid number of chromosomes. Polyploidy is far more common in plants than in animals and very high chromosome numbers may be found; for example in octaploids and decaploids (containing eight and ten times the haploid chromosome number). Polyploids are often larger and more vigorous than their diploid counterparts and the phenomenon is therefore exploited in plant breeding, in which the chemical colchicine can be used to induce polyploidy. Polyploids may contain multiples of the chromosomes of one species (autopolyploids) or combine the chromosomes of two or more species (allopolyploids). Polyploidy is rare in animals because the sex-determining mechanism is disturbed. For example a tetraploid XXXX would be sterile. *See* allopolyploidy; autopolyploidy.

polyribosome /pol-ee-**rÿ**-bŏ-sohm/ *See* ribosome.

polysaccharide /pol-ee-**sak**-ă-rÿd, -rid/ A polymer of monosaccharides joined by glycosidic links (*see* glycoside). They contain

Polysaccharide: the chain of amylose, one of the constituents of starch

many repeated units in their molecular structures and are of high molecular weight. They can be broken down to smaller polysaccharides, disaccharides, and monosaccharides by hydrolysis or by the appropriate enzyme. Important polysaccharides are inulin (hydrolyzed to fructose), starch (hydrolyzed to glucose), glycogen (also known as animal starch), and cellulose (hydrolyzed to glucose but not metabolized by humans). *See also* carbohydrates; sugar.

polysome /**pol**-ee-sohm/ *See* ribosome.

polysomy /**pol**-ee-soh-mee/ *See* aneuploidy.

polyspermy /**pol**-ee-sper-mee/ The penetration of several sperm into one ovum at fertilization; only one sperm actually fuses with the ovum nucleus. It occurs in a few animals with yolky eggs, e.g. birds. In most animals a fertilization membrane forms around the fertilized ovum, preventing polyspermy.

polystely /**pol**-ee-stee-lee/ Having many steles, as in the stems of some *Selaginella* species. *See also* distely.

polytene /**pol**-ee-teen/ Describing the chromosome condition caused by chromatids not separating after duplication. It leads to the formation of *giant chromosomes* consisting of numerous identical chromatids lying parallel to each other. Giant chromosomes have characteristic bands, caused by the degree of coiling of the DNA-histone fiber – highly coiled regions stain darker when prepared for microscopy. They are used to study gene activity and make chromosome maps. Polytene chromosomes are common in the salivary gland cells of dipterous insects, e.g. *Drosophila*.

pome /pohm/ A fleshy pseudocarpic fruit, characteristic of the family Rosaceae (e.g. apple). It is formed from an extremely perigynous ovary that subsequently becomes epigynous as the carpels fuse with the receptacle. The outer succulent portion of the fruit develops from the receptacle and encloses the pericarp or core.

pons /ponz/ (*pl.* **pontes**) (**pons cerebelli; pons Varolii**) A thick band of nerve fibers in the mammalian brain that passes across the medulla oblongata to link the two hemispheres of the cerebellum.

population A group of organisms of the same species (or other groups within which individuals may exchange genetic information) occupying a particular space. A population is continually modified by increases (birth and immigration) and losses (death and emigration), and is limited by the food supply and the effects of environmental factors such as disease.

Porifera /pŏ-**rif**-ĕ-ră/ A phylum of primitive multicellular animals, the sponges, that probably evolved a multicellular structure independently of the other multicellular animals. All are sessile and almost all are marine. The body of a sponge is a loose aggregation of cells, with minimal coordination between them, forming a vase-like structure. Flagellated cells (choanocytes) line the vase, and cause water currents to flow in through apertures (ostia) in the body wall and out through one or more openings (oscula) at the top. Sponges have an internal skeleton of chalk or silica spicules or protein fibers (as in the bath sponge).

porogamy /por-**og**-ă-mee/ The usual method of fertilization in angiosperms in which the pollen tube enters the ovule by the micropyle. *Compare* chalazogamy.

porphyrins /**por**-fă-rinz/ Cyclic organic structures that have the important characteristic property of forming complexes with metal ions. Examples of such *metalloporphyrins* are the iron porphyrins (e.g. heme in hemoglobin) and the magnesium porphyrin, chlorophyll, the photosynthetic pigment in plants. In nature, the majority of metalloporphyrins are conjugated to proteins to form a number of very important molecules, e.g. hemoglobin, myoglobin, and the cytochromes.

portal vein A vein connecting capillary networks of two particular regions that allows blood from one to be regulated by the other. *See* hepatic portal system; renal portal system.

postcaval vein /pohst-**kay**-văl/ *See* vena cava.

posterior /pŏ-**steer**-ee-er/ **1.** Designating the hind end of an animal: in bilaterally symmetrical animals this is the end directed backwards during locomotion. However, in bipedal animals, such as humans, the posterior side corresponds to the dorsal side of other animals. **2.** Designating the part of a flower or axillary bud facing towards the inflorescence axis or stem respectively. *Compare* anterior.

potassium /pŏ-**tass**-ee-ŭm/ One of the essential elements in plants and animals. It is absorbed by plant roots as the potassium ion, K+, and in plants is the most abundant cation in the cell sap. Potassium ions are required in high concentrations in the cell for efficient protein synthesis, and for glycolysis in which they are an essential cofactor for the enzyme pyruvate kinase. In animals the gradient of potassium and sodium ions across the cell membrane is responsible for the potential difference across the membrane, which is important for the transmission of nerve impulses.

potometer /pŏ-**tom**-ĕ-ter/ An apparatus for measuring the rate of water uptake by a cut shoot. This is normally closely related to the rate of water loss by transpiration, and the potometer can be used to compare transpiration rates under different conditions. The rate of uptake is measured by the progress of an air bubble in a capillary tube along a scale.

poxvirus /**poks**-vÿ-rŭs/ One of a group of large DNA-containing viruses that are responsible for smallpox, cowpox, and certain tumors in animals.

PPLO pleuropneumonia-like organism. *See* mycoplasmas.

P protein A protein found in the sieve elements of phloem in vascular plants that helps to plug leaks of sap from injured vessels. It normally forms a fine mesh underlying the plasma membrane, but following injury is swept by the surge of escaping sap into the center of the cell, creating a tangled mass that forms a plug at the sieve plate or sieve area.

P$_R$ (P660) *See* phytochrome.

Precambrian /pree-**kam**-bree-ăn/ The time in the earth's geological history that precedes the Cambrian period, i.e. from the origin of the earth, nearly 5 billion years ago, to the start of the Cambrian, around 570 Ma (million years ago). The term 'Precambrian' is now used mainly descriptively, and has been largely discarded as a geological term in the light of greater knowledge of the early evolution of life. Precambrian time is now divided into three eons: Hadean, from the earth's origin to about 3900 Ma: Archean, 3900–2390 Ma; and Proterozoic, 2390–570 Ma (the Cambrian marks the start of the Phanerozoic eon, which extends to the present day).

The oldest fossils discovered so far are remains of bacteria-like organisms, dating from about 3500 Ma. Indeed, there is abundant evidence of flourishing colonies of cyanobacteria and other bacteria throughout the Archean and Proterozoic eons. This takes the form of STROMATOLITES, rock structures representing the remains of sediment trapped or precipitated by bacterial communities. However, the earliest remains of single-celled eukaryotes are much later, dating from about 1400 Ma, while the first appearance of multicellular animals is in the so-called Ediacara fauna, in rocks dated to the last 100 million years of Precambrian time. *See also* Burgess shale.

precaval vein /pree-**kay**-văl/ *See* vena cava.

precipitin /pri-**sip**-ă-tin/ An antibody that combines with and precipitates soluble antigen, used in the *precipitin reaction* for identifying antigens.

precocial /pri-**koh**-shăl/ *See* nidifugous.

preformation /pree-for-**may**-shŏn/ The theory that the embryonic development of animals and plants consists merely of growth or extension of a preformed germ or program. Early adherents of the theory postulated the presence of tiny human figures (*homunculi*) in the heads of the sperms, while modern biochemical preformationists assume that the complexity of organisms is only the complexity of nucleic acid extended.

premaxilla /pree-maks-**il**-ă/ (*pl.* **premaxillae**) A paired membrane bone forming the anterior region of the upper jaw in most vertebrates. In mammals, it bears the incisor teeth.

premolar /pree-**moh**-ler/ A mammalian tooth situated between the canine teeth or incisors in front and the molars behind. Premolars are multiple-rooted and ridged and are used for grinding food. *See also* teeth.

pressure potential *See* water potential.

presumptive /pri-**zump**-tiv/ Describing embryonic tissues that are presumed to develop in a certain way. For example, presumptive neural plate of amphibians lies towards the animal pole of the blastula. *See* neural plate.

prickle A protective outgrowth from the surface of a plant. It may be a modified trichome and thus completely epidermal in origin, or it may also contain cortical and vascular tissue.

primary growth Growth derived solely from meristems present in the embryo, i.e. apical meristems. Such growth generally increases the length of plant organs. *Compare* secondary growth.

primary plant body The structure that is derived solely from meristems present in the embryo and their derivatives. *See* primary growth.

primary tissue Plant tissue that is derived solely from meristems present in the embryo and their derivatives. *See* primary growth.

Primates /**prÿ**-mayts, prÿ-**may**-teez/ The order of mammals that contains the monkeys, great apes, and humans. Most primates are relatively unspecialized arboreal mammals with a very highly developed brain, quick reactions, and large forward-facing eyes allowing binocular vision. The opposable thumb and (usually) big toe are used for grasping and the digits have nails. The young undergo a long period of growth and development, during which they learn from their parents. The New World monkeys have prehensile tails; the more advanced Old World monkeys lack prehensile tails, and great apes are larger tailless primates that typically swing from trees by their long arms.

primitive streak The first sign of embryo formation on the blastoderm of reptiles and birds and in the inner cell mass of mammals. It usually appears as a longitudinal wrinkle in the outer layer with a pit (*Hensen's node*) at the anterior end. This appearance is caused by convergence of the outer cells toward the streak and their sinking beneath the surface to become mesoderm; at the anterior end the cells sink into Hensen's node and move anteriorly to become notochord.

primordium /prÿ-**mor**-dee-ŭm/ (*pl.* **primordia**) A collection of cells that differentiates into an organ or tissue, e.g. the apical shoot and apical root primordia of the embryo. *See also* initial; leaf buttress.

prion /**pree**-on/ An infectious protein particle that causes various nervous diseases in humans and other animals, including bovine spongiform encephalopathy ('mad cow disease') in cattle, and a form of Creutzfeldt-Jakob disease (CJD) in humans. Prions are apparently unique in that unlike viruses, virions, and all other infectious agents they lack any form of genetic material (i.e. DNA or RNA). It is thought that the infectious prion is a variant of a membrane protein normally produced in cells. It can be transmitted in food, and induces changes in the folding of the normal proteins, causing them to form rod-shaped aggregates in the central nervous system, which are responsible for the symptoms of disease.

probiotic /proh-bӯ-**ot**-ik/ Any compound produced by a microorganism that promotes growth in other microorganisms. *Compare* antibiotic.

Proboscidea /proh-bŏ-**sid**-ee-ă/ The order that contains the largest terrestrial mammals – *Loxodonta* (African elephant) and *Elephas* (Indian elephant). Elephants are characterized by the trunk (*proboscis*), formed from the elongated nose and upper lip, which is used for bathing, drinking, and collecting vegetation. The single pair of upper incisor teeth grow into large ivory tusks. There are no lower incisors, canines, or premolars. The huge ridged molars are used for grinding vegetation. Only two pairs are used at a time and are replaced when worn down.

proboscis /proh-**boss**-iss/ Any tubular extension of the head or mouthparts, especially one used for feeding. An elephant's trunk, perhaps the best-known example, is a long muscular elongation of the upper lip and nose. It is used for transferring food to the mouth, and for sucking up water, which is then squirted into the mouth. The end of the trunk bears a pair of nostrils and also finger-like lobes, which can pick up small objects. The proboscis of a moth or butterfly is formed from certain mouthparts (the maxillae) and is used to suck nectar and other liquids. Other examples are common among the invertebrates.

procambial strand /proh-**kam**-bee-ăl/ The layer of cells that gives rise to the vascular tissue. It is discernable just below the apex as a strand of flattened cells which, if traced back along the shoot or root, may be seen to give rise to the primary vascular tissues. It is continuous with the intrafascicular cambium. In roots it may also be called plerome.

procaryote /proh-**ka**-ree-oht/ *See* prokaryote.

proctodeum /prok-tŏ-**dee**-ŭm/ (*pl.* proctodea) The posterior end of the alimentary tract of most animals, derived from an intucking of the embryonic ectoderm.

producer The first TROPHIC level in a food chain. Producers are those organisms that can build up foods from inorganic materials, i.e. green plants, algae, and some bacteria. Producers are eaten by herbivores, primary consumers. *Compare* consumer.

pro-estrus /proh-**es**-trŭs/ *See* estrous cycle.

profundal /prŏ-**fun**-dăl/ The deepwater zone of a lake beyond a depth of ten meters. Little light penetrates this zone and thus the inhabitants are all heterotrophic, depending on the LITTORAL and SUBLITTORAL organisms for basic food materials. Commonly found inhabitants include bacteria, fungi, mollusks, and insect larvae. Species found in the profundal zone are adapted to withstand low oxygen concentration, low temperatures, and low pH.

progesterone /prŏ-**jes**-tĕ-rohn/ A steroid hormone secreted by the corpus luteum in the ovary after ovulation. It initiates the preparation of the uterus for implantation of the ovum, the development of the placenta, and the development of the mammary gland in preparation for lactation. *See also* estrogen.

progestogen /prŏ-**jes**-tŏ-jĕn/ Any hormone whose effects resemble those of progesterone. Synthetic progestogens are used in therapy and oral contraceptives. *See also* estrogen.

proglottis /proh-**glot**-is/ (*pl.* proglottides) One of the many body segments of a cestode (tapeworm). As they mature they enlarge and develop sex organs. Mature proglottides, containing young embryos (hexacanths), become detached from the posterior end of the body and are passed out of the host in the feces. They may then be eaten by the intermediate host. *See* Cestoda.

programmed cell death *See* apoptosis.

prohormone /proh-**hor**-mohn/ The inactive form of a hormone: the form in which it is stored. Activation usually involves enzymatic removal of some part of the prohormone; for example, removal of amino acids from the polypeptide prohormone, proinsulin, to form insulin.

prokaryote /proh-**ka**-ree-oht/ (**procaryote**) An organism whose genetic material (DNA) is not enclosed by membranes to form a nucleus but lies free in the cytoplasm. Organisms can be divided into prokaryotes and EUKARYOTES, the latter having a true nucleus. This is a fundamental division because it is associated with other major differences. Prokaryotes comprise the two groups of bacteria – ARCHAEBACTERIA and EUBACTERIA. Some authorities place them in a single kingdom, Prokaryotae (Monera), whereas others regard them as equidistant from each other and the eukaryotes in evolutionary terms, and place them in separate DOMAINS. However, it is clear that prokaryote cells evolved first and gave rise to eukaryote cells. *See* endosymbiont theory.

prolactin /proh-**lak**-tin/ A hormone, formerly called luteotropic hormone, produced by the anterior pituitary gland. In mammals it stimulates and controls lactation after the mammary gland has been prepared for milk production by estrogens, progesterone, and other hormones. In birds prolactin stimulates secretion of crop milk from the crop glands. *See also* gonadotropin.

prolamellar body /proh-lă-**mel**-er, proh-**lam**-ĕ-ler/ *See* etioplast.

proline /**proh**-leen, -lin/ *See* amino acids.

promoter A specific DNA sequence at the start of a gene that initiates TRANSCRIPTION by binding RNA polymerase. In *Escherichia coli* the RNA polymerase has a protein 'sigma factor' that recognizes the promoter; in the absence of this factor the enzyme binds to, and begins transcription at, random points on the DNA strand. In eukaryotic cells, binding of RNA polymerase to the promoter involves proteins called transcription factors.

pronation /proh-**nay**-shŏn/ The rotational movement of the lower forelimb (forearm) so that the forefoot (hand) is twisted through 90 degrees in either direction in relationship to the elbow. Pronation in humans occurs when the palm of the hand faces downwards or backwards and the radius and ulna are crossed. Movement so that the palm of the hand faces upwards or forwards and the radius and ulna are parallel is *supination*.

pronephros /proh-**nef**-ros/ (*pl.* **pronephroi**) The first type of vertebrate kidney to develop; the functional kidney of larval fish and amphibians. It is comprised of a variable number of open-ended ducts, which are segmentally arranged, just behind the heart. They collect fluid waste from the coelom and are drained by a collecting tube (pronephric duct) leading to the cloaca. It is later replaced by the mesonephric or metanephric kidney.

prophage /**proh**-fayj/ *See* lysogeny.

prophase /**proh**-fayz/ The first stage of cell division in meiosis and mitosis. During prophase the chromosomes become visible and the nuclear membrane dissolves. Prophase may be divided into successive stages termed LEPTOTENE, ZYGOTENE, PACHYTENE, DIPLOTENE, and DIAKINESIS. The events occurring during these stages differ in meiosis and mitosis, notably in that bivalents (pairs of homologous chromosomes) are formed in meiosis, whereas homologous chromosomes remain separate in mitosis.

proplastid /proh-**plas**-tid/ A self-duplicating undifferentiated plastid, about 0.5–1 μm in diameter and found in the meristematic regions of plants. They grow and develop into plastids of different types.

They typically contain very rudimentary membrane systems, occasional starch grains, and DNA and ribosomes (sparse).

proprioceptor /proh-pree-ŏ-**sep**-ter/ A sensory nerve ending that, when stimulated by stretching or pressure, supplies information to the cerebellum of the brain about the position and movement of the various parts of the body. Proprioceptors, which occur in muscles, ligaments, tendons, joints, etc., have an important role in maintaining balance and posture. *See also* muscle spindle; Pacinian corpuscle.

prop root An adventitious root, found at the junction of stem and soil that serves to give additional support to the stem, as seen in maize.

prosencephalon /pros-en-**sef**-ă-lon/ *See* forebrain.

prosenchyma /pros-**enk**-ă-mă/ Any tissue consisting of elongated cells with tapering ends, e.g. much of the mechanical and conducting tissues in plants.

prostaglandin /pros-tă-**glan**-din/ (**PG**) One of a group of lipid-soluble hormones derived from arachidonic acid. They tend to act locally within tissues, being released from cells into the intercellular space to influence neighboring cells, and are rapidly broken down by enzymes. There are at least 16 types of prostaglandins in 9 different classes, designated PGA to PGI. Among their many physiological effects, they modulate the effects of other hormones, stimulate smooth muscle contraction in the uterus, influence blood pressure, enhance inflammation, and promote platelet adhesion and blood clotting. Different prostaglandins often have opposing actions; for example, PGE and PGA reduce blood pressure whereas PGF raises it.

prostate gland /pros-tayt/ A gland in male mammals surrounding the urethra in the region where it leaves the bladder. It releases a fluid containing various substances, including enzymes and an antiagglutinating factor, that contribute to the production of semen. Its size and secretory function is under the control of hormones (androgens).

prosthetic group The non-protein component of a conjugated protein. Thus the heme group in hemoglobin is an example of a prosthetic group, as are the coenzyme components of a wide range of enzymes.

protamine /**proh**-tă-meen, -min/ One of a group of polypeptides formed from a few amino acids. They are soluble in water, dilute acids, and bases. On heating they do not coagulate. When protamines are hydrolyzed they yield a large proportion of basic amino acids, particularly arginine, alanine, and serine. They occur in the sperm of vertebrates, packing the DNA into a condensed form.

protandry /proh-**tan**-dree/ The maturation of the anthers before the stigma, as seen in daisy. It is more common than protogyny. *See also* dichogamy.

protease /**proh**-tee-ayss/ (**proteinase**) An enzyme that catalyzes the hydrolysis of peptide bonds in proteins to produce peptide chains and amino acids. Individual proteases are highly specific in the type of peptide bond they hydrolyze.

proteasome /**proh**-tee-ă-sohm/ A protein complex found in the cytoplasm of cells that dismantles worn-out or unwanted proteins by breaking them up into peptide fragments. It consists of a hollow cylindrical array guarded by a protein cap at either end. Proteins destined for disposal are tagged by a small protein called UBIQUITIN and then unfolded before they enter the proteasome.

protein /**proh**-teen, -tee-in/ One of a large number of substances that are important in the structure and function of all living organisms. Proteins are polypeptides; i.e. they are made up of amino acid molecules joined together by peptide links. Their molecular weight may vary from a few thousand to several million. About 20 amino acids are present in proteins. Simple proteins contain only amino acids. In conjugated proteins, the amino acids are joined to other groups.

The *primary structure* of a protein is the particular sequence of amino acids present. The *secondary structure* is the way in which this chain is arranged; for example, coiled in an alpha helix or held in beta pleated sheets. The secondary structure is held by hydrogen bonds. The *tertiary structure* of the protein is the way in which the protein chain is folded. This may be held by cystine bonds and by attractive forces between atoms. The *quaternary structure* is the number and arrangement of the constituent polypeptides, in proteins that comprise two or more subunits (i.e. multimeric proteins).

proteinase /**proh**-tă-nayss, **proh**-tee-ă-nayss/ *See* protease.

protein kinase Any of a group of enzymes that catalyze the addition of a phosphate group to a protein. Such phosphorylation typically alters the activity or binding properties of the target protein in some way. Protein kinases are important components in cellular signaling pathways, and fall into two main categories, depending on which amino acids they phosphorylate: *tyrosine kinases* are directed toward tyrosine residues, whereas *serine/threonine kinases* are directed toward either serine or threonine residues.

protein sequencing The determination of the primary structure of proteins, i.e. the type, number, and sequence of amino acids in the polypeptide chain. This is done by progressive hydrolysis of the protein using specific proteases to split the polypeptides into shorter peptide chains. Terminal amino acids are labeled, broken off by a specific enzyme, and identified by chromatography. The first protein to be sequenced was insulin, by Frederick Sanger at Cambridge University in 1954.

protein synthesis The process whereby proteins are synthesized on the ribosomes of cells. The sequence of bases in MESSENGER RNA (mRNA), transcribed from DNA, determines the sequence of amino acids in the polypeptide chain: each codon in the mRNA specifies a particular amino acid. As the ribosomes move along the mRNA in the process of TRANSLATION, each codon is 'read', and amino acids bound to different TRANSFER RNA molecules are brought to their correct positions along the mRNA molecule. These amino acids are polymerized to form the growing polypeptide chain.

proteoglycan /proh-tee-oh-**glÿ**-kan/ (**mucoprotein**) A type of glycoprotein consisting of long branched heterogeneous chains of glycosaminoglycan molecules linked to a protein core of amino acids. Unlike more typical glycoproteins, they have a greater carbohydrate content, the protein core is rich in serine, and they have a higher molecular weight.

proteolysis /proh-tee-**ol**-ă-sis/ The hydrolysis of proteins into their amino acids. Enzymes that catalyze this are *proteases* or *proteolytic enzymes*.

proteolytic enzyme /proh-tee-ŏ-**lit**-ik/ *See* proteolysis.

proteome /**proh**-tee-ohm/ The full range of proteins synthesized by a particular cell, tissue, or organism. The proteome lies at the heart of cellular life; it determines the functions and interactions of any particular cell, and reflects the pattern of gene expression within the cell. It varies not only with the nature of the organism and the type of tissue from which a sample is obtained, but also changes over time with normal metabolic rhythms, and is affected by factors such as stage of development, aging, and disease. The study of proteomes, called *proteomics*, relies on fast-developing techniques, such as highly sensitive mass spectrometers and microarrays. Using automated high-throughput methods, it is possible to take a series of 'snapshots' of the proteome, and thus build up a picture of the numerous interactions between different proteins in the cell. The vast amounts of data generated by proteomics are stored on databases and analyzed by computer programs. Besides providing valuable insights into cell function, proteomics might identify new targets for drugs, and lead to the development of new diagnostic techniques, such as automated 'proteome readouts'. The Human Proteomics Organization (HUPO) is one body that is coordinating international efforts, so that contributions from scientists worldwide can be shared via the Internet. *See also* transcriptome.

proteoplast /**proh**-tee-ŏ-plast/ A colorless plastid (leukoplast) that stores protein. *See also* aleuroplast.

Proterozoic /prot-ĕ-roh-**zoh**-ik/ *See* Precambrian.

prothallus /proh-**thal**-ŭs/ (*pl.* **prothalli**) A flattened disk of cells that forms the free-living haploid gametophyte generation of certain pteridophytes, e.g. the fern *Dryopteris*. In homosporous plants, there is only one type of prothallus with both male and female sex organs. In heterosporous plants, the microspores give rise to small male prothalli bearing male sex organs (antheridia), and larger female prothalli bearing female sex organs (archegonia). The prothallus is greatly reduced in spermatophytes.

prothrombin /proh-**throm**-bin/ The inactive form of the enzyme thrombin in blood plasma. It is activated during blood clotting by another enzyme, thrombokinase, in the presence of calcium ions. *See* blood clotting.

Protista /proh-**tiss**-tă/ In some classifications, a kingdom of simple organisms including the bacteria, algae, fungi, and protozoans. It was introduced to overcome the difficulties of assigning such organisms, which may show both animal and plantlike characteristics, to the kingdoms Animalia or Plantae. Today the grouping is considered artificial and many taxonomists support a system whereby the bacteria and fungi are both assigned to separate kingdoms, while algae and protozoans consti-

tute various phyla of protoctists. *See* prokaryote; Protoctista.

Protochordata /proh-toh-kor-**day**-tă/ *See* acraniate.

protocooperation /proh-toh-koh-op-ĕ-**ray**-shŏn/ *See* mutualism.

Protoctista /proh-tok-**tiss**-tă/ A kingdom of simple eukaryotic organisms that includes the algae, slime molds, fungus-like oomycetes, and the organisms traditionally classified as protozoa, such as flagellates, ciliates, and sporozoans. Most are aerobic, some are capable of photosynthesis, and most possess undulipodia (flagella or cilia) at some stage of their life cycle. Protoctists are typically microscopic single-celled organisms, such as the amebas, but the group also has large multicellular members, for example the seaweeds, pondweeds, and other conspicuous algae. *See* Apicomplexa; Chlorophyta; Chrysomonada; Ciliophora; Euglenophyta; oomycete; Phaeophyta; Rhizopoda; Rhodophyta; Xanthophyta; Zoomastigina.

protoderm /**proh**-tŏ-derm/ The tissue that develops from the tunica initials of the apical meristem and gives rise to the epidermis and, in some plants, the root cap. In the root it may also be called dermatogen. *See* histogen theory.

protogyny /proh-**toj**-ă-nee/ 1. (*Botany*) The maturation of the stigma before the anthers as in figwort. It is less common than protandry though more effective in preventing self pollination. *See also* dichogamy.
2. (*Zoology*) The condition in hermaphrodite animals in which female gametes (ova) are produced before the male gametes (spermatozoa).

protonema /proh-tŏ-**nee**-mă/ (*pl.* **protonemata**) Most commonly, the young bryophyte gametophyte that develops following spore germination. In most mosses and liverworts it resembles the heterotrichous green algae but in *Sphagnum* the protonema filament is soon replaced by a thallose protonema. The mature gametophyte plants develop from buds that form at several points along the protonema. Secondary protonemata may arise from the leaves, stems, or rhizoids of the mature gametophyte, and wounded sporophyte tissue may also produce protonemata which then give diploid gametophytes. In the algal order Charales the erect filament formed when the zygote germinates is termed a protonema.

protonephridium /proh-toh-ni-**frid**-ee-ŭm/ (*pl.* **protonephridia**) The excretory organ of certain invertebrates (e.g. platyhelminths, rotifers, and some annelids). It consists of one or more flame cells connected by a tubule, which conducts waste products, collected in the flame cell cavities, to the exterior. *See also* nephridium.

proton pump Any transport protein that moves protons (hydrogen ions, H+) across a cell membrane against their concentration gradient. There are several different types, all of which power the proton pumping using energy released by the hydrolysis of ATP. In cells of plants, fungi, and bacteria, the export of protons by proton pumps in the plasma membrane (and also in the tonoplast membrane surrounding the vacuole in plant cells) is the principal mechanism for maintaining the RESTING POTENTIAL. Proton pumps also occur in the membranes of lysosomes and endosomes in animal cells, where they import protons to maintain the acidic conditions inside the organelle. Acid secretion by cells lining the mammalian stomach again involves proton pumps. ATP synthase, found in mitochondria and chloroplast membranes, is a proton pump working in reverse; i.e. it uses the movement of protons down a concentration gradient to power ATP synthesis. *See* electron-transport chain.

protophloem /proh-toh-**floh**-em/ The first formed primary PHLOEM, differentiated from the procambium in the region just behind the meristem.

protoplasm /**proh**-tŏ-plaz-ăm/ The living

contents of a cell, comprising the CYTOPLASM plus NUCLEUS.

protoplast /**proh**-tŏ-plast/ The protoplasm and plasma membrane of a cell (e.g. of a plant, alga, or bacterium) after removal of the cell wall. This can be achieved by physical means or by enzymic digestion. Protoplasts can be grown in culture and make possible certain observational or experimental work such as study of new cell wall formation, pinocytosis, and fusion of cells. Fusion of protoplasts of different species is being investigated by plant breeders as a means of crossing otherwise incompatible plants. Under suitable culture conditions the hybrid cell can develop to form a mature fertile plant. The ability to regenerate mature plants from single transformed protoplasts is essential for certain genetic engineering techniques.

protopodite /proh-**top**-ŏ-dÿt/ *See* biramous appendage.

protostele /**proh**-tŏ-steel, -stee-lee/ A simple form of stele, uninterrupted by leaf gaps, consisting of xylem in the center completely surrounded by phloem. Most roots have protosteles. The stems of some lower plants (e.g. *Lycopodium*) also have protosteles. The actinostele, haplostele, medullated protostele, mixed protostele, and plectostele are all modifications of the protostele.

Prototheria /proh-tŏ-**th'eer**-ee-ă/ *See* Monotremata.

protoxylem /proh-toh-**zÿ**-lĕm/ The primary xylem elements that are formed from the procambium first. The cells are usually annularly or spirally thickened and thus extensible.

protozoa /proh-tŏ-**zoh**-ă/ A group of single-celled heterotrophic often motile eukaryotic organisms, traditionally classified as animals and constituting a phylum, or subkingdom, Protozoa. In more recent classifications they are placed with other single-celled or simple multicelled eukaryotes in the kingdom Protoctista. They range from plantlike forms (e.g. *Euglena, Chlamydomonas*) to members that feed and behave like animals (e.g. *Amoeba, Paramecium*). There are over 30 000 species living universally in marine, freshwater, and damp terrestrial environments. Some form colonies (e.g. *Volvox*) and many are parasites (e.g. *Plasmodium*). Protozoa vary in body form but specialized organelles (e.g. cilia and flagella) are common. Reproduction is usually by binary fission although multiple fission and conjugation occur in some species. The main protozoan phyla are: Rhizopoda (rhizopods); Zoomastigina (flagellates); Apicomplexa (sporozoans); and Ciliophora (ciliates). *See also* Amoeba; Paramecium; Plasmodium; Apicomplexa; Ciliophora; Rhizopoda; Zoomastigina.

proventriculus /proh-ven-**trik**-yŭ-lŭs/ (*pl.* proventriculi) In birds, the anterior glandular part of the stomach, leading to the gizzard. In insects and crustaceans, the term is synonymous with the GIZZARD.

provirus /proh-**vÿ**-rŭs/ A viral chromosome that is integrated in a host chromosome and multiplies with it. Proviruses do not leave the host chromosome and begin a normal cycle of viral replication unless triggered to do so (*see* latent virus). *See also* retrovirus.

proximal /**proks**-ă-măl/ Denoting the part of an organ, limb, etc., that is nearest the origin or point of attachment. *Compare* distal.

pseudoallele /soo-doh-al-**eel**/ A mutation in a gene that produces an effect identical to another mutation at a different site in the same gene locus. The two pseudoalleles thus act as a single gene but do not occupy the same position, as evidenced by the occasional rare recombinations between them that result in the *cis-trans* effect.

pseudocarp /**soo**-dŏ-karp/ (**false fruit**) A fruit that includes other parts of the flower, e.g. the bracts, inflorescence, or receptacle, in addition to the ovary. *See also* composite fruit; pome; sorosis.

pseudogene /**soo**-doh-jeen/ A DNA sequence comparable to a gene that does not produce a polypeptide product, and hence is nonfunctional. Pseudogenes contain MUTATIONS that either prevent transcription, or prevent processing of any messenger RNA produced by transcription. They are thought to represent ancestral duplicates of functional genes that have acquired mutations and drifted into nonfunctionality, with no harmful effects to the organism. *Processed pseudogenes* differ in that they lack the intron and flanking sequences characteristic of functional genes. These are thought to represent DNA copies of cellular RNA, produced by reverse transcription, and then integrated at random into the genome, like a retrotransposon. Pseudogenes have potential to form new genes by further mutation as they already have useful sequences, such as those signaling transcription.

pseudoparenchyma /soo-doh-pă-**renk**-ă-mă/ A fungal or algal tissue resembling parenchyma but made up of interwoven hyphae (fungi) or filaments (algae). The stipe of the mushroom and the thallus of red algae (e.g. *Porphyra*) are pseudoparenchymatous tissues.

pseudopodium /soo-dŏ-**poh**-dee-ŭm/ (*pl.* pseudopodia) A temporary finger-like projection or lobe on the body of an ameboid cell. It is formed by a flowing action of the cytoplasm and functions in locomotion and feeding. *See* ameboid movement.

pseudopregnancy /soo-doh-**preg**-năn-see/ A physiological state resembling pregnancy that occurs in some female mammals such as rabbits, but without the formation of embryos. It is caused by secretion of hormone by the corpus luteum, either when a pronounced luteal phase of the estrous cycle occurs or when corpus luteum formation is triggered by copulation that does not result in fertilization.

Psilophyta /si-**lof**-ă-tă/ A phylum of vascular seedless plants comprising the whisk ferns. It contains only two living subtropical genera (*Psilotum* and *Tmesipteris*).

They lack shoots and roots; instead they have rhizoids on subterranean parts of the dichotomously branching axis, which bears tiny alternate scalelike or leaflike outgrowths. Whisk ferns may be descendants of *Rhynia* and similar leafless plants known from fossils of the late Silurian and early Devonian – some of the earliest of all plants.

psychrophilic /sȳ-krŏ-**fil**-ik/ Describing microorganisms that can live at temperatures below 20°C. *Compare* mesophilic; thermophilic.

pteridophyte /**te**-ră-dŏ-fȳt/ Any vascular non-seed-bearing plant. Pteriodophytes include the club mosses (phylum Lycophyta), horsetails (Sphenophyta), whisk ferns (Psilophyta), and ferns (Filicinophyta). In older classifications these groups constitute classes of a single phylum (or division), the Pteridophyta. In all pteridophytes, as in seed-bearing plants but unlike the non-vascular plants, he diploid sporophyte is the more conspicuous phase of the life cycle.

pterodactyl /te-tŏ-**dak**-tăl/ *See* Pterosauria.

Pterosauria /te-rŏ-**sor**-ee-ă/ An extinct order of flying reptiles, the pterodactyls, which were particularly common in the Jurassic and survived until the Cretaceous. Fossils are always found in marine deposits. They had long forelimbs with a very elongated fourth finger, which supported the delicate leathery wing membrane. Although they had some structural adaptations to flight, they were probably incapable of the same sort of flight as birds. Their very weak hind legs suggest that they were unable to stand upright on land and their mode of life probably involved swooping or gliding over the sea to catch fish. They had beaked jaws and primitive forms, such as *Rhamphorhynchus*, had teeth, but teeth are absent in more advanced forms, such as *Pteranodon*.

pteroylglutamic acid /te-roh-ăl-gloo-**tam**-ik/ *See* folic acid.

ptyalin /tȳ-ă-lin/ (**salivary amylase**) An enzyme present in the saliva of humans and some animals. It belongs to the group of carbohydrate-hydrolyzing enzymes known as AMYLASES. It catalyzes the conversion of starch to maltose.

ptyxis /tiks-is/ The way in which young leaves are folded or rolled in the bud. *See also* vernation.

pubic symphysis A joint formed by the union of the two pubic bones at the mid-ventral line of the pelvic girdle of mammals and many reptiles. It is slightly deformed in mammals during labor, in order to ease passage of the fetus. *See also* pubis; symphysis.

pubis /pyoo-bis/ (*pl.* **pubes**) (**pubic bone**) One of a pair of bones forming the anterior ventral portion of the tetrapod pelvic girdle. They are sometimes joined at the pubic symphysis.

puff (**Balbiani ring**) A swelling that is seen in certain areas of the giant (POLYTENE) chromosomes found in the salivary glands and other tissues of certain dipterous insects. Puffs originate in different regions of the chromosome in a certain sequence and their occurrence can be correlated with specific developmental events. Others occur only in certain tissues. The puffs are sites of active transcription of probably just a single gene, albeit present as numerous copies. The Balbiani ring is named for the French embryologist Édouard-Gérard Balbiani (1823–99).

pulmonary artery /**pul**-mŏ-nair-ee/ In mammals, a paired artery that carries deoxygenated blood from the right ventricle of the heart to the lungs. It is derived from the sixth aortic arch and also occurs in lungfish and other tetrapods. *See also* artery.

pulmonary vein A paired vein that carries oxygenated blood from the lungs to the left atrium of the heart in lungfish and tetrapods. *See also* vein.

pulp cavity The central core of a mammalian tooth, surrounded by dentine. It contains jelly-like connective tissue with blood, lymph vessels, and nerves, all of which originate in branches passing through the pulp canal. The outer layer of the pulp contains specialized cells (*odontoblasts*), which have fine cytoplasmic branches that penetrate the dentine. When growth is complete the pulp canal becomes constricted, allowing only sufficient blood supply to maintain the living cells. *See illustration at* teeth.

pulse A series of waves of dilation that spread outward from the heart along the main arteries. Each wave is caused by high pressure, which is produced when blood is discharged on contraction of the left ventricle. The pulse can be felt where the arteries pass near the surface, e.g. the wrist, but diminishes as it proceeds towards the capillaries. In humans, the number of pulsations per minute (the pulse rate) normally varies from 70 to 72 in men and 78 to 80 in women. The pulse travels at a much higher velocity than the blood flow.

pulvinus /pul-vy̆-nŭs/ (*pl.* **pulvini**) A specialized group of cells with large intercellular spaces that are located at the bases of leaves or leaflets in certain plants. They are involved in non-growth nastic movements, bringing these about by rapid changes in turgor through loss of water to the intercellular spaces.

punctuated equilibrium A theory of evolution proposing that there have been long periods of geological time, lasting for several million years, when there is little evolutionary change, punctuated by short periods of rapid speciation of less than 100 000 years. This is in contrast to the traditional theory (*see* neo-Darwinism) in which it is postulated that species have evolved continually and gradually throughout geological time.

pupa /pyoo-pă/ (*pl.* **pupae**) The third stage in the life cycle of insects that have complete metamorphosis, which follows the larval stage. During the pupal stage the insect does not feed, it is usually immobile, and internally it undergoes complete reorganization of its structure. At the end of the pupal stage the insect undergoes its final molt and the imago emerges. The pupa is often formed in a cocoon or made inconspicuous in some other way, as in the chrysalis of butterflies and moths. Mosquito pupae are fully mobile, although they do not feed. In mayflies the pupa gives rise to the *sub-imago*, which can fly but soon molts again to give the true imago.

pupil /pyoo-păl/ The hole (aperture) in the center of the iris of vertebrates and cephalopods through which light enters the eye. In humans, the pupil is round, but in many nocturnal animals, such as the cat, it is a slit. The size of the pupil can be altered by contraction of the muscles of the iris. *See also* iris. *See illustration at* eye.

pure line The succession of descendants of a homozygous individual that are identical to each other and continue to breed true, i.e. they produce genetically identical offspring. Pure lines cannot be improved by selection since all variation within them, barring the occasional mutation, is environmental. In plants pure lines are obtained by selfing, which halves the heterozygosity each generation, while in animals inbreeding tends to increase homozygosity.

purine /pyoor-een, -in/ A simple nitrogenous organic molecule with a double ring structure. Members of the purine group include adenine and guanine, which are constituents of the nucleic acids, and certain plant alkaloids, e.g. caffeine and theobromine.

Purkinje fibers /per-**kin**-jee/ (**Purkyne fibers**) A bundle of specialized cardiac muscle fibers that occurs along the midline of the heart ventricles in some mammals. It receives rhythmical impulses of electrical excitation from the pacemaker and spreads waves of contraction through the ventricle walls. The fibers are named for the Czech physiologist Johannes Evangelista Purkinje (1787–1869). *See also* pacemaker.

pyloric sphincter /pÿ-**lô**-rik, pi-/ *See* pylorus.

pylorus /pÿ-**lor**-ŭs, -**loh**-rŭs, pi-/ (*pl.* **pylori**) (**pyloric sphincter**) A ring of involuntary muscle in vertebrates that surrounds the opening from the stomach to the duodenum. It regulates the passage of food between stomach and duodenum.

pyramid of biomass A type of ecological pyramid based on the total amount of living material at each trophic level in the community, which is normally measured by total dry weight, and shown diagrammatically. The pyramid of biomass slopes more gently than the pyramid of numbers because organisms at successively higher levels in the pyramid tend to be larger than those below.

pyramid of numbers A type of ecological pyramid in which the number of individual organisms at each stage in the food chain of the ecosystem is depicted diagrammatically. The producer level forms the base, and successive levels the tiers. The shape of the pyramid of numbers depends upon the community considered; generally, the organism forming the base of a food chain is numerically very abundant, and each succeeding level is represented by fewer individual organisms, culminating with the final large predator. The pyramid of numbers may be partly inverted (base smaller than one or more of the upper tiers) if the producing organisms are large.

Theoretically the higher the level in the pyramid, the fewer organisms it can support. This has important implications regarding human food supplies as it illustrates that more people can be supported in a given area if their diet is mainly vegetarian – more land is needed per head if meat forms a large part of the diet.

pyranose /pÿ-ră-nohs, pi-**ran**-ohs/ A sugar that has a six-membered ring form (five carbon atoms and one oxygen atom). *See also* sugar.

pyrenocarp /pÿ-**ren**-ŏ-karp, pi-/ **1.** *See* drupe.

2. *See* ascocarp.

pyrenoid /**pÿ**-rĕ-noid/ A protein structure found in the chloroplasts of green algae and hornworts (*Anthoceros*). Pyrenoids are associated with the storage of starch.

Pyridine

pyridine /pi-ră-deen/ (C_5H_5N) An organic liquid of formula C_5H_5N. The molecules have a hexagonal planar ring and are isoelectronic with benzene. Pyridine is an example of an aromatic heterocyclic compound, with the electrons in the carbon–carbon pi bonds and the lone pair of the nitrogen delocalized over the ring of atoms. The compound is extracted from coal tar and used as a solvent and as a raw material for organic synthesis.

pyridoxine /pi-ră-**doks**-een, -in/ (**vitamin B₆**) A vitamin of the water-soluble VITAMIN B COMPLEX. Good sources include yeast and certain seeds (e.g. wheat and corn), liver, and to a limited extent, milk, eggs, and leafy green vegetables. There is also some bacterial synthesis of the vitamin in the intestine. Pyridoxine gives rise to the coenzyme pyridoxal phosphate, which is involved in various aspects of amino acid metabolism.

pyrimidine /pi-**rim**-ă-deen, pi-ră-mă-/ A simple nitrogenous organic molecule whose ring structure is contained in the pyrimidine bases cytosine, thymine, and uracil, which are constituents of the nucleic acids, and in thiamine (vitamin B_1).

pyruvic acid /pÿ-**roo-vik**, pi-/ An organic acid, $CH_3COCOOH$, that is an intermediate in several important metabolic pathways in living cells, including glycolysis and gluconeogenesis. The initial oxidation of glucose in glycolysis culminates in the conversion of phosphoenolpyruvate to pyruvic acid. Under aerobic conditions, the pyruvic acid is then converted to acetyl CoA, which enters the Krebs cycle for further oxidation. If oxygen is in short supply, most microorganisms and the cells of higher organisms convert pyruvic acid to lactate. However, organisms such as brewer's yeast can convert the pyruvic acid to ethanol by fermentation. Lactate is formed by intense muscle activity, and subsequently is transported to the liver, where it is converted back to pyruvic acid, and thence by the reactions of gluconeogenesis to glucose. Pyruvic acid is also the precursor for the synthesis in the body of certain amino acids.

Q

quadrat /**kwod**-rat/ A square area (standard size is one meter square) taken at random, within which the composition of organisms is noted. The quadrat sampling technique is mostly used in plant ecology to study plant communities but quadrats are also used as a sampling unit to count and weigh animals for an estimate of density or to discover animal distribution in a selected area. Permanent quadrats can be established that are examined at given intervals as a means of assessing changes in species composition in an area over a period of time. *See also* transect.

quadrate /**kwod**-rayt/ One of a pair of bones of the upper jaw in bony fish (Osteichthyes), amphibians, reptiles, and birds that form the points of articulation with the lower jaw. The quadrate bones are homologous with the palato-pterygoquadrate bar, a paired cartilage forming the upper jaw in cartilaginous fish (Chondrichthyes).

qualitative variation (**discontinuous variation**) A form of variation in which a character has two or more distinct forms. Examples are human blood groups and Mendel's pea characters. Qualitative variation generally occurs when there are two or more allelic forms of a major gene in a population.

quantitative inheritance (**polygenic inheritance; multifactorial inheritance**) The pattern of inheritance shown by traits, such as height in humans or grain yield in wheat, that show continuous variation within a certain range of values. Such traits are typically controlled by several or many different genes (polygenes) distributed among the genome. The identification and manipulation of genetic loci determining quantitative traits (*quantitative trait loci,* QTLs) is of great importance in plant and animal breeding, where the goal is often to improve such traits.

Quaternary /kwă-**ter**-nă-ree/ (**Neogene**) The most recent period of the Cenozoic era from about two million years ago to the present day and composed of the Pleistocene and Holocene epochs. Literally the 'fourth age', it is characterized by the emergence of humans. *See also* Ice Age.

quiescent center A group of cells in the center of the apical meristem in which mitotic divisions are rare or absent. The cells may begin dividing if another part of the meristem is damaged.

capitulum corymb umbel spike raceme

Racemose infloresence: various types

race **1.** (*Biology*) A term sometimes used for a genetically distinct group of individuals that form an identifiable population of a species but remain capable of breeding with members of other races. A race can be characterized by the geographical distribution, ecology, behavior, physiology, or other aspect of the members' phenotype. The term is often used to mean SUBSPECIES. *See also* variety.
2. (*Anthropology*) Any of several different types of human defined according to certain physical traits such as skin color, hair, and facial features. For example, one system divides humans into Mongolians, Ethiopians, and Caucasians. However, such races are often scientifically ill defined and arbitrary, and can be used to bolster prejudices and reinforce stereotypes.

raceme /ray-seem, ră-/ A type of racemose inflorescence in which stalked flowers are borne on an undivided peduncle, as in foxgloves (*Digitalis*).

racemose inflorescence /rass-ĕ-mohs/ (**indefinite inflorescence**) An inflorescence in which the growing point of the axis continues to develop and produce lateral branches; i.e. shows monopodial growth. This results in the older flowers being towards the base of the inflorescence, or, in flat-topped clusters, to the outside of the flower head. Types of racemose inflorescence include the capitulum, corymb, raceme, spike, and umbel. *Compare* cymose inflorescence. *See also* monopodial.

rachis /ray-kis/ *See* feathers.

rachis /ray-kis/ **1.** The main stalk on which the leaflets are borne in a compound leaf, or on which the pinnae are attached in the fern leaf.
2. The central axis of certain inflorescences (e.g. wheat) that bears the spikelets.
3. *See* feathers.

radial symmetry The arrangement of parts in an organism in such a way that cutting in any plane across the diameter splits the structure into similar halves (mirror images). Radial symmetry is characteristic of many sedentary animals, e.g. Cnidaria and Echinodermata. The term *actinomorphy* is generally used to describe radial symmetry in plants, particularly flowers. *See also* bilateral symmetry.

radicle /rad-ă-kăl/ The embryonic root. It is the first organ to emerge from the seed on germination. The radicle is joined to the hypocotyl and both tissues are derived from the four octants nearest the suspensor. The root cap over the tip of the radicle is derived from the closest cell of the suspensor.

radioactive dating Any method of dating that uses the decay rates of naturally occurring radioactive isotopes to assess the age of a specimen. Organic matter less than 7000 years old can be dated using radiocarbon dating. This uses the fact that the isotope carbon-14 is found in the atmosphere and taken in by plants when they photosynthesize, and subsequently assimilated by the animals that feed on them. When plants and animals die, no more carbon is taken in and the existing ^{14}C decays to the nonradioactive isotope carbon-12. If the proportion of ^{14}C to ^{12}C in the atmosphere and the decay rate of ^{14}C to ^{12}C are both known, as they are, then the sample may be dated by finding the present proportion of ^{14}C to ^{12}C. Specimens more than 7000 years old can be dated by other radioisotope methods such as potassium–argon dating.

radioimmunoassay /ray-dee-oh-i-myoo-noh-a-**say**, -**ass**-ay/ (**RIA**) A type of immunoassay used for finding the concentration of a particular substance, for example a protein, in a biological sample. The substance acts as an antigen, binding to specific antibodies. Also required is a preparation of the substance labeled with a radioisotope. Two series of mixtures are prepared, each using the same known concentrations of antibody and labeled (antigenic) substance; in one series various standard solutions of the substance are added; in the second series, various dilutions of the sample are added. In each case, the unlabeled antigen competes with the labeled antigen for binding sites on the antibody. When the antigen–antibody complexes are separated from the reaction mixture, the ratio of labeled antigen/unlabeled antigen can be measured, and the concentration of substance in the sample

found by comparison with the series of standard solutions. The technique is highly sensitive, and is widely used in medicine and research to make accurate determinations of biochemical substances, including a huge range of enzymes, hormones, drugs, and other substances.

radius /**ray**-dee-ŭs/ (*pl.* **radii**) One of the two long bones of the lower forelimb (forearm) in tetrapods. In humans, it forms the anterior (preaxial) border of the forearm, extending from the upper side of the elbow joint to the thumb side of the wrist (carpus). It is able to twist against the larger ulna bone in a pivot joint, to turn the palm of the hand upward or downward when the arm is extended. The radius and ulna may be fused in some species. *See illustration at* pentadactyl limb.

radula /**raj**-û-lă/ (*pl.* **radulae**) A ribbon-like strip on the tongue of most mollusks. It is covered with tiny horny teeth that act like a file to scrape away the surface of the vegetation on which it feeds. Continual growth from its origin enables the radula to be replaced as it wears away.

ragworms /**rag**-wermz/ *See* Nereis.

rainforest Natural forest that grows in regions of high rainfall, typically receiving more than 200–300 cm rain per year. The plentiful year-round supply of moisture sustains these highly diverse and productive ecosystems, which are repositories of vast numbers of animal and plant species. *Tropical rainforest* grows in equatorial regions, such as the Amazon basin and central Africa, and is dominated by tall flowering evergreen trees, which form a dense canopy some 30–40 m above ground. Epiphytes abound in the humid conditions, and there are large populations of insects, birds, and arboreal mammals. *Temperate rainforest* tends to be restricted to temperate seaboard regions, such as the Pacific northwest coast of North America. Here, conifers predominate, along with epiphytes and also a rich ground flora of herbs and shrubs. The many high-value timber trees in the rainforests has resulted in commercial exploitation and indiscriminate large-scale felling.

Rana /ran-ă, rah-nă/ (frog) *See* Anura.

rank The hierarchical status of a taxon in a classification scheme. For example, the taxon Annelida has the rank of phylum, while the taxon Oligochaeta has the rank of class.

raphe /ray-fee/ **1.** In an anatropous ovule, the portion of the funicle that is fused with the integument.
2. A slit found in the valve of motile diatoms.

raphides /raf-ă-deez/ Bunches of needle-like crystals of calcium oxalate found in certain plant cells.

rapid eye movement (REM) Describing a stage of sleep, also known as dreaming or desynchronized sleep (D-sleep), associated with dreaming in humans and activation of the autonomic nervous system in a wide range of animals. It is characterized by a range of physiological signs, such as involuntary muscle movements (including those of the eyes), irregularity in pulse and blood pressure, and penile erection in males.

ratites /rat-ÿts/ A group (formerly considered a subclass or superorder) that contains the flightless birds, such as *Struthio* (ostrich) of Africa – the largest living bird, *Dromaius* (emu) of Australia, *Rhea* of South America, and the recently extinct *Diornis* (moa) of New Zealand. Ratites are large heavy fast-running birds with long powerful hind limbs and reduced wings. They are confined to open lands in the southern hemisphere, where carnivores are few and their approach is easily seen. They have no keel on the sternum. The soft curly feathers lack barbs and the palate differs from that of other birds. The feathers and palate are thought to have evolved by neotony. Ratites are descended from flying birds by several different evolutionary lines and are probably not closely related to each other. *See also* Aves.

Rattus /rat-ŭs/ (rat) *See* Rodentia.

Raunkiaer's plant classification /rown-kee-airz/ A classification of growth forms based on the persistence of the shoots and the position of the resting buds. The system simplifies assessment of the percentages of different plant forms in any given type of vegetation. The classification is named for the Danish botanist Christian C. Raunkiaer (1860–1938). *See also* chamaephyte; cryptophyte; helophyte; hemicryptophyte; phanerophyte; therophyte.

ray *See* medullary ray.

ray initial *See* initial.

reaction time The time that elapses between stimulation of a whole organism and the production of a detectable response. Reflex arcs shorten reaction time to a minimum, by means of direct connections between receptor and effector, with only a very few intervening synapses. Non-reflex responses are slower, due to the delay in impulse transmission at each synapse. *Compare* latent period.

reading frame A sequence of codons in a messenger RNA (mRNA) molecule that runs from a start codon to a stop codon, as recognized by the ribosome during translation. It thus contains the genetic information required for the assembly of a specific polypeptide. However, a few mRNAs contain two or more overlapping reading frames, in which the same base sequence can be read as different sequences of codons, depending on where the ribosome starts translation. Hence, two or more different polypeptides can be encoded by the same mRNA. *See also* open reading frame.

recapitulation /ree-kă-pich-û-**lay**-shŏn/ The theory proposed by the German biologist Ernst Heinrich Haeckel (1834–1919) that the embryological development of an organism summarizes the evolutionary history of the species. The theory is now regarded as a gross oversimplification, though it is true that the embryos of related species resemble each other more closely than do the adults.

Recent *See* Holocene.

receptacle /ri-**sep**-tă-kăl/ **1.** (thalamus; torus) The tip of the angiosperm flower stalk upon which the other organs are inserted. The way in which the receptacle develops determines the position of the gynoecium relative to the other floral parts. If the receptacle is dome- or saucer-shaped the gynoecium is superior, while a gynoecium inserted on a flask-shaped receptacle is termed inferior. *See also* epigyny; hypogyny; perigyny.
2. The swollen portion of the lamina (blade) bearing the conceptacles in certain algae (e.g. *Fucus*).

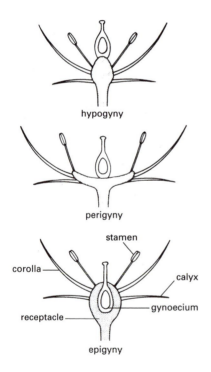

hypogyny

perigyny

stamen

corolla

calyx

receptacle

gynoecium

epigyny

Receptacle: the various arrangements
of the floral parts on the receptacle

receptor /ri-**sep**-ter/ A cell or organ that is specialized to receive and respond to stim-

uli from outside or inside the body of an organism. The eyes, ears, and nose are receptors that respond to light, sound, and airborne chemicals, respectively.

recessive /ri-**sess**-iv/ An allele that is only expressed in the phenotype when it is in the homozygous condition. *Compare* dominant. *See also* double recessive.

recipient /ri-**sip**-ee-ĕnt/ A person or animal that receives blood, tissues, or organs from another person or animal. *Compare* donor.

reciprocal cross A cross that tests whether the inheritance of a particular character is affected by the sex of the parent. The cross is thus made both ways, i.e. the character under consideration is carried by the female in one cross and by the male in the second cross. The procedure can demonstrate which characters are controlled by sex-linked genes.

recombinant DNA /ri-**kom**-bă-nănt/ **1.** Any DNA fragment or molecule that contains inserted foreign DNA, whether from another organism or artificially constructed. Recombinant DNA is fundamental to many aspects of GENETIC ENGINEERING, particularly the introduction of foreign genes to cells or organisms. There are now many techniques for creating recombinant DNA, depending on the nature of the host cell or organism receiving the foreign DNA. Particular genes or DNA sequences are cut from the parent molecule using specific type II RESTRICTION ENDONUCLEASES, or are assembled using a messenger RNA template and the enzyme REVERSE TRANSCRIPTASE. In GENE CLONING using cultures of bacterial or eukaryote tissue cells, the foreign gene is inserted into a VECTOR, e.g. a bacterial plasmid or virus particle, which then infects the host cell. Inside the host cell the recombinant vector replicates and the foreign gene is expressed. Plasmids are also used to insert foreign DNA into plants. One of the most common is the Ti (tumor-inducing) plasmid of the bacterium *Agrobacterium tumefaciens*. This causes crown gall tumors in

plants, and its plasmid has been used on a range of crop plants. Some animal and plant cells take up foreign DNA directly from their environment. For example, mouse embryos can be injected with DNA. The embryos are then implanted into receptive mothers, which give birth to transgenic offspring. The same principle is used with other mammals, including sheep. Another technique, used in transfecting certain plant cells or cell organelles, for example, is to shoot DNA-coated microprojectiles, such as tungsten or gold particles, at the host cell target. This is termed *biolistics. See* transfection.
2. DNA formed naturally by recombination, e.g. by crossing over in meiosis or by conjugation in bacteria. *See* recombination.

recombination /ree-kom-bă-**nay**-shŏn/ The regrouping of genes that regularly occurs during meiosis as a result of the independent assortment of chromosomes into new sets, and the exchange of pieces of chromosomes (crossing over). Recombination results in offspring that differ both phenotypically and genotypically from both parents and is thus an important means of producing variation.

rectum /**rek**-tŭm/ The last part of the alimentary canal, in which feces are stored and released at intervals to the exterior through an anus or cloaca. In mammals it is closed by a sphincter. In insects it may be used for reabsorption of water.

red algae *See* Rhodophyta.

red blood cell (**red blood corpuscle**) *See* erythrocyte.

redia /**ree**-dee-ă/ (*pl.* **rediae**) *See* cercaria.

redox *See* oxidation–reduction.

redox potential /**ree**-doks/ (**oxidation–reduction potential**) The tendency of a substance to accept electrons (i.e. be reduced) or donate electrons (i.e. be oxidized) in a chemical reaction. It is measured in volts using an electrode placed in a standard so-

lution of the oxidized and reduced forms of the substance, with reference to a hydrogen electrode. Compounds with large negative redox potentials tend to donate electrons and are strong reducing agents; examples are NADH and NADPH (*see* NAD; NADP). Compounds with large positive redox potentials are strong oxidizing agents, for example water. *See* oxidation–reduction.

reduction division The first division of MEIOSIS, including prophase, metaphase I, and anaphase I. It results in a haploid number of chromosomes gathering at each end of the nuclear spindle.

reflex /**ree**-fleks/ An innate and stereotyped response to a stimulus. In vertebrates, the receptor and effector involved are connected by neurones running to and from the central nervous system, forming a reflex arc. The functions of reflexes are to perform repetitive actions, such as swimming in fishes, and to provide very fast response to harmful stimuli, by connecting receptor and effector with a minimum of synaptic delay.

Peripheral nerves emerge from the central nervous system segmentally; body reflexes are mediated in the appropriate sector of the spinal cord and head reflexes in the brain. However, the reflex arc can be, and usually is, connected to interneurones, which transmit impulses to other body levels, so that reflex action can involve many levels at once. There may also be quite complex nervous integration of reflexes, especially those concerning difficult learnt activities, such as speech and walking. *See also* conditioned reflex.

refractory period The period following passage of an impulse along a nerve when either no stimulus, however large, will evoke a further impulse (the absolute refractory period) or only an abnormally large stimulus will evoke further impulse (the relative refractory period). During this time the resting potential of the cell membrane is recovered by the active pumping of sodium ions out of the cell. *See also* nerve impulse.

regeneration /ri-jen-ĕ-**ray**-shŏn/ The regrowth by an organism of an organ or tissue that has been lost through injury, autotomy, etc. The powers of regeneration vary between different groups; they are best seen in plants and lower animals. In some cases a complete organism can sometimes be regenerated from a few cells. Regeneration in mammals is limited to wound-healing and regrowth of peripheral nerve fibers.

regma /**reg**-mă/ A type of capsular fruit, found for example in geranium, that resembles the carcerulus except that there is an explosive splitting into segments.

reinforcement The strengthening of a response during learning. The achievement of a goal by performing a particular action reinforces the behavior carried out to achieve the goal. In experiments a reward for carrying out a desired action is a positive reinforcement to encourage learning of the action; punishment to teach avoidance of a particular type of behavior is a negative reinforcement.

relative molecular mass (**molecular weight**) Symbol: M_r The ratio of the mass of an atom or molecule to one-twelfth of the mass of an atom of carbon-12. It is thus a mass ratio, and has no units. It is commonly used to indicate the size of proteins and other large biomolecules. For example, a protein of M_r 20 000 would have an equivalent molecular mass of 20 000 daltons (or 20 kDa) (*see* dalton).

relaxin /ri-**laks**-in/ A hormone produced by the corpus luteum and responsible for the inhibition of uterine contraction. During birth, relaxin stimulates dilation of the cervix and relaxes the pubic symphysis enabling the pelvic girdle to widen.

release factor A protein that is instrumental in terminating TRANSLATION of messenger RNA (mRNA) by ribosomes and releasing the newly formed polypeptide from the ribosome. It recognizes a stop codon in the mRNA molecule, and binds to the ribosomal A site.

releaser A stimulus that elicits an instinctive behavioral pattern by an animal. A social releaser is produced by a member of the same species as the reactor, and is commonly used in courtship and threat displays.

releasing hormone Any hormone that triggers the release of another hormone by a target tissue, particularly any of the hormones released by the hypothalamus that stimulate secretion of specific hormones by the pituitary gland. For example, corticotropin-releasing hormone causes the adenohypophysis of the pituitary to release corticotropin.

rem /rem/ (*radiation equivalent man*) A unit for measuring the effects of radiation dose on the human body. One rem is equivalent to an average adult male absorbing one rad of radiation. The resultant biological effects depend on the type of radiation as well as the energy deposited per kilogram. *See also* gray.

REM *See* rapid eye movement.

renal capsule /**ree**-năl/ *See* Bowman's capsule.

renal pelvis *See* pelvis.

renal portal system A venous pathway comprising the renal portal veins, which carry blood from the capillary beds in the tail and/or hind limbs to the capillary bed in the kidneys. Blood then leaves the kidneys via the renal veins and returns to the heart. A renal portal system is found in animals with mesonephric kidneys, i.e. most fish, amphibians, and some reptiles.

renal tubule *See* kidney tubule.

renin /**ree**-nin/ A proteolytic enzyme produced in the kidney and released into the bloodstream that splits angiotensin from its precursor, angiotensinogen. *See* angiotensin.

rennin /**ren**-in/ *See* chymosin.

repetitive DNA DNA that consists of multiple repeats of the same nucleotide sequences. Unlike prokaryotic cells, eukaryotic cells contain large amounts of repetitive DNA – in mammals it makes up roughly half of the total DNA. Most repetitive DNA does not encode any product and its function is unknown; it constitutes part of the so-called 'junk DNA'. One major component of this consists of very long stretches of tandemly repeated short DNA sequences (usually less than 10 bp) that flank the centromeres of each chromosome. When the total DNA is subject to centrifugation, these *centromeric DNA* repeats separate out as a distinct band called SATELLITE DNA. Tandemly repeated short sequences also occur at the tips of chromosomes (*telomeric DNA*); this safeguards functional regions against loss of material from the tips. Other types of repetitive DNA occur throughout the genome, e.g. as noncoding introns within genes and as *spacer DNA* between genes. VARIABLE NUMBER TANDEM REPEATS (VNTRs) consist of hundreds or thousands of repeats of particular sequences (each sequence typically 15–100 nucleotides long), occurring at numerous sites within the genome. These make up *minisatellite DNA*, whereas *microsatellite DNA* comprises repeats of shorter sequences (2–10 bp). Many TRANSPOSONS also contribute to repetitive DNA. Functionally important repetitive DNA is represented by multiple copies of particular genes or gene sequences (*see* gene family). For example, genes encoding histones or ribosomal RNAs are often repeated to form tandem arrays. *See also* selfish DNA.

replacing bone *See* cartilage bone.

replica /rep-lă-kă/ A thin detailed copy of a biological specimen, obtained by spraying the surface with a layer of plastic and carbon. Replicas are used in electron-microscope work.

replication /rep-lă-**kay**-shŏn/ The mechanism by which exact copies of the genetic material are formed. Replicas of DNA are made when the double helix unwinds and the separated strands serve as templates along which complementary nucleotides are assembled through the action of the enzyme DNA polymerase. The result is two new molecules of DNA each containing one strand of the original molecule, and the process is termed *semiconservative replication*. In eukaryotes, replication proceeds simultaneously in both directions at many points along the DNA molecule – the *origins of replication*. At each site the two new daughter strands are assembled in different ways, because of the orientation of the parent strands: one new strand (the *leading strand*) is assembled continuously from its growing end, while the other new strand (the *lagging strand*) is assembled from short segments, which are joined together in a process termed *discontinuous replication*. In RNA viruses an RNA polymerase is involved in the replication of the viral RNA.

reporter gene A gene whose expression can be readily detected and is used to study the regulation of gene activity, and the behavior of foreign genes (transgenes) in their host cells. For example, a reporter gene can be fused with the promoter of a particular gene of interest to identify in which tissues the gene is expressed. Reporter genes are selected because the reporter proteins they encode are easy to identify in tissue samples or cell cultures. A favorite example is the *lacZ* gene of *E. coli*. This encodes the enzyme β-galactosidase, which normally breaks down lactose, but is also able to degrade the synthetic lactose analog X-Gal, forming a bright-blue dye. Hence, expression of the *lacZ* reporter gene is revealed by a blue stain. Another popular reporter protein is the green fluorescent protein (GFP), encoded by a gene from the jellyfish *Aequorea victoria*. GFP reveals itself by emitting green fluorescent light.

reporter molecule A molecule having a characteristic property, e.g. fluorescence, ultraviolet absorbance, that is sensitive to polarity. The molecule is introduced into a protein so that changes in the property can be monitored in order to measure changes in the environment of the protein.

repressor /ri-**press**-er/ A protein molecule that prevents protein synthesis by binding to the operator sequence of the gene and preventing transcription. The molecule is produced by a regulatory gene and may act either on its own or in conjunction with a *corepressor*. In some cases another molecule, an *inducer*, may bind to the repressor, weakening its bonds with the operator and derepressing the gene, allowing transcription to proceed. *See* operon.

reproduction /ree-prŏ-**duk**-shŏn/ *See* asexual reproduction; sexual reproduction.

Reptilia /rep-**til**-ee-ă/ The class of vertebrates that contains the first wholly terrestrial tetrapods, which are adapted to life on land by the possession of a dry skin with horny scales, which prevents water loss by evaporation. Fertilization is internal and there is no larval stage. The young develop directly from an amniote egg that has a leathery shell and is laid on land, i.e. it is cleidoic. Respiration is by lungs only and the heart has four chambers, although oxygenated and deoxygenated blood usually mix. Other advanced features are the clawed digits and the metanephric kidney. Like amphibians, but unlike birds and mammals, reptiles are poikilothermic.

Reptiles, notably the DINOSAURS, were the dominant tetrapods in the Mesozoic period. Modern forms include the predominantly terrestrial lizards and snakes (order SQUAMATA), as well as the aquatic crocodiles and turtles. Reptiles evolved from primitive Amphibia and Mesozoic reptiles included aerial members (e.g. *Pteranodon*) and aquatic forms (e.g. *Ichthyosaurus*), as well as the terrestrial dinosaurs (e.g. *Tyrannosaurus*). Some groups gave rise to birds and mammals. Primitive reptiles had a pineal eye but this is lost in most modern forms. *See also* Ichthyosauria; Pterosauria.

residual body /ri-**zij**-û-ăl/ *See* lysosome.

residual volume The volume of air that cannot be forcibly expelled from the lungs. In humans it amounts to approximately 1200 ml. *See* vital capacity.

resin /**rez**-in/ One of a group of acidic substances occurring in many trees and shrubs (e.g. conifers and some poplars) either as sticky glassy solids or in solution in essential oils, i.e. as balsams, such as turpentine. Resins may be phenolic derivatives or oxidation products of terpenes, a group of substances which have branched-chain carbon skeletons consisting of 5-carbon units. They are usually secreted by special cells into long resin ducts or canals. Sometimes they are produced in response to injury or infection. They sometimes form a sticky covering to buds (e.g. horse chestnut), reducing transpiration and giving protection. Some are important commercially, for example oleoresin from pine tree bark is a source of rosin which, on distillation, yields turpentine.

resolution (of racemates) The separation of a racemate into the two optical isomers. This cannot be done by normal methods, such as crystallization or distillation, because the isomers have identical physical properties. The main methods are:
1. Mechanical separation. Certain optically active compounds form crystals with distinct left- and right-handed shapes. The crystals can be sorted by hand.
2. Chemical separation. The mixture is reacted with an optical isomer. The products are then not optical isomers of each other, and can be separated by physical means.

For instance, a mixture of D- and L-forms of an acid, acting with a pure L-base, produces two salts that can be separated by fractional crystallization and then reconverted into the acids.
3. Biochemical separation. Certain organic compounds can be separated by using bacteria that feed on one form only, leaving the other.

resolving power (resolution) The ability of an optical system to form separate images of closely spaced objects. *See* microscope.

respiration /res-pă-**ray**-shŏn/ The oxidation of organic molecules to provide energy in plants and animals. In animals, food molecules are respired, but autotrophic

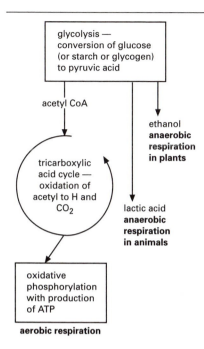

glycolysis —
conversion of glucose
(or starch or glycogen)
to pyruvic acid

acetyl CoA

ethanol
**anaerobic
respiration
in plants**

tricarboxylic
acid cycle —
oxidation of
acetyl to H and
CO_2

lactic acid
**anaerobic
respiration
in animals**

oxidative
phosphorylation
with production
of ATP

aerobic respiration

Respiration: some of the principal
biochemical pathways

plants respire molecules that they have themselves synthesized by photosynthesis. The energy from respiration is used to attach a high-energy phosphate group to ADP to form the short-term energy carrier ATP, which can then be used to power energy-requiring processes within the cell. The actual chemical reactions of respiration are known as *internal* (*cell* or *tissue*) *respiration* and they normally require oxygen from the environment (*see* aerobic respiration). Some organisms are able to respire, at least for a short period, without the use of oxygen (*see* anaerobic respiration), although this process produces far less energy than aerobic respiration. Respiration usually involves an exchange of gases with the environment; this is known as *external respiration*. In small animals and all plants exchange by diffusion is adequate, but larger animals generally have special respiratory organs with large moist and ventilated surfaces (e.g. lungs, gills) (*see* ventilation) and there is often a circulatory system to transport gases internally to and from the respiratory organs.

The complex reactions of cell respiration fall into two stages, GLYCOLYSIS and the KREBS CYCLE. Glycolysis occurs within the cytoplasm, but the Krebs cycle enzymes are localized within the mitochondria of eukaryotes; cells with high rates of respiration (e.g. insect flight muscles) have many mitochondria. Glycolysis results in partial oxidation of the respiratory substrate to the 3-carbon compound pyruvate, but it can also occur in anaerobic conditions, when the pyruvate is converted in animals to lactic acid and in plants to ethanol. The net yield of glycolysis is 2 molecules of ATP plus 2 molecules of the reduced coenzyme $NADH_2$ for each molecule of glucose respired.

In the Krebs cycle, which requires free oxygen, pyruvate is converted into the 2-carbon acetyl group, which becomes attached to a coenzyme forming acetyl coenzyme A. This then enters a cyclic series of reactions during which carbon dioxide is evolved and hydrogen atoms are transferred to the coenzymes NAD and FAD. The energy released by the Krebs cycle is transferred via the reduced coenzymes $NADH_2$ and $FADH_2$ to an ELECTRON-TRANSPORT CHAIN embedded in the inner mitochondrial membrane. According to the chemiosmotic theory, electrons flow through the various components of the chain, while a gradient of hydrogen ions is created across the membrane. The osmotic and electrical force of this gradient drives the formation of ATP. The overall net yield from each molecule of glucose oxidized during aerobic respiration is traditionally quoted as 38 (or 36) molecules of ATP. However, current opinion is that aerobic respiration gives a net yield of 31 or 29.5 ATP per glucose molecule, depending on the shuttle mechanism by which NADH in the cytoplasm is transported into the mitochondrion. This compares with only 2 ATP during anaerobic respiration. ATP is not transported between cells, but is made in the cell where it is required.

respiratory chain /res-pă-ră-tor-ee, -toh-ree/ The ELECTRON-TRANSPORT CHAIN in aerobic respiration.

respiratory movement *See* ventilation.

respiratory organ An organ by which oxygen is absorbed into the body of an animal and by which carbon dioxide is released. Gaseous exchange at the organ is by diffusion, so the exchanging surface of the organ must be folded or subdivided to provide a large area that is thin and moist. In terrestrial animals the surface is kept moist by special mucus-secreting cells. There is usually a mechanism to ventilate the surface on one side and a flow of fluid (tissue fluid or blood) on the other side to transport gases between the surface and the other tissues of the body. *See also* bronchus; bronchiole; gill; gill books; lung; lung book; trachea; ventilation.

respiratory pigments Colored compounds that can combine reversibly with oxygen. Hemoglobin is the blood pigment in all vertebrates and a wide range of invertebrates. Other blood pigments, such as hemerythrin (containing iron) and hemocyanin (containing copper), are found in lower animals, and in many cases are dissolved in the plasma rather than present in cells. Their affinity for oxygen is comparable with that of hemoglobin, though oxygen capacity is generally lower.

respiratory quotient (**RQ**) The ratio of the volume of carbon dioxide expired by an organism compared to the volume of oxygen consumed during the same period of respiration. A theoretical RQ can be calculated for the various foodstuffs used in respiration, giving a value of 1 for carbohydrates, 0.7 for fats, and 0.8 for proteins. However, in practice, more than one foodstuff is respired at one time and other metabolic processes may produce carbon dioxide or use oxygen, so an RQ measurement for an organism gives unreliable information about the type of foodstuff respired.

response A change in an organism or in part of an organism that is produced as a reaction to a stimulus.

resting potential The potential difference that exists across the plasma membrane of a cell in a resting, or nonexcited state. Animal cells, including neurons (nerve cells), normally maintain a resting potential in the range -20 to -100 mV. In plant cells the value is typically about -120mV in the cytosol; this becomes more positive in the vacuole, by some 20–30mV. The resting potential results from differences in concentrations of ions and electrically charged molecules in the cell's interior compared with the exterior. For example, the concentration of potassium ions (K^+) inside a typical mammalian cell is 20–40 times greater than that in blood, whereas the internal concentration of sodium ions (Na^+) is 8–12 times lower than that in blood. Cells regulate their internal ion concentrations by selectively exporting and/or importing ions. In animal cells, the resting potential is achieved chiefly by the outward diffusion of K^+ through K^+ channels in the plasma membrane, while largely blocking the entry of Na^+ by keeping Na^+ channels closed. The K^+ concentration gradient, on which this outward diffusion depends, is maintained by the active pumping of K^+ ions into the cell, in exchange for Na^+. The chief mechanism in plant cells is the active pumping of H^+, across the plasma membrane out of the cell, and across the tonoplast into the vacuole, by a proton pump. The resting potential is important in various aspects of a cell's life, including the regulation of pH and osmotic pressure; uptake or expulsion of materials; growth and movement (of plant cells); the conduction of nerve impulses by neurons, and initiation of contraction in muscle cells.

restriction endonuclease (**restriction enzyme**) Any of a large group of enzymes, found mainly in bacteria, that can cleave and fragment DNA internally (*see* endonuclease). They are so named because they cut only foreign DNA entering the cell, and so act to 'restrict' the replication of the DNA of an invading virus; hence their natural function is protection of the bacterial cell. Some restriction endonucleases cleave DNA at random, but a particular group of enzymes, known as *class II restriction enzymes*, cleave DNA at specific sites, called *restriction sites*. These sites consist of a sequence of four to eight base pairs. The same sites on the host cell's own DNA are protected from this enzyme activity by methylation, which is controlled by another type of site-specific enzyme.

Two properties of restriction enzymes have made them indispensable tools for producing recombinant DNA molecules for use in genetic engineering. First, digestion of DNA with a particular restriction enzyme produces a characteristic and reproducible set of double-stranded DNA fragments, called *restriction fragments*. Second, many restriction enzymes make a staggered cut across the two strands of the DNA molecule, giving the resulting DNA fragment identical single-stranded 'sticky' ends. The ends readily form hydrogen bonds with the complementary ends on all other fragments generated by the same restriction enzyme. This enables restriction fragments from quite different sources (e.g. different organisms) to be easily joined together, forming chimeric molecules. Over 2500 type II restriction endonucleases have been discovered, with some 200 different cleavage site specificities. They have been named according to the organism in which they occur. For example, *Eco*R1 is obtained from *Escherichia coli*, strain R, and was the first enzyme to be isolated in this bacterium.

restriction fragment length polymorphism (RFLP)

Variation among the members of a population in the sites at which restriction enzymes cleave the DNA, and hence in the size of the resulting DNA fragments. It results from differences between individuals in nucleotide sequences at the cleavage sites (restriction sites). The presence or absence of particular restriction sites can be ascertained using DNA probes in the technique called SOUTHERN BLOTTING. Restriction sites vary enormously, and this variation is exploited in analyzing and comparing the genomes of different individuals, e.g. to establish how closely related they may be. Restriction sites are also invaluable as genetic markers in chromosome mapping, and can be used to track particular genes. *See* chromosome map.

restriction map A map of a segment of DNA showing the cleavage sites of RESTRICTION ENDONUCLEASES and their physical distance apart, usually measured in base pairs. It can be used to reveal variations in restriction sites between individuals of the same species or between different species (*see* restriction fragment length polymorphism). This variation serves as a key to the organism's genes, since the restriction sites can be used as markers to identify closely linked genetic loci and allow investigation of deletions, insertions, or other mutations. They are an essential tool in chromosome mapping.

The first step in constructing a restriction map is to label the ends of the DNA with a radioisotope. Then the DNA is subjected to a series of total and partial digests with one or more restriction enzymes. The fragments resulting from each digest are separated according to size by gel electrophoresis, and the order of fragments, and hence restriction sites, deduced from the various fragment sizes and the labeled ends. *See also* genetic fingerprinting.

restriction point A point late in G_1 phase of the CELL CYCLE when a cell becomes committed to enter the S phase and complete the cycle.

reticular activating system /ri-**tik**-yŭ-ler/ (reticular formation) A network of short-axoned nerve cells in the brain stem that receives stimuli from peripheral sense organs and transmits them to the cerebral cortex, so regulating the level of consciousness. Destruction of the system or deprivation of sensory input results in hallucination or coma.

reticulate thickening /ri-**tik**-yŭ-lit, -layt/ The type of secondary-wall formation in which an irregularly branching mass of lignified deposits covers the inner wall of the cell. It is usually found in metaxylem tracheids and vessels. Reticulate thickening is often difficult to distinguish from the ladder-like scalariform thickening. *See* xylem.

reticulocyte /ri-**tik**-yŭ-lŏ-sÿt/ An immature red blood cell. It develops in the red marrow in bones from a *proerythroblast* (red cell precursor) that gradually accumulates hemoglobin until it is a fully formed reticulocyte. In mammals it then loses its nucleus and is released into the blood as an erythrocyte.

reticuloendothelial system /ri-**tik**-yŭ-loh-en-doh-**th'ee**-lee-ăl/ *See* mononuclear phagocyte system.

reticulum /ri-**tik**-yŭ-lŭm/ (*pl.* **reticula**) The second region of the specialized stomach of ruminants (e.g. the cow). The cud passes back into the reticulum after it has been regurgitated and chewed. It is lined with tough cornified stratified squamous epithelium and is formed in sectional folds like a honeycomb.

retina /**ret**-ă-nă/ The innermost light-sensitive layer of the vertebrate eye. It consists of two types of photosensitive cells (RODS and CONES), adjacent to, and at right angles to, the choroid. The rods and cones are connected by synapses to bipolar and ganglion nerve cells. From the ganglion cells, nerve fibers pass over the inner surface of the retina to the optic nerve. Light entering the eye through the cornea and pupil has to pass through all the layers of the retina before it reaches the sensitive ends of the rods and cones, except at the FOVEA. *See illustration at* eye.

retinal /**ret**-ă-năl/ (**retinene**) An aldehyde derivative of retinol (vitamin A). Retinal is a constituent of the light-sensitive conjugated protein, RHODOPSIN, which occurs in the rod cells of the retina.

retinene /**ret**-ă-neen/ *See* retinal.

retrotransposon /ret-roh-tran-**spoh**-zon/ A type of TRANSPOSON found in the DNA of eukaryotes that moves to new sites within the genome by making an RNA version of itself. This mechanism is analogous to the replication mechanism of retroviruses, hence the name. Like retroviruses integrated in the DNA of a host cell, retrotransposons use the RNA polymerase and other enzymes of the cell to assemble an RNA intermediate. This is converted into a DNA version by a reverse transcriptase enzyme encoded by the retrotransposon. The DNA then inserts at a new site within the host DNA.

retrovirus /**ret**-roh-vÿ-rŭs/ An RNA-containing virus whose genome becomes integrated into the host DNA (by means of the enzyme reverse transcriptase) and then replicates with it. These viruses can cause cancerous changes in their host cells (i.e. they are oncogenic) by means of the activity of one or more of their genes (*see* oncogene). Examples of retroviruses are the Rous sarcoma virus (RSV), which was discovered in 1911 and causes cancer in chickens, and HIV (the virus causing AIDS). *See* HIV.

reverse transcriptase /tran-**scrip**-tayss/ An enzyme that catalyzes the synthesis of DNA from RNA (i.e. the reverse of transcription, in which mRNA is synthesized from a DNA template). The enzyme occurs in certain RNA viruses (*see* retrovirus) and enables the viral RNA to be 'transcribed' into DNA, which is then integrated into the host DNA and replicates with it. It is also used in genetic engineering to make complementary DNA (cDNA) from an RNA template.

RFLP *See* restriction fragment length polymorphism.

R$_f$ value In paper chromatography, the distance traveled by the solute divided by the distance traveled by the solvent front, the latter always being taken as one. The R$_f$ (relative front) value is thus always between zero and one and is characteristic of a particular molecule. In this way various amino acids, chlorophylls, etc., may be identified. *See* paper chromatography.

rhesus factor /**ree**-sŭs/ (**Rh factor**) An antigen attached to human red blood cells, so named because it is also present in the rhesus monkey. The antigen is present in most people, who are therefore described as rhesus-positive (Rh-positive), but absent in others (Rh-negative).

Normally, neither type of blood contains the anti-Rh antibody. However, it may be present in the blood of Rh-negative women who have borne Rh-positive children: during pregnancy, the red cells of the fetus, carrying the Rh factor, can diffuse across the placenta and stimulate anti-Rh antibody production in the mother's blood. If the woman subsequently becomes pregnant with another Rh-positive fetus, the anti-Rh antibody could diffuse into the circulation of the fetus, producing a serious condition called hemolytic disease of the newborn, in which the red cells of the fetus are destroyed. To prevent this, Rh-negative mothers are injected soon after delivery with a concentrated immunoglobulin that destroys any Rh-positive fetal red cells in her circulation, thus preventing the production of anti-Rh antibodies.

Rhizobium /rÿ-**zoh**-bee-ŭm/ A spherical or rod-shaped bacterium that can live either freely in the soil or symbiotically in the root nodules of leguminous plants and a few other species, such as alder. The bacteria can move slowly through the soil by means of flagella and are attracted to and infiltrate the root hairs of leguminous plants. They produce infection threads that penetrate the cells of the root cortex, which are stimulated to divide rapidly and form a swollen mass of tissue, the root nodule. The central region of the nodule consists of enlarged cells containing large numbers of bacteria. The outer region of the nodule contains vascular strands linking with the vascular bundles of the root. The bacteria in the nodules use atmospheric nitrogen and the resulting nitrates can be passed to the plant. In return, the bacterium is supplied with carbohydrates, such as sugars. *See also* nitrogen cycle; nitrogen fixation.

rhizoid /rÿ-zoid/ A unicellular rootlike structure found in certain algae and in the gametophyte generation of the bryophytes and some pteridophytes. It serves to anchor the plant and to absorb water and nutrients.

rhizome /rÿ-zohm/ A stem that grows horizontally below ground. Rhizomes may be fleshy (e.g. iris) or wiry (e.g. couch grass) and may serve as an organ of perennation or vegetative propagation, or occasionally both. *Compare* stolon. *See also* rootstock.

rhizomorph /rÿ-zŏ-morf/ A structure found in certain fungi (e.g. *Agaricus*) consisting of a mass of hyphae forming a complex strand of tissue that serves to transport food from one part of the mycelium to another, thus aiding propagation.

rhizophore /rÿ-zoh-for, -fohr/ An organ found only in *Selaginella* species that arises just beyond each fork of the stem. It resembles a root in being positively geotrophic and having a similar internal anatomy. However, it arises exogenously and lacks a root cap. True roots develop endogenously from the tip of the rhizophore.

Rhizopoda /rÿ-**zop**-ŏ-dă/ A phylum of ameboid protozoan protoctists that lack undulipodia (cilia and flagella). They have an irregular shape due to the formation of pseudopodia, used for locomotion and food capture. Most are solitary, occurring in marine and fresh waters, but a few are parasites (e.g. *Entamoeba*). Many (e.g. *Amoeba*) have no internal skeleton, but some (e.g. *Difflugia*) have a protective outer test or shell.

rhizosphere /rÿ-zŏ-sfeer/ The area of soil surrounding plant roots, in which any microorganisms present are affected by the presence of the roots; it usually extends a few millimeters from the surface of the root. Such substances as carbohydrates, amino acids, and vitamins pass from the root into the soil and are often of nutrient value to microorganisms. Consequently, the rhizosphere may contain 10–50 times more bacteria than the surrounding soil.

Rhodophyta /roh-**doff**-ă-tă/ (**red algae**) A phylum of protoctists comprising aquatic, mainly marine, algae characterized by their red color. The color results from photosynthetic pigments, phycoerythrin and phycocyanin (*see* phycobilins). Food is stored as floridean starch and sugar and glycerol compounds. The cell walls contain sulfated polysaccharides in addition to cellulose and hemicellulose. The Rhodophyta, in common with the conjugating green algae (Gamophyta), lack undulipodia throughout their life cycle. Examples of red algae are *Porphyra* and *Gelidium*, the source of agar, used in bacterial cultures.

rhodopsin /roh-**dop**-sin/ (**visual purple**) A light-sensitive pigment in the retina. It has a protein component, opsin, linked to a non-protein molecule, retinal, which is a derivative of vitamin A. It is localized in the ROD cells. When light strikes the retina, rhodopsin is split into its separate components, opsin and retinal, from which it is subsequently regenerated. The biochemical mechanism for cone vision is analogous to rod vision, retinal being used as the chromophore, but the protein component being different.

rhombencephalon /rom-ben-**sef**-ă-lon/ *See* hindbrain.

rhytidome /**rit**-ă-dohm/ A form of BARK consisting of dead cortex and phloem as well as cork.

rib One of a series of slender curved paired bones, in the anterior body region of most vertebrates, that are attached to the vertebral column. In humans there are 12 pairs; they articulate posteriorly with the thoracic vertebrae and extend anteriorly around the thorax and connect with the breast bone (sternum) to form a cage protecting the heart and lungs. Only the first 7 pairs connect directly with the sternum (*true ribs*). Ribs 8, 9, and 10 are attached anteriorly to the rib above (*false ribs*). The ribs are also involved in respiratory movements. In many fish there are ribs of two kinds: dorsal ribs between the muscle blocks (myotomes) and ventral ribs – not

connected with the vertebrae – that support the viscera.

riboflavin /rÿ-boh-**flay**-vin/ (**vitamin B₂**) A water-soluble vitamin of the VITAMIN B COMPLEX. It is found in cereal grains, peas, beans, liver, kidney, and milk. Riboflavin is a constituent of several enzyme systems (flavoproteins), acting as a coenzyme for hydrogen transfer in the reactions catalyzed by these enzymes.

Two different forms of phosphorylated riboflavin are known to exist in various enzyme systems. These are FMN (flavin mononucleotide) and FAD (flavin adenine dinucleotide).

ribonuclease /rÿ-boh-**new**-klee-ayss/ (**RNase**) Any nuclease enzyme that cleaves the phosphodiester bonds between adjacent nucleotides in RNA. *Exoribonucleases* cleave nucleotides from one or both ends of the RNA molecule, whereas *endoribonucleases* only cleave bonds within the molecule.

ribonucleic acid /rÿ-boh-new-**klee**-ik/ *See* RNA.

ribose /**rÿ**-bohs/ A monosaccharide, $C_5H_{10}O_5$; a component of RNA.

ribosomal RNA /rÿ-bŏ-**soh**-măl/ (**rRNA**) The form of RNA that is found in the RIBOSOMES.

ribosome /**rÿ**-bŏ-sohm/ A small organelle found in large numbers in all cells that acts as a site for protein synthesis. Ribosomes are often bound to endoplasmic reticulum or may occur free in the cytoplasm. In most species they are composed of roughly equal amounts of protein and RNA. The ribosome consists of two unequally sized rounded subunits arranged on top of each other. Eukaryotic cells have larger ribosomes than prokaryotic cells but the ribosomes in mitochondria and chloroplasts are about the same size as prokaryotic ribosomes.

During TRANSLATION the ribosome moves along the messenger RNA (mRNA), enabling the peptide linkage of amino acids delivered to the site by transfer RNA molecules according to the code in mRNA. Several ribosomes may be actively engaged in protein synthesis along the same mRNA molecule, forming a polyribosome, or polysome.

ribozyme /rȳ-bŏ-zȳm/ Any RNA molecule that acts as an enzyme. It is now known that certain introns (non-coding RNA sequences) can catalyze their own removal from the primary RNA transcript and the splicing together of the cleaved ends, in a process called *self-splicing*. Ribosomal RNA is also believed to have ribozyme activity by catalyzing peptide bond formation between each incoming amino acid and the end of the growing polypeptide chain (*see* translation). Ribozymes have remarkable similarities with viroids, the minute plant pathogens consisting simply of RNA circles, and it has been proposed that viroids are escaped introns.

ribulose bisphosphate (RUBP) A 5-carbon compound that accepts carbon dioxide during PHOTOSYNTHESIS, in a reaction catalyzed by the enzyme RIBULOSE 1,5-BIS-PHOSPHATE CARBOXYLASE. Each molecule is then converted to two molecules of 3-carbon phosphoroglyceric acid. The ribulose bisphosphate is regenerated in the Calvin cycle when the carbon dioxide is converted into carbohydrate.

ribulose 1,5-bisphosphate carboxylase (**rubisco**) The enzyme that catalyzes the fixation of carbon dioxide (CO_2) into carbohydrates, the key step of the Calvin cycle during PHOTOSYNTHESIS. One molecule of CO_2 combines with the 5-carbon sugar ribulose bisphosphate (RuBP) to form two molecules of phosphoglyceric acid (PGA). Rubisco also catalyzes the key reaction of PHOTORESPIRATION, a seemingly wasteful metabolic process that competes with photosynthesis by consuming oxygen and liberating CO_2. Rubisco is a large enzyme, comprising 16 subunits, and is located in the chloroplast stroma. It accounts for some 50% of all chloroplast protein, and is thought to be the most abundant protein in nature.

rickettsiae /rik-et-see-ee/ (*sing.* rickettsia) Obligate intracellular parasitic bacteria found in certain arthropods. They are not pathogenic to the arthropod, but if an infected animal bites a mammal severe infection can result. Diseases caused by rickettsiae include typhus, psittacosis, and Rocky Mountain spotted fever.

Ringer's solution *See* physiological saline.

ring-porous Describing wood in which the largest vessels are in the early season (spring) wood, producing distinct growth rings, as seen in elm. *Compare* diffuse-porous. *See also* annual ring.

ritualization /rich-û-ă-li-zay-shŏn/ The process by which certain patterns of animal behavior are modified to form easily recognized social signals. Ritualized movements form part of the displays performed in threat and courtship and are usually developed from intention movements or from displacement activities.

RNA (**ribonucleic acid**) A nucleic acid found mainly in the cytoplasm and involved in protein synthesis. It is a single polynucleotide chain similar in composition to a single strand of DNA except that the sugar ribose replaces deoxyribose and the pyrimidine base uracil replaces thymine. RNA molecules can display various types of folds, turns, and loops due to pairing between complementary bases in different regions of the molecule. These features are important in determining the three-dimensional structure, and hence the function, of the various types of RNA. RNA is synthesized on DNA in the nucleus and exists in three forms (*see* messenger RNA; transfer RNA; ribosome). Some types of RNA have catalytic activity (*see* ribozyme). In certain viruses, RNA is the genetic material. *See also* heterogeneous nuclear RNA.

RNA interference (RNAi) The phenomenon by which small RNA molecules within a cell can induce cleavage of complementary sequences in messenger RNAs

(mRNAs), thereby preventing translation and protein synthesis, and hence effectively 'silencing' expression of the corresponding gene(s). It is brought about most effectively by short double-stranded RNA molecules, typically 19–21 bp long and called *small interfering RNAs* (siRNA). First discovered in the early 1990s, this phenomenon is now known in a wide range of organisms, including plants, fungi, nematodes, insects, and mammals. It might have evolved as a defense mechanism against RNA viruses or transposons. The mechanism is still unclear, but it is thought that the siRNA binds to an enzyme complex and partially unwinds. This enables the complex to undergo base pairing with a complementary region of an mRNA molecule. The complex then cleaves the mRNA, making subsequent translation of the mRNA impossible. RNAi is now widely used as an experimental technique to investigate how the loss of function of particular genes affects cultured cells or entire organisms (e.g. nematode worms and fruit flies). The siRNAs are custom-made with a sequence complementary to that of the target mRNA, and introduced to cells by microinjection or other techniques. *See also* gene silencing.

RNA polymerase *See* polymerase.

RNase /ar-en-**ayss**/ (**ribonuclease**) An enzyme that catalyzes the hydrolysis of the sugar–phosphate bonds of RNA. There are several types, each having a specific action. For example ribonuclease T1 degrades RNA to mono- and oligonucleotides terminating in a 3′-guanine nucleotide, while those produced by ribonuclease T2 terminate in a 3′-adenine nucleotide.

rod One of the two types of light-sensitive cells in the RETINA of the vertebrate eye. Rods are concerned with vision in dim light; they are found chiefly in the periphery of the retina and are absent from the fovea. They contain a pigment, visual purple (rhodopsin), that is bleached by light energy. This photochemical reaction breaks down rhodopsin into a protein (opsin) and retinal (a derivative of vitamin A) and causes nerve impulses to pass from the rod cells to the brain. The rhodopsin is continually reformed from retinal, using energy from ATP in the mitochondria of the rod. In bright light, the reformation does not keep pace with the destruction, so that rods can only function in dim light. Several rods connect with the same bipolar cell (retinal convergence) so that sharp images are not seen. They have low visual acuity, but high sensitivity due to summation of the impulses from several rods starting an impulse from the bipolar cell. *See also* cone.

Rodentia /roh-**den**-shă/ The largest and most successful order of mammals, including *Rattus* (rat), *Mus* (mouse), *Sciurus* (squirrel), and *Castor* (beaver). Rodents are herbivorous or omnivorous mammals with one pair of chisel-like incisor teeth projecting from each jaw at the front of the mouth and specialized for continuous gnawing. The incisors, which grow throughout life, have enamel only on the front. The wearing down of the softer dentine behind produces a sharp cutting edge. Skin folds can be inserted into the gap (diastema) between the incisors and ridged grinding molars so that inedible material (such as wood) need not be swallowed. Rodents are found universally and are mostly nocturnal and terrestrial. They are noted for their rapid breeding.

root 1. The organ that anchors a plant to the ground and that is responsible for the uptake of water and mineral nutrients from the soil. Roots develop from the radicle of the embryo and, according to the nature of branching from the seedling root, a fibrous or tap root system develops. Roots differ from shoots in lacking chlorophyll and not producing buds or leaves. In this way a root may be distinguished from an underground stem. Roots also differ from shoots in the arrangement of xylem and phloem, having a solid central strand of vascular tissue rather than a hollow cylinder of conducting tissue.

The growing point of the root is protected by a *root cap* to withstand the abrasion that occurs as the root grows through the soil. The direction of root growth is controlled by both gravity and water supply. The main absorptive region of the root is just beyond the zone of elongation behind the root tip, where the *root hairs* are formed. The fine root hairs arise from single epidermal cells. Further back along the root, lateral roots are formed, which develop from within the vascular tissue and grow out through the cortex, this being termed endogenous root formation. All roots whose derivation can eventually be traced back to the radicle are called *primary roots*. Roots that arise in any other way are termed *adventitious roots*, for example those that develop from the stem in bulbs and corms. In many plants, particularly biennials, the root may become swollen with carbohydrates, and act as an underground food store during the winter. Many other root modifications are seen such as buttress roots, contractile roots, prop roots, pneumatophores, etc.

The roots of many plant species exist in association with fungi as mycorrhiza or, as in the Leguminosae, with nitrogen-fixing bacteria to form *root nodules* (*see* nitrogen fixation; Rhizobium). Both these associations are important in the nutrition of the plants affected.
2. The part of a tooth inside the gum, held in the jaw bone. *See* teeth.

root cap A conical-shaped structure that is formed by the activity of the meristem at the root apex and forms a protective cap around the root tip. It is constantly replaced by newly formed cells as the older tissue is sloughed off with growth of the root through the soil.

root nodule *See* nitrogen fixation.

root pressure The positive hydrostatic pressure in the xylem vessels of a mature root that forces water upward into the stem of a plant. It can be demonstrated by cutting off the stem of a herbaceous plant just above the soil line. Sap continues to exude from the stump due to the root pressure. This pressure arises because the roots take up mineral ions from soil and actively transport them to the xylem vessels within

the central core (stele) of the root. Water follows into the stele by osmosis, but is prevented from diffusing back into the root cortex because of the watertight endodermis. Hence a positive pressure is established within the stele.

rootstock A vertical, usually short, underground stem. It is found in many angiosperms, e.g. rhubarb and strawberry, and in certain pteridophytes, e.g. *Osmunda*. It can reproduce vegetatively.

Rotifera /roh-**tif**-ĕ-ră/ A phylum of microscopic aquatic invertebrates that are widely distributed, usually in fresh waters. Rotifers are bilaterally symmetrical and unsegmented with a body divided into head, trunk, and tail regions. They are characterized by a ciliated crown on the head (*corona*), used in feeding and locomotion, which appears like a rotating wheel when beating. The muscular pharynx has well-developed jaws. Excretion is carried out by protonephridia. Males are often degenerate and parthenogenesis is common.

round window (fenestra rotunda) A membrane that lies between the middle and inner ear in higher vertebrates. It moves back and forth to compensate for the pressure changes in the perilymph, caused by vibrations of the oval window. *See illustration at* ear.

roundworms *See* Nematoda.

RQ *See* respiratory quotient.

r selection A form of natural selection that favors individuals or species that have a high BIOTIC POTENTIAL, and whose populations can grow rapidly where resources are plentiful. For example, following a fire, flood, or other catastrophe, a habitat often contains few competitors, and becomes available for *r*-strategists, which are typically fast growing and produce large numbers of offspring. Good examples are annual herbs and small shrubs. *Compare* K strategists.

rubisco /roo-**biss**-koh/ *See* ribulose 1,5-bisphosphate carboxylase.

rumen /**roo**-měn/ The first region of the specialized stomach of ruminants (e.g. the cow). It is sometimes called the *paunch*. Here the food is degraded and fermented by millions of symbiotic bacteria and protoctists, some of which produce the enzyme cellulase. Mammals are unable to produce this enzyme themselves, and since plant food consists mainly of cellulose, they cannot digest it without the help of these bacteria. After some time in the rumen the partly digested food, now called the *cud*, is regurgitated back to the mouth and chewed again, before being swallowed and passed into the reticulum.

ruminants /roo-mă-nănts/ *See* Artiodactyla.

runner A branch, formed from an axillary bud, that grows horizontally along the ground. The axillary buds of the runner may develop into daughter plants with adventitious roots growing from the node, as seen in the creeping buttercup. Conversely, only the terminal bud may form a plant, as in the strawberry. A new runner then develops from a branch of this daughter plant. *See also* offset; stolon.

rusts Parasitic basidiomycete fungi of the order Uredinales. The name derives from the characteristic yellow-brown streaks that appear on the host plant following the eruption of masses of spores (uredospores) through the host epidermis. The complicated life cycle of rusts involves the formation of a series of spore types – uredospores, teleutospores, basidiospores, and aecidiospores – and some need two hosts to complete their life cycle. An economically important rust is *Puccinia*, which infects many crops, notably cereals. Occasionally the phycomycete fungus *Albugo* is called rust or white rust.

saccharide /**sak**-ă-rўd, -rid/ *See* sugar.

Saccharomyces /sak-ă-roh-**mў**-seez/ (**yeasts**) A genus of unicellular ascomycete fungi that can live in both aerobic and anaerobic conditions. They are important in the brewing and baking industries, respectively, for the alcohol and carbon dioxide they produce by anaerobic respiration. Reproduction is generally asexual by budding although in adverse conditions sexual spores may be formed. These yeasts are also used in gene cloning as convenient eukaryotic hosts to express effects of inserted foreign genes.

sacculus /**sak**-yŭ-lŭs/ (*pl.* **sacculi**) The lower chamber of the labyrinth of the vertebrate inner ear, which bears the hearing organ – the cochlea of reptiles, birds, and mammals. There is a patch of sensory epithelium in the wall lining and granules of calcium carbonate in the cavity, which together are responsible for the detection of changes in position of the head with respect to gravity and the rate of change. *See also* macula; otolith. *See illustration at* ear.

sacral vertebrae /**say**-krăl/ Large strong vertebrae between the lumbar and coccygeal regions of the vertebral column that articulate with the pelvic girdle of tetrapods. There is one sacral vertebra in amphibians and two or more in reptiles, birds, and mammals.

sacrum /**say**-krŭm/ (*pl.* **sacra**) One or more fused sacral vertebrae in the lower back region that are attached to the ilia and give support to the pelvic girdle of tetrapods. In humans the sacrum is a large triangular bony mass comprising five fused sacral vertebrae.

safranin /**saf**-ră-nin/ *See* staining.

sagittal /**saj**-ă-tăl/ (*Anatomy*) Describing a section or plane that runs lengthwise along a body or part, dividing it into right and left sides. A *mid-sagittal* plane produces equal and symmetrical right and left halves.

saliva /să-**lў**-vă/ A secretion produced by the salivary glands of animals, consisting mainly of mucus. It is used to moisten and lubricate the food and in some animals contains enzymes. For example, in humans and certain insects, the enzyme amylase is present and starts off the process of starch digestion. In some insects (e.g. mosquito) the saliva contains an anticoagulant.

salivary glands /**sal**-ă-vair-ee/ Glands that secrete a watery secretion (saliva) into the buccal cavity. In the rabbit there are four pairs: parotid, infra-orbital, submaxillary, and sublingual; in humans there are three pairs: parotid, submandibular, and sublingual.

saltatory conduction /**sal**-tă-tor-ee, -toh-ree/ The mode of transmission of a NERVE IMPULSE along a myelinated nerve fiber whereby the impulse leaps between the nodes of Ranvier, considerably speeding up its passage. The fastest nerve impulses known, traveling up to 120 m/s, occur in vertebrate myelinated fibers.

The myelin sheath insulates against loss of local currents between the nodes; they are therefore transmitted along the fiber axis to a node, where in the absence of myelin they generate an action potential.

samara /**sam**-ă-ră, să-**mair**-ă/ A dry one-seeded indehiscent fruit with the pericarp extended into a membranous wing, e.g. the ash fruit.

sand Mineral particles consisting mainly of quartz, felspar, and mica, and measuring between 1.00 and 0.05 mm in diameter. Sandy soils are light and drain well. However, retention of nutrients is poor and such soils are often structureless and lose water quickly.

saprophyte /**sap**-rŏ-fÿt/ An organism that derives its nourishment by absorbing the products or remains of other organisms. Many fungi and bacteria are saprophytes and are important in food chains in returning nutrients to the soil by means of putrefaction and decay.

saprozoic /sap-rŏ-**zoh**-ik/ Describing an organism that feeds on organic material in solution, rather than on solid organic material.

sapwood (**alburnum**) The outer living xylem cells in a tree trunk, consisting of xylem elements, parenchyma, and medullary rays, that are actively involved in water transport and food storage. *Compare* heartwood.

SAR *See* systemic acquired resistance.

sarcolemma /sar-koh-**lem**-ă/ *See* skeletal muscle.

sarcomere /**sar**-kŏ-meer/ The contractile element in a SKELETAL MUSCLE fibril (MYOFIBRIL). Each sarcomere is joined to the next one by a fibrous lattice called the Z disk (Z line). Thick filaments of the protein myosin form a dark central *A band*. On either side of this is a light area, the *I band*, in which are thin filaments of another protein, actin. The two types of filament overlap in the dark band except in the center, leaving a slightly lighter *H band* (Hensen's disk). According to the *sliding filament theory* of muscle contraction, projecting parts of myosin molecules form crossbridges to connect with binding sites on adjacent actin molecules. ATP from mitochondria in the myofibril provides energy for the bridges to oscillate and pull the actin filaments along in a ratchet action, thus making the whole sarcomere shorter. The shortening of all the sarcomeres of all the myofibrils results in the contraction of a muscle fiber when it is stimulated by nerve impulses.

sarcoplasm /**sar**-koh-plaz-ăm/ The cytoplasm of the fibers of striated muscle, excluding the myofibrils.

sarcoplasmic reticulum /sar-koh-**plaz**-mik/ (SR) A modified form of smooth endoplasmic reticulum found in skeletal and cardiac muscle. It forms a network of tubules in the muscle-cell cytosol, into which calcium ions (Ca^{2+}) are continually pumped from the cytosol by calcium ion pumps in the membrane of the reticulum. Muscle contraction is initiated by the release of Ca^{2+} from the SR via calcium ion release channels. This release of Ca^{2+} is triggered by the arrival of a nerve impulse at the muscle cell's plasma membrane. The impulse penetrates deep within the muscle cell, between the myofibrils, via invaginations of the plasma membrane known as transverse tubules. Within about 30 milliseconds, the Ca^{2+} ions are pumped back into the SR, thus restoring the resting concentration of Ca^{2+} in the cytosol.

satellite DNA DNA that can be separated by centrifugation from the main DNA fraction. Such DNA forms distinct 'satellite' bands along the centrifuge tube. The bulk of satellite DNA is from chromosomal regions adjacent to the centromeres, which consists of multiple tandem repeats of short base sequences, stretching to hundreds of kilobases in length. *Minisatellite DNA* consists of another type of repetitive DNA, namely variable number tandem repeats (VNTR). These are slightly longer (15–100 bp) base sequences, repeated in tandem arrays extending typically to 1–5 kb. VNTRs are dispersed throughout the genome, and are used to identify individuals in genetic fingerprinting. *Microsatellite DNA* consists of dinucleotide tandem repeats dispersed throughout the genome. The uniformity of base composition of these different types of repetitive DNA gives each a characteristic density on centrifugation, hence their tendency to separate into satellite bands.

scalariform thickening /skă-**la**-ră-form/ A type of secondary wall formation consisting of interlaced spiral bands of thickening giving a ladder-like formation. It is found in metaxylem tracheids and vessels. *See* xylem.

scale 1. A small horny or bony plate, numbers of which cover the body of reptiles and fishes. Reptiles have horny epidermal scales (*corneoscutes*), which may be fused to underlying bony dermal plates (*osteoscutes*). Different groups of fishes have different types of scales. Bony fishes have one of two types of bony scales: *ctenoid scales*, which are rounded with serrated exposed edges; and *cycloid scales*, which are smooth and rounded. Cartilaginous fishes, such as sharks and rays, have toothlike scales made of dentine (*see* denticle). *See also* cosmoid scales; ganoid scales.
2. A modified hair found on the wing of a butterfly or moth. The numerous tiny scales overlap like roof tiles to cover the wing, and are often pigmented and intricately sculpted, sometimes producing vivid iridescent colors.

scanning electron microscope *See* microscope.

scapula /**skap**-yŭ-lă/ One of two large flat triangular bones forming the dorsal portion of the pectoral girdle on each side in most vertebrates; they form the shoulder blades in humans. Each articulates with a humerus bone at a concave articular surface (the glenoid cavity) on its outer lateral angle and provides attachment for muscles of the forelimbs. They overlap parts of the second to seventh ribs and, although in most vertebrates they are fused to the vertebral column, in mammals they are bound by muscles to the back of the thorax to allow free movement of the shoulders.

schizocarp /**skiz**-ŏ-karp/ A dry fruit, formed from two or more carpels, that di-

vides at maturity into one seeded achene-like segments termed mericarps. Such fruits are seen in hollyhock.

schizogeny /skiz-**oj**-ĕ-nee/ The separation of plant cells at the middle lamellae to give intercellular spaces, which may have special functions. An example is seen in the resin ducts of conifers. *Compare* lysogeny.

Schultze's solution /**shûlt**-sĕz/ A solution of zinc chloride, potassium iodide, and iodine used mainly for testing for cellulose and hemicellulose. Both materials stain a blue color with the reagent, that of hemicellulose being weaker.

Schwann cell /shvahn, shwahn/ A cell that makes a section of the MYELIN SHEATH of a medullated nerve fiber. During development, the cell becomes spirally wrapped around the nerve fiber. The cell is named for the German physiologist Theodor Schwann (1810–82).

scion /**sȳ**-ŏn/ In a plant GRAFT, the part that is grafted on to the stock. The scion may be a bud or shoot.

sclera /**skleer**-ă/ (**sclerotic**) The tough outer protective opaque coat of the vertebrate eye. It consists of fibrous connective tissue, with the bundles running in all directions, and maintains the shape of the eye. It is continuous with the transparent cornea in front of the eye. *See illustration at* eye.

sclereid /**skleer**-ee-id/ Any sclerenchyma cell, excluding the fibers. The various forms of sclereid include the star-shaped astrosclereid, the rod-shaped macrosclereid, and the isodiametric stone cell.

sclerenchyma /skli-**renk**-ă-mă/ The main supporting tissue in plants, made up of cells with heavily thickened often lignified walls and empty lumina. Unlike collenchyma, it is not very extensible and is thus not formed in quantity until after the young tissues have fully differentiated. Sclerenchyma is often found associated with vascular tissue and exists as two distinct types of cell: the fiber and the sclereid.

scleroprotein /skleer-oh-**proh**-teen, -tee-in/ One of a group of proteins obtained from the exoskeletal structures of animals. They are insoluble in water, salt solutions, dilute acids, and alkalis. This group exhibits a wide range of both physical and chemical properties. Typical examples of scleroproteins are keratin (hair), elastin (elastic tissue), and collagen (connective tissue).

sclerotic /skli-**rot**-ik/ *See* sclera.

sclerotium /skli-**roh**-shee-ŭm/ (*pl.* sclerotia) The resting body of certain fungi, e.g. ergot, formed from a mass of hyphae. *See also* stroma.

sclerotome /**skleer**-ŏ-tohm/ The part of each of the SOMITES of vertebrates that contributes to the axial skeleton.

scolex /**skoh**-leks/ (*pl.* scolices) The head of a cestode (tapeworm). It is spherical, with a narrow neck leading to the region where proglottids are produced. It has a crown of hooks and four lateral suckers for attaching the worm to the lining of the gut of the final host.

scorpions *See* Arachnida.

scrotum (scrotal sac) (*pl.* scrota) A pouch of skin that hangs external to the body directly behind the penis in most male mammals. It is divided into two compartments, each containing a TESTIS. The testes are thus maintained at a temperature lower than that of the body in order to insure optimum development of sperm.

scutellum /skoo-**tel**-ŭm/ (*pl.* scutella) The part of the embryo of Gramineae (grasses) that lies next to the endosperm. Some believe it to be the modified cotyledon while others think that the scutellum and coleoptile together represent the cotyledon.

Scyliorhinus /sil-ee-ŏ-**ry**-nŭs/ A genus of dogfish (order SELACHII) common univer-

sally in coastal waters. They have a streamlined body and a well-developed heterocercal tail. The lateral line contains the neuromast sense organs, which are sensitive to water pressure and current direction. The five gill slits and spiracle, situated laterally, are not covered by an operculum. Dogfish justify their common name by hunting in packs, relying on their highly developed sense of smell. The male bears claspers for internal fertilization and the eggs are enclosed in horny capsules.

scyphistoma /sy̆-**fis**-tŏ-mă/ (*pl.* scyphistomae) The polyp stage in the life cycle of scyphozoans (jellyfish), which develops from the planula. It undergoes transverse splitting (strobilization) to produce small free-swimming ephyrae, which develop into adult jellyfish.

Scyphozoa /sy̆-fŏ-**zoh**-ă/ A class of cnidarians, the jellyfish, in which the medusa is the only or dominant form and the polyp is absent or restricted to a small larval stage (*see* scyphistoma). The medusae are highly organized with the mouth at the end of a tube (manubrium) hanging down underneath and leading to the coelenteron, which is divided into four pouches and contains a canal system (gastrovascular cavity) for food distribution. The tentacles around the rim bear stinging cnidocytes. Jellyfish are found universally and range in diameter from about 70 mm (e.g. *Aurelia*) to 2 m (e.g. *Cyanea*).

sea urchins *See* Echinoidea.

sebaceous gland /si-**bay**-shŭs/ A gland, situated at the upper end of a hair follicle, near the skin surface, that secretes an oily secretion, sebum, into the follicle. Sebum keeps hair and skin in good condition and has antiseptic properties.

sebum /**see**-bŭm/ A complex oily secretion produced by the sebaceous glands in the mammalian skin. It is secreted onto the skin preventing desiccation, and onto the hairs and skin making them water-repellent. It also contains an antiseptic ingredient to kill bacteria. *See also* skin.

secondary growth Plant growth derived from secondary or lateral meristems, i.e. the vascular and cork cambia. It is usually absent in monocotyledons. In dicotyledons, the result of secondary growth is termed *secondary thickening* since there is usually an increase in width rather than length. The activity of the vascular cambium gives rise to the secondary xylem and phloem. The cork cambium gives rise to the periderm, a protective layer of tissue on the outside of the stem or root, which consists of the cork, the phellogen, and the phelloderm. *Compare* primary growth.

secondary plant body The parts of the plant formed by secondary growth, e.g. secondary vascular tissue produced by the intrafascicular cambium and cork cells produced by the phellogen.

secondary sexual characteristic A characteristic that develops in male and female animals at the onset of sexual maturity in association with masculinity and femininity. Most secondary sexual characteristics result from the effects of hormones (e.g. androgens or estrogens) secreted by special cells in the gonads at this time. For example, in male humans hair begins to grow on the face and the voice becomes deeper, whilst in female humans the breasts develop and the hip girdle enlarges. *See also* sexual selection.

secondary thickening *See* secondary growth.

second messenger A molecule or ion that relays signals from cell-surface receptors to enzymes and other components of signal pathways inside cells. For example, binding of a hormone to an external binding site on a receptor activates intracellular enzymes that bring about a rapid rise in the concentration of particular second messengers, which in turn activate other components of the signal pathway and influence cell metabolism in some way. Second messengers include cyclic AMP, cyclic GMP (cyclic guanosine monophosphate), inositol trisphosphate, and calcium ions.

secretin /si-**kree**-tin/ A polypeptide hormone secreted by the mucosa of the duodenum and jejunum when the stomach empties its contents into the intestine. It stimulates alkaline pancreatic secretions to neutralize the acidic chyme from the stomach. Its secretion is also stimulated by the release of bile.

secretion 1. A substance or fluid produced in cells and released to the surrounding medium. The secretion may be a fluid (e.g. sweat) or molecules (e.g. enzymes, hormones).
2. The process by which cells release proteins and other molecules to their surroundings. Secretory proteins are sorted and packaged by the GOLGI APPARATUS, in a sequence of events called the *secretory pathway*. Some proteins are packaged in transport vesicles for *constitutive* (i.e. continuous) *secretion* at the cell surface, or for sorting to lysosomes or insertion in the plasma membrane. Other proteins are stored in secretory vesicles for *regulated secretion*, i.e. to await a specific stimulus, such as a hormonal signal. A vesicle releases its contents by fusing with the plasma membrane, or membrane of the target organelle within the cell.

seed The structure that develops from the ovule following fertilization in angiosperms or gymnosperms. In flowering plants one or more seeds are contained within a fruit developed from the ovary wall. The individual seeds are composed of an embryo and, in those seeds in which food is not stored in the embryo cotyledons, a nutritive endosperm tissue. This difference enables seeds to be classified as nonendospermic or endospermic. The whole is surrounded by a testa developed from the integuments of the ovule. In gymnosperms the seeds do not develop within a fruit but are shed 'naked' from the plant. Following dispersal from the parent plant, seeds may germinate immediately to form a seedling or may remain in a relatively inactive dormant state until conditions are favorable for germination. In annual plants, seeds provide the only mechanism for surviving the cold or dry seasons. Seeds may be formed asexually in certain plants by apomixis, e.g. in dandelion.

The development of the seed habit, which makes water unnecessary for fertilization, is one of the most significant advances in plant evolution. It has enabled gymnosperms and angiosperms to colonize dry terrestrial habitats where lower plants are unable to establish themselves.

seed ferns Seed-bearing plants represented only by fossil forms that flourished in the Carboniferous but became extinct in the Cretaceous. The plant body resembled a fern and did not produce flowers, the seeds developing from megasporangia borne on the fronds.

seed plants *See* spermatophyte.

segment One of a series of repeated parts of the body. *See* metameric segmentation.

segmentation 1. *See* metameric segmentation.
2. *See* cleavage.

segregation The separation of the two alleles of a gene into different gametes, brought about by the separation of homologous chromosomes during meiosis. *See* Mendel's laws.

seismonasty /sȳz-moh-nas-tee/ (seismonastic movements) A NASTIC MOVEMENT in response to shock.

Selachii /si-layk-ee-ȳ/ The order of Chondrichthyes that contains the sharks. Sharks are fast aggressive predators with a widely gaping mouth and numerous sharp teeth that are continuously replaced. The streamlined torpedo-shaped body tapers into a well-developed heterocercal tail and the paired fins have narrow bases, making them mobile and effective in controlling motion through the water. The spiracle and gill slits are situated laterally.

Skates and rays are sometimes included in this order. They are specialized for living on the sea bed, having a dorsoventrally flattened body, dorsal eyes and spiracle,

ventral gill openings, and winglike pectoral fins. *See also* Scyliorhinus.

Selaginella /si-laj-ă-nel-ă/ A genus of club mosses (*see* Lycophyta) comprising the spike mosses. A typical representative of the genus is *S. kraussiana*, which is a creeping regularly branched plant with four rows of leaves arranged along the horizontal stem in opposite pairs. The roots develop from a unique structure (the rhizophore) and the strobili arise as vertical branches.

selection pressure The intensity with which the environment eliminates a particular phenotype, so causing the gene responsible for this to decrease in the population. It is thus a measure of the force of natural selection.

selfish DNA (junk DNA) A term sometimes used to describe DNA that is present within the genome of an organism but has no apparent function. This includes much of the REPETITIVE DNA, such as the variable number tandem repeats and the mobile genetic elements (i.e. TRANSPOSONS and retrotransposons), that is dispersed between genes and within introns. Some biologists also regard introns as selfish DNA. The hypothesis is that selfish DNA has somehow evaded the forces of natural selection and become a 'molecular parasite', existing solely to perpetuate itself in the genome. However, it may be that this seemingly functionless DNA is required for some as yet unknown purpose. For example, it might provide the bulk needed for normal chromosome separation during nuclear division, or act as spacer DNA to enable functioning genes to be controlled effectively.

self-sterility The condition in many hermaphrodite animals and plants whereby male gametes cannot fertilize female gametes from the same individual. *See* incompatibility.

Seliwanoff's test /sel-ee-vah-noffs/ A standard test for the presence of fructose in solution. A few drops of Seliwanoff's reagent, resorcinol in hydrochloric acid, are heated with the test solution. A red color or red precipitate indicates fructose. The test is named for the Russian chemist Feodor Feodorowitsch Seliwanoff (1859–1938).

semen /see-mĕn/ A fluid containing spermatozoa and nutritive substances, produced by male mammals. The testes produce the spermatozoa, and the other constituents of the semen are produced by the prostate gland and the seminal vesicles. Semen is placed in the body of the female during mating.

semicircular canals Three looped canals that form part of the labyrinth of the vertebrate inner ear and detect changes in the rate of movement of the head. They are positioned on the utriculus at right angles to each other and bear a swelling (ampulla) at one end of each canal, which contains sensory cells. Movement in a particular plane causes the endolymph to lag behind in the canal of the same plane and stimulate the cells. *See illustration at* ear.

semiconservative replication *See* replication.

semilunar valve /sem-ee-**loo**-ner/ Either of two valves in the mammalian heart that prevent backflow of blood from the major arteries into the ventricles. The pulmonary valve lies at the entrance to the pulmonary artery; the aortic valve is at the entrance to the aorta. Each consists of three half-moon-shaped flaps that under backward pressure fill with blood like small pouches and close the entrance. When blood is pumped out of the ventricles, the flaps are pressed against the vessel wall and the valve opens.

seminal receptacle /sem-ă-năl/ *See* spermatheca.

seminal vesicle (**vesicula seminalis**) **1.** One of a pair of small elongated glands in most male mammals that opens into the vas deferens. It secretes a thick alkaline fluid – containing substances such as fruc-tose, proteins, and various chemicals – that contributes to the semen. Its growth and activity are largely under the influence of hormones (androgens).
2. An organ in lower vertebrates and some invertebrates used for the storage of sperm.

seminiferous tubule /sem-ă-**nif**-ĕ-rŭs/ A mass of minute coiled tubules within the vertebrate testis in which spermatozoa are produced. In man, each is about 15 mm in diameter and 50 cm long and drains into small collecting ducts (the vasa efferentia).

semipermeable membrane /sen-ee-**per**-mee-ă-băl/ *See* osmosis.

senescence /sĕ-**ness**-ĕns/ **1.** (*Biology*) The advanced phase of the aging process of an organism or part of an organism that precedes natural death. It is usually characterized by a reduction in capacity for self-maintenance and repair of cells, and hence deterioration. The degree of senescence varies between groups and its mechanism remains largely uncertain. However, discoveries in cell biology and genetics have prompted certain theories. The *telomere theory of aging* proposes that senescence involves failure of chromosomes to replicate. This is based on the finding that the telomeres, which form the tips of an organism's chromosomes, shorten progressively with successive replication cycles, eventually making replication impossible. The *mitochondrial theory of aging* states that the accumulation of mutations in mitochondrial DNA are linked to a decline in mitochondrial function with age, which is responsible for the age-related decline in oxidative phosphorylation.
2. (*Botany*) The deterioration of the leaves of a deciduous plant toward the end of the growing season, culminating in ABSCISSION.

sense organ One or more sensory cells (receptors) and associated structures in an animal that are able to respond to a stimulus from inside or outside. The stimulus is converted into an electrical impulse and sent along nerve fibers to the brain for interpretation and response. In general, a sense organ can only respond to a specific stimulus. Hence there are different organs for touch, heat, pressure, etc. They may be distributed over the body or concentrated in certain regions, e.g. the taste buds in the mouth of terrestrial vertebrates. *See also* ear; eye.

sense strand *See* nontemplate strand.

sensillum /sen-**sil**-ŭm/ (*pl.* **sensilla**) A type of sense organ found in insects and other arthropods. Each consists of a basal cluster of receptor neurons, from which extend fine dendrites that penetrate a hair, peg, plate, or other cuticular structure. The sensory neurons are all derived from a single mother cell, but may respond to different types of stimuli, even within a single sensillum. Some are chemoreceptors, responding to different chemical substances present in the air or the substrate. The chemicals diffuse through pores in the hairshaft of the sensillum and stimulate the dendrites. Sensilla can also act as mechanoreceptors, for example, responding to the distortions of the hairshaft caused by movements of neighboring joints.

sensitization /sen-să-ti-**zay**-shŏn/ The increase in the reaction of an organism or cell to an antigen to which it has been previously exposed. It may occur naturally or be artificially induced, e.g. following vaccination. *See* immunity.

sepal /**see**-păl/ One of the structures situated immediately below the petals of a flower. Their collective name is the *calyx*. They are often green and hairy and enclose and protect the flower bud. Sometimes they are brightly colored (e.g. in orchids) and attract insects for pollination. *See illustration at* flower.

septum /**sep**-tŭm/ (*pl.* **septa**) A wall, partition, or membrane separating two cavities. For example a septum separates the coelom of one segment of an earthworm from that of the next segment; the capsule of a poppy is divided by septa.

seral stage /**seer**-ăl/ (**seral community**) *See* sere.

sere /seer/ A plant succession in which each community itself effects changes in the habitat that determine the nature of the following stage. The successive stages are known as *seral stages*. Seres result eventually in a climax community. *Hydroseres*, starting in water, and *xeroseres*, starting in dry conditions, both tend towards mesophytic conditions. *Microseres* occur in microhabitats and *subseres* are secondary seres, appearing when the biotic components of a primary sere are destroyed, e.g. by fire. *See also* succession.

serine /se-**reen**, -rin/ *See* amino acids.

serology /si-**rol**-ŏ-jee/ The *in vitro* study of reactions between antigens and antibodies in the blood serum. Various serological tests involving specific types of reaction enable the identification of blood groups, pathogens, diseases, etc. *See also* agglutination; complement fixation; precipitin.

serotonin /seer-oh-**toh**-nin/ (**hydroxytryptamine**) A substance that serves as a neurohormone that acts on muscles and nerves, and a neurotransmitter found in both the central and peripheral nervous systems. It controls dilation and constriction of blood vessels and affects peristalsis and gastrointestinal tract motility. Within the brain it plays a role in mood behavior. Many hallucinogenic compounds (e.g. LSD) antagonize the effects of serotonin in the brain.

serous membrane /**seer**-ŭs/ The tissue that lines cavities, in vertebrates, that do not open to the exterior, e.g. the pleural and peritoneal cavities. It consists of mesothelium and underlying connective tissue.

Sertoli cells /ser-**toh**-lee/ Large pillar-like cells in the germinal epithelium of the vertebrate testis, which protect and nourish developing spermatozoa. They also secrete the hormone inhibin, which inhibits secretion of follicle-stimulating hormone by the pituitary. The cells are named for the Italian histologist Enrico Sertoli (1842–1910). *See also* spermatogenesis.

serum /seer-ŭm/ (*pl.* **sera**) *See* blood serum.

sesamoid bone /sess-ă-moid/ A generally small oval bone (nodule) that develops as an ossification within a tendon of vertebrates, especially mammals. There may be numerous sesamoid bones in the body occurring in tendons subjected to friction; for example, those passing over an articular surface or bony ridge. *See also* patella.

sessile /sess-ăl/ **1.** (*Botany*) Describing any organ (e.g. an acorn) that is attached to the main body of the plant instead of being stalked.
2. (*Zoology*) Describing an animal that lives permanently attached to a substrate. Sponges, for example, live permanently attached to rocks.

seta /see-tă/ (*pl.* **setae**) **1.** (*Botany*) The part of the sporogonium that forms the stalk between the foot and the capsule in bryophytes, e.g. *Funaria* and *Pellia*.
2. (*Zoology*) *See* chaeta.

Sewall Wright effect /soo-ăl rÿt/ *See* genetic drift.

sex chromosomes The chromosomes that determine sex in most animals. There are two types: in mammals these are called the *X chromosome* and the *Y chromosome*. In the heterogametic sex (XY) they can usually be distinguished from the other chromosomes, because the Y chromosome is much shorter than the X chromosome with which it is paired (unlike the remaining chromosomes, which are in similar homologous pairs). *See* sex determination; sex linkage.

sex determination In species having almost equal numbers of males and females sex determination is genetic. Very occasionally a single pair of alleles determine sex but usually whole chromosomes, the sex chromosomes, are responsible. The 1:1 ratio of males to females is obtained by crossing of the homogametic sex (XX) with the heterogametic sex (XY). In most animals, including humans, the female is

XX and the male XY, but in birds, butterflies and some fishes this situation is reversed. In some species sex is determined more by the number of X chromosomes than by the presence of the Y chromosome, but in humans the Y chromosome is important in determining maleness. Many genes are involved in determining all aspects of maleness or femaleness, but in mammals one particular gene on the Y chromosome acts as sex switch to initiate male development. In the absence of this male switch gene (i.e. in XX individuals) the fetus develops as a female. Rarely, sex is subject to environmental control, in which case unequal numbers of males and females develop. In bees and some other members of the Hymenoptera, females develop from fertilized eggs and are diploid while males develop from unfertilized eggs and are haploid, the numbers of each sex being controlled by the queen bee.

sex hormone Any of several hormones responsible for the development and functioning of the reproductive organs. They are also involved in the development of secondary sex characteristics. They are secreted mainly by the gonads and include androgens in males and estrogens and progesterone in females.

sex linkage The coupling of certain genes (and therefore the characters they control) to the sex of an organism because they happen to occur on the X sex chromosome. The heterogametic sex (XY), which in humans is the male, has only one X chromosome and thus any recessive genes carried on it are not masked by their dominant alleles (as they would be in the homogametic sex). Thus in humans recessive forms of the sex-linked genes appear in the male phenotype far more frequently than in the female (in which they would have to be double recessives). Color blindness and hemophilia are sex linked. *See* carrier.

sex ratio The ratio of the number of males to females in a population. The *primary sex ratio* is is the ratio applying at conception, usually 1:1, whereas the *sec-*

sexual reproduction

ondary sex ratio is that applying at sexual maturity.

sexual reproduction The formation of new individuals by fusion of two nuclei or sex cells (gametes) to form a zygote. In unicellular organisms whole individuals may unite but in most multicellular organisms only the gametes combine. In organisms showing sexuality, the gametes are of two types: male and female (in animals, spermatozoa and ova). They are produced in special organs (carpel and anther in plants; ovary and testis in animals), which, with associated structures, form a reproductive system and aid in the reproductive process. Individuals containing both systems are termed monoecious or hermaphrodite.

Generally meiosis occurs before gamete formation, resulting in the gametes being haploid (having half the normal number of chromosomes). At fertilization, when the haploid gametes fuse, the diploid number of chromosomes is restored. In this way sexual reproduction permits genetic recombination, which results in greater variety in offspring and consequently provides a mechanism for evolution to take place by natural selection.

Apomixis and parthenogenesis are usually regarded as modified forms of sexual reproduction.

sexual selection The aspect of natural selection concerned with traits that contribute to mating success. The term was coined by Charles Darwin to account for the evolution of secondary sexual features such as the elaborate tail feathers of peacocks. Such features are costly in terms of resources, and may reduce the survival of the adult, for instance by making it more noticeable to predators. However, these costs are offset by the greater reproductive success enjoyed by the male, up to the point where survival of the male is so compromised that mating is affected. Competition for mates between members of the same sex – usually males – results in selection for traits that improve fighting prowess, such as the antlers of male deer. Competition to attract females results in traits such as long or gaudy tail feathers, elaborate courtship rituals, or mating calls. Some interpret such traits as means of signaling the individual's vigor to potential mates.

shadowing A method of preparation of material for electron microscopy enabling surface features to be studied. It can be used for small entire structures, subcellular organelles, or even large molecules (e.g. DNA). The specimen is supported on a plastic or carbon film on a small grid and sprayed with vaporized metal atoms from one side while under vacuum. The coated specimen appears blacker (more electron-opaque) where metal accumulates, and the lengths and shapes of 'shadows' cast (regions behind the objects not coated with metal) give structural information. It is often used in association with FREEZE FRACTURING.

shoot The aerial photosynthetic portion of a plant that generally consists of a stem upon which leaves, buds, and flowers are borne.

short-day plant (SDP) A plant that flowers in response to a light period shorter than a critical maximum. *See* critical day length; photoperiodism.

shoulder girdle *See* pectoral girdle.

siblings (sibs) Two or more offspring from the same cross. In animals, brothers and sisters are siblings. In plants the products of a self-pollination are termed sibs.

sieve elements /siv/ Elongated cells that, placed end to end, make up the sieve tubes in angiosperms. Most of the organelles, including the nucleus, break down during the development of the sieve element so all that remains of the cell contents is the cytoplasm, which runs from cell to cell through the pores in the perforated end walls or sieve plates.

sievert Symbol: Sv /**see**-vert/ The SI unit of dose equivalent. It is the dose equivalent when the absorbed dose produced by ionizing radiation multiplied by certain dimensionless factors is 1 joule per kilogram (1 J kg^{-1}). The dimensionless factors are used to modify the absorbed dose to take account of the fact that different types of radiation cause different biological effects. The unit is named for the Swedish physicist Rolf Sievert (1898–1966).

sieve tube A column of cells formed from sieve elements, in which food is translocated in plants.

silicon /sil-ă-kŏn/ A trace element found in many animals and plants, although not essential for growth in most organisms. It is found in large quantities in the cell walls of certain algae (e.g. desmids, diatoms) and horsetails, and in smaller amounts in the cell walls of many higher plants. It forms the skeleton of certain marine animals, e.g. the siliceous sponges. Silicon is also found in connective tissue.

silicula /să-lik-yŭ-lă/ (silicle) A capsular fruit typical of the Cruciferae (e.g. honesty) that is formed from a bicarpellary ovary. It is flattened, short, and broad and is divided into two loculi by a false septum.

siliqua /să-lee-kwă, sil-ă-kwă/ (silique) (pl. siliquae) A fruit of some Cruciferae, similar to the silicula but longer and thinner, for example the wallflower fruit.

Silurian /să-loor-ee-ăn, sÿ-/ The period, some 440–405 million years ago, between the Ordovician and the Devonian periods of the Paleozoic. It is characterized by early land plants, primitive jawless fish, and many invertebrates. *See also* geological time scale.

single-cell protein (SCP) Protein produced from microorganisms, such as bacteria, yeasts, mycelial fungi, and unicellular algae, used as food for humans and other animals.

single-factor inheritance The control of one character by one gene. This gives rise to discontinuous variation in such characters, and intermediates between the dominant and recessive forms of the gene do not usually occur; for example, a person either is or is not red-green color blind. *Compare* multifactorial inheritance.

sinoatrial node /sÿ-noh-ay-tree-ăl/ *See* pacemaker.

sinus /sÿ-nŭs/ An anatomical cavity, space, or channel. Examples are the nasal sinuses in the skull.

sinusoid /sÿ-ŭ-soid/ A small blood vessel or space within certain tissues, such as the liver, spleen, and bone marrow. Compared to a capillary it usually connects two veins, has a wider lumen, and an irregular wall containing macrophages, which in some places is incomplete and allows direct contact between blood and tissue.

sinus venosus /vee-noh-sŭs/ The thin-walled first chamber of the heart in fish and amphibians, which receives deoxygenated blood from the body. It is absorbed into the right atrium of other vertebrates.

Siphonaptera /sÿ-fŏ-nap-tĕ-ră/ The order of insects that contains the fleas, which are all ectoparasites of mammals and birds, e.g. *Pulex irritans* (human flea). Some transmit serious diseases; for example *Xenopsylla* (the rat flea) carries bubonic plague. Generally each species is limited to one host. Fleas are small wingless insects with laterally compressed bodies to ease movement over the host, legs adapted for jumping and clinging to the host, and mouthparts modified for piercing and sucking. The grublike larvae feed on organic detritus in the host's nest.

siphonostele /sÿ-fŏ-nŏ-steel, -stee-lee/ A term covering both medullated protosteles and solenosteles. *See* stele.

SI units (Système International d'Unités) The internationally adopted system of units used for scientific purposes. It has seven base units (the meter, kilogram, second, kelvin, ampere, mole, and candela) and two supplementary units (the radian and steradian). Derived units are formed

BASE AND DIMENSIONLESS SI UNITS

Physical quantity	Name of SI unit	Symbol for SI unit
length	meter	m
mass	kilogram(me)	kg
time	second	s
electric current	ampere	A
thermodynamic temperature	kelvin	K
luminous intensity	candela	cd
amount of substance	mole	mol
*plane angle	radian	rad
*solid angle	steradian	sr

(* = supplementary units)

DERIVED SI UNITS WITH SPECIAL NAMES

Physical quantity	Name of SI unit	Symbol for SI unit
frequency	hertz	Hz
energy	joule	J
force	newton	N
power	watt	W
pressure	pascal	Pa
electric charge	coulomb	C
electric potential difference	volt	V
electric resistance	ohm	Ω
electric conductance	siemens	S
electric capacitance	farad	F
magnetic flux	weber	Wb
inductance	henry	H
magnetic flux density	tesla	T
luminous flux	lumen	lm
illuminance (illumination)	lux	lx
absorbed dose	gray	Gy
activity	becquerel	Bq
dose equivalent	sievert	Sv

DECIMAL MULTIPLES AND SUBMULTIPLES USED WITH SI UNITS

Submultiple	Prefix	Symbol	Multiple	Prefix	Symbol
10^{-1}	deci-	d	10^{1}	deca-	da
10^{-2}	centi-	c	10^{2}	hecto-	h
10^{-3}	milli-	m	10^{3}	kilo-	k
10^{-6}	micro-	μ	10^{6}	mega-	M
10^{-9}	nano-	n	10^{9}	giga-	G
10^{-12}	pico-	p	10^{12}	tera-	T
10^{-15}	femto-	f	10^{15}	peta-	P
10^{-18}	atto-	a	10^{18}	exa-	E
10^{-21}	zepto-	z	10^{21}	zetta-	Z
10^{-24}	yocto-	y	10^{24}	yotta-	Y

by multiplication and/or division of base units; a number have special names. Standard prefixes are used for multiples and submultiples of SI units.

skeletal muscle (**striated muscle; striped muscle; voluntary muscle**) Muscle that moves the bones of the skeleton. Each muscle is made up of many microscopic *muscle fibers*, bound together with connective tissue and surrounded by a sheath (*epimysium*). Skeletal muscle has a typically striped appearance. The muscle fibers are long and narrow with tapering ends. Each has an outer membrane (*sarcolemma*) inside which are many oval nuclei. The cytoplasm (*sarcoplasm*) contains many large mitochondria and longitudinal MYOFIBRILS, which contain the contractile elements – the SARCOMERES – giving the striated appearance. The epimysium is continuous with the nonelastic fibers of the tendons attached to the tapering ends of the muscle. The tendons penetrate the tissues of the bones to which the muscle is joined, the *origin* of the muscle being on the stationary bone and the *insertion* on the movable one. When the muscle contracts it becomes shorter and fatter and the tendons pull on the bones, bringing about movement at the joint. All skeletal muscles are under the voluntary control of the central nervous system.

skeleton A hard structure that supports and maintains the shape of an animal. It may be external to the body (EXOSKELETON) or within the body (ENDOSKELETON).

skin The outer layer of the body of an animal. In vertebrates it protects the animal from excessive loss of water, from the entry of disease-causing organisms, from damage by ultraviolet radiation, and from mechanical injury. It contains numerous nerve endings and therefore also acts as a peripheral sense organ. In warm-blooded animals it plays a part in the regulation of body temperature. It consists of two layers, the inner DERMIS and the outer EPIDERMIS. The former originates from the mesoderm and the latter from the ectoderm. *See* hair; sebaceous gland; sweat gland.

skull The part of the vertebrate skeleton that encloses and protects the head. It consists of the braincase (cranium) and bones of the face and upper jaw, which provide articulation for the lower jaw. The base of the skull articulates with the vertebral column, and contains an opening (foramen magnum) for the spinal cord to pass from the brain into the vertebral column.

sleep A normal recurrent state of reduced responsiveness to external stimuli in vertebrates. Sleep in humans is characterized by

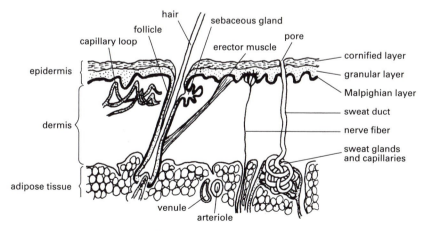

Skin: section through the skin

typical brain wave patterns, recorded as electroencephalograms (EEGs), which demonstrate the existence of different phases of sleep. Other mammalian species show comparable EEG patterns, but in lower vertebrates (frogs, fish) the characteristic signs of sleep may vary and a formal definition of the state becomes more difficult. Sleep can be distinguished from HIBERNATION and similar states by the fact that it is easily reversible.

sliding filament theory *See* sarcomere.

sliding growth A pattern of plant growth seen, for example, in many epidermal cells where, in order to accommodate growth by adjoining cells, expanding cell walls slide along each other. Thus growth is achieved without disruption of neighboring cells. *Compare* intrusive growth; symplastic growth.

slime bacteria *See* myxobacteria.

slime molds (**slime fungi**) Simple unicellular or multicellular eukaryotic organisms that display distinct changes in form during their life cycle. They have an ameboid stage, which lacks cell walls and feeds by engulfing bacteria by phagocytosis; and they can form multicellular differentiated reproductive structures that resemble the fruiting bodies of fungi or primitive plants. The ameboid stages aggregate in masses, often visible as slimy masses on rotting logs, vegetation, etc. The slime molds are divided into two groups. The 'true' slime molds, or *plasmodial slime molds*, form a large mass of multinucleate cytoplasm (*plasmodium*), which pulsates internally and moves only as it grows. In contrast, the *cellular slime molds* form a mass of cells, or slug (*pseudoplasmodium*), that moves around leaving a slime trail. In both cases the mass of cells eventually matures to the fruiting stage, in which a stalked fruiting body (*sporophore*) is formed by differentiation. This produces spores, which give rise to individual amebas. In the plasmodial slime molds, haploid amebas of distinct mating types fuse to produce a diploid

zygote, whereas sexual reproduction is rare or absent in the cellular slime molds.

The classification of the slime molds has always been contentious. They exhibit features of animals, fungi, and plants, and at some time have been classified as all three. For example, in one scheme they are placed in the phylum Myxomycota as part of the kingdom Fungi. This phylum contains two classes: Myxomycetes (plasmodial slime molds) and Acrasiomycetes (cellular slime molds). However, certain recent classifications place them respectively in the phyla Myxomycota and Rhizopoda in the kingdom Protoctista.

small calorie *See* calorie.

small intestine The narrow tube between the stomach and the colon. It consists of the duodenum, jejunum, and ileum, and is lined with villi. Here digestion and absorption of food takes place.

smooth muscle (**involuntary muscle**) The muscle of all internal organs (viscera) and blood vessels (except the heart). Usually it is in the form of tubes or sheets, up to several layers thick. The cells are long, narrow, and tapering, with a single nucleus and cytoplasm containing loose bundles of thick and thin contractile filaments, composed of the contractile proteins myosin and actin, respectively. These bundles connect to plaques in the muscle cell plasma membrane and to dense bodies in the cytoplasm, which act like the Z disks of skeletal muscle SARCOMERES. The calcium ions that trigger contraction are stored not in a sarcoplasmic reticulum but in the extracellular space. In vertebrates, smooth muscle is not under voluntary control, being supplied by the autonomic nervous system. It contracts when stretched, may have spontaneous rhythmic contractions, and can remain in a state of continuous contraction (tonus) for long periods without fatigue.

smuts Basidiomycete fungi of the order Ustilaginales. Many are important parasites of cereals and form a mass of sooty black spores in place of the grain. Examples of smuts are *Tilletia* and *Ustilago*.

snails *See* Gastropoda.

snakes *See* Squamata.

sodium /**soh**-dee-ŭm/ An element essential in animal tissues, and often found in plants although it is believed not to be essential in the latter. It is found in bones, and is the most abundant ion in the blood and cell fluids, being extremely important in maintaining the osmotic balance of animal tissues. *See* sodium pump.

sodium pump An active transport system within a cell membrane, by which sodium ions entering the cell are transported back into the extracellular medium against a concentration gradient. It maintains the differential concentration of ions either side of the membrane (the outside being kept positive) necessary to produce a resting potential. It consists of an integral membrane protein that hydrolyzes ATP to ADP to provide the energy required for the pumping mechanism.

soft palate *See* palate.

soil The accumulation of mineral particles and organic matter that forms a superficial layer over large parts of the earth's surface. It provides support and nutrients for plants and is inhabited by numerous and various microorganisms and animals. A section down through the soil is termed a *soil profile* and this can characteristically be divided into three main layers or *horizons*. Horizon A, the top soil, is darker than the lower layers due to the accumulation of organic matter as HUMUS. It is the most fertile layer and contains most of the soil population and a high proportion of plant roots. Horizon B, the subsoil, contains materials washed down from above and may be mottled with various colors depending on the iron compounds present. Horizon C is relatively unweathered parent material from which the mineral components of the above layers are derived. The depth and content of the horizons are used to classify soils into various types, e.g. podsols and brown earths. The texture, structure, and porosity of soil depends largely on the sizes of the mineral particles it contains and on the amount of organic material present. Soils also vary depending on environmental conditions, notably rainfall. *See also* peat; brown earth; podsol.

solenocyte /sŏ-**lee**-nŏ-sÿt/ *See* flame cell.

solenostele /sŏ-**lee**-nŏ-steel, -stee-lee/ A type of stele with leaf gaps separated vertically to the extent that only one gap is cut through in a cross section. *Compare* dictyostele.

solute potential *See* water potential.

somatic /soh-**mat**-ik/ Describing the cells of an organism other than germ cells. Somatic cells divide by mitosis producing daughter cells identical to the parent cell. A somatic mutation is a mutation in any cell not destined to become a germ cell; such mutations are therefore not heritable.

somatic cell hybridization The fusion of cultured human cells with hamster or mouse cells to create a hybrid cell containing nuclei from both cell types. The nuclei then fuse, and during subsequent divisions, some human chromosomes are lost, eventually resulting in a cell line in which just one or several human chromosomes are stably inherited. The technique is used to assign genes to chromosomes. For example, if the hybrid cell expresses a particular human protein, the corresponding gene must be located on the human chromosome present in the cell. *See also* chromosome map.

somatic motor nerve A nerve carrying impulses to a skeletal muscle.

somatomedin /soh-mă-tŏ-**mee**-din/ *See* insulin-like growth factor.

somatostatin /soh-mă-tŏ-**stay**-tin/ A peptide hormone produced by the hypothalamus that inhibits the release of GROWTH HORMONE from the pituitary, and also the release of insulin and glucagon from the pancreas in fasted animals. It is found in various types of tissues.

somatotropin /soh-mă-tŏ-**troh**-pin/ *See* growth hormone.

somite /soh-**mÿt**/ Any of the blocks of tissue into which the mesoderm of vertebrate embryos is divided lateral to the notochord and neural tube. The segmentation of this mesoderm into somites usually starts at about the level of the hindbrain and continues both anteriorly into the head and posteriorly into the trunk. Each somite later forms a muscle block (*myotome*), a portion of kidney (*nephrotome* or *intermediate cell mass*), and contributions to the axial skeleton (*sclerotome*) and dermis (*dermatome*).

soredium /sŏ-**ree**-dee-ŭm/ (*pl.* soredia) A small fragment of a LICHEN that contains components of both partners – fungal hyphae and cells of the alga or cyanobacterium. Soredia are dispersive structures, light enough to be carried by wind, water, or animals, and capable of producing new lichens.

sorosis /sŏ-**roh**-sis/ A type of composite fruit incorporating a spike, as seen in the pineapple.

sorus /**sor**-ŭs, **soh**-rŭs/ (*pl.* sori) 1. A reproductive structure found in ferns comprising a collection of sporangia borne on a cushion of tissue termed the placenta. The placenta develops over a vein ending on the underside of the leaf. The sorus is covered by a flap of tissue, the indusium. *See also* placenta, def. 3.
2. The reproductive area of certain algal thalli, e.g. *Laminaria*.

Southern blotting A technique for transferring DNA fragments from an electrophoretic gel to a nitrocellulose filter or nylon membrane, where they can be fixed in position and probed using DNA probes. Named for its inventor, the British molecular biologist E. M. Southern (1938–), it is widely used in genetic analysis. The DNA is first digested with restriction enzymes and the resulting mixture of fragments separated according to size by electrophoresis on agarose gel. The double-

stranded DNA is then denatured to single-stranded DNA using sodium hydroxide, and a nitrocellulose filter pressed against the gel. This transfers, or blots, the single-stranded DNA fragments onto the nitrocellulose, where they are permanently bound by heating. The DNA probe can then be applied to locate the specific DNA fragment of interest, while preserving the electrophoretic separation pattern. *See* DNA probe. *Compare* Western blotting.

spadix /**spay**-diks/ (*pl.* spadices) A type of inflorescence found in the family Araceae, e.g. cuckoopint. It is a modified spike with a large fleshy axis on which are borne small hermaphrodite or, more usually, unisexual flowers. The inflorescence is enclosed by a large bract, the spathe, which gradually unfurls to reveal the flowers.

spathe /spay*th*/ A large bract that encloses the spadix. It may be foliose or petalloid and has been shown to attract insects in certain species.

specialization 1. *See* adaptation.
2. *See* physiological specialization.

speciation /spee-shee-**ay**-shŏn/ The formation of one or more new species from an existing species. Speciation occurs when an isolated population develops distinctive characteristics as a result of natural selection, and cannot then reproduce with the rest of the population, even if there are no geographical or other physical reasons to prevent them from doing so. *See also* adaptive radiation.

species /**spee**-sheez/ 1. A group of organisms that resemble each other and cannot be subdivided into two or more species. Such a group can be defined in various ways. According to the *biological species concept*, a species is a population of organisms whose members can breed amongst themselves and produce fertile offspring. Two or more related species unable to breed because of geographical separation are called *allopatric species*. Related species which are not geographically isolated, and which could interbreed, but in

practice do not because of differences in behavior, breeding season, etc., are called *sympatric species*. Broad uniformity among the members of a biological species is maintained by sexual reproduction, which ensures genetic recombination and sharing of a common gene pool. Members of different species are prevented from breeding by ISOLATING MECHANISMS, which block gene flow between different groups, thus promoting genetic divergence between them. However, such a concept cannot be applied to exclusively asexual organisms, such as certain groups of fungi and bacteria. Also, it is contradicted by instances of interspecies mating, especially in plants, fungi, and bacteria. These problems are overcome by the *phylogenetic species concept*, which views a species as an irreducible group of organisms whose members share certain defining, or derived traits (*see* apomorphy) by virtue of descent from an immediate common ancestor. Hence, this concept defines a species as a group having a shared and unique evolutionary history. It is less restrictive than the biological species concept: breeding between members of different species does not undermine the concept. However, increasingly fine-scale discrimination among defining traits can lead to division of species into ever-smaller groups.
2. A category, or RANK, used in the classification of organisms. It can be subdivided into, for example, subspecies or races. Related species are grouped into genera (*see* genus). *See also* binomial nomenclature.

spectrophotometer /spek-troh-foh-**tom**-ĕ-ter/ An instrument for measuring the amount of light of different wavelengths absorbed by a solution. It gives information about the identity or amount of the specimen and can be used to plot absorption spectra.

sperm *See* spermatozoon.

spermatheca /sper-mă-**th'ee**-kă/ (*pl.* spermathecae) (seminal receptacle) A saclike organ in some female or hermaphrodite invertebrates, e.g. the earthworm. It acts as a store for sperm received during copulation, which are held by the female until required for fertilizing ova.

spermatid /**sper**-mă-tid/ A reproductive cell resulting from the second meiotic division of a spermatocyte. It matures and undergoes a series of changes, which transform it into a spermatozoon. *See also* spermatogenesis.

spermatocyte /sper-**mat**-ŏ-sÿt/ A reproductive cell, within the seminiferous tubules of the testis, that develops during the formation of spermatozoa. A primary spermatocyte develops from a spermatogonium, which has undergone a period of multiplication and growth. It divides by meiosis and the first meiotic division produces two secondary spermatocytes with haploid nuclei. Each secondary spermatocyte undergoes a second meiotic division to produce two spermatids. One primary spermatocyte thus forms four spermatids, which later become spermatozoa. *See also* spermatogenesis.

spermatogenesis /sper-mat-ŏ-**jen**-ĕ-sis, sper-mă-tŏ-/ The formation of spermatozoa within the testis in male animals. Precursor cells in the germinal epithelium lining the seminiferous tubules begin to multiply by mitosis and form spermatogonia, even before the animal is born. However, the production of spermatogonia is most significant from the onset of sexual maturity. They give rise to huge numbers of spermatozoa; in man, the process of producing one mature spermatozoon takes up to 90 days.

A spermatogonium destined to form spermatozoa migrates inward towards the lumen of the tubule and enters a growth phase, which results in the formation of a primary spermatocyte. The primary spermatocyte then undergoes meiosis and the first meiotic (or reduction) division results in the formation of two secondary spermatocytes, containing the haploid number of chromosomes. Each secondary spermatocyte undergoes the second meiotic division and produces two spermatids. By a series of changes the spermatids then become transformed into spermatozoa, during which time they are attached to Sertoli cells. When mature, the spermatozoa pass from the seminiferous tubules into the epididymis for temporary storage.

spermatogonium /sper-mat-ŏ-**goh**-nee-ŭm, sper-mă-tŏ-/ (*pl.* **spermatogonia**) A reproductive cell in the testis, situated in the germinal epithelium that lines the seminiferous tubules. It undergoes a period of multiplication and growth to give rise to spermatocytes. *See also* spermatogenesis.

spermatophore /sper-**mat**-ŏ-for, -fohr, sper-mă-tŏ-/ A gelatinous packet containing spermatozoa produced by some animals with internal fertilization. It may be transferred directly to the female (as in cephalopods and insects) or deposited in water or moist soil to be taken up by the female (as in salamanders and newts).

spermatophyte /sper-**mat**-ŏ-fÿt, sper-mă-tŏ-/ (**seed plant**) Any seed-bearing plant. In many older classifications they constituted the division (phylum) Spermatophyta, subdivided into the classes Angiospermae and Gymnospermae. *See also* tracheophyte.

spermatozoid /sper-mă-tŏ-**zoh**-id/ *See* antherozoid.

spermatozoon /sper-mă-tŏ-**zoh**-on/ (*pl.* **spermatozoa**) (**sperm**) The small motile mature male reproductive cell (gamete) formed in the testis. It differs in form and size between species; in humans it is about 52–62 μm long and comprises a head region containing a haploid nucleus, a middle region containing mitochondria, and a long tail region containing an undulipodium (flagellum). It is covered by a small amount of cytoplasm and a plasma membrane. They remain inactive until they pass from the testis during coitus, when secretions from the prostate gland and seminal vesicles stimulate undulating movements to pass along the tail and effect locomotion. About 200–300 million spermatozoa may be released in a single ejaculation, although only one may fertilize each ovum.

S phase *See* cell cycle.

Sphenophyta /sfe-**noff**-ă-tă/ A phylum of vascular, non-seed-bearing plants that contains one living order, the Equisetales, comprising one genus, *Equisetum* (horsetails), along with three extinct orders, the Calamitales, Sphenophyllales, and Pseudoborniales. Horsetails have jointed stems and leaves arranged in whorls. The Sphenophyta were particularly abundant during the Carboniferous, when the genus *Calamites* formed a large proportion of the forest vegetation.

spherosome /**sfeer**-ŏ-sohm/ A small spherical organelle of plant cells, about 0.8–1.0 μm in diameter, bounded by a single membrane and storing lipid.

sphincter /**sfink**-ter/ A muscle that surrounds an opening or tube and constricts or completely closes the tube when it contracts. Examples are the pyloric sphincter between the stomach and the duodenum and the muscle around the urethra where it leaves the urinary bladder.

spiders *See* Arachnida.

spike A type of racemose inflorescence having sessile flowers borne on an elongated axis, as in wheat. The catkin and spadix are modifications of the spike.

spinal column *See* vertebral column.

spinal cord The longitudinal nerve tract of the vertebrate central nervous system. It connects the brain and the nerve cells that supply the organs and muscles of the body via the series of paired spinal nerves along its length. It is contained in the protective vertebral canal of the backbone. In cross section, an outer region of white matter containing ascending and descending nerve fibers surrounds a roughly H-shaped region of gray matter, which consists of nerve cell bodies. In the center is a narrow canal filled with cerebrospinal fluid. *See also* spinal nerves.

spinal nerves The paired nerves that arise at intervals along the length of the spinal cord to supply each segment of the body. Each spinal nerve is connected to the spinal cord by both a *dorsal root*, which carries sensory nerve fibers and bears a ganglion containing sensory nerve cell bodies, and a *ventral root*, which carries the fibers of motor nerve cells. These two roots combine before emerging from the vertebral column to form the nerve trunk. This later divides into smaller trunks that supply the skin, muscles, and internal organs of a particular body region. The spinal nerves in the region of each limb are connected to form a nerve plexus.

spindle The structure formed during mitosis and meiosis that is responsible for moving the chromatids and chromosomes to opposite poles of the cell. The spindle consists of a longitudinally orientated system of protein microtubules whose synthesis starts late in interphase under the control of a microtubule-organizing center. In plants and animals this is the CENTROSOME. A special region of the centromeres of each pair of sister chromatids, the *kinetochore*, becomes attached to one or a bundle of spindle microtubules. During anaphase, the kinetochore itself acts as the motor, disassembling the attached microtubules and hauling the chromatid towards the spindle pole. Later in anaphase, the unattached interpolar microtubules actively slide past each other, thus elongating the entire spindle.

spine 1. (*Botany*) A modified leaf reduced to a sharply pointed structure as a protection against predators, as in barberry. In some species only a part of the leaf is modified, as in holly. *Compare* prickle; thorn. **2.** (*Zoology*) *See* vertebral column.

spinneret A paired appendage on the abdomen of a spider for spinning silk, used for making its web or egg cocoons or for binding prey caught on the web. The silk is secreted as a liquid by silk glands but hardens as it passes out through the spinnerets.

spiracle /spȳ-ră-kăl/ **1.** An opening leading to a trachea in an insect, isopod, centipede, millipede, or arachnid. Spiracles occur in pairs on each side of the body. In adult insects, one pair occurs on each of the posterior two thoracic segments and the anterior eight abdominal segments. The arrangement in larval and pupal stages of insects may vary. **2.** The opening from the gill chambers (atria) of the frog tadpole, on the left side only. **3.** The anterior gill cleft of cartilaginous fish, usually much reduced.

spiral thickening (helical thickening) The type of secondary wall formation in which a spiral band of lignified deposits is formed on the inner wall of the cell. It is found in protoxylem and first metaxylem tracheids and vessels. Like annular thickening, it allows for continued elongation of the XYLEM.

spiral valve 1. A spirally arranged fold of epithelium in the intestine of all fish except teleosts. It delays the passage of food and provides an increased surface area for secretion and absorption. **2.** *See* conus arteriosus.

spirillum /spȳ-ril-ŭm/ (*pl.* spirilla) The appearance of helically shaped bacteria. Cells are usually found singly and possess flagella at one or more poles.

spirochetes /spȳ-rŏ-keets/ Long spirally twisted bacteria surrounded by a flexible wall. An axial filament, with a similar structure to a bacterial flagellum, is spirally wound about the protoplast inside the cell wall. They swim actively by flexing the cell. They are found in mud and water and can withstand low oxygen concentrations. Many spirochetes are pathogens causing yaws, syphilis, and relapsing fever.

Spirogyra /spȳ-rŏ-jȳ-ră/ A genus of filamentous green algae found in freshwater and having a characteristic spiral chloroplast. Reproduction is by conjugation: two filaments become aligned and pairs of cells, one from each filament, become joined by a conjugation tube. The contents of one cell pass through the tube and fuse with the contents of the second cell to form a zygote. *See* Chlorophyta.

spleen A lymphoid organ situated just beneath the stomach in vertebrates. It produces lymphocytes and destroys and stores red blood cells. The spleen consists of loose connective tissue containing lymphoid tissue (Malpighian bodies), which surrounds a network of sinuses. The circulation of blood through the spleen is slow and blood can leak out of the sinuses into the lymphoid tissue. Thus, there is ample opportunity for phagocytosis of red blood cells and bacteria.

sponges *See* Porifera.

spongy mesophyll *See* mesophyll.

spontaneous generation The erroneous belief that modern living organisms can be formed from inorganic material, given the right conditions. This belief, disproved by Redi and Pasteur in the 17th and 19th centuries, should not be confused with the concept of gradual inorganic evolution and abiogenesis. *See also* abiogenesis; evolution; origin of life.

spontaneous movements *See* autonomic movements.

sporangium /spŏ-**ran**-jee-ŭm/ (*pl.* **sporangia**) The plant reproductive body in which asexual spores are formed.

spore /spor, spohr/ A uni- or multicellular plant reproductive body. Generally the term is applied to reproductive units produced asexually, such as the spores of bryophytes and pteridophytes. However certain sexually formed structures (e.g. the oospore) are also called spores. A prefix is often added, providing information as to the nature of the spore, for example conidiospores arise on a conidium, zoospores are motile, etc.

spore mother cell (**sporocyte**) A cell that gives rise to four haploid spores by meiosis.

In heterosporous species, many of the potential megaspores often abort.

sporocarp /**spor**-ŏ-karp, **spoh**-rŏ-/ **1.** A hard spore-containing structure found in water ferns (e.g. *Marsilea* and *Pilularia*). The sori become enclosed by the growing together of fertile fronds, and the spores are not liberated until the sporocarp decays and ruptures.
2. *See* ascocarp.

sporocyst /**spor**-ŏ-sist, **spoh**-rŏ-/ **1.** The tough covering of a spore, such as is found around the spores of sporozoan protoctists (e.g. *Monocystis* and *Plasmodium*).
2. In parasitic flatworms (e.g. *Fasciola*), a sac in which redia larvae are produced. It develops from a miracidium. *See* cercaria.

sporocyte /**spor**-ŏ-sÿt, **spoh**-rŏ-/ *See* spore mother cell.

sporogonium /spor-ŏ-**goh**-nee-ŭm, spoh-rŏ-/ (*pl.* **sporogonia**) The sporophyte generation in mosses and liverworts. It develops from the zygote and comprises the foot, seta, and capsule. The sporogonium is parasitic on the gametophyte generation.

sporophore /**spor**-ŏ-for, **spoh**-rŏ-fohr/ The aerial spore-producing body of certain fungi, e.g. the mushroom of *Agaricus*.

sporophyll /**spor**-ŏ-fil, **spoh**-rŏ-/ The sporangium-bearing structure of vascular plants. In some ferns sporangia develop on the normal foliage leaves but in higher plants, the sporophylls are highly modified leaves. They may be grouped together in a strobilus, or, in the angiosperms, located in the flowers.

sporophyte /**spor**-ŏ-fÿt, **spoh**-rŏ-/ The diploid generation giving rise asexually to haploid spores. In vascular plants, the sporophyte is the dominant generation, while in bryophytes it is parasitic on the gametophyte. *See* alternation of generations; gametophyte.

sporozoan /spor-ŏ-**zoh**-ăn, spoh-rŏ-/ Any of various parasitic protoctists with a complicated life cycle involving the alternation of sexual and asexual reproduction and the production of spores to insure dispersal to another host. In older classifications they constituted a class of Protozoa, the Sporozoa, but in more recent schemes they are distributed among several phyla, especially the Apicomplexa.

Squamata /skway-mă-tă/ The order that contains the most successful living reptiles, the lizards and their descendants, the snakes, characterized by a body covering of overlapping horny scales. Lizards typically have a long tail, four limbs – although some, e.g. *Anguis* (slowworm), are limbless – an eardrum, and movable eyelids. Snakes lack an eardrum and their eyes are covered by transparent spectacle eyelids. They have an elongated body lacking limbs and girdles, a deeply forked protrusible sensory tongue, and an extremely wide jaw gape made possible by the loose articulations of the skull bones. The prey is swallowed whole. Primitive snakes (e.g. *Python*) suffocate their prey; the more advanced types use their fanglike teeth. Some, e.g. *Vipera* (viper), inject fast-acting poisons through their fangs to kill large animals.

squamosal /skway-**moh**-săl/ One of a pair of bones on the side of the skull of most vertebrates. In mammals, each has a process anterior to the ear, which curves forward and fuses in an arch with the jugal bone to form the cheek bones. The process also articulates at its posterior end with the dentary of the lower jaw.

staining A procedure that is designed to heighten contrast between different structures. Normally biological material is lacking in contrast, protoplasm being transparent, and therefore staining is essential for an understanding of structure at the microscopic level. *Vital stains* are used

COMMON STAINS FOR LIGHT MICROSCOPY		
Stains	*Final Color*	*Suitable for*
aniline (cotton) blue	blue	fungal hyphae and spores
aniline sulfate or hydrochloride	yellow	lignin
borax carmine	pink	nuclei; particularly for whole amounts (large pieces) of animal material
eosin	pink	cytoplasm; *see* hematoxylin
	red	cellulose
Feulgen's stain	red/purple	DNA; particularly to show chromosomes during cell division
hematoxylin	blue	nuclei; mainly used for sections of animal tissue with eosin as counterstain for cytoplasm; also for smears
iodine	blue–black	starch; therefore for plant storage organs
Leishman's stain	red–pink	blood cells
	blue	white blood cell nuclei
light green or fast green	green	cytoplasm and cellulose; *see* safranin
methylene blue	blue	nuclei; suitable as a vital stain
phloroglucinol	red	lignin
safranin	red	nuclei. Lignin and suberin. Mainly used for sections of plant tissue with light green as counterstain for cytoplasm

to stain and examine living material. Most stains require dead or nonliving material. Staining is done after fixation and either during or after dehydration. *Double staining* involves the use of two stains; the second is called the *counterstain*. *Acidic stains* have a colored anion, *basic stains* have a colored cation. Some stains are neutral. Materials can be described as *acidophilic* or *basophilic* depending on whether they are stained by acidic or basic dyes respectively. Basic stains are suitable for nuclei, staining DNA. Stains for light microscopy are colored dyes; those for electron microscopy contain heavy metals, e.g. uranyl acetate, lead citrate, and osmium tetroxide.

stamen /stay-měn/ The male reproductive organ in flower plants consisting of a fine stalk, the *filament*, bearing the pollen producing *anther*. It is equivalent to the microsporophyll present in the gymnosperms and heterosporous pteridophytes. The collective term for the stamens is the *androecium*.

staminode /stam-ă-nohd/ A sterile stamen. It may be rudimentary or partly developed, consisting of only the filament, as in figwort, or it may form a conspicuous part of the flower, as in iris.

standing crop The nutritional portion of the biomass in an area at a given moment.

stapes /stay-peez/ (*pl.* **stapes**) (**stirrup**) The stirrup-shaped bone attached to the oval window (fenestra ovalis) of the ear, which forms the innermost ear ossicle in mammals. It is homologous with the hyomandibular of fishes. *See illustration at* ear. *See also* incus; malleus.

Staphylococcus /staf-ă-lŏ-**kok**-ŭs/ A genus of Gram-positive spherical nonmotile bacteria. They are facultative anaerobes and do not form spores. Many species are parasites or pathogens of animals and some cause wound infections, abscesses, and a type of food poisoning. They are killed by pasteurization and many common disinfectants.

starch A polysaccharide that occurs exclusively in plants. Starches are extracted commercially from maize, wheat, barley, rice, potatoes, and sorghum. They exist in the plant cells as granules dispersed in the cytoplasm. The starches are storage reservoirs for plants; they can be broken down by enzymes to simple sugars and then metabolized to supply energy needs. Starch is a dietary component of animals. In humans it is digested by salivary and pancreatic amylase then further degraded by maltase to yield glucose, which may be stored as glycogen (animal starch). Excess starch, i.e. above the maximum liver and muscle storage capacity, is converted to lipids and stored as fat. Starch is not a single molecule but a mixture of amylose (water-soluble, blue color with iodine) and amylopectin (not water-soluble, violet color with iodine). The composition is amylose 10–20%, amylopectin 80–90%.

starch sheath The innermost layer of cells of the cortex replacing the endodermis in some stems, especially young herbaceous dicotyledonous stems. The starch sheath contains prominent starch grains and is thought by some to be involved in the perception of gravity.

starch–statolith hypothesis A hypothesis concerning the mechanism of gravity perception in plants. *See* geotropism.

starfish *See* Asteroidea.

statoblast /stat-ŏ-blast/ An internal bud produced asexually by freshwater bryozoans. Statoblasts have a thick protective covering, enabling them to survive extremes of drought and temperature after they have broken away from the parent colony.

statocyst /stat-ŏ-sist/ An organ concerned with the perception of gravity in invertebrates. Statocysts are seen, for example, in the tentacles of *Obelia* and in the antennules of the crawfish. It is similar in structure and function to the macula of vertebrates.

statocyte /stat-ŏ-sÿt/ A plant cell containing STATOLITHS, and thought to be involved in the plant's perception of gravity. *See* geotropism.

statolith /stat-ŏ-lith/ One of a number of large starch grains found in the statocytes, plant cells that are thought to be gravity sensitive. They move through the cytoplasm to the lowermost cell surface, enabling the plant to detect the direction of gravity. *See* geotropism.

stearic acid /stee-a-rik/ (octadecanoic acid) A saturated carboxylic acid, which is widely distributed in nature as the glyceride ester. It is present in most fats and oils of animal and vegetable origin, particularly the so-called hard fats, i.e. those of higher melting point.

stele /steel, -stee-lee/ The vascular tissue and (if present) the surrounding pericycle and endodermis of a stem or root. Stelar arrangements vary considerably in plants from the simple protostele, in which leaf gaps are absent, to the complex dictyostele, in which there are many closely spaced leaf gaps. Intermediate between these is the solenostele in which the leaf gaps are more widely spaced vertically so that only one is seen per cross section of the stem. The different types of stele are shown in the illustration.

stem A longitudinal axis upon which are borne the leaves, buds, and reproductive organs of the plant. The stem is generally aerial and erect but various modifications are found, for example underground stems like rhizomes, bulbs, and corms, and horizontal structures, such as runners. The stem serves to conduct water and food materials up and down the plant and, particularly in young plants, it may serve as a photosynthetic organ. The stem is generally cylindrical and consists of regularly arranged conducting (vascular), strengthening, and packing cells, the whole being surrounded by a protective epidermis.

stem cell Any cell that remains undifferentiated and capable of unlimited division in order to provide new cells for growth or replacement of tissues. After division it may produce more stem cells or cells that differentiate into specialized tissue cells. An early embryo contains stem cells with the potential to develop into any of the differentiated cells found in the body. Throughout adult life various types of stem cells remain active in many of the body's

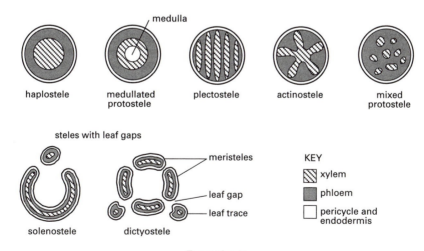

Types of stele

tissues. For example, intestinal stem cells continually divide to replenish the cells of the gut lining, whereas stem cells in bone marrow give rise to differentiated blood cells, such as lymphocytes and erythrocytes. Stem cells that divide to produce just one type of differential cell are called *unipotent* ('having a single capability'). Stem cells that are capable of producing two, three, or more types of differentiated cells are called *pluripotent* ('having many capabilities').

stenohaline /sten-ŏ-**hay**-lin, -lўn, -**hal**-in, -ўn/ Describing organisms that are unable to tolerate wide variations of salt concentrations in the environment. *Compare* euryhaline.

sterigma /sti-**rig**-mă/ (*pl.* **sterigmata**) A finger-like projection upon which spores are formed in many fungi. In basidiomycete fungi, four sterigmata, each bearing one basidiospore, are usually borne on each basidium. In certain ascomycete fungi, e.g. *Penicillium*, the sterigmata give rise to chains of conidiophores.

sternum /**ster**-nŭm/ **1.** A shield or rod-shaped bone (the breastbone) in the midline of the ventral side of the thorax of tetrapods, to which the ventral ends of the ribs are usually attached. It is important for wing-muscle attachment in birds and bats. In humans it is an elongate flattened bone articulating with the clavicles at its upper end and with ribs along each side. *See also* keel.
2. The plate, stiffened with chitin, that forms the protective ventral covering to each segment of the thorax and abdomen of an insect. *See also* pleuron; tergum.

steroid /**ste**-roid/ Any member of a group of compounds having a complex basic ring structure. Examples are corticosteroid hormones (produced by the adrenal gland), sex hormones (progesterone, androgens, and estrogens), bile acids, and sterols (such as cholesterol). *See also* anabolic steroid; sterol.

sterol /**ste**-rol, -rohl/ A steroid with long aliphatic side chains (8–10 carbons) and at least one hydroxyl group. They are lipid-soluble and often occur in membranes (e.g. cholesterol and ergosterol).

stigma /**stig**-mă/ **1.** The receptive tip of the carpel of plants. *See illustration at* flower. **2.** *See* eyespot.

stilt root An enlarged form of prop root, seen in some mangroves, that helps support plants in unstable soils.

stimulus (*pl.* **stimuli**) A change in the external or internal environment of an organism that elicits a response in the organism. The stimulus does not provide the energy for the response.

stipe /stўp/ **1.** In the more highly differentiated algae (e.g. wracks and kelps), the stalk between the holdfast and the blade. **2.** The stalk of the fruiting body of certain basidiomycete fungi (e.g. *Agaricus*) that bears the cap or pileus.

stipule /**stip**-yool/ A modified leaf found as an outgrowth from the petiole or leaf base. Stipules are seen around the base of the petiole in the garden pea, in which they serve as additional photosynthetic organs, and growing along the length of the petiole in the rose.

stolon /**stoh**-lon/ **1.** (*Zoology*) A branched stemlike structure in some invertebrates (e.g. cnidarians and tunicates) from which new individuals develop.

Steroid ring structure

2. (*Botany*) An initially erect branch that, due to its great length, eventually bends over towards the ground. Where a node touches the soil adventitious roots develop and the axillary bud at that node grows out to form a new plant. The stolon thus acts as an organ of vegetative reproduction. *See also* offset; runner.

stoma /stoh-mă/ (*pl.* **stomata**) One of a large number of pores in the epidermis of plants through which gaseous exchange occurs. In most plants stomata are located mainly in the lower epidermis of the leaf. Each stoma is surrounded by two crescent-shaped GUARD CELLS, which regulate the opening and closing of the pore by changes in their turgidity.

stomach The part of the alimentary canal in vertebrates that lies between the esophagus and the duodenum. It acts as a storage organ so that food can be eaten at intervals instead of continuously. It is a large sac with thick muscular walls and is closed at each end by a ring of muscle (sphincter). It expands to hold a meal for several hours, during which time the food is churned by muscular contractions, mixed with hydrochloric acid, and the protein in it is partly digested by the enzymes of gastric juice. When the food has been reduced to semiliquid chyme, it is passed, a little at a time, through the pylorus to the duodenum. *See* duodenum.

The stomach is lined with mucous membrane containing simple tubular glands (gastric pits). These contain oxyntic cells, which secrete hydrochloric acid; peptic or chief cells, which secrete the enzymes pepsin and rennin; and goblet cells, which secrete mucus. There are three muscle layers in the stomach: circular, longitudinal, and oblique.

In birds, the posterior part of the stomach is the gizzard. In some herbivores, there are several compartments.

stomium /stoh-mee-ŭm/ A structure involved in the dispersal of spores in certain plant structures. For example in the sporangium of the fern *Dryopteris* the stomium is seen below the annulus and is gradually ruptured as the annulus dries out. In the stamen of angiosperms the cells in the groove between the two pollen sacs form a stomium, and rupture as the anther dries out.

stomodeum /stoh-mŏ-dee-ŭm, stom-ŏ-/ (*pl.* **stomodea**) An invagination of ectoderm of the animal embryo that forms the anterior region of the alimentary canal – the mouth, pharynx, and esophagus in mammals.

stone cell (**brachysclereid; sclereid**) An isodiametric sclereid found either singly or in groups in the parenchyma and phloem of stems and some fruits (e.g. pear).

stratum corneum /stray-tŭm kor-nee-ŭm, strat-ŭm kor-nee-ŭm/ The outermost layer of the epidermis of vertebrates. Its cells are dead, flattened, and dry, containing a high proportion of the horny protein keratin. It provides the main protection of the body against water loss and the entry of disease-causing organisms. *See also* Malpighian layer.

Streptococcus /strep-tŏ-kok-ŭs/ A genus of spherical Gram-positive bacteria that usually occur in pairs or chains; most strains are nonmotile. Most species are parasites or pathogens of animals, often occurring in the respiratory or alimentary tracts. Some species are hemolytic (i.e. they destroy red blood cells) and cause such diseases as scarlet fever and rheumatic fever. Streptococci are killed by pasteurization and common disinfectants; penicillin, tetracycline, and other antibiotics are effective against hemolytic strains.

stretch receptor *See* muscle spindle; proprioceptor.

stretch reflex An involuntary response (*see* reflex) that tends to maintain a constant muscle length when the load is increased. Loading a muscle causes it to stretch, which is detected by stretch receptors (MUSCLE SPINDLES) inside the muscle. These receptors send sensory impulses to the spinal cord, where the information is integrated and impulses are sent out via motor neurons to increase contraction of the muscle and counteract the stretching. Such reflexes are essential in maintaining balance and posture, as well as in performing voluntary muscle actions.

striated muscle /strȳ-**ay**-tid/ *See* skeletal muscle.

stridulation /strij-û-**lay**-shŏn/ The production of sounds by some insects, usually males, by rubbing together parts of the body. Grasshoppers stridulate by rubbing the hindlimbs against the forewing; crickets by rubbing forewings together. Stridulation has an important role in attracting and stimulating a female during courtship.

strobila /strŏ-**bȳ**-lă/ (*pl.* strobilae) The form or part of an animal in which asexual reproduction occurs by transverse division (*strobilization*) into a number of separate individuals. Examples include the scyphistoma stage in the life cycle of a jellyfish and the chain of proglottids in a tapeworm.

strobilus /strŏ-**bȳ**-lŭs/ (*pl.* strobili) **1.** The reproductive structure of the gymnosperms and certain pteridophytes. In the club mosses, horsetails, and *Selaginella* only one type of strobilus is formed, while in the gymnosperms both female megastrobili and male microstrobili develop. Strobili are also termed *cones*.
2. A type of dry composite fruit, the individual fruits being achenes. Strobili are found in hops.

stroke volume The volume of blood pumped by the ventricles of the heart during a single heartbeat. It is affected by several factors, notably by the degree to which the ventricles are stretched by filling with venous blood. The more blood that fills the ventricles during diastole, the more the ventricles are stretched, and the greater the force of contraction of the ventricles during systole. Hence increased venous filling pressure causes increased stroke volume, up to a point where overstretching causes decreased contraction. This relationship is called the *Frank–Starling law*, named for

the German physiologist Otto Frank (1865–1944) and the British physiologist Ernest Henry Starling (1866–1927). *See* cardiac cycle.

stroma /**stroh**-mă/ **1.** (*Botany*) The colorless ground matter between the grana lamellae in a chloroplast.
2. (*Botany*) A mass of fungal hyphae, sometimes including host tissue, in which fruiting bodies may be produced. An example is the compact black fruiting body of the ergot fungus, *Claviceps purpurea*.
3. (*Zoology*) A tissue that acts as a framework; for example, the connective tissue framework of the ovary or testis that surrounds the cells concerned with gamete production.

stromatolite /stroh-**mat**-ŏ-lȳt/ A layered cushion-like mass of chalk formed by the actions of certain bacteria, notably blue-green bacteria (*see* Cyanobacteria). Communities of these organisms trap and bind lime-rich sediments. Modern stromatolite-building communities are confined to salt flats or shallow salty lagoons where bacterial predators cannot survive, but the fossil record demonstrates a much more widespread distribution in the Precambrian period. Some stromatolites date back nearly 4000 million years, making them the oldest known fossils.

style The stalklike portion of a carpel, joining the ovary and the stigma. The style may be elongated in plants relying on wind, insect, or animal pollination so that the stigma has a greater chance of coming into contact with the pollinating agent. *See illustration at* flower.

subarachnoid space /sub-ă-**rak**-noid/ The space between the middle arachnoid membrane and the inner pia mater, which surrounds and protects the brain and spinal cord in vertebrates. It is crossed by delicate tissue strands and filled with cerebrospinal fluid, which cushions the central nervous system against external shocks.

subclavian artery /sub-**klay**-vee-ăn/ In mammals, a large paired artery that carries oxygenated blood to the forelimbs. It arises on the left side from the aorta and on the right side from the innominate artery.

subcutaneous tissue /sub-kyoo-**tay**-nee-ŭs/ A layer of tissue beneath the dermis, which contains stored fat. It is important in restricting heat loss in aquatic mammals and in some hibernating mammals; in the latter it also acts as an essential food store.

suberin /**soo**-bĕ-rin/ A mixture of substances produced in the walls of cork tissue. It is similar in properties and functions to CUTIN. The Casparian band in roots and some stems contains suberin, lignin, or similar substances.

subhymenium /sub-hȳ-**mee**-nee-ŭm/ (*pl.* subhymenia) The intermediate layer of tissue between the trama and hymenium in the gills of basidiomycete fungi.

sublittoral /sub-**lit**-ŏ-răl/ **1.** The marine zone that extends from low tide to a depth of around 200 m. Large algae (e.g. kelps) are found in shallower waters, while certain red algae may be found in deeper water. Numerous animals are found in this zone, including mollusks, echinoderms, arthropods, cnidarians, etc. *Compare* littoral; benthic.
2. The zone in a lake or pond between the LITTORAL and PROFUNDAL zones, extending from a depth of about six to ten meters. Its depth is limited by the *compensation level* – the depth at which the rate of photosynthesis is equalled by the rate of respiration, and below which plants cannot live. The sublittoral zone contains plankton, a large mollusk population, and freshwater crustaceans.

subsidiary cell *See* accessory cell.

subspecies /**sub**-spee-sheez, sub-**spee**-/ The taxonomic group below the species level. Crosses can generally be made between subspecies of a given species but this may be prevented in the wild by various isolating mechanisms, e.g. geographical isolation or different flowering times.

substitution *See* point mutation.

substrate /**sub**-strayt/ **1.** The substance upon which an enzyme acts.
2. The nonliving material upon which an organism lives or grows.

succession A progressive series of changes in vegetation and animal life of an area from initial colonization to the final stabilized stage, or climax. The climax is stable because the succession can progress no further under the climatic, edaphic, and other environmental factors present at the time. *See* sere.

succinic acid /suk-**sin**-ik/ A dicarboxylic acid formed by fermentation of sugars. It occurs in algae, lichens, sugars, and other plant substances. Succinate ions have an important role in the Krebs cycle.

succulent /**suk**-yŭ-lĕnt/ Describing a plant with thick fleshy leaves that are rounded, cylindrical, or spherical. Such leaves are an adaptation to life in dry habitats, by minimizing the ratio of surface area to volume and so reducing water loss by evaporation from leaf surfaces. Examples include ice plants and *Sedum*. *See* xerophyte.

succus entericus /**suk**-ŭs en-te-ră-kŭs/ The viscous alkaline secretion produced by Brunner's glands in the wall of the duodenum. It consists of water, hydrogencarbonate ions, and mucoproteins, and serves to protect the walls of the small intestine from the corrosive effects of the acidic and proteolytic chyme entering from the stomach. It was formerly thought to contain digestive enzymes, but all duodenal enzymes are now known to be localized on or within the epithelial cells.

sucker An underground shoot that at some stage emerges above the soil surface and gives rise to a new plant, which initially is nourished by the parent plant until it becomes established. Suckers are troublesome in certain ornamentals (e.g. roses) in which the plant has been grafted onto a wild root stock, since sucker shoots develop from the root stock and grow at the expense of the ornamental.

sucrose /soo-krohs/ (**cane sugar**) A sugar that occurs in many plants. It is extracted commercially from sugar cane and sugar beet. Sucrose is a disaccharide formed from a glucose unit and a fructose unit. It is hydrolyzed to a mixture of fructose and glucose by the enzyme invertase. Since this mixture has a different optical rotation (levorotatory) from the original sucrose, the mixture is called *invert sugar*.

suction pressure (**SP**) *See* osmosis.

sugar (**saccharide**) One of a class of sweet-tasting carbohydrates that are soluble in water. Sugar molecules consist of linked carbon atoms with –OH groups attached, and either an aldehyde or ketone group. The simplest sugars are the *monosaccharides*, such as GLUCOSE and FRUCTOSE, which cannot be hydrolyzed to sugars with fewer carbon atoms. They can exist in a chain form or in a ring formed by reaction of the ketone or aldehyde group with an –OH group on one of the carbons at the other end of the chain. It is possible to have a six-membered (*pyranose*) ring or a five-membered (*furanose*) ring. Monosaccharides are classified according to the number of carbon atoms: a *pentose* has five carbon atoms and a *hexose* six. Monosaccharides with aldehyde groups are *aldoses*; those with ketone groups are *ketoses*. Thus, an *aldohexose* is a hexose with an aldehyde group; a *ketopentose* is a pentose with a ketone group, etc.

Two or more monosaccharide units can be linked in *disaccharides* (e.g. SUCROSE), *trisaccharides*, etc. *See also* polysaccharide.

sugar acid An acid formed from a monosaccharide by oxidation. Oxidation of the aldehyde group (CHO) of the aldose monosaccharides to a carboxyl group (COOH) gives an *aldonic acid*; oxidation of the primary alcohol group (CH_2OH) to COOH yields *uronic acid*; oxidation of both the primary alcohol and carboxyl groups gives an *aldaric acid*. The uronic acids are biologically important, being

components of many polysaccharides, for example glucuronic acid (from glucose) is a major component of gums and cell walls, while galacturonic acid (from galactose) makes up pectin. Ascorbic acid or vitamin C is an important sugar acid found universally in plant tissues, particularly in rose hips and citrus fruits.

sugar alcohol (**alditol**) An alcohol derived from a monosaccharide by reduction of its carbonyl group (CO) so that each carbon atom of the sugar has an alcohol group (OH). For example, glucose yields sorbitol, common in fruits, and mannose yields mannitol.

sulfonamide /sul-**fon**-ă-mÿd/ One of a group of bacteriostatic drugs having a sulfonamide group (SO_2NH_2). The bacteriostatic action is due to the similarity in chemical structure of sulfonamides to para-aminobenzoic acid, required by some bacteria to manufacture the vitamin folic acid. The bacteria are unable to distinguish the two and take up the sulfonamide if it is present in higher concentration, which prevents folic acid synthesis, and hence development and reproduction. Examples of sulfonamides are sulfanilamide, sulfafurazole, and sulfamerazine.

sulfur /**sul**-fer/ An essential element in living tissues, being contained in the amino acids cysteine and methionine and hence in nearly all proteins. Sulfur atoms are also found bound with iron in ferredoxin, one of the components of the electron transport chain in photosynthesis. Plants take up sulfur from the soil as the sulfate ion SO_4^{2-}. The sulfides released by decay of organic matter are oxidized to sulfur by sulfur bacteria of the genera *Chromatium* and *Chlorobium*, and further oxidized to sulfates by bacteria of the genus *Thiobacillus*. There is thus a cycling of sulfur in nature.

sulfur bacteria Filamentous autotrophic chemosynthetic bacteria that derive energy by oxidizing sulfides to elemental sulfur and build up carbohydrates from carbon dioxide. An example is *Beggiatoa*. *See also* photosynthetic bacteria.

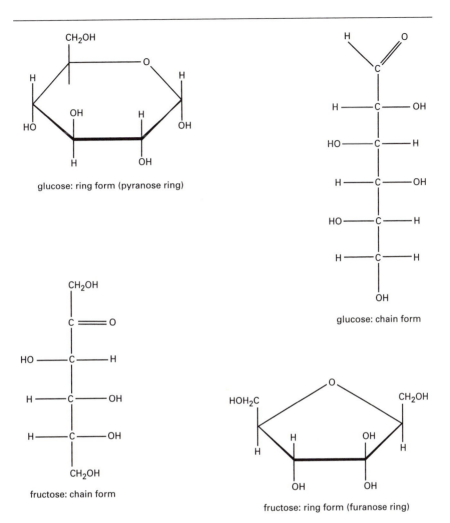

glucose: ring form (pyranose ring)

glucose: chain form

fructose: chain form

fructose: ring form (furanose ring)

glycosidic link

sucrose: a disaccharide

Sugars

sulfur dioxide (**sulfur(IV) oxide**) A pungent gas, SO_2, formed by burning sulfur in air, and also as a by-product of the combustion of fossil fuels, such as oil, coal, and gas. It is a major pollutant, combining with moisture in the atmosphere to form sulfurous and sulfuric acids, which are important components of ACID RAIN.

summation 1. The additive effect of several impulses arriving at a synapse of a nerve and/or muscle cell, when individually the impulse cannot evoke a response. The impulses either arrive simultaneously at different synapses at the same cell (*spatial summation*) or in succession at one synapse (*temporal summation*). Stimulation of the synapse elicits a graded postsynaptic potential and if the potential exceeds the threshold level, a postsynaptic impulse is triggered. Summation is one of the major mechanisms of integration in the nervous system. *Compare* facilitation.
2. The interaction of two substances with similar effects in a given system, such that the combined effect is greater than their separate effects.

supercooling /soo-per-kool-ing/ The ability of a liquid to cool below its freezing point and remain in a liquid state. Many plants, animals, and other organisms that live in cold environments survive only because of various adaptations that permit supercooling of their tissue fluids below freezing point, in some cases as low as –40°C. Above this temperature, ice crystals will form only in the presence of nucleating agents that enable the crystals to 'seed'. So for an organism to survive, the tissue fluids are kept free of nucleating agents, and any ice crystals that do form are restricted to extracellular spaces. This draws cytoplasmic water from cells, concentrating the cytoplasm still further and lowering its freezing point.

supergene /soo-per-jeen/ A collection of closely linked genes that tend to behave as a single unit because crossing over between them is very rare.

superior Above. In botany, the term is used with reference to the position of the ovary in relation to the other parts of the flower. When the ovary is superior, the petals, sepals, and stamens are inserted at the base of the ovary where it joins the flowerstalk, as in buttercup flowers. In a floral formula a superior ovary is denoted by a line below the carpel number. *Compare* inferior. *See also* hypogyny; perigyny.

supernormal stimulus /soo-per-**nor**-măl/ A stimulus presented in animal behavior experiments that is more effective than the natural stimulus. For example, an oyster-catcher will incubate a clutch of five eggs in preference to a normal clutch of three; also, if presented with a normal egg and one twice its size, it usually chooses to incubate the large egg.

supination /soo-pă-**nay**-shŏn/ *See* pronation.

survivorship curve /ser-**vÿ**-ver-ship/ A graph that plots the number of surviving members of a population against the age of the population. The shape of the curve reflects the distribution of mortality over time. A straight diagonal line indicates fairly constant mortality over time, with equal numbers dying in each time period, whereas an initial steep decline shows that large numbers die during early life, with relatively few surviving to maturity. A fairly flat curve ending in a steep decline reflects a population, such as that of stable human societies, in which the number of offspring is small but most survive to maturity.

suspension culture A method of growing free-living single cells or small clumps of cells in a liquid medium. Microorganisms or cells of plant callus tissue may be grown in this way; the liquid medium is agitated to keep the cells in suspension. Individual cells from plant suspension cultures may be carefully isolated and grown into entire plants, the process being regulated by hormone treatment. In the case of callus, it can be demonstrated that differentiated plant cells can be dissociated into single cells and subsequently regenerated to reproduce an entire plant.

suspension feeding (microphagous feeding) A type of feeding in which minute food particles are removed from dispersion in a liquid medium surrounding the animal. *See also* ciliary feeding; filter feeding.

suspensor /sŭ-**spen**-ser/ A temporary stalklike structure, found in angiosperms, that pushes the embryo into the nutritive endosperm after fertilization. It is also seen in certain pteridophytes (e.g. *Selaginella*) and in gymnosperms (e.g. *Cycas*), in which it pushes the embryo into the female gametophyte tissue. The suspensor may be uni- or multicellular and develops from the outer of the two cells arising from the first division of the zygote.

suture /**soo**-cher/ (*Anatomy*) A line marking the fusion or junction between adjacent parts. Examples are the sutures between the bones of the skull, and, in plants, the margins of the carpel.

swallowing *See* deglutition.

sweat The watery fluid secreted by the sweat glands. It contains small amounts of sodium, chloride, and potassium ions and urea; certain bacteriological and fungicidal substances are also present and contribute to the body's defense against disease. Water in the sweat evaporates from the skin surface and helps to keep the body cool. In humans, *thermoregulatory sweating* (*thermal sweating*) takes place over the entire body surface, whereas *emotional sweating* is confined to the palms, soles, and armpits. *Sensible sweating*, which occurs in hot weather or during muscular exercise, results in a large quantity of sweat production accompanied by dilatation of the blood capillaries in the skin initiated by bradykinin in the sweat. In a humid atmosphere, evaporation is impeded, thus preventing heat loss (the presence of sweat on the skin inhibits further sweating).

sweat gland A coiled tubular gland in the dermis of mammals that produces sweat. A tube runs from the gland, through the epidermis, and sweat is secreted on to the outer surface of the skin (*see* sweat). Modified sweat glands (*ceruminous glands*) in the external auditory meatus of the ear produce wax.

swim bladder A large thin-walled cavity found in bony fish, by which the fish is able to adjust its buoyancy as it swims at different depths. It contains oxygen and nitrogen. A network of blood vessels surrounds the bladder and oxygen can be extracted from the blood to adjust the volume of gas in the bladder. The bladder is also responsible for the detection of sounds, which cause high-frequency changes in the tension in the bladder walls, and in some teleost fish it acts as a resonator or sound producer. In certain fish, the swim bladder is connected by a tube to the pharynx and in lungfish it functions as a lung. *See also* Weberian ossicles.

syconus /sȳ-**koh**-nŭs/ (**synconium**) A composite succulent fruit that develops from a hollow capitulum, as in the fig. The small individual flowers borne on the inside of the capitulum each form a single drupe, these being the 'pips' of the fig.

symbiosis /sim-bȳ-**oh**-sis, -bee-/ (*pl.* **symbioses**) Any close association between two or more different organisms, as seen in parasitism, mutualism, and commensalism. The term is usually used more narrowly to mean MUTUALISM.

sympathetic nervous system (thoracolumbar nervous system) One of the two divisions of the autonomic nervous system, which supplies motor nerves to the smooth muscles of internal organs and to heart muscle. Sympathetic nerve fibers arise via spinal nerves in the thoracic and lumbar regions. Their endings release mainly norepinephrine, which increases heart rate and breathing rate, raises blood pressure, and slows digestive processes, thereby preparing the body for 'fight or flight' and antagonizing the effects of the parasympathetic nervous system. The medulla of the adrenal gland is supplied only by sympathetic fibers, which trigger the release of epinephrine into the bloodstream, thus enhancing the effects of the sympathetic system. *See also* autonomic nervous system; parasympathetic nervous system.

sympatric species /sim-**pat**-rik/ *See* species.

symphysis /sim-fă-sis/ (*pl.* **symphyses**) A type of joint in which two bones are connected by a fibrocartilaginous disk and fibrous ligaments. It allows for slight movement under deformation. *See also* pubic symphysis.

symplast /**sim**-plast/ The living system of interconnected protoplasts extending through a plant body. Cytoplasmic connections between cells are made possible by the plasmodesmata. The symplast pathway is an important transport route through the plant. *Compare* apoplast.

symplastic growth /sim-**plas**-tik/ A form of plant growth in which neighboring cell walls stay in contact and grow at the same rate. *Compare* intrusive growth; sliding growth.

sympodial /sim-**poh**-dee-ăl/ Describing the system of branching in plants in which the terminal bud of the main stem axis stops growing and growth is taken over by lateral buds. These in turn lose their dominance and lateral buds take over their role. This process may repeat itself to form a multiple branching system. Sympodial growth is also called definite or cymose branching and is typical of the formation of a cymose inflorescence. *Compare* monopodial.

synapomorphy A homologous derived trait (APOMORPHY) that is shared by two or more species because they inherited the trait from a recent shared ancestor. Synapomorphies are thus indicators of closely related evolutionary groups, and according to the cladistics approach, are the only valid criteria for constructing phylogenetic trees. For example, possession of an amnion (an embryonic membrane) is a synapomorphy that defines the group of vertebrates called the amniotes, which include mammals, birds, and the reptile groups. *Compare* plesiomorphy.

synapse /si-**naps**, sÿ-naps/ The junction between two neurons (nerve cells), between a neuron and a muscle cell, or between two muscle cells, across which impulses can be transmitted. There are two main types: *chemical synapses*, which are most common and relay impulses using chemical messengers (NEUROTRANSMITTERS); and *electrical synapses*, which occur between certain neurons in the central nervous system and between cardiac muscle cells.

Chemical synapses are formed between the knoblike axon endings of one neuron and the dendrites or cell body of another. One neuron may have many synapses with other neurons. Each synapse consists of adjacent specialized regions in the plasma membranes of both neurons, separated by a narrow gap (synaptic cleft), typically 20–30 nm wide in vertebrate synapses. A nerve impulse arriving at the axon ending of the presynaptic cell causes small vesicles to fuse with the presynaptic membrane and release neurotransmitter. This diffuses across the cleft and binds to receptors in the plasma membrane of the postsynaptic cell, which triggers the temporary opening of various ion channels in the postsynaptic membrane. Depending on the ion channels involved, this may either start a nerve impulse in the postsynaptic cell (excitation) or prevent impulses from other neurons being transmitted (inhibition). Most chemical synapses will only transmit nerve impulses in one direction.

In electrical synapses, adjoining cells are directly connected by numerous gap junctions, and impulses are transmitted between cells by the flow of ions through channels created by these junctions. Hence, transmission is virtually instantaneous, compared with the typical delay of 0.5 ms in chemical synapses. *See* facilitation; summation.

synapsis /si-**nap**-sis/ (**pairing**) The association of homologous chromosomes during the prophase stage of meiosis that leads to the production of a haploid number of bivalents. Homologous chromosomes pair point to point so that corresponding regions lie in contact.

syncarpy /sin-kar-pee/ The condition in which an ovary is made up of fused carpels, as in the primrose. *Compare* apocarpy.

synchronous culture /sink-rŏ-nŭs/ A culture of cells in which all the individuals are at approximately the same point in the cell cycle. Cells can be synchronized by a variety of means, e.g. temperature, shock, or drugs. Such cultures are of great value in some physiological and biochemical investigations.

syncytium /sin-**sish**-ee-ŭm, -**sit**-/ (*pl.* **syncytia**) An area of animal cytoplasm containing many nuclei, the whole being bounded by a continuous cell membrane. This gives rise to a multinucleate condition. The term may be applied to an area of cytoplasm partially divided by membranes into discrete cells but with extensive cytoplasmic continuity. Such structures are to be found in striped and cardiac muscle, insect eggs, and some protoctists. *Compare* coenocyte; symplast.

synecology /sin-i-**kol**-ŏ-jee/ The study of all the living and relevant nonliving components of a natural community and their relationships with each other. *Compare* autecology.

synergid cells /si-**ner**-jid/ Two haploid cells located near the egg cell at the micropylar end of the embryo sac, in flowering plants. They do not participate in the fertilization process and abort soon afterward.

synergism /sin-er-jiz-ăm/ **1.** The interaction of two substances, e.g. drugs or hormones, which have similar effects in a given system, such that the effect produced is greater than the sum of their separate effects. *Compare* antagonism; summation.

2. The coordinated action of muscles to produce a particular movement. *Compare* antagonism.

syngamy /sing-gă-mee/ *See* fertilization.

syngraft /sing-graft, -grahft/ *See* graft.

synovial membrane /si-**noh**-vee-ăl/ The membrane that forms the capsule surrounding a joint. It consists of tough connective tissue with a high proportion of white collagen fibers and it secretes the viscous lubricating *synovial fluid.*

syrinx /si-rinks/ (*pl.* **syringes**) The sound-producing organ in birds. It is similar to the vocal cords but positioned at the base of the trachea.

systematics The area of biology that deals with the diversity of living organisms, their relationships to each other, and their classification. The term may be used synonymously with *taxonomy.*

systemic acquired resistance (SAR) The generalized increased resistance of a plant to attack by a pathogen following a localized infection by a pathogen. It is thought that a signal molecule is mobilized by the initial lesion, and that this spreads throughout the rest of the plant via the phloem, to induce the synthesis of antipathogen enzymes in healthy tissues.

systemic arch A blood vessel found in adult tetrapods that carries usually oxygenated blood from the heart to the dorsal aorta, which serves the body. It is derived from the fourth aortic arch. Amphibians and reptiles retain both arches, birds only the right arch, and mammals only the left arch. *See also* aorta.

systole /sis-tŏ-lee/ The phase of the CARDIAC CYCLE when the cardiac muscle contracts. Contraction of the atria (atrial systole) propels blood into the ventricles; contraction of the ventricles (ventricular systole) expels blood into the aorta and pulmonary artery.

T

tactic movement *See* taxis.

Taenia /tee-nee-ă/ (tapeworm) *See* Cestoda.

tagma /tag-mă/ (*pl.* **tagmata**) One of the sections of an arthropod body formed by fusion of body segments. For example, the tagmata of an insect are the head, thorax, and abdomen.

tannin /tan-in/ One of a mixed group of substances which, as defined by industry, combine with hide to form leather. Tannins are also used in dyeing and ink manufacture. Many plants accumulate tannins, particularly in leaves, fruits, seed coats, bark, and heartwood. Their astringent taste may deter animals from eating the plant and they may discourage infection. Tannins precipitate proteins and hence inactivate enzymes; they are therefore segregated in cell vacuoles, organelles, or cell walls. Chemically they are polymers derived either from carbohydrates and phenolic acids by condensation reactions, or from flavonoids.

tapetum /tă-pee-tŭm/ (*pl.* **tapeta**) **1.** (*Zoology*) A light-reflecting layer in the interior of the eyeball of many vertebrates, especially nocturnal animals and deep-water fishes. It usually consists of glistening connective tissue or guanine crystals lying within the choroid coat. Light entering the eye is reflected back onto the retina by the tapetum. Some of the reflected light passes out through the pupil, so that the eyes seem to glow in the dark (e.g. when caught in the headlights of a car).
2. (*Botany*) A food rich layer surrounding the spore mother cells in the anthers of vascular plants. These cells usually disintegrate, liberating food substances that are subsequently absorbed by the spore mother cells and the developing spores.

tapeworms *See* Cestoda.

tarsal bones /tar-săl/ Bones in the distal region of the hind limb of tetrapods; in humans they form the ankle and heel bones. In the typical pentadactyl limb there are 12 tarsal bones arranged in three rows. However, there are various modifications and reductions to this pattern; in humans there are only seven. They articulate with each other and with the metatarsal bones distally. One tarsal bone, the *talus*, forms the hinge joint of the ankle with the tibia and fibula. *Compare* carpal bones.

tarsus /tar-sŭs/ (*pl.* **tarsi**) **1.** The collection of tarsal bones, which form the ankle and heel in humans. *Compare* carpus.
2. The fifth segment of an insect leg; it is often divided into a number of portions.

taste bud A small bulblike group of chemical receptor cells in vertebrates that is responsible for the sense of taste. In terrestrial vertebrates, taste buds are usually embedded in small projections (papillae) of the epithelium of the throat and mouth, especially the tongue. Chemical substances in solution stimulate the cells to send nerve impulses to the brain for interpretation as taste. Humans are considered to have four kinds of taste buds, which distinguish sweet, sour, salt, and bitter chemicals. Aquatic vertebrates may have taste buds anywhere on the body surface.

taxis /taks-iss/ (*pl.* **taxes**) (tactic movement) Movement of an entire cell or organism (i.e. locomotion) in response to an external stimulus, in which the direction of movement is dictated by the direction of the stimulus. Movement towards the stimulus is positive taxis and away from the stimulus is negative taxis. It is achieved by protoplasmic streaming, extrusion of cell substances, or by locomotory appendages, such as cilia and flagella. *See* aerotaxis; chemotaxis; phototaxis. *See also* nastic movements; tropism.

taxometrics /taks-ŏ-**met**-riks/ *See* numerical taxonomy.

taxon /**taks**-on/ (*pl.* **taxa**) A group of any rank in taxonomy. Ranunculaceae (a family) and *Triticum* (a genus) are examples.

taxonomy /taks-**on**-ŏ-mee/ The area of SYSTEMATICS that covers the principles and procedures of CLASSIFICATION.

Taxus /**taks**-ŭs/ (**yew**) *See* Coniferophyta.

TCA cycle *See* Krebs cycle.

T-cell (**T-lymphocyte**) A lymphocyte involved primarily in cell-mediated immunity. Like B-CELLS, T-cells originate in the bone marrow, but during embryological development they migrate to the thymus where they mature and differentiate (B-cells are believed to mature in the bone marrow). There are several types of T-cell, which act collaboratively in identifying and destroying virus-infected body cells, and possibly cancerous cells. All of them possess on their surface special receptors, T-cell receptors, that are structurally related to the antibody molecules secreted by B-cells, and that bind to specific viral antigens on the surface of infected body cells. Unlike antibodies, they do not bind soluble antigens. However, a T-cell will generally only recognize its specific antigen if the antigen is bound to a marker protein (an MHC protein; *see* major histocompatibility complex) on the surface of a cell. Binding of the T-cell receptor to the MHC–antigen complex acts as the stimulus for the T-cell's rapid division, to form a clone with four types of T-cell.

Cytotoxic T-cells (T_C cells) recognize other body cells that display the same antigens on their surface, and lyse the target cells by releasing a protein called perforin. This makes a hole in the cell membrane of the infected cell, causing it to burst. Hence, viral replication by that cell ceases. *T-helper cells* (T_H cells) release substances, called lymphokines, that stimulate the growth and maturation of both T_C cells and B-cells. In fact, T_H cells are essential for antibody production by B-cells. But T_H cells will only release lymphokines if their receptor sites bind to an antigen-MHC complex on the surface of *antigen-presenting cells*. These are cells, such as macrophages and B-cells, that possess class II MHC proteins on their surface. T_H cells are destroyed by HIV to produce the symptoms of AIDS. A third type of T-cell, the *T-suppressor cell* (T_S cell) suppresses the activities of T_H cells and thus limits antibody production by B-cells. *See* immunity; killer cell.

tectum /**tek**-tŭm/ (*pl.* **tecta**) The dorsal roof region of the midbrain that is the dominant brain center in fishes and amphibians. It is also important in reptiles and birds and gives rise to a pair of prominent optic lobes. In mammals the tectum is much less important, being concerned only with certain visual and auditory reflexes.

teeth Hard dense structures growing on the jaws of vertebrates and used for seizing, biting, and chewing. In mammals each tooth consists of dentine, covered by enamel, enclosing a pulp cavity. It has a crown above the gum (gingiva) and a root embedded in a socket of the jaw bone. Fish, amphibians, and many reptiles have teeth that are modified denticles distributed over the palate. *See also* dentition.

Teleostei /tel-ee-**ost**-ee-ÿ/ The largest order of bony fishes and the most numerous group of living vertebrates. They are found in most types of aquatic environment and show great variety of form. Teleosts have thin rounded bony scales, a symmetrical (homocercal) tail, and shortened jaws with reduced cheek bones,

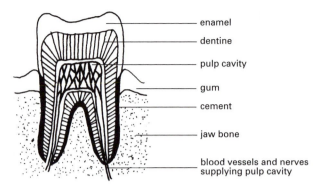

enamel

dentine

pulp cavity

gum

cement

jaw bone

blood vessels and nerves
supplying pulp cavity

Teeth: section through a molar tooth

which allow the mouth to gape widely. In most the fins are supported by a few strong movable spines, and the pelvic fins, at the anterior end of the body, assist the pectorals. Internally there is a swim bladder, which is hydrostatic in function and confers buoyancy and thus great maneuverability and is a major contribution to their success. Fertilization is external and the eggs are unprotected. *See also* Osteichthyes.

telomere /tel-ŏ-meer/ The tip of a eukaryotic chromosome. It consists of numerous tandemly repeated short nucleotide sequences that do not encode any product. Each replication cycle results in the loss of nucleotides from the ends of the DNA molecule, but although repeated cell division causes gradual shortening of the telomere, this does not normally result in loss of functional genes. However, it has been proposed that excessive telomere shortening might be the cause of certain degenerative changes associated with old age. Germ-line cells are protected against telomere shortening by a telomerase enzyme, which repeatedly adds nucleotides to the ends of the DNA molecule during the replication cycle.

telophase /tel-ŏ-fayz/ The final stage in mitosis and meiosis before cells enter interphase. During this stage chromosomes uncoil and disperse, the nuclear spindle degenerates, and a new nuclear membrane

forms. The cytoplasm may also divide during this phase.

telson /tel-sŏn/ The last segment of the abdomen of crustaceans, flattened to form a plate. In the lobster, the telson and the uropods form a fan that can be used for swimming backwards for short distances.

temperate phage A bacteriophage that becomes integrated into the bacterial DNA and multiplies with it, rather than replicating independently and causing lysis of the bacterium. *See* lysogeny. *Compare* virulent phage.

template Any molecule that serves as a pattern for the synthesis of another molecule. *See* replication.

template strand (**anticoding strand; noncoding strand**) The strand of the DNA double helix that serves as the template for the synthesis of RNA during TRANSCRIPTION of a gene. Hence, the base sequence of the template strand is complementary to that of the corresponding messenger RNA molecule that, by convention, contains the genetic code (with uracil instead of thymine). The template strand can switch from one DNA strand to the other depending on which gene is being transcribed. *Compare* nontemplate strand.

tendinous cords /ten-dă-nŭs/ (**chordae tendineae**) Tough strands of connective tissue that run from the undersides of the bicuspid and tricuspid valves to muscles in the walls of the left and right ventricles of the heart respectively. They prevent the valves from turning inside out under pressure of ventricular contraction.

tendon The tough nonelastic connective tissue that joins a muscle to a bone. It consists of a mass of parallel white collagen fibers, which are continuous with those of the muscle sheath (epimysium) and the periosteum of the bone. When the muscle contracts, the tendon pulls on the bone, causing movement at the joint.

tendril A slender structure found in plants, used for twining and support. It may be a modified terminal bud, as in the grape vine, a modified lateral branch, as in the passion flower, or a modified leaflet, as seen in many of the pea family. Tendrils may be branched, or unbranched, and may have terminal adhesive disks, as in the Virginia creeper. Each tendril is slightly curved and the cells of the concave surface respond to specific tactile stimuli (i.e. show haptotropism) by losing water. This reduction in cell volume causes the twining of the tendril about the support.

teratogen /te-ră-tŏ-jĕn, tĕ-**rat**-ŏ-jĕn/ Any environmental factor that causes physical defects (teratomas) in a fetus. Teratogens include various drugs (e.g. thalidomide), infections (e.g. German measles), and irradiation. Teratogens interfere with essential growth mechanisms, causing arrested or distorted growth; the fetus is particularly sensitive during the first two months, when rudimentary growth patterns are being established. In later life, when growth patterns are well established, teratogens have no effect.

teratology /te-ră-**tol**-ŏ-jee/ The study of plant and animal abnormalities (teratomas).

tergum /**ter**-gŭm/ (*pl.* **terga**) The plate, stiffened by chitin, that forms the protective dorsal covering to each segment of the thorax and abdomen of an insect. *See also* pleuron; sternum.

terminalization /ter-mă-nă-li-**zay**-shŏn/ The movement of chiasmata to the end of the bivalent arms, a process that may occur during late prophase I of meiosis. The chiasmata can slip off the ends of the bivalents, and thus chiasma frequency may be reduced by terminalization.

terpene /**ter**-peen/ One of a complex group of lipids based on the hydrocarbon skeleton C_5H_8 (isoprene). *Monoterpenes* are built from two C_5 residues ($C_{10}H_{16}$), *diterpenes* from four, etc. The C_{10} to C_{20} terpenes are present in essential oils, giving the characteristic scent of some plants (e.g. mint). Some terpenoid substances are physiologically active, e.g. vitamins A and E.

territory An area occupied and defended by an animal for such purposes as mating, nesting, and feeding. The type and size of territory depends on its function – a nesting territory may be small but a feeding territory may be very large – and on the size and nature of the animal and its requirements. Territories are common among vertebrates, particularly birds, and also occur occasionally in certain invertebrates.

Tertiary /**ter**-shee-air-ee, -shă-ree/ The larger and older period of the Cenozoic, being composed of the Paleocene, Eocene, Oligocene, Miocene, and Pliocene epochs (65–2 million years ago). Literally the 'third age', it is characterized by the emergence of mammals. *See also* geological time scale.

testa /**tes**-tă/ (*pl.* **testae**) The hard dry protective covering of a seed, formed from the integuments of the ovule. After fertilization the layers of the integuments fuse and become thickened and pigmented. In some species the testa may be modified in some way, as in *Gossypium* (cotton) in which long cotton fibers develop on the seed coat surface.

test cross *See* back cross.

testicle *See* testis.

testis /**tes**-tis/ (*pl.* **testes**) (**testicle**) The male reproductive organ of animals, which produces spermatozoa. In vertebrates there is a pair of testes, which also produce sex hormones (androgens). They develop in the abdominal cavity near the kidneys but in most mammals migrate downwards during fetal development and come to lie outside the body cavity, within a pouch of skin (scrotum) situated behind the penis. They usually remain there throughout life but in a few cases only during the breeding season. In man, the testes are oval structures about 4–5 cm long. Each is comprised of a fibrous capsule (the *tunica albuginea*), which surrounds a mass of seminiferous tubules separated into compartments by fibrous tissue. Between the tubules lie interstitial (or Leydig's) cells, which produce androgens. Spermatozoa are continuously produced within the seminiferous tubules from the onset of sexual maturity to the age of about 70. They migrate via efferent ducts to the epididymis for temporary storage. *See also* vas deferens.

testosterone /tes-**tos**-tĕ-rohn/ A naturally occurring ANDROGEN that is secreted by the testis under the influence of luteinizing hormone. Its secretion during adult life is responsible for the development, function, and maintenance of secondary male sexual characteristics, male sex organs, and spermatogenesis. Testosterone is also secreted from the adrenal cortex and the ovaries. It is metabolized in the liver, its metabolites (e.g. androsterone) being excreted in the urine.

tetrad /**tet**-rad/ 1. A group of four spores formed as a result of meiosis in a spore mother cell.
2. In meiosis, the association of four homologous chromatids seen during the pachytene stage of prophase.

tetraploid /**tet**-ră-ploid/ A cell or organism containing four times the haploid number of chromosomes. Tetraploid organisms may arise by the fusion of two diploid gametes that have resulted from the nondis-junction of chromosomes at meiosis. Tetraploids may also arise through nondisjunction of the chromatids during the mitotic division of a zygote. *See also* polyploid; allotetraploid; autotetraploid.

tetrapod /**tet**-ră-pod/ A vertebrate with four limbs; i.e. an amphibian, bird, mammal, or reptile.

thalamus /**thal**-ă-mŭs/ (*pl.* **thalami**) A part of the midbrain that relays sensory and motor impulses to and from the appropriate centers in the cerebral hemispheres.

thallus /**thal**-ŭs/ (*pl.* **thalli**) A simple plant body showing no differentiation into root, leaf, and stem and lacking a true vascular system. It may be uni- or multicellular, and is found in the algae, lichens, bryophytes, and the gametophyte generation of the pteridophytes.

thermogenesis /th'er-moh-**jen**-ĕ-sis/ The generation of heat from an animal's energy stores to raise its body temperature. Although performed chiefly in birds and mammals (*see* endotherm), thermogenesis also occurs in some 'cold-blooded' vertebrates and invertebrates. One way of producing heat is *shivering*, in which sets of skeletal muscles contract repeatedly and rapidly to produce heat, rather than mechanical work, from the breakdown of ATP. Another way is *nonshivering thermogenesis*, which takes place in fat-storage cells (adipose tissue). The fat is oxidized inside the cell to generate heat, instead of being transported to the liver for ATP production. Certain mammals have special adipose tissue known as *brown fat*, which can provide bursts of intense heat. Such tissue is useful in assisting the arousal of hibernating animals, or in protecting newborn animals against hypothermia, for example.

thermonasty /**ther**-moh-nas-tee/ (**thermonastic movements**) A NASTIC MOVEMENT in response to change in temperature.

thermoperiodism /ther-moh-**peer**-ee-ŏ-diz-ăm/ The phenomenon shown by certain plants (e.g. chrysanthemum and tomato) in which there is a response to alternating periods of low and high temperatures. Such plants will flower earlier and more profusely if subjected to low night and high day temperatures. *See also* photoperiodism; vernalization.

thermophilic /ther-moh-**fil**-ik/ Describing microorganisms that require high temperatures (around 60°C) for growth. It is exhibited by certain bacteria that grow in hot springs or compost and manure. *Compare* mesophilic; psychrophilic.

therophyte /**theer**-ŏ-fÿt/ A plant that survives the winter as a seed and completes its life cycle between the spring and autumn (i.e. an annual). *See also* Raunkiaer's plant classification.

thiamine /**th'ÿ**-ă-min, -meen/ (**thiamin**; **vitamin B**$_1$) A water-soluble vitamin of the VITAMIN B COMPLEX. Good sources of thiamine are unrefined cereal grains, liver, heart, and kidney. Thiamine deficiency predominantly affects the peripheral nervous system, the gastrointestinal tract, and the cardiovascular system. Thiamine has been shown to be of value in the treatment of beriberi. Thiamine, in the form of thiamine diphosphate, is the coenzyme for the decarboxylation of some acids such as pyruvic acid.

thigmotropism /thig-**mot**-rŏ-piz-ăm/ (**haptotropism**) A tropism in which the stimulus is touch. The tendrils of climbing plants are thigmotropic. *See* tropism.

thin-layer chromatography A chromatographic method in which a glass plate is covered with a thin layer of inert absorbent material (e.g. cellulose or silica gel) and the materials to be analyzed are spotted near the lower edge of the plate. The base of the plate is then placed in a solvent, which rises up the plate by capillary action, separating the constituents of the mixtures. The principles involved are similar to those of paper chromatography and, like paper chromatography, two dimensional methods can also be employed.

thoracic duct /thor-**ass**-ik, -thoh-**rass**-/ A dorsal longitudinal lymphatic vessel – the major vessel of the lymphatic system – that collects lymph from most of the body. In mammals, it begins below the diaphragm, ascends in front of the vertebral column, and drains into the innominate vein at the base of the neck.

thoracic vertebrae The vertebrae of the upper back region; there are 12 in humans. Each is distinguished by articular surfaces on the sides of the body (centrum) and on the transverse processes for the upper ends of a pair of ribs.

thorax /**thor**-aks, **thoh**-raks/ **1.** The section of the body cavity of vertebrates that contains the heart and lungs. It is protected by the sternum and ribs and in mammals is separated from the abdomen by the diaphragm.
2. In arthropods, the part of the body between the head and the abdomen. In insects there are three thoracic segments each covered by four cuticular plates (a tergum, sternum, and two pleurae). They bear the walking legs and wings. *See also* cephalothorax.

thorn A stiff sharply pointed woody process that may be found on the stems of vascular plants. It is a modified branch and is supplied with vascular tissue. *Compare* prickle; spine.

threonine /**three**-ŏ-neen, -nin/ *See* amino acids.

threshold The minimum stimulus intensity that will initiate a response in excitable tissue, such as a muscle or nerve cell.

thrombin /**throm**-bin/ An enzyme that converts the soluble protein fibrinogen into the fibrous fibrin during BLOOD CLOTTING. It is formed from prothrombin under the influence of thromboplastin, calcium ions, and other clotting factors, which are activated when blood is removed from the circulation, usually by injury.

thrombocyte /**throm**-bŏ-sÿt/ *See* platelet.

thylakoid /**th'ÿ**-la-koid/ An elongated flattened fluid-filled sac forming the basic unit of the photosynthetic membrane system in chloroplasts and photosynthetic bacteria. *See* chloroplast.

thymidine /**th'ÿ**-mă-deen, -din/ The nucleoside formed when thymine is linked to D-ribose by a β-glycosidic bond.

thymine /**th'ÿ**-meen, -min/ A nitrogenous base found in DNA. It has a pyrimidine ring structure. *See illustration at* DNA.

thymus gland /**th'ÿ**-mŭs/ A gland consisting of two lobes, which lie in the lower part of the neck and upper part of the chest. The lobules, which are found in each lobe, are composed of an outer cortex and an inner medulla portion. The thymus controls the development of lymphocytes, particularly T-cells, which are produced in the thymus by stem cells derived from bone marrow. Newly formed T-cells are influenced by extracellular proteins and thymic peptides to differentiate into subpopulations with distinct immunological identities and roles. The thymus degenerates from the time the animal reaches sexual maturity, or earlier.

thyroid gland /**th'ÿ**-roid/ A gland situated in the neck, consisting of two lateral lobes on either side of the trachea and larynx, giving the gland a butterfly appearance. Its main function is the regulation of metabolic rate by production of thyroid hormones.

thyroid hormones Hormones produced by the thyroid gland, which increase cell metabolism. The most important is thyroxine. *See also* triiodothyronine.

thyrotropin /th'ÿ-roh-**troh**-pin/ (**thyroid-stimulating hormone**) A hormone, produced by the anterior pituitary gland, that stimulates the thyroid gland to release thyroxine. The level of thyroxine controls thyrotropin release by a negative feedback mechanism.

thyroxine /th'ÿ-**roks**-een, -in/ (**thyroid hormone**; T_4) An iodine-containing polypeptide hormone that is secreted by the thyroid gland and is essential for normal cell metabolism. Its many effects include increasing oxygen consumption and energy production. It is used therapeutically to treat hypothyroidism (cretinism and myxedema).

tibia /**tib**-ee-ă/ (*pl.* **tibiae**) **1.** The large inner long bone of the lower hindlimb of tetrapods; it forms the shinbone in humans. The articular surfaces (condyles) on its upper end articulate with those of the femur as the hinge joint of the knee. The lower end articulates laterally with the fibula and distally with one of the tarsal bones of the heel to form the ankle joint. A downward projection, the *medial malleolus*, forms a prominence on the inner side of the ankle.
2. The fourth segment of an insect leg, between the femur and the tarsus.

tidal volume The volume of air exhaled from the lungs after a normal intake of breath. It is about 500 ml in an adult human. *See also* vital capacity.

tight junction (**zonula occludens**) A structure that encircles many animal epidermal cells holding adjacent ells tightly together with no intercellular space. Tight junctions prevent material at the epidermal surface from penetrating between the cells, particularly at sites of absorption, e.g. the gut and kidney. Below them are DESMOSOMES.

tissue A group of cells that is specialized for a particular function. Examples are connective tissue, muscular tissue, and nervous tissue. Several different tissues are often incorporated in the structure of each organ of the body.

tissue culture The growth of cells, tissues, or organs in suitable media *in vitro*. Such media must normally be sterile, correctly pH balanced, and contain all the necessary micro and macronutrients, carbohydrates, vitamins, and hormones for growth. Studies of such cultures have shed light on physiological processes that would be difficult to follow in the living organism. The cytokinins were discovered through work on tobacco pith tissue culture.

Plant tissue-culture techniques also have important practical applications, enabling, for example, large-scale multiplication of plants by micropropagation and the generation of disease-free material by meristem-tip culture.

Tissue engineering is a relatively new area of biomedical technology concerned with producing synthetic or semisynthetic tissues that resemble natural body tissues. Implants can be constructed using combinations of cultured human cells, artificial bipolymers, and growth factors. Products such as synthetic skin and cartilage are already being used to supplement or replace natural tissue grafts in surgery. In future, a wide range of tissues and organs might be grown *in vitro* starting from cultures of STEM CELLS.

titer /tȳ-ter, tee-/ A measure of the concentration of an antibody in serum. It is estimated from the highest dilution that will still produce a detectable reaction with the appropriate antigen and is expressed as the reciprocal of this dilution.

toads *See* Anura.

tocopherol /toh-**kof**-ĕ-rol, -rohl/ *See* vitamin E.

tolerance *See* immunological tolerance.

tone (**tonus**) The state of continuous but partial contraction of muscle tissue, due to steady nervous stimulation. It enables an animal to maintain its posture.

tongue An organ of taste situated in the buccal cavity of vertebrates. In mammals it consists of smooth and striped muscle covered with stratified epithelium, beneath which is loose connective tissue. The upper surface of the tongue is covered with projections (papillae) containing special chemoreceptors (taste buds), which are sensitive to chemicals in solution. The tongue can be divided into regions corresponding to different taste sensations. In most animals it is attached to the back of the mouth and is often protrusive.

In fish the tongue is a pad of tissue used in swallowing. The tongue in some amphibians, especially frogs, is attached to the front of the buccal cavity and is forked at the end. It can be flicked out to catch insects on its sticky surface.

tonoplast /**tohn**-ŏ-plast/ The membrane that surrounds the large central vacuole of plant cells.

tonsils Small bodies of lymphoid tissue in tetrapods, situated near the back of the throat. They produce lymphocytes and are therefore concerned with defense against bacterial invasion.

tooth *See* teeth.

topoisomerase /top-oh-ȳ-**som**-ĕ-rayss/ An enzyme that catalyzes the twisting or coiling of large molecules, such as DNA.

tornaria /tor-**nair**-ee-ă/ The free-swimming larva of hemichordates, such as the acorn worms. Tornaria larvae, which have folded bands of cilia for swimming, resemble the bipinnaria and pluteus larvae of echinoderms. This suggests a close evolutionary relationship between the echinoderms and the chordate group through the hemichordates.

torus /**tor**-ŭs, toh-rŭs/ (*pl.* **tori**) 1. A disk-shaped structure formed from lignin on the middle lamella of a bordered pit. The structure is found mainly in the conifers. *See illustration at* pit.
2. *See* receptacle.

totipotency /toh-**tip**-ŏ-tĕn-see/ The ability, shown by many, if not all, living cells, to form all the types of tissues that constitute the mature organism. This may be achieved, even if the cells have completely differentiated, provided that the appropriate balance of nutrients and hormones is given. The best example of this phenomenon is the formation of adventitious em-

IMPORTANT TRACE ELEMENTS IN PLANTS AND ANIMALS

Trace element	Compounds containing	Metabolic role
copper	cytochrome oxidase	oxygen acceptor in respiration
	plastocyanin	electron carrier in photosynthesis
	hemocyanin	respiratory pigment in some marine invertebrates
	tyrosinase	melanin production – absence causes albinism
zinc	alcohol dehydrogenase	anaerobic respiration in plants – converts acetaldehyde to ethanol
	carbonic anhydrase	CO_2 transport in vertebrate blood
	carboxy peptidase	hydrolysis of peptide bonds
cobalt	vitamin B_{12}	red blood cell manufacture – absence causes pernicious anaemia
molybdenum	nitrate reductase	reduction of nitrate to nitrite in roots
	a nitrogen-fixing enzyme	nitrogen fixation
manganese	enzyme cofactor	oxidation of fatty acids
		bone development
fluorine	associated with calcium	tooth enamel and skeletons
boron		mobilization of food in plants

bryos in carrot tissue cultures. Totipotency demonstrates that each cell retains the full genetic potential of the species.

toxin /toks-in/ A chemical produced by a pathogen (e.g. bacteria, fungi) that causes damage to a host cell in very low concentrations. Toxins are often similar to the enzymes of the host and interfere with the appropriate enzyme systems. *See also* endotoxins; exotoxins.

trabecula /trǎ-**bek**-yŭ-ler/ (*pl.* **trabeculae**) An elongated cell or a line of cells across a cavity. In plants, trabeculae may be found in *Selaginella* stems, where they suspend the steles in large air spaces. They also divide the mature *Isoetes* sporangium into compartments, and connect the spore sac of *Funaria* with the capsule.

trace element An element required in trace amounts (a few parts per million of food intake) by an organism for health. A list of the more important trace elements with examples of their uses is given in the table above.

trachea /tray-kee-ǎ, trǎ-**kee**-/ (*pl.* **tracheae**) 1. (*Zoology*) The windpipe: a wide tube leading from the throat to the bronchi in land vertebrates. Its walls are stiffened with incomplete rings of cartilage, preventing collapse yet retaining flexibility. *See illustration at* alimentary canal.
2. (*Zoology*) A tube leading from each spiracle in insects and most other land arthropods. It branches into fine *tracheoles*, which penetrate the muscles and organs. Pumping actions ventilate the main tracheae; oxygen then diffuses through to the tracheoles, where it dissolves in fluid (which fills the finest tracheoles), and into the surrounding tissues.
3. (*Botany*) *See* vessel.

tracheid /tray-kee-id/ An elongated xylem conducting element with oblique end walls. Tracheids have heavily lignified walls and the only connection between adjacent tracheids is through paired pits. Tracheids form the only xylem conducting tissue of pteridophytes and most gymnosperms. *Compare* vessel.

tracheophyte /**tray**-kee-ŏ-fÿt/ Any plant with a differentiated vascular system; i.e. all plants except the bryophytes (liverworts and mosses). In some classifications they constitute the division (phylum) Tracheophyta, comprising the various subdivisions of vascular plants. *Compare* pteridophyte; spermatophyte.

trama /**tram**-ă, **trah**-mă/ The inner tissue of the gills in basidiomycete fungi that is made up of loosely packed hyphae.

transamination /trans-am-ă-**nay**-shŏn/ The transfer of an amino group from an amino acid to an α-keto acid, producing a new α-keto acid and a new amino acid. This is catalyzed by a transaminase enzyme in conjunction with the cŏenzyme pyridoxal phosphate. The amino group becomes attached to the coenzyme to form pyridoxamine phosphate, and is then transferred to the α-keto acid, which is usually pyruvic acid, oxaloacetic acid, or α-ketoglutaric acid.

transcription /tran-**skrip**-shŏn/ The process in living cells whereby RNA is synthesized according to the template embodied in the base sequence of DNA, thereby converting the cell's genetic information into a coded message (messenger RNA, mRNA) for the assembly of proteins, or into the RNA components required for protein synthesis (ribosomal RNA and transfer RNA). The term is also applied to the formation of single-stranded DNA from an RNA template, as performed by the enzyme reverse transcriptase, for example in retrovirus infections. Details of DNA transcription differ between prokaryote and eukaryote cells, but essentially it involves the following steps. Firstly, in the stage called initiation, the enzyme RNA polymerase binds to a site near the coding region of the gene, starts to unwind the two DNA strands, and begins transcription. During elongation, RNA polymerase moves along one of the DNA strands, the transcribed strand (or anticoding strand, since the code is carried by the complementary base sequence of the RNA), and nucleotides are assembled to form a complementary RNA molecule. The polymerase enzyme proceeds until reaching a stop signal, when formation of the RNA strand is terminated. Behind the enzyme, the DNA double helix re-forms, stripping off the newly synthesized RNA strand. In eukaryotes, transcription is initiated and regulated by a host of proteins called *transcription factors*; in prokaryotes an accessory *sigma factor* is essential for transcription.

In many eukaryotic genes the functional message is contained within discontinuous segments of the DNA strand (EXONS), interrupted by non-functional segments (INTRONS). Initially, both exons and introns are transcribed to form HETEROGENEOUS NUCLEAR RNA (hnRNA). Subsequently the non-coding intron sequences are spliced out to form the fully functional mature mRNA transcript, which then leaves the nucleus to direct protein assembly in the cytoplasm, in the process called TRANSLATION. *See also* messenger RNA; polymerase.

transcriptome /tran-**skrip**-tohm/ The set of RNA transcripts that is present in a particular cell, tissue, or organism at any given time. The transcriptome is thus a 'snapshot' reflecting the pattern of gene expression at any instant in time. The transcripts, in the form of messenger RNAs (mRNAs), are sampled from cell contents and their types and relative abundance determined using techniques such as the polymerase chain reaction and DNA MICROARRAYS. The study of sets of RNA transcripts is called *transcriptomics*. This looks at how various factors, such as stage of development, hormones, drugs, or disease affect the transcriptome. It complements and overlaps with study of the PROTEOME.

transduction /trans-**duk**-shŏn, tranz-/ The transfer of part of the DNA of one bacterium to another by a bacteriophage. The process does occur naturally but is mainly known as a technique in recombinant DNA technology, and has been used in mapping the bacterial chromosome.

transect /tran-**sekt** (*vb.*); **tran**-sekt (*n.*) / A line or belt designed to study changes in species composition across a particular area. The transect sampling technique is most often used in plant ecology to study changes in the composition of vegetation. A long tape marked at set intervals is laid across the area to be studied, for example, the edge of a wood, and the species found growing at each interval recorded. *See also* quadrat.

transfection /trans-**fek**-shŏn/ The uptake and expression of foreign DNA by a eukaryotic cell. Various methods are used to introduce novel genes into eukaryotic cell cultures in order to modify the cells' genotype and pattern of gene expression. Most widely used are DNA carriers such as cationic polymers, liposomes, modified viruses, and calcium phosphate. Cationic polymers, for instance, form complexes with the foreign DNA and enter the cell by endocytosis. Other methods include biolistics, in which the DNA is coated on tungsten microprojectiles, which are then shot at high velocity into the cell; and electroporation, in which uptake of the DNA is induced by subjecting cells to brief pulses of high-voltage electric current. *Compare* transformation.

transferase /**trans**-fer-ayss/ An enzyme that catalyzes reactions in which entire groups or radicals are transferred from one molecule to another. Hexokinase catalyzes the transfer of a high energy terminal phosphate group from ATP to glucose to give glucose-6-phosphate and ADP.

transfer cell A specialized type of plant cell in which the cell wall forms protuberances into the cell, thus increasing the surface area of the wall and plasma membrane. They are active cells, containing many mitochondria, and are concerned with short-distance transport of solutes. They are common in many situations, for example as gland cells and epidermal cells, and in xylem and phloem parenchyma, where they are concerned with active loading and unloading of vessels and sieve tubes.

transfer RNA (**tRNA**) A type of RNA that participates in protein synthesis in living cells. It attaches to a particular amino acid – forming an *amino-acyl tRNA* – and imports the amino acid to the site of polypeptide assembly at the RIBOSOME when the appropriate codon is reached during TRANSLATION of messenger RNA (mRNA). Each tRNA molecule consists of 70–80 nucleotides; some regions of the molecule undergo base pairing and form a double helix, while in others the two strands separate to form loops. When flattened out the tRNA molecule has a characteristic 'cloverleaf' shape with three loops; the base of the 'leaf' carries the amino-acid binding site, and the middle loop contains the anticodon, whose base triplet pairs with the complementary codon in the mRNA molecule. Bacterial cells typically contain 30–40 different tRNAs, whereas in plant and animal cells the number is generally 50–100. Hence, some cells have fewer tRNAs than the 61 possible amino acid-specifying codons in the genetic code. This is because some tRNAs can recognize two or more different codons for their particular amino acid due to nonstandard pairing between the third bases of the codon and anticodon – a phenomenon known as 'wobble'. In eukaryotic cells, moreover, some different tRNAs will bind the same amino acid.

The correct amino acid is attached to a tRNA molecule by an enzyme called an *aminoacyl-tRNA ligase* (or aminoacyl-tRNA synthetase). There are 20 of these, one for each type of amino acid. This attachment also involves the transfer of a high-energy bond from ATP to the amino acid, which provides the energy for peptide bond formation during translation.

transformation /trans-fer-**may**-shŏn/ 1. The uptake by a cell, especially a prokaryotic cell, of DNA from its environment and the subsequent expression of the DNA. Some bacteria can naturally take up fragments of DNA and incorporate them into their chromosome, thereby acquiring new traits, such as resistance to a particular antibiotic. This can be demonstrated by growing bacteria in the presence of dead

cells, culture filtrates, or extracts of other strains. Transformation is also performed using recombinant DNA techniques in cloning DNA and in creating transgenic bacteria that express foreign proteins, such as human insulin. *Compare* transfection.
2. The conversion of normal cells in tissue culture to cells having properties of tumor cells. The change is permanent and transformed cells are often malignant. It may be induced by certain viruses or occur spontaneously.

transgenic /trans-**jen**-ik, tranz-/ Describing organisms, especially eukaryotes, containing foreign genetic material – *transgenes*. Genetic engineering has created a wide range of transgenic animals, plants, and other organisms, for both experimental and commercial purposes. Examples include dairy cows that secrete drugs in their milk, and herbicide-resistant crop plants. *See* recombinant DNA.

transition state (**activated complex**) Symbol: ‡ A short-lived high-energy molecule, radical, or ion formed during a reaction between molecules possessing the necessary activation energy. The transition state decomposes at a definite rate to yield either the reactants again or the final products. The transition state can be considered to be at the top of the energy profile.

For the reaction,
$$X + YZ = X...Y...Z^{\ddagger} \rightarrow XY + Z$$
the sequence of events is as follows. X approaches YZ and when it is close enough the electrons are rearranged producing a weakening of the bond between Y and Z. A partial bond is now formed between X and Y producing the transition state. Depending on the experimental conditions, the transition state either breaks down to form the products or reverts back to the reactants.

transition zone The zone in a vascular plant where the root and shoot structures merge, and where arrangement of the vascular tissue is intermediate between that of the root and shoot. *See* hypocotyl.

translation /trans-**lay**-shŏn, tranz-/ The process whereby the genetic code of messenger RNA (mRNA) is deciphered by the machinery of a cell to make proteins. Molecules of mRNA are in effect coded messages based on information in the cell's genes and created by the process of TRANSCRIPTION. They relay this information to the sites of protein synthesis, the RIBOSOMES. In eukaryotes these are located in the cytoplasm, so the mRNA must migrate from the nucleus. The first stage in translation is *initiation*, in which the two subunits of the ribosome assemble and attach to the mRNA molecule near the *initiation codon*, which signals the beginning of the message. This also involves various proteins called *initiation factors*, and the initiator TRANSFER RNA (tRNA), which always carries the amino acid N-formyl methionine.

The next stage is *elongation*, in which the peptide chain is built up from its component amino acids. tRNA molecules successively occupy two sites on the larger ribosome subunit in a sequence determined by consecutive codons on the mRNA. As each pair of tRNAs occupies the ribosomal sites, their amino acids are joined together by a peptide bond. As the ribosome moves along the mRNA to the next codon, the next tRNA enters the first ribosomal site, and so on, leading to elongation of the peptide chain. Having delivered its amino acid, the depleted tRNA is released from the second ribosomal site, which is then occupied by the tRNA with the growing chain. This process continues until the ribosome encounters a *termination codon*. Then the polypeptide chain is released and the ribosome complex dissociates, marking the *termination* of translation.

Following its release, the polypeptide may undergo various changes, such as the removal or addition of chemical groups, or even cleavage into two. This *post-translational modification* produces the fully functional protein. Folding of the protein is assisted by a class of molecules called *chaperones*.

translocation /trans-loh-**kay**-shŏn/ **1.** The movement of mineral nutrients, elaborated food materials, and hormones through the plant. In vascular plants, the xylem and phloem serve to translocate such substances. Carbohydrates, amino acids, and other organic compounds are moved both upwards and downwards in the phloem, whereas water moves from roots to leaves through the xylem. There is evidence that mineral salts are moved in both the xylem and phloem. *See* mass flow; transpiration.

2. *See* chromosome mutation.

transpiration /tran-spă-**ray**-shŏn/ The loss of water vapor from the surface of a plant. Most is lost through stomata when they are open for gaseous exchange. Typically, about 5% is lost directly from epidermal cells through the cuticle (*cuticular transpiration*) and a minute proportion through lenticels. A continuous flow of water, the *transpiration stream*, is thus maintained through the plant from the soil via root hairs, root cortex, xylem, and tissues such as leaf mesophyll served by xylem. Water evaporates from wet cell walls into intercellular spaces and diffuses out through stomata. Transpiration may be useful in maintaining a flow of solutes through the plant and in helping to cool leaves through evaporation, but is often detrimental under conditions of water shortage, when wilting may occur. It is favored by low humidity, high temperatures, and moving air. *Compare* guttation.

transplantation /trans-plan-**tay**-shŏn/ The transfer of a tissue or organ from one part of an animal to another part or from one individual to another. *See also* graft.

transplantation antigen *See* histocompatibility antigen.

transposon /trans-**poh**-zon/ A segment of an organism's DNA that can insert at various sites in the genome – a process called *transposition* – either by physically moving from place to place, or by producing a DNA copy that inserts elsewhere. Transposons represent one of the two main categories of mobile elements occurring in DNA. The most rudimentary, called *insertion sequences* (IS), occur in bacteria, these are typically 1–2 kb long and comprise just one or sometimes two protein-coding genes flanked at either end by inverted repeats. The gene encodes a transposase enzyme, which catalyzes the cleavage of the insertion sequence from the chromosome and its insertion at another site. Other insertion sequences form a DNA replica of themselves, which then inserts at a new site. *Bacterial transposons* are larger than insertion sequences; besides the IS components they also contain antibiotic-resistance genes. Such transposons are used in bacterial genetics to insert within and disrupt particular genes. The antibiotic resistance conferred by the transposon serves to identify cells containing it.

The first mobile elements to be discovered were *eukaryotic transposons*, whose existence was postulated by Barbara McClintock in the 1940s, based on observations of kernel colour in corn (maize). She proposed the existence of genetic entities that could enter a gene, so creating mutants in which kernel color was changed from purple to white. The entity could subsequently leave the gene, causing reversion to the wild type. These entities were later identified as being similar to bacterial insertion sequences and encoding transposase. However, transposons are rare in eukaryotes, compared to the much more common RETROTRANSPOSONS.

transverse process A lateral projection on each side of a vertebra of tetrapods. It sometimes forms an articular surface for the upper end of a rib, especially in the upper back region. *See also* thoracic vertebrae.

transverse tubules (T tubules) Invaginations of the plasma membrane of a muscle cell that permit the propagation of action potentials deep within the cell. They enter the muscle cell (fiber) along the Z lines and make close contact with the SARCOPLASMIC RETICULUM, an arrangement that enables virtually simultaneous stimulation of contraction of myofibrils throughout the cell.

Trematoda /trem-ă-toh-dă, tree-mă-/ A class of parasitic Platyhelminthes, the flukes, including *Fasciola* (the liver fluke of cattle and sheep) and *Schistosoma* (the human blood fluke that causes schistosomiasis). Flukes have an oval or elongated body covered by a thick cuticle to prevent digestion by the host, suckers for attachment, and a forked gut.

Fasciola and *Schistosoma* are endoparasites with a complex life cycle and more than one host. *Schistosoma* eggs hatch into miracidia larvae in fresh water and enter a freshwater snail, in which they produce CERCARIA larvae, which penetrate human skin.

Some flukes are ectoparasites (e.g. *Gyrodactylus*) and have a rapid life cycle with only one host; they are placed in the order Monogenea. The endoparasitic flukes form the order Digenea. These orders are sometimes regarded as two separate classes.

Triassic /trÿ-**ass**-ik/ The oldest period of the Mesozoic, 250–215 million years ago. It is marked by a decrease in the number and variety of cartilaginous fishes and an increase in primitive amphibians and reptiles. *See also* geological time scale.

tribe In plant classification, a group of closely related genera. A number of tribes comprise a subfamily. The tribe is only introduced in classifications of very large families such as the grasses in which the tribes Oryzeae, Triticeae, and Aveneae are examples in the subfamily Pooideae. Tribe names generally end in *eae*.

tricarboxylic acid cycle /trÿ-boks-**il**-ik/ *See* Krebs cycle.

triceps A muscle that runs along the back of the humerus in the upper arm. When it contracts it extends the elbow joint, and hence works as an antagonist of the BICEPS muscle.

trichome /**trÿ**-krohm/ (**hair**) A uni- or multicellular outgrowth arising solely from the epidermis. Root and leaf hairs and glandular hairs are types of trichome.

tricuspid valve /trÿ-**kus**-pid/ A valve consisting of three membranous flaps or cusps

situated between the atrium and ventricle of the right side of the heart in mammals and birds. It prevents the backflow of blood into the atrium when the ventricle contracts. *See also* tendinous cords. *See illustration at* heart.

trigeminal nerve /trÿ-**jem**-ă-năl/ (**cranial nerve V**) One of the pair of nerves that arises from the anterior end of the vertebrate hindbrain to suply the mouth and jaws. It carries sensory nerve fibers from the head surface and mouth cavity and motor nerve fibers serving the jaw muscles. *See* cranial nerves.

triglyceride /trÿ-**gliss**-ĕ-rÿd/ (**triacylglycerol**) An ester of glycerol in which all the –OH groups are esterified; the acyl groups may be the same or different. Many LIPIDS are triglycerides in which the parent acid(s) of the acyl group(s) are long-chain fatty acids. In animals the triglycerides are more frequently saturated and have higher melting points than the triglycerides of plant origin, which are generally unsaturated. The triglycerides that are synthesized by organisms act as supporting material for internal organs, cell walls, etc., as transport mechanisms for nonpolar material, and as a food reserve. *See also* carboxylic acid; glyceride.

triiodothyronine /trÿ-ÿ-ŏ-doh-**th'ÿ**-rŏ-meen, -min/ (T_3) A hormone secreted by the thyroid gland that has similar actions to but is metabolically more potent than THYROXINE. Its levels in the blood are much lower than those of thyroxine. It is also produced in tissues, such as the kidney and liver, by conversion of thyroxine.

Trilobita /trÿ-lŏ-**bÿ**-tă/ A class of extinct marine bottom-dwelling arthropods abundant in the Cambrian and Silurian and thought to be closely related to the ancestors of other arthropods, particularly the Crustacea. Trilobites had an oval flattened body divided longitudinally into three lobes and transversely into a head, thorax, and abdomen. The head bore compound eyes, antennae, and four pairs of jointed forked appendages with an inner projection used as an aid in feeding. There were numerous paired appendages on the body.

triploblastic /trip-**loh**-blas-tik/ Describing an animal whose body is made from three embryonic germ layers: ectoderm on the outside, endoderm lining the gut and allied structures, and mesoderm between these two layers. Each of the three layers gives rise to a particular set of tissues and organs. Most animals are triploblastic; exceptions are the sponges and cnidarians.

triploid /**trip**-loid/ A cell or organism containing three times the haploid number of chromosomes. Triploid organisms arise by the fusion of a haploid gamete with a diploid gamete that has resulted from the nondisjunction of chromosomes at meiosis. Triploids are usually sterile because one set of chromosomes remains unpaired at meiosis, which disrupts gamete formation. In flowering plants the endosperm tissue is usually triploid, resulting from the fusion of one of the pollen nuclei with the two polar nuclei.

trisomy /**trÿ**-soh-mee/ *See* aneuploidy.

trochanter /troh-**kan**-ter/ One of several projections on the upper end of the femur of tetrapods, to which the muscles of the hindleg are attached.

trochlear nerve /**trok**-lee-er/ (**cranial nerve IV**) One of a pair of nerves that emerges from the dorsal midbrain in vertebrates to supply the superior oblique muscle of each eyeball. It contains chiefly motor nerve fibers. *See* cranial nerves.

trochophore /**trok**-ŏ-for, -fohr/ The free-swimming ciliated larva of mollusks, annelid worms, and several minor groups of invertebrates. Its rounded body is encircled by a band of cilia and it has other bands and tufts of long cilia on its surface. It has a digestive canal opening by a mouth and anus. The presence of this larva indicates a close evolutionary relationship between the groups that possess it. Since it is unlike the dipleurula larva, it is thought that the mollusks, annelids, and related groups are

only distantly related to the echinoderms. *Compare* dipleurula.

trophic level /**troff**-ik/ In complex natural communities, organisms whose food is obtained from plants by the same number of steps are said to belong to the same trophic or energy level. The first and lowest trophic level contains the producers, green plants that convert solar energy to food by photosynthesis. Herbivores occupy the second trophic level and are primary consumers. At the third level carnivores eat the herbivores (the secondary consumer level), and at the fourth level secondary carnivores eat the primary carnivores (the tertiary consumer level). These are general categories as many organisms feed on several trophic levels, for example omnivores eat both plants and animals. A separate trophic level is occupied by decomposers or transformers, and consist of organisms, such as fungi and bacteria, that break down dead organic matter into nutrients usable by the producers. *See* food chain.

trophoblast /**trof**-ŏ-blast/ The cells of the outer wall of the mammalian blastocyst. It is the part of the blastocyst that is attached to the wall of the uterus and it forms the part of the early placenta that is in closest contact with the maternal tissues.

tropism /**troh**-piz-ăm/ (**tropic movement**) A directional growth movement of part of a plant in response to an external stimulus. Tropisms are named according to the stimulus. The organ is said to exhibit a positive or negative tropic response, depending on whether it grows towards or away from the stimulus respectively, e.g. shoots are positively phototropic but negatively geotropic. Growth straight towards or away from the stimulus (0° and 180° orientation respectively) is called *orthotropism*. Primary roots and shoots are orthotropic to light and gravity. By contrast, growth at any other angle to the direction of the stimulus as by branches or lateral roots is called *plagiotropism*. The mechanism involved in the latter is poorly understood. Since the receptor for the stimulus is often separate from the region of growth, tropic move-

TROPISMS	
Stimulus	*Type of tropism*
light	photo- or heliotropism
gravity	geotropism
chemical	chemotropism
water	hydrotropism
solid surface (touch)	thigmo- or haptotropism

ments are often mediated by hormones. *See* phototropism; geotropism. *See also* nastic movements; taxis.

tropomyosin /troh-pŏ-**mȳ**-ŏ-sin, trop-ŏ-/ A protein found in skeletal muscle that regulates contraction of the muscle filaments. It forms a rodlike molecule that, in the resting stage, blocks the myosin binding sites on the actin filaments. Another protein, *troponin*, is associated with tropomyosin. A nerve impulse triggers contraction by causing an influx of calcium ions from the sarcoplasmic reticulum. The troponin molecules bind calcium ions and in so doing cause the tropomyosin molecules to shift position slightly, so exposing the myosin binding sites and allowing contraction to proceed. Tropomyosin, but not troponin, is also found in smooth muscle. *See* sarcomere.

troponin /**troh**-pŏ-nin, **trop**-ŏ-/ *See* tropomyosin.

trypsin /**trip**-sin/ An enzyme that catalyzes the partial hydrolysis of peptides. It catalyzes the hydrolysis of peptide bonds formed from the carbonyl group of lysine and arginine residues. Trypsin is found in pancreatic tissue and in pancreatic juices in an inactive form, trypsinogen.

trypsinogen /trip-**sin**-ŏ-jĕn/ The inactive form of trypsin, found in pancreatic tissues and pancreatic juices. Trypsinogen is converted into the active form by action of the enzyme enteropeptidase (enterokinase).

tryptophan /**trip**-tŏ-fan/ *See* amino acids.

T tubules *See* transverse tubules.

tuber A swollen underground stem or root that contains stored food, and acts as an organ of perennation and vegetative propagation. Stem tubers (e.g. potato) develop at the end of underground stems by swelling of nodes and internodes. There is an increase in pith tissue to form a round tuber that bears buds in the axils of greatly reduced scale leaves. The stem connecting the tuber to the parent plant then severs. Root tubers may develop in the same way from adventitious roots, as in dahlias.

tubulin /tew-byŭ-lin/ *See* microtubule.

tumor /**tew**-mer/ A swelling or abnormal growth of tissue. A *malignant tumor* is one caused by CANCER, whereas a *benign tumor* is noncancerous.

tumor suppressor gene Any of a class of genes whose products normally perform functions that reduce the likelihood of cancer. Many of them act to negatively control (retard) the cell cycle or positively control (enhance) programmed cell death (APOPTOSIS). Hence mutations of such genes can 'take the brakes off' the cell cycle, allowing defective DNA to go unrepaired for instance, or reduce the effectiveness of apoptosis in destroying cancerous cells when they arise. For example, mutations of the p53 tumor suppressor gene are associated with 50% of human cancers. *See also* oncogene.

tunica–corpus theory /**tew**-nă-kă **kor**-pŭs/ A theory of apical organization and development that distinguishes two separate tissue zones, the tunica and the corpus, in the apex of a flowering plant. The *tunica* is made up of one or more peripheral layers in which cell division is mostly anticlinal. The *corpus* is the inner area of tissue in which cell arrangement and division is irregular. The epidermis originates in the tunica region and the other stem tissues arise from either the tunica or the corpus, depending on the species.

Tunicata /tew-nă-**kay**-tă/ *See* Urochordata.

Turbellaria /ter-bĕ-**lair**-ee-ă/ A class of small mostly aquatic free-living platyhelminths, e.g. *Planaria*. Turbellarians have a flat body covered with cilia for locomotion, a ventral mouth and protrusible pharynx, and often have tentacles and eyes on the head. *See also* Planaria.

turgor /**ter**-ger, -gor/ The state, in a plant or prokaryote cell, in which the protoplast is exerting a pressure on the cell wall owing to the intake of water by OSMOSIS. The cell wall, being slightly elastic, bulges but is rigid enough to prevent water entering to the point of bursting. The cell is then said to be turgid. Turgidity is the main means of support of herbaceous plants. *See* plasmolysis.

turion /**tewr**-ee-ŏn/ A swollen detached winter bud that contains stored food and is protected by an outer layer of leaf scales and mucilage. It is an organ of perennation or vegetative propagation, and is characteristic of various water plants (e.g. *Sagittaria*).

Turner's syndrome A condition in the human female caused by partial or complete lack of an X chromosome. This causes undeveloped ovaries, reduced stature, webbing of the neck, deafness, and mental deficiency. In less severe forms the normal number of chromosomes may be present but one of the X chromosomes is structurally abnormal. The syndrome is named for the American endocrinologist Henry H. Turner (1892–1970).

tylose /**tÿ**-lohs/ An ingrowth from a parenchyma cell into an adjacent tracheid or vessel through a paired pit. Tyloses are often found in injured tissue, older wood, and below an abscission layer, and can completely block the conducting vessel.

tympanic cavity /tim-**pan**-ik/ *See* middle ear.

tympanum /**tim**-pă-nŭm/ (*pl.* **tympana**) (**tympanic membrane; eardrum**) A thin membrane that separates the outer and middle ear of vertebrates. Sound waves cause vibrations of the tympanum, which are then transmitted to the inner ear via the ear ossicles of the middle ear. *See illustration at ear.*

type The material used to define a species. It is usually a dried specimen stored in a herbarium but may also be a drawing. The term type is also used to describe the representative species of a genus, the representative genus of a family, etc. For example the genus *Solanum* is the type genus of the family Solanaceae.

tyrosine /**tÿ**-rŏ-seen, -sin, **ti-**/ *See* amino acids.

U

ubiquinone /yoo-bă-kwi-**nohn**, -**kwin**-ohn/ (**coenzyme Q**) A coenzyme that is an essential component of the ELECTRON-TRANS-PORT CHAIN.

ubiquitin /yoo-**bik**-wă-tin/ A small protein found in cells of all tissues that tags cellular proteins for degradation by PRO-TEASOMES. An enzyme complex recognizes certain amino-acid sequences in target proteins and attaches a string of ubiquitin molecules – a process called *ubiquitination* – which is thought to direct the protein to the proteasome.

ulna /**ul**-nă/ One of the two long bones of the lower forelimb (forearm) in tetrapods. In humans, it forms the posterior (postaxial) border of the forearm, extending from the back of the elbow to the wrist and lying parallel to the smaller radius bone. Its hooklike upper end forms the point of the elbow and its inner curved surface articulates over the lower end of the humerus.

ultracentrifuge /ul-tră-**sen**-tră-fyooj/ A high-speed centrifuge, operating at up to a million revolutions per second, that is used to sediment protein and nucleic acid molecules. Ultracentrifuges operate under refrigeration in a vacuum chamber and forces 50 million times gravity may be reached. The rate of sedimentation depends on the molecular weight of the molecule and thus the ultracentrifuge can be used to separate a mixture of large molecules, and estimate sizes.

ultrafiltration /ul-tră-fil-**tray**-shŏn/ The separation of solvent from a solution by forcing it through a suitable membrane by the application of hydrostatic pressure. A generalized ultrafiltration occurs through the semipermeable walls of blood capillaries; blood pressure forces fluid and inorganic ions through the walls into intercellular spaces, but leaves proteins and cells behind in the plasma. The resultant tissue fluid is eventually returned to the venous system via the lymphatic system. The kidney glomerulus is specialized for ultrafiltration as a way of removing excess water, ions, and waste products as urine.

ultramicrotome /ul-tră-**mÿ**-kroh-tohm/ *See* microtome.

ultrastructure /ul-tră-**struk**-cher/ (**fine structure**) The detailed structure of biological material as revealed, for example, by electron microscopy, but not by light microscopy.

Ulva /**ûl**-vă/ (**sea lettuce**) A genus of marine green algae with a fine leafy thallus two cells thick. It has a diplobiontic life cycle with similar haploid and diploid generations. *See* Chlorophyta.

umbel /**um**-bĕl/ A type of inflorescence in which the stem axis is not elongated and individually stalked flowers arise from the same point on the stem. These flowers are massed on one plane, giving the appearance of an umbrella, with the oldest flowers on the outside and the youngest in the middle. The umbel is typical of the carrot family (Umbelliferae).

umbilical cord /um-**bil**-ă-kăl/ A cord of tissue connecting the abdomen of the embryo to the placenta in pregnant mammals. It contains two arteries (*umbilical arteries*) and a vein (*umbilical vein*) through which blood, containing useful and waste substances, is transported to and from the embryo. It is severed at birth and atrophies leaving a scar, the *umbilicus* (navel in humans). The umbilical cord is formed mainly of mesodermal tissue and embryonic membranes.

undulipodium /un-jŭ-lă-**poh**-dee-ŭm/ (*pl.* **undulipodia**) A whiplike organelle that protrudes from a eukaryotic cell and is used chiefly for locomotion (e.g. sperm) or feeding (e.g. ciliate protoctists). Undulipodia include all eukaryotic cilia and flagella, which share the same essential structure, and differ markedly from bacterial flagella (*see* flagellum). The shaft comprises a cylindrical array of nine doublet microtubules surrounding a central core of two single microtubules. The outer wall of the shaft is an extension of the cell membrane. Movement of the shaft is produced by projecting arms of a protein (dynein) on the microtubules, which cause adjacent microtubule doublets to slide past each other. This requires energy from ATP. At the base of the shaft near the cell surface is a basal body, or kinetosome. This consists of a cylindrical array of nine triplet microtubules, from which the shaft grows. Its structure is similar to that of a centriole. Flagella tend to be larger than cilia, and produce successive waves of bending that are propagated to the tip of the shaft, as in a sperm. Cilia characteristically beat in a different manner, with a power stroke and a recovery stroke. *See* cilium.

ungulate /**ung**-gyŭ-lit, -layt/ A hoofed grazing mammal; one belonging to the order Perissodactyla or Artiodactyla.

unguligrade /**ung**-gyŭ-lă-grayd/ The mode of progression in some mammals in which only the very tips of the fingers or toes are in contact with the ground. It is typical of hoofed mammals (ungulates), such as the horse. *Compare* digitigrade; plantigrade.

unicellular /yoo-nă-**sell**-yŭ-ler/ Describing organisms that exist as a single cell. Such a state is characteristic of protoctists and bacteria and is also found in many

algae and fungi. *Compare* multicellular; acellular.

unisexual /yoo-nă-**sek**-shoo-ăl/ Describing organisms that have either male or female sex organs, but not both. Unisexual plants may be monoecious or dioecious. *Compare* hermaphrodite.

universal indicator *See* indicator.

uracil /**yoor**-ă-săl/ A nitrogenous base that is found in RNA, replacing the thymine of DNA. It has a pyrimidine ring structure.

urea /yû-**ree**-ă/ A water-soluble nitrogen compound, $H_2N.CO.NH_2$. It is the main excretory product of catabolism of amino acids in certain animals (ureotelic animals). *See also* ornithine cycle.

urea cycle *See* ornithine cycle.

ureotelic /yoor-ee-oh-**tel**-ik/ Excreting nitrogen in the form of urea. Amphibians and mammals are ureotelic. *Compare* ammonotelic; uricotelic.

ureter /yû-**ree**-ter/ One of a pair of ducts in reptiles, birds, and mammals that transports urine from the kidneys to the cloaca (in reptiles and birds) or bladder (in mammals). It is associated with the metanephric kidney and functionally replaces the Wolffian duct of fish and amphibians.

urethra /yû-**reeth**-ră/ A duct in mammals that takes urine from the bladder to the exterior. In males, it also transports spermatozoa and passes through the penis. *See also* vas deferens.

uric acid /**yoor**-ik/ A nitrogen compound produced from purines. In certain animals (uricotetic animals), it is the main excretory product resulting from breakdown of amino acids. In humans, uric acid crystals in the joints are the cause of gout.

uricotelic /yoor-ă-koh-**tel**-ik/ Excreting nitrogen in the form of uric acid. Reptiles and birds are uricotelic. *Compare* ammonotelic; ureotelic.

uridine /**yoor**-ă-deen, -din/ The nucleoside formed when uracil is linked to D-ribose by a β-glycosidic bond.

urine /**yoor**-in/ The liquid excreted through the urethra or cloaca. It is produced in the kidneys and contains urea or uric acid, and numerous other substances in small amounts.

Urochordata /yoor-ŏ-kor-**day**-tă/ (**Tunicata**) A marine subphylum of chordates including the sessile sea squirts (e.g. *Ciona*) and pelagic forms (e.g. *Oikopleura*). The chordate characters of notochord, dorsal nerve cord, and gill slits are clearly seen in the tadpole-like free-swimming larva. In the adult the gill slits are modified for filter feeding, the notochord is absent, and the nerve cord reduced (except in *Oikopleura*, in which the notochord and nerve cord are retained). The unsegmented globular body is enclosed in a protective tunic with two openings, an inhalent mouth and an exhalent atriopore. *Compare* Cephalochordata.

urostyle /**yoor**-ŏ-stÿl/ A pointed rod of bone at the hind end of the vertebral column of frogs and toads (Anura). It results from the fusion of caudal vertebrae.

uterus /yoo-tĕ-rŭs/ (*pl.* **uteri**) (**womb**) A thick-walled muscular structure between the bladder and rectum in female mammals. It may be paired but in humans forms a single inverted pear-shaped organ about 6–8 cm long. It leads anteriorly to the two Fallopian tubes and posteriorly to the vagina and the outside. The uterus lining (*see* endometrium) undergoes cyclical changes in thickness under the influence of reproductive hormones and, if fertilization of an ovum occurs, it provides a source of attachment and nourishment for the fetus. The uterus expands and becomes increasingly muscular to accommodate the growing fetus, although it returns to normal about 6–8 weeks after the birth of young.

utriculus /yû-**trik**-yŭ-lŭs/ (*pl.* **utriculi**) (**utricle**) The upper chamber of the labyrinth of the vertebrate inner ear, which bears the semicircular canals. There is a patch of sensory epithelium in the wall lining and granules of calcium carbonate in the cavity, which together are responsible for the detection of changes in position of the head with respect to gravity and the rate of change. *See also* macula. *See illustration at* ear.

V

vaccination /vak-să-**nay**-shŏn/ The introduction of antigens into the body to induce production of specific antibodies, either to confer immunity against subsequent infection by the same antigen or, less commonly, to treat a disease. The various types of antigens used include attenuated or dead microorganisms or harmless microorganisms that are closely related to pathogenic types. They are made into a suspension (a vaccine), which is usually injected or ingested.

vacuole /**vak**-yoo-ohl/ A spherical fluid-filled organelle of variable size found in plant and animal cells, bounded by a single membrane and functioning as a compartment to separate a variety of materials from the cytoplasm. Vacuoles have a variety of specialized functions, for example as food vacuoles, CONTRACTILE VACUOLES, and autophagic vacuoles. Many mature plant cells have a single large central vacuole that confines the cytoplasm to a thin peripheral layer. It is bounded by a membrane called the tonoplast and contains cell sap. This contains substances in solution, e.g. sugars, salts, and organic acids, often in high concentrations resulting in a high osmotic pressure. Water therefore moves into the vacuole by osmosis making the cell turgid. Vacuoles may also contain crystals and waste substances.

vagina /vă-**jy**-nă/ A distensible muscular duct in most female mammals that extends from the uterus (or uteri), between the bladder and rectum, to the exterior. It receives the male penis during mating. A similar structure occurs in some invertebrates. The vagina is usually singular, having formed from the lower part of the Müllerian ducts, which fuse in the embryo.

vagus nerve /**vay**-gŭs/ (**cranial nerve X**) One of the pair of nerves that arises from the medulla oblongata in the vertebrate brain and extends into the body to supply the major internal organs. It is the major nerve of the parasympathetic system, carrying motor nerve fibers to the heart, lungs, stomach, intestine, liver, kidneys, etc. It also contains sensory nerve fibers running from the viscera to the brain. *See* cranial nerves.

valine /**val**-een, -in/ *See* amino acids.

valve 1. A structure in an opening or tube that can open and close to restrict flow to one direction. There are several types of valves in the heart and veins, all of which consist of flaps of tissue (cusps) that can seal off the passageway to prevent backflow of blood. *See* mitral valve; semilunar valve; tricuspid valve.
2. Either of the halves of the shell of a bivalve mollusk.
3. Either of the two parts of the cell wall of a diatom.
4. One of the parts that make up the wall of dry fruit, such as a capsule.

variable number tandem repeats (**VNTRs**) A type of REPETITIVE DNA consisting of nucleotide sequences, typically 15–100 bp long, that are repeated in tandem arrays of varying length (1–5 kb) at sites scattered throughout the genome. All humans, for example, have a particular 16-base sequence that is repeated hundreds of times among the different chromosomes. However, the number of repeats at each site is unique for each individual, which makes variable number tandem repeats useful as genetic markers and for GENETIC FINGERPRINTING.

variation The extent to which the characteristics of a species can vary. Variation can be caused by environmental and genetic factors. Environmental variation (phenotypic plasticity) results in differences in the appearance of individuals of a species because of differences in nutrition, disease, population density, etc. Genetic variation is caused by recombination and occasionally mutation.

variety The taxonomic group below the subspecies level. The term is often loosely used to describe breeds of livestock or various cultivated forms of agricultural and horticultural species. *See also* cultivar.

vascular bundle /vas-kyŭ-ler/ A strand of conducting tissue found in vascular plants. In plants showing secondary thickening (e.g. dicotyledons) the vascular bundles contain meristematic cambium tissue between the xylem and phloem, but this is generally lacking in plants without secondary thickening (e.g. monocotyledons). *See also* vascular tissue.

vascular cambium *See* intrafascicular cambium.

vascular plants Plants containing differentiated cells forming vascular tissue, which comprises the xylem and phloem. Vascular tissue transports water and nutrients through the plant and also provides strength and support. It is characteristic of the tracheophytes (pteridophytes and spermatophytes). Vascular plants are able to achieve considerable vertical growth upwards and into the soil. They have thus been able to colonize the drier habitats that are inaccessible to the more primitive nonvascular bryophytes.

vascular system 1. (*Botany*) The system of conducting tissue in angiosperms, gymnosperms, and pteridophytes, that is responsible for the transport of water, mineral salts, and foods to and from the roots and aerial parts of the plant. It also gives mechanical support, especially in older stems of perennial plants, which largely consist of vascular tissue. *See also* vascular tissue.
2. (*Zoology*) A continuous fluid-filled system of vessels in animals. An example is the blood vascular system.

vascular tissue (fascicular tissue) A tissue found in seed plants and pteridophytes consisting principally of XYLEM and PHLOEM (water- and food-conducting tissues respectively). It also contains strengthening tissue (sclerenchyma) and packing tissue (parenchyma). The arrangement of vascular tissue in the stem is very varied, giving a number of different types of stele.

Primary vascular tissue, which is found in all vascular plants, is formed from the procambium. Secondary vascular tissue, found only in plants with secondary thickening, develops from the vascular cambium. The vascular cambium extends to form a complete ring of meristematic tissue around the stem, the separate vascular bundles being linked by interfascicular cambium.

vas deferens /vas def-ĕ-renz/ (*pl.* vasa deferentia) One of the main pair of ducts in animals that convey sperm from the testis to the exterior. In male mammals, it leads from the epididymis and opens into the urethra just after it leaves the bladder.

vas efferens /ef-ĕ-renz/ (*pl.* vasa efferentia) In reptiles, birds, and mammals, one of a number of small ducts that convey spermatozoa from the seminiferous tubules of the testis to the epididymis. They derive from tubules of the embryonic mesonephros. In invertebrates, vasa efferentia conduct spermatozoa directly from the testis to the vas deferens.

vasoconstriction /vas-oh-kŏn-**strik**-shŏn/ The reduction in diameter of small blood vessels due to contraction of the smooth muscle in their walls. It results from stimulation by vasoconstrictor nerve fibers or from secretion (or injection) of epinephrine in response to decreased blood pressure, low external temperature, pain, etc. *See also* vasomotor nerves.

vasodilatation /vas-oh-dil-ă-**tay**-shŏn/ (**vasodilation**) The increase in diameter of small blood vessels due to relaxation of the smooth muscle in their walls. It results from stimulation by vasodilator nerve fibers or inhibition of vasoconstrictor nerve fibers in response to increased blood pressure, exercise, high external temperature, etc. *See also* vasomotor nerves.

vasomotor nerves /vas-oh-**moh**-ter/ Nerve fibers of the autonomic nervous system that control the diameter of blood vessels. They transmit impulses from the vasomotor center in the medulla oblongata of the brain to the smooth muscle in the vessel walls, causing them to become constricted (vasoconstrictor nerve fibers) or dilated (vasodilator nerve fibers).

vasopressin /vas-oh-**pres**-in/ (**antidiuretic hormone; ADH**) A peptide hormone that is synthesized in the mammalian hypothalamus and released from nerve endings in the posterior pituitary gland (*see* neurosecretion). Its chief effect is to increase the permeability to water of the kidney tubules, thereby stimulating reabsorption of water and making urine more concentrated. Release of vasopressin is triggered by raised osmotic pressure of the blood, which is detected by osmoreceptors in the hypothalamus. It thus acts to increase blood volume, and hence blood pressure. Vasopressin also stimulates the contraction of smooth muscle in the walls of arterioles, which further contributes to raised blood pressure. Birds, fishes, and reptiles produce vasotocin, a hormone very similar to vasopressin and producing similar effects. *See also* oxytocin.

vector /**vek**-ter/ **1.** An animal, often an insect or tick, that carries a disease-causing organism from an infected to a healthy animal or plant, causing the latter to become infected; for example, the mosquito transmits malaria and other diseases to humans. **2.** (**cloning vector**) An agent used as a vehicle for introducing foreign DNA, for example a new gene, into host cells. Several types of vector are used in gene cloning, notably bacterial plasmids and bacterio-

phages. The segment of DNA is first spliced into the DNA of the vector, then the vector is transferred to the host cell (e.g. the bacterium, *E. coli*), where it replicates along with the host cell. The result is a clone of cells, all of which contain the foreign gene, which is expressed by the host cell machinery. *See* gene cloning.

vegetal pole /**vej**-ĕ-tăl/ (**vegetative pole**) The end of a spherical animal egg that is opposite the animal pole. The vegetal pole contains most of the yolk and is furthest from the nucleus. *Compare* animal pole.

vegetative /**vej**-ĕ-tay-tiv/ Describing or relating to an involuntary function, such as digestion or the autonomic nervous activity, or to a structure or stage in development that is concerned with nutrition and growth rather than with sexual reproduction. Vegetative reproduction is asexual reproduction.

vegetative nucleus (**tube nucleus**) One of the two or three nuclei in a young pollen grain that are formed after division of the haploid nucleus. After the pollen grain germinates on the stigmatic surface, the tube nucleus is the first to migrate down the pollen tube, and is thought to regulate its growth and development. The tube nucleus disintegrates as the pollen tube grows down the style.

vegetative propagation (**vegetative reproduction**) **1.** (*Botany*) Asexual reproduction in plants by means of large multicellular structures that become detached from the parent and develop into new individuals, e.g. gemmae, corms, bulbs, stolons, tubers. Many of these also serve as perennating organs. Vegetative propagation is a common natural process but many artificial techniques of vegetative propagation have also been developed. *See also* budding; cutting; graft. **2.** (*Zoology*) Asexual reproduction in animals, as by budding.

vein 1. (*Zoology*) A blood vessel that conveys blood from the capillary network in the tissues to the heart. All veins except the

pulmonary vein carry deoxygenated blood and most contain valves, which maintain direction of flow. Compared with arteries they have a larger lumen and thinner walls. **2.** (*Zoology*) One of numerous chitinous tubes that support and strengthen the wing of an insect.
3. (*Botany*) One of the vascular bundles in a leaf.

velamen /vĕ-**lay**-mĕn/ (*pl.* **velamina**) A layer surrounding the aerial roots of epiphytic plants (e.g. orchids) which, due to the spongy nature of the cells, is able to soak up surface water. It is made up of several layers of dead empty cells situated external to an exodermis. The cells are spirally thickened, and are translucent to allow light through to the photosynthetic tissue beneath.

veliger /**vel**-ă-jer/ The second larval stage of aquatic mollusks (except cephalopods), which develops from the trochophore. During this stage the shell and foot develop and the viscera become rotated so as to produce the asymmetry characteristic of the adult mollusk.

velum /**vel**-ŭm/ (*pl.* **vela**) *See* annulus.

vena cava /**vee**-nă **kay**-vă/ (*pl.* **venae cavae**) Either of two main veins that convey deoxygenated blood from the body to the right atrium of the heart in tetrapods. The anterior vena cava (or *precaval vein*) is a paired vein serving the head and forelimbs. The posterior vena cava (or *postcaval vein*) is a single vein serving most of the body and the hindlimbs. The anterior vena cava is homologous with the anterior cardinal vein in fish. The posterior vena cava is derived from the posterior cardinal vein and the renal portal system.

venation /vee-**nay**-shŏn/ **1.** (*Botany*) The distribution of veins (vascular bundles) in a leaf. Dicotyledons usually show a netlike arrangement whereas monocotyledons show a parallel distribution of veins.
2. (*Zoology*) The distribution of veins in an insect's wing. There is a great variety of patterns between species, which is useful as a means of identification.

venter /**ven**-ter/ The swollen base of an archegonium that contains the egg cell (oosphere).

ventilation The process by which an organism maintains a flow of air or water over its respiratory surfaces in the lungs, gills, or other respiratory organs. This increases the rate at which oxygen enters and carbon dioxide leaves the blood. Most active animals possess some form of ventilation mechanism, typically involving muscular movements (*respiratory movements*). For example, bony fishes can actively pump water over their gills by alternately expanding and contracting the mouth and gill cavities. In mammals, air is alternately drawn into and expelled from the lungs by muscular movements of the rib cage and diaphragm. *See* respiration.

ventral /**ven**-trăl/ **1.** (*Zoology*) Designating the side of an animal nearest the substrate, i.e. the lower surface. However, in bipedal animals, such as humans, the ventral side is directed forwards corresponding to the anterior side of other animals.
2. (*Botany*) Designating the upper or adaxial surface of the lateral organs of plants, e.g. leaves.
Compare dorsal.

ventral aorta The blood vessel in fish and embryo tetrapods that carries deoxygenated blood from the anterior end of the heart and divides to form the six paired aortic arches. In adult tetrapods it is equivalent to the ascending limb of the aorta.

ventral root *See* spinal nerve.

ventricle /**ven**-tră-kăl/ **1.** In mammals, either of the two thick-walled muscular lower chambers of the HEART. When they contract, the bicuspid and tricuspid valves close and blood is forced into the aorta and pulmonary artery, respectively. When they relax, semilunar valves in these arteries close to prevent blood returning to the ventricles. The right ventricle pumps deoxygenated blood to the lungs and the left ventricle pumps oxygenated blood around the body. The hearts of reptiles and birds also have two ventricles, amphibians and fish have one.

2. One of four fluid-filled interconnecting cavities within the brain. Ventricles I and II within the cerebral hemispheres are each connected via a small hole (foramen) with ventricle III, lying in the mid-line. This is linked through the narrow cerebral aqueduct of the midbrain to ventricle IV of the hindbrain, which is connected to the central canal of the spinal cord. In two thin-walled areas – CHOROID PLEXUSES – cerebrospinal fluid is filtered from the blood and enters the ventricles.

venule /**ven**-yool/ A small vein that collects deoxygenated blood from the capillary networks in tissues.

vermiculite /ver-**mik**-yŭ-lÿt/ A very light mica-based substance used as a support material for plant growth in some experiments where soil would be unsuitable. It can, for instance, be sterilized and mineral solutions of known composition can be fed to the plants.

vermiform appendix /**ver**-mă-form/ *See* appendix.

vernalization /ver-nă-li-**zay**-shŏn/ The cold treatment of partially germinated seeds. Certain plants will only flower if exposed to low temperatures (1–2°C) at an early period of growth, i.e. they have a chilling requirement. Thus winter varieties of cereals will only flower in summer if sown the previous autumn. Spring sown winter varieties remain vegetative throughout the season unless they have been vernalized. Vernalization is an important technique in countries where severe winters can kill autumn sown crops.

vernation /ver-**nay**-shŏn/ The arrangement of leaves in relation to each other in the bud. *See also* ptyxis.

versatile Describing an anther that is attached dorsally to the tip of the filament. This allows the anther to turn freely in the wind, thus aiding pollen dispersal. *Compare* basifixed; dorsifixed.

vertebra (*pl.* **vertebrae**) One of a series of bones or cartilages forming the VERTEBRAL COLUMN of vertebrates. Tetrapods have bony vertebrae, each consisting typically of a ventral main body (centrum) and a dorsal neural arch of bone, which forms an opening (neural canal) for the spinal cord. Arising from each arch are various processes for muscle attachment. This general arrangement is modified for different regions of the body. Some fishes have vertebrae made of cartilage. *See* caudal vertebrae; cervical vertebrae; lumbar vertebrae; thoracic vertebrae; sacral vertebrae.

vertebral column (**backbone; spinal column**) A series of bones or cartilages (vertebrae) that run along the dorsal side of the vertebrate body from the head to the tail region. There are 26 vertebrae in the adult human – 7 cervical vertebrae, 12 thoracic vertebrae, 5 lumbar vertebrae, a sacrum, and a coccyx – each separated by a disk of fibrocartilage (the intervertebral disk) and attached to muscles. The whole column provides a flexible axial support to the body and forms a protective channel (neural canal) for the spinal cord. The vertebrae become larger and stronger toward the major weight-bearing region, i.e. where the pelvic girdle is attached to the sacrum; the ribs and pectoral girdle articulate with the column in the thoracic region.

vertebrate *See* craniate.

verticillaster /ver-tă-si-**las**-ter/ (**false whorl**) A type of inflorescence in which two oppositely placed cymes are inserted either side of a stem giving the impression of a whorl. This flower structure is typical of the deadnettle family.

vesicle /**ves**-ă-kăl/ A small vacuole of variable origin, such as a Golgi vesicle or endocytotic vesicle. *See* endocytosis; Golgi apparatus.

vessel (**trachea**) An advanced form of xylem conducting tissue composed of vertically arranged vessel elements. Vessels are only found in the angiosperms and the gymnosperm order Gnetales. *Compare* tracheid.

vessel element One of the cells that makes up a xylem vessel. In contrast to tracheids, the most advanced vessel elements are often more broad than long and have horizontal rather than slanting perforation plates.

vestibular apparatus /ves-**tib**-yŭ-ler/ The parts of the EAR concerned with sensing the position and motion of the body, and thus maintaining balance. The apparatus consists of the UTRICULUS and SACCULUS (which together comprise the vestibule of the inner ear) and the three SEMICIRCULAR CANALS positioned on the utriculus.

vestibulocochlear nerve /ves-tib-yŭ-loh-**kok**-lee-er/ *See* auditory nerve.

vestigial organ An organ that is functionless and generally reduced in size but bears some resemblance to the corresponding fully functioning organs found in related organisms. Examples include the wings of flightless birds, the limb girdles of snakes, the appendix and the ear muscles of humans, and the scale leaves of parasitic flowering plants. The presence of vestigial organs is thought to indicate that the ancestors of the organism possessed fully functioning organs, which, because of gradual changes in the environment or their lifestyle, became of less use and so did not develop fully in modern forms.

vibrio /**vib**-ree-oh/ Any comma-shaped bacterium.

villus /**vil**-ŭs/ (*pl.* **villi**) One of the microscopic finger-like projections in the lining of the small intestine. Millions of these villi give the appearance of velvet to the lining and enormously increase the surface area for absorption. Each villus is covered by a single layer of columnar cells through which the soluble products of digestion can readily pass into the blood or lymph. Each also contains a network of blood capillaries and a lacteal. Strands of muscle contract rhythmically, to shorten the villus and empty the lacteal and capillaries, so keeping up the diffusion gradient. The surface area is further increased by *microvilli*, visible only through an electron microscope, which cover the surface of each cell of the columnar epithelium.

Similar structures also occur in the chorion (*chorionic villi*), especially in the placenta where they provide a large surface area for exchange of materials between fetal and maternal blood.

violaxanthin /vÿ-ŏ-lă-**zan**-thin/ A xanthophyll pigment found in the brown algae. *See* photosynthetic pigments.

virion /**vÿ**-ree-on, **vi**-/ The extracellular inert phase of a virus. A virion consists of a protein coat surrounding one or more strands of DNA or RNA. Virions may be polyhedral or helical and vary greatly in size, though all are submicroscopic.

viroid /**vÿ**-roid/ A tiny infectious agent, about 20 nm in length, found in plants that is similar to a virus but lacks a capsid, consisting simply of a circle of RNA, 300–400 nucleotides long. Viroids replicate within the plant cell and cause characteristic disease symptoms; examples are the potato spindle tuber viroid, the avocado sunblotch viroid, and the hop stunt viroid.

virulence /**vi**-ryŭ-lĕns, -rŭ-/ The relative pathogenicity of an organism. Virulence depends on the invasiveness and toxicity of the pathogen. Virulence may vary between strains of the same organism.

virulent phage A bacteriophage that infects a bacterial cell and immediately replicates, causing lysis of the host cell. *Compare* temperate phage.

virus /**vÿ**-rŭs/ An extremely small infectious agent, generally 30–300 nm in diameter, that requires a living cell in order to replicate itself. Viruses infect virtually all types of eukaryotic organisms as well as bacteria, and cause a variety of diseases in plants and animals, such as smallpox, the common cold, and tobacco mosaic disease. Each type of virus is usually restricted to a fairly narrow range of hosts. Viruses that infect only bacteria are called PHAGES (or bacteriophages). Outside their living host cell viruses exist as inactive particles consisting of a core of DNA or RNA surrounded by a protein coat (CAPSID). The inert extracellular form of the virus, termed a VIRION, penetrates the host membrane and liberates the viral nucleic acid into the cell. Viral growth cycles inside the cell are of two types. Most viruses undergo a *lytic cycle*, in which the viral nucleic acid is transcribed and translated by the host cell's enzymes and ribosomes to produce the proteins needed for the formation of daughter virions. The virions are released by rupture (lysis) of the host cell. In contrast, some phages enter a quiescent state inside the cell called LYSOGENY. Here the viral DNA is integrated with the host's genome. Some animal viruses, notably the RETROVIRUSES, can also integrate their genome with that of the host; these viruses can introduce ONCOGENES, which cause cancer in the host.

visceral arch /viss-ĕ-răl/ One of a series of bony or cartilaginous skeletal arches, in fish and other vertebrate embryos, that occur in the lateral walls of the pharynx and support the tissue behind the mouth and between the gill slits. Typically there are seven, each forming an incomplete ring of elements – a mid-ventral element from which a series of elements extends upwards on each side, almost to the mid-dorsal line. They are often modified for different functions. *See* branchial arch; hyoid arch; mandibular arch.

visceral cleft *See* gill cleft.

visceral pouch *See* gill pouch.

visual purple *See* rhodopsin.

vital capacity The volume of air that is moved in and out of the lungs when taking the deepest possible intake of breath and forcing air out to the maximum possible extent. It is affected by factors such as size of the individual, posture, smoking, and disease. *See also* residual volume.

vital stains Nontoxic coloring materials that can be used in dilute concentrations to stain living material without damaging it. Examples of vital stains include Janus green, which selectively stains mitochondria and nerve cells, and trypan blue, which has an affinity for the macrophages of the reticuloendothelial system. *See also* staining.

vitamin A (vitamin A_1; retinol) A fat-soluble vitamin (a derivative of the yellow pigment, carotene) occurring in milk, butter, cheese, liver, and cod-liver oil. It can also be formed in the body by oxidation of carotene, which is present in fresh green vegetables and carrots. Vitamin A is a component of the visual pigment RHODOPSIN, and is also required for normal growth and development, and for reproduction. Deficiency in vitamin A can result in a reduced resistance to disease and in night blindness.

vitamin B complex A group of ten or more water-soluble vitamins, which tend to occur together. They can be obtained from whole grains of cereals and from meat and liver. Since the B vitamins are present in most unprocessed food, deficiency diseases only occur in populations living on restricted diets. Many of the B vitamins act as coenzymes involved in the normal oxidation of carbohydrates during respiration.

The vitamins of the B complex include thiamine (vitamin B_1), riboflavin (vitamin B_2), nicotinic acid (niacin), pantothenic acid (vitamin B_5), pyridoxine (vitamin B_6), cyanocobalamin (vitamin B_{12}), biotin, lipoic acid, and folic acid.

vitamin C (ascorbic acid) A water-soluble vitamin, which is widely required in metabolism. The major sources of vitamin C are fresh fruit and vegetables. Humans and other primates are among the few mammalian groups that are unable to synthesize ascorbic acid. It is needed for the formation of collagen, and severe deficiency results in inability to form connective tissue, hence the symptoms of scurvy. It is also an antioxidant.

vitamin D A fat-soluble vitamin found in fish-liver oil, butter, milk, cheese, egg yolk, and liver. Its principal action is to increase the absorption of calcium and phosphorus from the intestine. The vitamin also has a direct effect on the calcification process in bone. Deficiency results in inadequate deposition of calcium in the bones, causing rickets in young children and osteomalacia in adults.

The term vitamin D refers, in fact, to a group of compounds, all sterols, of very similar properties. The most important are vitamin D_2 (*calciferol*) and vitamin D_3 (cholecalciferol). Precursors of these are converted to the vitamins in the body by the action of ultraviolet radiation.

vitamin E (tocopherol) A fat-soluble vitamin found in wheat germ, dairy products, and in meat. It is thought to function primarily as an antioxidant, trapping free radicals and preventing oxidation of lipids. Severe deficiency in infants may lead to high rates of red-blood cell destruction and hence to anemia. However, there are very few deficiency effects apparent in humans.

vitamin K (phylloquinone; menaquinone) A fat-soluble vitamin that is required to catalyze the synthesis of prothrombin, a blood-clotting factor, in the liver. Intestinal microorganisms are capable of synthesizing considerable amounts of vitamin K in the intestine and this, together with dietary supply, insures that deficiency is unlikely to occur in any but the newborn. A newborn child may be deficient as the intestine is sterile at birth and the level supplied by the mother during gestation is limited. Thus during the first few days of life blood-clotting deficiency may be observed, but this is readily rectified by a small injection of the vitamin.

vitamins Organic chemical compounds that are essential in small quantities for metabolism. The vitamins have no energy value; most of them seem to act as catalysts for essential chemical changes in the body, each one influencing a number of vital processes. Vitamins A, D, E, and K are the fat-soluble vitamins, occurring mainly in animal fats and oils. Vitamins B and C are the water-soluble vitamins. If a diet lacks vitamins, this results in the breakdown of normal bodily activities and produces disease symptoms. Such deficiency diseases can usually be remedied by including the necessary vitamins in the diet. Plants can synthesize vitamins from simple substances, but animals generally require them in their diet, though there are exceptions to this. These include vitamins synthesized by bacteria in the gut, and some that can be manufactured by the animal itself. A precursor of vitamin D_2 (ergosterol), for example, can be converted in the skin by ultraviolet radiation.

vitelline membrane /vi-**tel**-in, vÿ-/ *See* egg membrane.

vitreous humor The firm semifluid jelly that fills the space in the vertebrate eye behind the lens. It contains a delicate network of fibers resembling collagen and helps to maintain the shape of the eyeball. *See illustration at* eye.

viviparity /viv-ă-**pa**-ră-tee/ **1.** (*Zoology*) A type of sexual reproduction in animals in which the embryo develops within the mother's body and derives continuous nourishment by close contact with maternal tissues, usually through a placenta. It results in the birth of live young and occurs in most mammals. *Compare* oviparity; ovoviviparity.
2. (*Botany*) The production of young plants instead of flowers, as in some grasses.
3. (*Botany*) The germination of seeds or spores that are still attached to the parent plant.

vocal cords A pair of mucous membrane folds stretched across the anterior opening of the LARYNX. Vibration of the cords by expelled air produces vocal sounds. The pitch of the sound varies according to the length and tension of the cords, which is controlled by muscles of the larynx. *See also* syrinx.

Volkmann's canals /fol'ks-mahn/ Vascular channels found in bone material. They contain blood vessels that connect with those of the Haversian canals. The canals are named for the German physiologist Alfred W. Volkmann (1800–77).

voluntary muscle *See* skeletal muscle.

vulva (*pl.* **vulvae**) The exterior opening of the vagina.

W

Wallace's line /wol-ăs/ A hypothetical boundary in Indonesia between the islands of Bali and Lombok. It separates the Australasian and Oriental zoographical regions and was drawn by the British naturalist Alfred Russel Wallace (1823–1913), co-founder with Darwin of the theory of evolution.

warm-blooded *See* homoiothermy.

warning coloration (**aposematic coloration**) A conspicuous coloring or marking by which a noxious or dangerous animal can be recognized by potential attackers. Since these animals have developed their foul taste or dangerous nature to protect themselves from predation, it is important for them to warn potential predators of this. Warning coloration is common in insects, for example, the bright black-and-yellow stripes of many wasps. Venomous snakes often advertize their dangerous nature by their conspicuous markings. Warning coloration, may, however, be mimicked (*see* mimicry).

water A colorless odorless liquid, H_2O, occurring ubiquitously in nature and the most abundant constituent of living organisms. It is a powerful solvent and hydrating agent, a strong oxidizing agent, and serves as both a medium and a reactant for metabolic reactions. The structure of liquid water is still not fully understood, but its properties mean that its maximum density occurs at nearly $-4°C$, consequently ice floats on water, which prevents lakes and seas from freezing solid and permits aquatic life to continue in freezing conditions. Pure water contains low concentrations of hydroxonium (H_3O^+) and hydroxyl (OH^-) ions; acids increase the concentration of H_3O^+ ions, whereas bases increase the concentration of OH^- ions.

water fleas *See* Daphnia.

water potential Symbol: ψ The chemical potential of water in a biological system compared to the chemical potential of pure water at the same temperature and pressure. It is measured in kilopascals (kPa), and pure water has the value 0 kPa; solutions with increasing concentrations of solute have more negative values of ψ, since the solute molecules interfere with the water molecules. This effect is termed the *solute potential* (or osmotic potential), denoted by ψ_s; it is measured in kPa and always has a negative value, with increasing concentrations of solute having increasingly negative values of ψ_s. In turgid plant cells there is also a pressure exerted by the walls of the cell; this is called the *pressure potential*; it is denoted by ψ_p, and has a positive value (although in xylem cells it may be negative due to water movement in the transpiration stream). Hence for a plant cell the water potential of the cell, ψ_{cell}, is given by:

$$\psi_{cell} = \psi_s + \psi_p$$

If the water potential of a cell is less (i.e. more negative) than its surroundings the cell will take up water by OSMOSIS, until the solute potential is just balanced by the pressure potential, when the cell is described as being fully turgid. If a cell's water potential is greater (i.e. less negative) than its surroundings, the cell will lose water to its surroundings.

water vascular system In echinoderms, a system of canals filled with sea water. Water is taken in through the *madreporite* (sieve plate) on the upper surface and passes down through the *stone canal* to the *ring canal*. From this, five radial canals lead outward (into the arms in starfish) and carry the water to the tube feet, which operate by hydrostatic pressure.

wax One of a group of water-insoluble substances with a very high molecular weight; they are esters of long-chain alcohols with fatty acids. Waxes form protective coverings to leaves, stems, fruits, seeds, animal fur, and the cuticles of insects, serving principally as waterproofing. For example, waxy deposits on some plant organs add to the efficiency of the cuticle in reducing transpiration, as well as cutting down airflow over the surface and forming a highly reflective surface, thus reducing energy available for evaporation. They may also occur in plant cell walls, e.g. leaf mesophyll. They are used in varnishes, polishes, and candles.

Weberian ossicles /vay-**beer**-ee-ăn **oss**-ă-kălz/ A paired chain of three or four small bones in certain fish (e.g. carps and catfishes) that connect the air bladder with the auditory capsule. They are modified from the first four vertebrae and are comparable in function with the ear ossicles in higher vertebrates in that they conduct pressure changes from the air bladder to the inner ear. The ossicles are named for the German anatomist Ernst Heinrich Weber (1795–1878).

Weismannism /**vÿss**-mahn-iz-ăm/ The ideas put forward by the German biologist August Friedrich Leopold Weismann (1834–1914) criticizing the theory of the inheritance of acquired bodily characteristics implicit in Lamarckism and certain aspects of Darwinism. Weismann synthesized his ideas into the 'Theory of the Continuity of the Germ Plasm', which emphasized the distinction between the somatic cells and the germ cells and stated that inheritance was effected only by the germ cells.

Western blotting A technique analogous to SOUTHERN BLOTTING used to separate and identify proteins instead of nucleic acids. The protein mixture is separated by

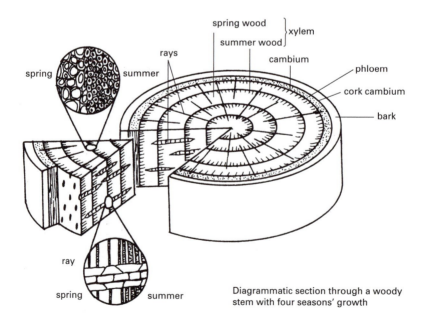

Diagrammatic section through a woody stem with four seasons' growth

electrophoresis and blotted onto a nitrocellulose filter. Antibodies specific to the protein of interest are applied and bind to their target proteins. A second antibody, specific to the first antibody, is then applied. This carries a radioactive label, so enabling it to be located by autoradiography wherever it binds to the antibody-protein complex.

whalebone *See* baleen.

whales *See* Cetacea.

white blood cell (**white blood corpuscle**) *See* leukocyte.

white matter Nerve tissue that consists chiefly of the fibers (axons) of nerve cells and their whitish myelin sheaths. It forms the outer region of the spinal cord and occurs in many parts of the brain.

wild type The most commonly found form of a given gene in wild populations. Wild-type alleles, often designated +, are usually dominant and produce the 'normal' phenotype.

windpipe *See* trachea.

Wolffian duct /wûl-fee-ăn/ One of a pair of ducts in fish and amphibians that transports urine from the kidney to the cloaca.

In the male it is a urinogenital duct, also transporting spermatozoa from the testes. In reptiles, birds, and mammals, it is functionally replaced by the ureter and persists only in the male, forming the epididymis and vas deferens. The duct is named for the German embryologist in Russia Kaspar F. Wolff (1733–94). *See also* mesonephros.

womb *See* uterus.

wood The hard fibrous structure found in woody perennials such as trees and shrubs. It is formed from the secondary XYLEM and thus only found in plants that show secondary thickening, namely the gymnosperms and dicotyledons. Water and nutrients are only transported in the outermost youngest wood, termed the sapwood. The nonfunctional compacted wood of previous seasons' growth is called the heartwood and it is this that is important commercially. Wood is classified as hardwood or softwood depending on whether it is derived from dicotyledons (e.g. oak) or conifers (e.g. pine). Hardwood is generally harder than softwood but the distinction is actually based on whether or not the wood contains fibers and vessels in addition to tracheids and parenchyma. *See also* annual ring; cambium; cork cambium.

woody perennial *See* perennial.

X

xanthophyll /**zan**-thŏ-fil/ One of a class of yellow to orange pigments derived from carotene, the commonest being lutein. *See* carotenoids; photosynthetic pigments.

Xanthophyta /zan-**th'off**-ă-tă/ A phylum of photosynthetic protoctists comprising the yellow-green algae, which are found chiefly in freshwater as pond scum. Some are unicellular, some form filamentous, branched, or bunched colonies, and yet others have large multinucleate cells. Their photosynthetic organelles contain chlorophylls *a* and *c*, and also xanthin pigments, which give the yellow-green coloration. Photosynthetic products are stored as oils, never as starch. The xanthophytes form motile cells (zoospores) equipped with two undulipodia (flagella) of unequal length and capable of germinating to form new colonies. Sexual reproduction also occurs in certain species, such as the filamentous *Vaucheria*.

xanthoproteic test /zan-thoh-proh-**tee**-ik/ A standard test for proteins. Concentrated nitric acid is added to the test solution. A yellow precipitate produced either immediately or on gentle heating indicates a positive result.

X chromosome The larger of the two types of SEX CHROMOSOME in mammals and certain other animals. It is similar in appearance to the other chromosomes and carries many sex-linked genes. *See* sex determination.

xenograft /**zen**-ŏ-graft, -grahft/ *See* graft.

xeromorphic /zeer-ŏ-**mor**-fik/ Structurally adapted to withstand dry conditions. *See* xerophyte.

xerophyte /**zeer**-ŏ-fÿt/ Any plant adapted to growing in dry conditions or in a physiologically dry habitat, such as an acid bog or a salt marsh, by storing available water, reducing water loss, or possessing deep root systems. Succulents, such as cacti and agaves, have thick fleshy stems or leaves to store water. Features associated with reducing water loss include: shedding or dieback of leaves; waxy leaf coatings coupled with closure or plugging of stomata; sunken or protected stomata; folding or repositioning of leaves to reduce sunlight absorption; and the development of a dense hairy leaf covering. *Compare* hydrophyte; mesophyte.

x-ray crystallography The use of *x-ray diffraction* by crystals to give information about the 3-D arrangement of the atoms in the crystal molecules. When x-rays are passed through a crystal a *diffraction pattern* is obtained as x-rays, whose wavelength is comparable with the distances between atoms, are diffracted by the atoms, rather as light is diffracted by a diffraction grating. The technique has been successful in helping to determine the structures of some large biological molecules such as DNA, RNA, viruses, and a variety of proteins, e.g. hemoglobin, myoglobin, and lysozyme.

x-ray diffraction *See* x-ray crystallography.

xylem /**zÿ**-lĕm/ The water-conducting tissue in vascular plants. It consists of dead hollow cells (the tracheids and vessels), which are the conducting elements. It also contains additional supporting tissue in the form of fibers and sclereids and some living parenchyma. The proto- and metaxylem,

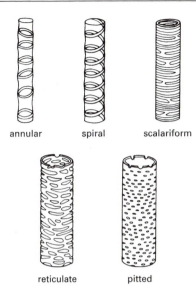

annular   spiral   scalariform

reticulate   pitted

Xylem: stages of xylem thickening

which together constitute the primary xylem, are formed from the procambium of the apical meristems, while the secondary xylem is differentiated from the cambium or lateral meristem.

The secondary cell walls of xylem vessels and tracheids become thickened with lignin to give greater support. The pattern of thickening varies according to the position and age of the xylem. *Annular* and *spiral thickening* is seen in protoxylem where extensibility is an important factor. More extensive *reticulate* and *scaliform thickening*, which prevents extension, is found in the metaxylem. The most extreme form of thickening is seen in the *pitted thickening* of the secondary xylem where all the secondary wall is thickened except for the small areas termed pits. Movement of water from roots to leaves via the xylem is termed the transpiration stream.

Y

Y chromosome The smaller of the two types of SEX CHROMOSOME in mammals and certain other animals. It is found only in the heterogametic sex. *See* sex determination.

yeasts *See* Saccharomyces.

yocto- Symbol: y A prefix denoting 10^{-24}. For example, 1 yoctometer (ym) = 10^{-24} meter (m).

yolk The food store, consisting of proteins and fats, in eggs.

yolk sac The sac, connected to the gut of the embryo and developed from it, that contains the yolk in reptiles, sharks, and birds. When birds hatch from the egg, the yolk sac is drawn into the abdomen of the newly hatched chick. The yolk gives the chick food for the first few days, until it becomes able to feed itself.

yotta- Symbol: Y A prefix denoting 10^{24}. For example, 1 yottameter (Ym) = 10^{24} meter (m).

Z

zepto- Symbol: z A prefix denoting 10^{-21}. For example, 1 zeptometer (zm) = 10^{-21} meter (m).

zetta- Symbol: Z A prefix denoting 10^{21}. For example, 1 zettameter (Zm) = 10^{21} meter (m).

zinc *See* trace elements.

Z line *See* sarcomere.

zona pellucida /zoh-nă pe-loo-să-dă/ (*pl.* zonae pellucidae) The thick clear membrane surrounding the mammalian egg. It is surrounded by *cumulus cells* in the freshly ovulated egg, but these disperse as the sperms pass between them and usually only one sperm penetrates the zona by enzyme action.

zonula occludens /zohn-yŭ-lă ŏ-kloo-denz/ *See* tight junction.

zoogeography /zoh-ŏ-jee-og-ră-fee/ The study of the geographical distribution of animal species. Such study shows that the earth can be divided into distinct geographical regions (*faunal regions*), each having its own unique collection of animal species (*see* Australasia; Ethiopian; Nearctic; Neotropical; Oriental; Palaearctic). For example, the continents of the southern hemisphere – Australasia, Africa (south of the Sahara), and South America – each have a characteristic fauna not found elsewhere. Anteaters, sloths, and armadillos are native to South America; marsupial and monotreme mammals are characteristic of Australasia; while Africa shows a greater diversity of fauna than any other zoogeographical region. *See also* phytogeography; Wallace's line.

zoology /zoh-ol-ŏ-jee/ The scientific study of animals.

Zoomastigina /zoh-ŏ-măs-tij-ă-nă/ A phylum of heterotrophic protoctists that possess one or more undulipodia (flagella) for locomotion. It includes both free-living forms (e.g. *Naegleria*) and parasites (e.g. *Trypanosoma*, which is responsible for sleeping sickness), and aerobic and anaerobic forms. In some recent classifications this phylum has been disbanded and its members dispersed to other taxa according to whether or not they possess mitochondria.

zooplankton /zoh-ŏ-plank-tŏn/ *See* plankton.

zoosporangium /zoh-ŏ-spŏ-ran-jee-ŭm/ (*pl.* zoosporangia) A sporangium that produces zoospores, as in some phycomycetes, and green and brown algae.

zoospore /zoh-ŏ-spor, -spohr/ An asexual motile spore produced by a zoosporangium. It has one or more flagella. These motile spores may encyst in adverse conditions, or may be the means by which the fungus penetrates a new host.

zwitterion /tsvit-er-ÿ-on, -ŏn, zwit-/ An ion with both a positive and a negative charge. Amino acids can form zwitterions: the amino group has the form $-NH_3^+$ and the acid group is ionized as $-COO^-$.

zygomorphy /zÿ-gŏ-mor-fee/ *See* bilateral symmetry.

zygomycete /zÿ-goh-mÿ-seet/ Any fungus belonging to a phylum (Zygomycota) whose members have non-septate hyphae (i.e. lacking cross walls). They are mostly saprophytic, absorbing nutrients from decaying vegetation and other organic matter; the bread molds *Rhizopus* and *Mucor* are examples. Others are symbiotic mycorrhizal fungi (e.g. *Glomus*), while some parasitize animals or protoctists; for instance, *Cochlonema* is a parasite of amebas. Zygomycetes can reproduce both asexually and sexually. In asexual reproduction, spores develop inside reproductive structures (sporangia), and when released can germinate to form new mycelia. In sexual reproduction hyphae of compatible mating types undergo conjugation, their tips fusing to form a thick-walled zygosporangium, inside which nuclei from each mating type fuse together. The zygosporangium can germinate to form a new mycelium, or produce sporangia.

zygospore /zȳ-gŏ-spor, -spohr/ A resistant sexual spore formed when a zygote develops a thick wall. Zygospores are characteristic of the Mucorales (such fungi as *Mucor* and *Rhizopus*) and the Conjugales (such algae as *Spirogyra* and *Zygnema*).

zygote /zȳ-goht/ The diploid cell resulting from the fusion of two haploid gametes. A zygote usually undergoes cleavage immediately. *See also* embryo; gamete.

zygotene /zȳ-gŏ-teen/ In meiosis, the stage in mid-prophase I that is characterized by the active and specific pairing (synapsis) of homologous chromosomes, leading to the formation of a haploid number of bivalents.

zymogen granule /zȳ-mŏ-jĕn/ A secretory granule found in large numbers in enzyme-secreting cells. The granules are vesicles containing an inactive precursor of the enzyme, the zymogen, e.g. trypsinogen in exocrine cells of the pancreas. It is activated after secretion via exocytosis at the plasma membrane. The vesicles are usually derived from the GOLGI APPARATUS, where the enzyme is processed and concentrated after synthesis on the rough endoplasmic reticulum.

APPENDIXES

Appendix I

The Animal Kingdom

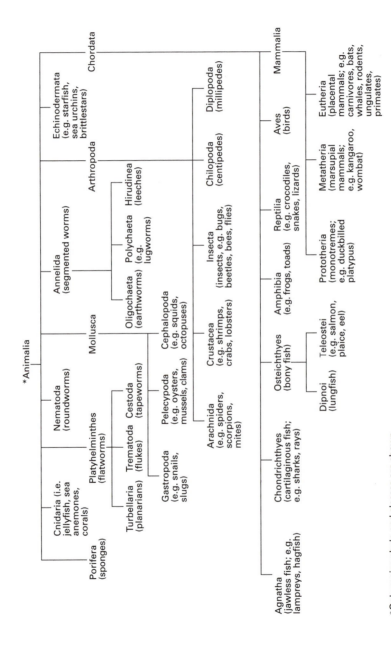

*Only major phyla and classes are shown

Appendix II

The Plant Kingdom

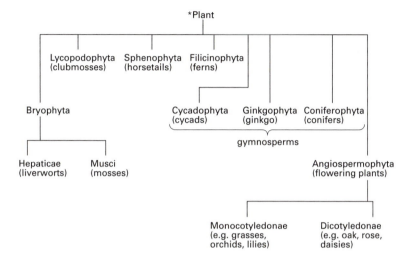

*Extinct and mostly extinct groups are excluded

Appendix III

Amino Acids

amino acid	symbol	R group
alanine	Ala	$-CH_3$
valine	Val	$-CH\Big\langle{{CH_3}\atop{CH_3}}$
leucine	Leu	$-CH_2-CH\Big\langle{{CH_3}\atop{CH_3}}$

isoleucine	Ile	$-CH-CH_2-CH_3$
		     $\mid$
		     CH_3

proline	Pro	$\boxed{\begin{array}{c}COO^-\\ {}^{H_2}\,{}_{'}C\\ H\end{array}}\ \begin{array}{c}H_2\\ C-CH_2\\ N-CH_2\\ H\end{array}$
methionine	Met	$-CH_2-CH_2-S-CH_3$
phenylalanine	Phe	$-CH_2-\hexagon$
tryptophan	Trp	$-CH_2-C$ ... HC ... N ... H

amino acids with nonpolar R groups

The **amino acids** commonly found in proteins are alpha amino acids. In all except proline, where the complete structure is given, the remainder of the molecule has the structure

$$HOOC-\underset{\underset{NH_2}{|}}{\overset{\overset{H}{|}}{C}}-$$

402

Amino Acids
(continued)

amino acid	symbol	R group
glycine	(Gly)	$-H$
serine	(Ser)	$-CH_2-OH$
threonine	(Thr)	$-\overset{\displaystyle OH}{\underset{\displaystyle H}{C}}-CH_3$
cysteine	(Cys)	$-CH_2-SH$
asparagine	(Asn)	$-CH_2-C\overset{\displaystyle NH_2}{\underset{\displaystyle O}{}}$
glutamine	(Gln)	$-CH_2-CH_2-C\overset{\displaystyle NH_2}{\underset{\displaystyle O}{}}$
tyrosine	(Tyr)	$-CH_2-\!\!\bigcirc\!\!-OH$

amino acids with uncharged polar R groups

amino acid	symbol	R group	
aspartic acid	(Asp)	$-CH_2-C\overset{\displaystyle O^-}{\underset{\displaystyle O}{}}$	acidic amino acids
glutamic acid	(Glu)	$-CH_2-CH_2-C\overset{\displaystyle O^-}{\underset{\displaystyle O}{}}$	
lysine	(Lys)	$-CH_2-CH_2-CH_2-CH_2-\overset{+}{N}H_3$	basic amino acids
arginine	(Arg)	$-CH_2-CH_2-CH_2-NH-\overset{\displaystyle NH_2}{\underset{\displaystyle \overset{+}{N}H_2}{C}}-NH_2$	
histidine	(His)	$-CH_2-\overset{\displaystyle C=CH}{\underset{\displaystyle HN\quad NH}{}}$ $\underset{\displaystyle \overset{\displaystyle \overset{+}{C}}{H}}{}$	

amino acids with charged polar R groups

403

Appendix IV

Web Sites

General sites for biology resources:

The US Geological Survey	http://biology.usgs.gov
National Biological Information Infrastructure	www.nbii.gov

Museum sites:

American Museum of Natural History	www.amnh.org
Canadian Museum of Nature	www.nature.ca
Harvard Museum of Natural History	www.hmnh.harvard.edu
Natural History Museum of Los Angeles County, California	www.nhm.org
Smithsonian Institution National Museum of Natural History	www.mnh.si.edu
The Natural History Museum, London	www.nhm.ac.uk
University of Michigan Museum of Zoology	www.ummz.lsa.umich.edu
Yale Peabody Museum of Natural History	www.peabody.yale.edu

Annotated evolution links:

Nearctica	www.nearctica.com/evolve/evolve.htm

Links for Palaeontology:

UK Palaeontological Association	www.palass.org
University of Calfornia, Berkeley	www.ucmp.berkeley.edu

Biochemistry and on cell and molecular biology:

European Bioinformatics Centre	http://ebi.ac.uk
Harvard University Department of Molecular and Cellular Biology	http://mcb.harvard.edu/BioLinks.html
Human Genome Project	www.ornl.gov/hgmis
International Union of Biochemistry and Molecular Biology (Nomenclature)	www.chem.qmul.ac.uk/iubmb
Leeds University Bioinformatics Links	www.bioinf.leeds.ac.uk/ bioinformatics.html
Oxford University Bioinformatics Centre	www.molbiol.ox.ac.uk

Web Sites
(continued)

Princeton University	www-hhmi.princeton.edu/hhmi/index.html
Southwest Biotechnology and Informatics Center	www.swbic.org
UK Human Genome Mapping Project	www.hgmp.mrc.ac.uk
University of Arizona, Biochemistry	www.biology.arizona.edu/biochemistry/biochemistry.html
University of Arizona, Molecular Biology	http://www.biology.arizona.edu/molecular_bio/molecular_bio.html
University of Southern Maine, Biochemistry Resources	http://www.usm.maine.edu/~rhodes/Biochem

Plant science:

American Society of Plant Biologists	www.aspb.org/education
Botanical Society of America	www.botany.org
Internet Directory for Botany	www.botany.net/IDB/info.html
Missouri Botanical Garden	www.mobot.org
New York Botanical Garden	www.nybg.org
Royal Botanic Gardens, Kew	www.rbgkew.org.uk
University of California Botanical Garden	www.mip.berkeley.edu/garden

The environment:

The US Environmental Protection Agency	www.epa.gov/enviroed
Environmental Sciences Division, Oak Ridge National Laboratory	www.esd.ornl.gov
Smithsonian Environmental Research Center	www.serc.si.edu

Biographies:

Nobel Prizes	http://nobelprize.org
University of Calfornia, Berkeley	www.ucmp.berkeley.edu/help/topic/history.html

Bibliography

Alberts, Bruce; Alexander Johnson; Julian Lewis; Martin Raff; Keith Roberts ; & Peter Walter. *Molecular Biology of the Cell.* 4th ed. New York: Garland Science, 2002.

Audesirk, Gerald; Teresa Audesirk; & Bruce E. Byers. *Biology: Life on Earth.* 7th ed. NewYork: Prentice Hall, 2004.

Campbell, Neil A. & Jane B. Reece. *Biology.* 7th ed. New York: Benjamin Cummings, 2004.

Gould, James L. & William T. Keeton. *Biological Science.* 6th ed. New York: W. W. Norton, 1996.

Gould, Steven Jay. *The Structure of Evolutionary Theory.* Cambridge, Mass.: Harvard University Press, 2002.

Krough, David. *Biology: A Guide to the Natural World.* 3rd ed. New York: Prentice Hall, 2004.

Margulis, L. & K. V. Schwartz. *Five Kingdoms: An Illustrated Guide to the Phyla of Life on Earth.* 3rd ed. New York: W. H. Freeman, 1998.

Nelson, David L. & Michael M. Cox. *Lehninger Principles of Biochemistry.* 3rd. ed. New York: Worth Publishers, 2000.

Purves, William K.; David Sadava; Gordon H. Orians; & H. Craig Heller. *Life: The Science of Biology.* 6th ed. New York: W. H. Freeman, 2001.

Raven, Peter H.; George B. Johnson; Susan Singer; & Jonathon Losos. *Biology.* 7th ed. New York: Prentice Hall, 2004.

Roberts, Michael; Michael Reiss; & Grace Monger. *Advanced Biology.* Walton-on Thames, UK: Nelson, 2000.

Taylor, Martha R. *Student Study Guide for Biology.* 5th ed. New York: Benjamin Cummings, 1999.